Gmelin Handbook of Inorganic and Organometallic Chemistry

8th Edition

Gmelin Handbook of Inorganic and Organometallic Chemistry

8th Edition

Gmelin Handbuch der Anorganischen Chemie

Achte, völlig neu bearbeitete Auflage

PREPARED AND ISSUED BY

Gmelin-Institut für Anorganische Chemie
der Max-Planck-Gesellschaft
zur Förderung der Wissenschaften
Director: Ekkehard Fluck

FOUNDED BY — Leopold Gmelin

8TH EDITION — 8th Edition begun under the auspices of the Deutsche Chemische Gesellschaft by R. J. Meyer

CONTINUED BY — E. H. E. Pietsch and A. Kotowski, and by Margot Becke-Goehring

Springer-Verlag Berlin Heidelberg GmbH 1993

Volumes published on "Phosphorus" (Syst. No. 16)

Main Volume

A: History. Occurrence – 1965

B: Technology. Toxicity. The Element – 1964

C: The Compounds of Phosphorus – 1965

Formula and Heading Index – 1967

Supplement Volume

C 1: Mononuclear Compounds with Hydrogen – 1993 (present volume)

Gmelin Handbook of Inorganic and Organometallic Chemistry

8th Edition

P

Phosphorus

Supplement Volume C 1

Mononuclear Compounds with Hydrogen

With 13 illustrations

AUTHORS Werner Behrendt, Ulrich W. Gerwarth †, Reinhard Haubold, Jörn v. Jouanne, Hannelore Keller-Rudek, Dieter Koschel, Hans Schäfer, Joachim Wagner

EDITORS Werner Behrendt, Ulrich W. Gerwarth †, Reinhard Haubold, Jörn v. Jouanne, Hannelore Keller-Rudek, Dieter Koschel, Hans Schäfer, Joachim Wagner

CHIEF EDITORS Dieter Koschel, Hans Schäfer

System Number 16

Springer-Verlag Berlin Heidelberg GmbH 1993

LITERATURE CLOSING DATE: MAY 1992
IN SOME CASES MORE RECENT DATA HAVE BEEN CONSIDERED

Library of Congress Catalog Card Number: Agr 25-1383

ISBN 978-3-662-08849-4 ISBN 978-3-662-08847-0 (eBook)
DOI 10.1007/978-3-662-08847-0

Originally published by Springer-Verlag Berlin in 1993.
Softcover reprint of the hardcover 8th edition 1993

Preface

This volume C 1 is the first supplement volume to "Phosphor" C which was published in 1965 and covers the compounds of phosphorus. Starting with the binary species formed between phosphorus and hydrogen, the present volume deals with the neutral mononuclear compounds PH through PH_5; the ions featuring the same stoichiometric composition are covered in separate sections.

PH and PH_2 are the major initial gaseous decomposition products of PH_3 and, thus, also intermediates in many of its gas-phase reactions. Both molecules and their ions have been thoroughly investigated by a variety of modern, high-resolution spectroscopic methods during the last three decades. The coverage of their physical, and mostly molecular, properties represents the largest part of the first two chapters (PH and ions pp. 2 to 47; PH_2 and ions pp. 47 to 111).

PH_3 is the only compound described in this volume which is thermally stable under normal conditions. It is the phosphorus analog of ammonia, but exhibits, however, a quite different chemical behavior towards most elements and compounds. The majority of its physical, and in particular spectroscopic, properties have been determined in great detail since the sixties, partially in regard to spectroscopic investigations of the atmospheres of the outer planets. From an industrial point of view, PH_3 plays a major part in the manufacture of a variety of electronic devices, mostly based on III-V semiconductor materials, for the microelectronics industry; PH_3 is a starting material for the production of many organophosphorus chemicals and is the effective compound formed during the fumigation of, e.g., grain silos. The critical coverage of the large number of papers considered to be relevant for the description of PH_3 in this volume amounts to about 197 pages (PH_3 and ions pp. 112 to 312).

The last chapters deal, in addition to other ionic species, with the description of PH_4 which was detected in low-temperature matrices by ESR spectroscopy, the cation PH_4^+ which can be obtained, e.g., in strongly acidic, liquid media (PH_4 and ions pp. 312 to 320), and the hypothetical parent phosphorane PH_5 (PH_5 and ions pp. 321 to 324).

Di- and polynuclear phosphorus-hydrogen compounds will be covered along with their organophosphorus derivatives as model compounds in the forthcoming Supplement Volume C 2.

Frankfurt am Main, October 1993 — Dieter Koschel, Hans Schäfer

Table of Contents

Phosphorus and Hydrogen

General Remarks

In addition to the mononuclear species, predominantly PH, PH_2, and PH_3 (see below), described in the following chapters of the present volume, most of the compounds formed between phosphorus and hydrogen can be classified according to several homologous series of compounds. The chemical composition of these series can be expressed by the general formulas P_nH_{n+2}, P_nH_n, and P_nH_{n-2m} where m = 1 to about 9. All of these polyphosphanes, which will be covered in the forthcoming "Phosphorus" Suppl. Vol. C 2, contain trivalent pyramidal phosphorus atoms. Since the thermal and photolytic stability of most of the polyphosphanes, in particular the hydrogen-richest series, stongly decreases with increasing n, chain-like or monocyclic compounds are generally realized only when n is relatively small. With an increasing number of phosphorus atoms in the molecule, there is a marked tendency to undergo condensation reactions releasing PH_3 and forming hydrogen-poor, polycyclic phosphanes exhibiting cage-like molecular structures which resemble segments of the molecular framework of Hittorf's violet phosphorus.

To date only very few polyphosphanes such as P_2H_4, P_3H_5, P_5H_5, and P_7H_3 have been isolated and fully characterized. Most of the phosphorus hydrides belonging to the above mentioned series have only been detected either by NMR spectroscopy in complex mixtures of phosphanes or by mass spectrometry. However, during the last two decades many examples of organyl-substituted derivatives, even of compounds with higher phosphorus content, have been obtained. These polycyclic organophosphanes can serve as models for the corresponding PH-containing parent species and will therefore also be dealt with in this context in the forthcoming "Phosphorus" Suppl. Vol. C 2.

1 Monophosphorus Compounds

General Remarks

Of the neutral mononuclear phosphorus-hydrogen compounds, PH_n where n = 1 to 5, described in this volume, only PH_3 is thermally and photolytically stable under normal conditions. Since this molecule is quite interesting scientifically, e.g., as the phosphorus analog of ammonia, and is also important from an industrial point of view, it is not surprising that the coverage of its extended chemistry and physical properties amounts to the largest part of the present volume. PH and PH_2 which are the two major primary decomposition products of PH_3 have been thoroughly investigated by a variety of spectroscopic methods, which now permits a detailed description of their physical, and in particular their molecular properties. While the PH_4 radical could be detected spectroscopically by ESR in low-temperature matrices, the hypothetical PH_5 molecule has only been investigated by theoretical methods.

The different cations and anions of each molecule are covered in separate chapters. Most of the information on these ionic species results from gas-phase studies or quantum-chemical calculations. Only PH_2^- and PH_4^+ have been obtained to date as ions of salt-like compounds, either in solution or in the solid state: PH_2^-, e.g., in the alkali metal phosphides MPH_2, where M = Li to Cs; and PH_4^+, e.g., in the phosphonium halides PH_4X, where X = Cl, Br, I.

1.1 PH and Ions

1.1.1 PH Phosphanediyl, Phosphanylidene, Phosphinidene

CAS Registry Numbers: PH *[13967-14-1]* Phosphinidene; PD *[53120-25-5]*, Phosphinidene, hydro-*d*-

1.1.1.1 Formation. Occurrence

The phosphanediyl molecule is unstable under normal conditions; investigations of its properties are usually restricted to the determination of spectroscopic data. PH and PD are obtained from reactions of phosphorus with H or D atoms and from the decomposition of PH_3 or PD_3; for the spectroscopically identified states of PH or PD formed under various experimental conditions, see Section 1.1.1.4, p. 23.

The contact of the H or D atoms (from the exposure of H_2 or D_2 to an electric [1] or an MW discharge [2 to 4]) at ambient temperature with red phosphorus [2, 3] or the vapor of white phosphorus [4, 5] yields phosphanediyl. The reactions are accompanied by a bright green chemiluminescence [4]; PH_2 [3] and PH_3 [1] were noted as additional products. An attempt to bypass the use of D_2 by forming PD via the reaction of red phosphorus with MW-discharged D_2O was only partially successful, because spectra of the product formed were of low quality with respect to bands of PD [6].

The formation of PH or PD during UV photolysis of PH_3 or PD_3 is described in Section 1.3.1.5.1.3.1, p. 206. The molecules were also obtained from PH_3 or PD_3 via three-photon excitation and dissociation upon laser irradiation in the range 383 to 394 nm [7]. Formation of phosphanediyl from decomposing gaseous phosphane was observed in focused IR laser beams (cf. 1.3.1.5.1.3.2, p. 210), in shock waves [8] and during thermal decomposition (cf. 1.3.1.5.1.2, p. 201), and in discharges (cf. 1.3.1.5.1.4, p. 214). For the intermediate formation (and decomposition) of PH upon the sorption of PH_3 on metal surfaces, see Section 1.3.1.5.11, p. 287.

Bombarding PH_3 with electrons generates electronically excited PH for spectroscopic investigations; an electron beam of, e.g., 60-eV [9] and periodic pulses of, e.g., 20-keV electrons were used [10]. PH forms via $PH_3 \rightarrow PH + H^+ + H$, where the appearance potential of the additional product H^+ of 20.5±1 eV was determined by photoionization time-of-flight mass spectrometry. The formation of PH via $PH_3 \rightarrow PH + H_2^+$ was not observed; an appearance potential of 17.95 eV for H_2^+ was calculated from thermodynamic data [11].

The formation of PH or PD was observed in reactions of PH_3 or PD_3 with atomic H, O, N, and F (cf. 1.3.1.5.5, p. 233). Formation of the molecules was also discussed in the mechanisms of the reactions of PH_3 with CO or SiH_4 upon UV irradiation (cf. 1.3.1.5.6, p. 249), with SiH_4 upon IR irradiation in the presence of the photosensitizer SiF_4 (cf. 1.3.1.5.6, p. 250), and with recoil ^{32}P atoms generated by neutron irradiation (cf. 1.3.1.5.1.5, p. 215). Intermediate formation of PH was suggested to occur during the pyrolysis of $SiH_3PH_2/(CH_3)_2SiD_2$ mixtures at 570 K [12].

The kinetics of the hypothetical hydrogen transfer reactions $P + HA \rightarrow PH + A$ with A = H, Li, Na, K, Be, Mg, B, Al, C, Si, N, O, S, F, Cl, Br, I at 1000 K was computed using the bond energy bond order (BEBO) method and the activated complex theory or the simple collision theory [13]. However, the values reported have partially been questioned [14].

Bands of PH were assigned in the spectrum of the sun photosphere [8]; calculated band widths for estimated PH band positions in the near-UV range using various models of the solar atmosphere, are given in [15]. The occurrence of PH in the atmosphere of cool red giant stars was modeled by [16].

References:

[1] Anacona, J. R.; Davies, P. B.; Hamilton, P. A. (Chem. Phys. Lett. **104** [1984] 269/71).
[2] Davies, P. B.; Russell, D. K.; Smith, D. R.; Thrush, B. A. (Can. J. Phys. **57** [1979] 522/8).
[3] Davies, P. B.; Russell, D. K.; Thrush, B. A. (Chem. Phys. Lett. **36** [1975] 280/2).
[4] My, L. T.; Peyron, M. (J. Chim. Phys. Phys.-Chim. Biol. **59** [1962] 688/95).
[5] Ram, R. S.; Bernath, P. F. (J. Mol. Spectrosc. **122** [1987] 275/81).
[6] Uehara, H.; Hakuta, K. (J. Chem. Phys. **74** [1981] 4326/9).
[7] Ashfold, M. N. R.; Dixon, R. N.; Stickland, R. J. (Chem. Phys. Lett. **111** [1984] 226/33).
[8] Guenebaut, H.; Pascat, B. (C. R. Hebd. Seances Acad. Sci. **255** [1962] 1741/3).
[9] Fink, E.; Welge, K. H. (Z. Naturforsch. **19a** [1964] 1193/201).
[10] Gustafsson, O.; Kindvall, G.; Larsson, M.; Senekowitsch, J.; Sigray, P. (Mol. Phys. **56** [1985] 1369/80).

[11] Zarate, E. B.; Cooper, G.; Brion, C. E. (Chem. Phys. **148** [1990] 277/88).
[12] Elliott, L. E.; Estacio, P.; Ring, M. A. (Inorg. Chem. **12** [1973] 2193/4).
[13] Mayer, S. W.; Schieler, L. (J. Phys. Chem. **72** [1968] 236/40).
[14] Truhlar, D. G. (J. Chem. Phys. **56** [1972] 3189/90).
[15] Sinha, K.; Tripathi, B. M.; Atalla, R. M.; Singh, P. D. (Sol. Phys. **115** [1988] 221/7; C. A. **109** [1988] No. 240016).
[16] Johnson, H. R.; Sauval, A. J. (Astron. Astrophys. Suppl. Ser. **49** [1982] 77/87; C. A. **97** [1982] No. 136174).

1.1.1.2 Molecular Properties

1.1.1.2.1 Electronic Structure

Quantum-Chemical Calculations

Some sixty ab initio (and a few semiempirical) studies of varying scope and accuracy have been published including the ground state and, in a few cases, the excited states of the PH molecule and its positive and negative ions. More than half of the calculations went beyond the Hartree-Fock limit generally applying various configuration-interaction and perturbation-theory methods, and, more recently, combinations of both. Studies that deal with the evaluation of molecular properties of PH per se (instead of only testing a computational method) are quoted in the following sections. For a more complete compilation of ab initio calculations on PH, the reader is referred to the various bibliographies on quantum-chemical calculations listed on p. 176.

The following abbreviations for the quantum-chemical methods will be used in the subsequent sections:

abbreviation	method
SCF	self-consistent field
UHF	unrestricted Hartree-Fock
(Q)CI	(quadratic) configuration interaction
"full" CI	extrapolation of MRD CI calculation
MC	multiconfiguration
CAS	complete active space

abbreviation	method
MRD	multireference double excitation
PNO	pair natural orbital
CEPA	coupled electron pair approximation
MP2, MP4	Møller-Plesset perturbation theory of 2nd, 4th order
G1, G2	Gaussian-1, Gaussian-2 (=MP4+QCI)
GVB	generalized valence bond
FSGO	floating spherical Gaussian orbitals
CNDO	complete neglect of differential overlap
MINDO	modified intermediate neglect of differential overlap

Electron Configuration. Term Values

Term values for electronic states of PH which have been studied experimentally and theoretically are compiled in Table 1. For details see below.

Lowest Energy Configuration

The ten core and six valence electrons of the PH molecule combine to the lowest energy configuration KL $(4\sigma)^2$ $(5\sigma)^2$ $(2\pi)^2$, where KL $=(1\sigma)^2$ $(2\sigma)^2$ $(3\sigma)^2$ $(1\pi)^4$ represents the P1s, 2s, 2p core orbitals and 4σ, 5σ, and 2π are essentially the P3s, P3pσ-H1s bonding, and P3pπ orbitals, respectively. This configuration gives rise to the ground state X $^3\Sigma^-$, which correlates with the ground state atoms $P(^4S)+H(^2S)$, and to two excited metastable states a $^1\Delta$ and b $^1\Sigma^+$, which correlate with $P(^2D)+H(^2S)$ and $P(^2P)+H(^2S)$; see e.g. [1, pp. 341, 369/71].

The **ground state** X $^3\Sigma^-$ has been observed by laser magnetic resonance (LMR) in the far and mid IR and by IR absorption and emission spectroscopy. Moreover, it has been identified with the lower state of the visible-UV emission and absorption systems b $^1\Sigma^+ \rightarrow$ X $^3\Sigma^-$, A $^3\Pi_i \leftrightarrow$ X $^3\Sigma^-$, and (tentatively) $^3\Pi$(Rydberg)$\leftarrow$X $^3\Sigma^-$ (see p. 29).

Some experimental information concerning orbital population in the ground state is provided by the LMR spectra: The ^{31}P hyperfine coupling constants (cf. p. 11) yielded a value of 86.6% for the spin density on the P atom and show that the unpaired electron orbital 2π is essentially the pure P3pπ orbital having only 0.98% s character [2], which confirms the earlier conclusions based on the ^{31}P and ^1H hf constants [3].

The total molecular energy of the ground state was estimated to be $E_t^{exp}=-342.648$ E_h [4, 5]; the so far best theoretical value, $E_t=-341.42843$ E_h, resulted from a recent G2 calculation using a method that treats the electron correlation by Møller-Plesset perturbation theory (MP4) and quadratic configuration interaction (QCI) [6].

The **metastable** states a $^1\Delta$ and b $^1\Sigma^+$ first predicted to lie about 0.95 and 1.88 eV above the ground state [7], both have been detected experimentally.

The a $^1\Delta$ state has been identified with the lower state of two absorption systems in the vacuum UV, $^1\Phi\leftarrow$a $^1\Delta$ and $^1\Pi\leftarrow$a $^1\Delta$ [8]. A term value of $T_0=0.950\pm0.010$ eV ($=7660\pm$ 80 cm^{-1}) was derived from the energy separation of the PH(X $^3\Sigma^-$, v=0)$\leftarrow$PH$^-$(X $^2\Pi$, v=0) and PH(a $^1\Delta$, v=0)$\leftarrow$PH$^-$(X $^2\Pi$, v=0) peaks in the laser (488 nm=2.54 eV) photoelectron spectrum of PH$^-$ [9]. The (less precise) value of $T_0\approx0.9$ eV could be derived from the second threshold energy for photodetachment (hν=0.8 to 2.8 eV) of PH$^-$ [10]. The a $^1\Delta$ state has also been observed by LMR in the far IR; the hf constants indicate that the unpaired electron orbital 2π is almost exclusively the P3pπ atomic orbital [11].

Table 1

Electronic States of PH. Experimental and Calculated Term Values T_e (T_0) in eV.

valence states				Rydberg states			
state	configuration	T_e (T_0) exp.	T_e calc.[a)]	state	configuration	T_e (T_0) exp.	T_e calc.[a)]
X $^3\Sigma^-$	$\ldots(5\sigma)^2(2\pi)^2$	0	0	$^3\Pi$	$\{\ldots(5\sigma)^2(2\pi)^1\}$ 4s	—	6.28
a $^1\Delta$	$\ldots(5\sigma)^2(2\pi)^2$	(0.950±0.010) [9]	1.12	$^1\Pi$	$\{\ldots(5\sigma)^2(2\pi)^1\}$ 4s	—	6.59
b $^1\Sigma^+$	$\ldots(5\sigma)^2(2\pi)^2$	1.77603±0.00001 [13]	1.98	$^3\Pi$	$\{\ldots(5\sigma)^2(2\pi)^1\}$ 4pσ	—	7.38
A $^3\Pi_i$	$\ldots(5\sigma)^1(2\pi)^3$	3.6571 (PH) [24, 25]	3.86	$^1\Pi$	$\{\ldots(5\sigma)^2(2\pi)^1\}$ 4pσ	—	7.75
		3.6579 (PD) [24, 25]	3.6924 [26]	$^3\Sigma^{+}$[c)]	$\{\ldots(5\sigma)^2(2\pi)^1\}$ 4pπ	(8.1548) [14]	—
c $^1\Pi$	$\ldots(5\sigma)^1(2\pi)^3$	—	5.02	$^3\Phi$	$\{\ldots(5\sigma)^2(2\pi)^1\}$ 3dδ	—	8.60
$^5\Sigma^-$	$\ldots(5\sigma)^1(2\pi)^2(6\sigma)^1$	—	repulsive[b)]	$^1\Phi$	$\{\ldots(5\sigma)^2(2\pi)^1\}$ 3dδ	(8.5803) [8]	8.61
d $^1\Sigma^+$	$\ldots(4\sigma)^2(5\sigma)^0(2\pi)^4$	—	9.76	$^3\Pi$	$\{\ldots(5\sigma)^2(2\pi)^1\}$ 3dδ	(8.6273) [8]	8.63
				$^1\Pi$	$\{\ldots(5\sigma)^2(2\pi)^1\}$ 3dδ	(8.7262) [8]	8.63
				$^5\Sigma^-$	$\{\ldots(5\sigma)^1(2\pi)^2\}$ 4s	—	8.37
				$^5\Sigma^-$	$\{\ldots(5\sigma)^1(2\pi)^2\}$ 4pσ	—	9.27

[a)] If no other reference is given, the vertical excitation energies ($\Delta E \approx T_e$, see p. 6) are from the "full" CI estimate following MRD CI calculations [21]. – [b)] Vertical excitation energy is 7.09 eV at r = 1.43 Å [21]. – [c)] Possibly identical with the $^1\Sigma^+ \ldots(4\sigma)^2(5\sigma)^0(2\pi)^4$ valence state [16].

The b $^1\Sigma^+$ state was identified with the upper state of the forbidden b $^1\Sigma^+ \rightarrow$ X $^3\Sigma^-$ transition, attributed to the weak emission system at 695 nm. The term value T_0=14340 cm^{-1}=1.78 eV derived experimentally for the first time by [12] has been refined to T_e=14325.5±0.1 cm^{-1} by a rotational-vibrational analysis of the highly resolved b→X spectrum [13]. Moreover, the b $^1\Sigma^+$ state was found to be the lower state of the two-photon resonance enhancement, $^3\Sigma^+$(Rydberg)←b $^1\Sigma^+$ or $^1\Sigma^+$(valence)←b $^1\Sigma^+$, observed in the multiphoton, (3+2+1) or (3+2+2), excitation, fragmentation, and ionization process $PH_3 \rightarrow PH_3^* \rightarrow$ PH(b $^1\Sigma^+$)→PH^+ with near-UV radiation [14 to 16].

Term values T_e for a $^1\Delta$ and b $^1\Sigma^+$ or vertical excitation energies ΔE(a←X, b←X) (the T_e and ΔE values nearly coincide, because the bond length of PH remains practically the same in the three lowest states, cf. p. 19) have been calculated by various theoretical methods: T_e (in eV)=0.87 [17], 0.95±0.01 [18] (SCF+semiempirical E_{corr} estimate), 0.94 (MINDO/3+CI) [19, 20], 1.10 (MRD CI) [21], 1.127 (CEPA) [22] for a $^1\Delta$ and T_e (in eV)=1.88 (SCF+semiempirical E_{corr} estimate) [17], 1.76 (MINDO/3+CI) [19, 20], 1.62 (MP4) [23], 1.96 (MRD CI) [21] for b $^1\Sigma^+$.

Excited Configurations

Valence States. The lowest excited electron configuration results from 2π←5σ excitation and gives rise to the valence states A $^3\Pi_i$ and $^1\Pi$, which correlate with P(^{2}D)+H(^{2}S). The first of them, identified with the upper state of the prominent A $^3\Pi_i \leftrightarrow$ X $^3\Sigma^-$ absorption and emission system at 340 nm (cf. pp. 27/8), has been studied in detail. Analysis of the fine structure by considering spin-spin and second-order spin-orbit interactions leads to the following subband origins for PH and PD (three standard deviation in parentheses) [24]:

substate	T_0(PH) in cm^{-1}	T_0(PD) in cm^{-1}
$^3\Pi_{0^+}$	29434.61(5)	29495.66(2)
$^3\Pi_{0^-}$	29434.28(5)	29495.37(2)
$^3\Pi_1$	29316.81(5)	29377.90(2)
$^3\Pi_2$	29203.21(5)	29264.21(2)

Term values for A $^3\Pi_i$, T_e=29498 cm^{-1} for PH and T_e=29505 cm^{-1} for PD, were calculated by Huber and Herzberg [25] using the $T_0(^3\Pi_1)$ (=T_0(eff)) values and vibrational constants given by [24]. The lower values (T_e=29484 and 29498 cm^{-1} for PH and PD) given in the original paper [24] obviously have been calculated neglecting the anharmonicity terms $\omega_e x_e$.

The experimental term value quoted by [26] with reference to [24] seems to be erroneous.

Theoretical results are: T_e=29783 cm^{-1} (MCSCF-CI) [26] and ΔE(A←X) (≥T_e)=3.86 eV (MRD CI) [21] and 3.83 eV (SCF+semiempirical E_{corr} estimate) [17].

The $^1\Pi$ (...(5σ)^1 (2π)^3) state predicted at 5.06 eV (MRD CI), 5.02 eV ("full" CI) [21], and 5.22 eV (SCF+semiempirical E_{corr} estimate) [17] above the ground state has not been observed; this in accordance with the rather flat potential curve indicating quasi-dissociative character (MRD CI) [21].

Most recently, the d $^1\Sigma^+$ (...(4σ)^2 (2π)^4) state, by analogy with the NH radical (cf. "Nitrogen" Suppl. Vol. B 1), has been assumed to be the resonance-enhancing state in the multiphoton excitation, fragmentation, and ionization process $PH_3 \rightarrow PH_3^* \rightarrow$ PH(b $^1\Sigma^+$)→PH^+ rather than the $^3\Sigma^+$ Rydberg state (see below) [16].

Among the various states arising from 6σ←5σ excitation (cf. [1, p. 337]), the repulsive $^5\Sigma^-$ state which correlates with the ground-state atoms P(^{4}S)+H(^{2}S) and which is supposed to be responsible for the predissociation of the A $^3\Pi_i$ state (cf. p. 28) has been studied theoretically (MRD CI) [21].

A semiempirical study (MINDO/3 + CI) deals with the known valence states of PH and a number of unobserved bound and repulsive states with vertical excitation energies up to ~8 eV [20].

Rydberg States. Among the various Rydberg states associated with excitations from the 2π MO into P4s, 4p, or 3d levels, the states $^1\Phi$, $^1\Pi$, and $^3\Pi$, all arising from the $\{PH^+(X\ ^2\Pi)\}$ 3dδ configuration, have been identified with the upper states of three absorption transitions in the vacuum UV, namely $^1\Phi$, $^1\Pi \leftarrow a\ ^1\Delta$ and (tentatively) $^3\Pi \leftarrow X\ ^3\Sigma^-$; the term values T_0(in cm^{-1}) = x + 61548.68, x + 62725.28 (with $x = T_0(a\ ^1\Delta) = 7660 \pm 80\ cm^{-1}$ [9]), and 69587.9 were determined [8].

The states $\{PH^+(X\ ^2\Pi)\}$ 4pπ $^3\Sigma^+$ at T_0(in cm^{-1}) = y + 51431.6 ± 1.0 for PH and y + 51433.5 ± 1.0 for PD (with $y = T_0(b\ ^1\Sigma^+) = 14345.2 \pm 0.2\ cm^{-1}$ for PH as derived from the results of [13]) were suggested to be the upper states of the two-photon excitation $^3\Sigma^+ \leftarrow b\ ^1\Sigma^+$ responsible for the resonance enhancement observed in the (3+2+1)-photon fragmentation, excitation, and ionization process $PH_3 \rightarrow PH(b\ ^1\Sigma^+) \rightarrow PH(^3\Sigma^+) \rightarrow PH^+$ with near-UV radiation [14, 15]. A reassignment of the resonance-enhancing state to the valence state d $^1\Sigma^+$, ...$(4\sigma)^2 (5\sigma)^0 (2\pi)^4$, however, was proposed recently (cf. p. 29) [16].

The PH^+ ion yield curve resulting from photoionization (125 to 104 nm) of PH exhibits an extensive autoionization structure; twenty-four features could be attributed to three Rydberg series of PH, nsσ (n = 3 to 10, quantum defect δ = 0.04), ndσ (n = 3 to 5, δ = −0.07), and ndπ (n = 3 to 14, δ = −0.14), all converging to $PH^+(a\ ^4\Sigma$, ...$(4\sigma)^2 (5\sigma)^1 (2\pi)^2)$ at 11.852 ± 0.002 eV [27].

References:

[1] Herzberg, G. (Molecular Spectra and Molecular Structure, Vol. 1: Spectra of Diatomic Molecules, Van Nostrand, Princeton, N. J., 1961).
[2] Ohashi, N.; Kawaguchi, K.; Hirota, E. (J. Mol. Spectrosc. **103** [1984] 337/49).
[3] Davies, P. B.; Russell, D. K.; Smith, D. R.; Thrush, B. A. (Can. J. Phys. **57** [1979] 522/8).
[4] Cade, P. E.; Huo, W. M. (J. Chem. Phys. **47** [1967] 649/72).
[5] Cade, P. E.; Huo, W. M. (At. Data Nucl. Data Tables **12** [1973] 415/66).
[6] Curtiss, L. A.; Raghavachari, K.; Trucks, G. W.; Pople, J. A. (J. Chem. Phys. **94** [1991] 7221/30).
[7] Jordan, P. C. (J. Chem. Phys. **41** [1964] 1442/9).
[8] Balfour, W. J.; Douglas, A. E. (Can. J. Phys. **46** [1968] 2277/80).
[9] Zittel, P. F.; Lineberger, W. C. (J. Chem. Phys. **65** [1976] 1236/43).
[10] Rackwitz, R.; Feldmann, D.; Kaiser, H. J.; Heinicke, E. (Z. Naturforsch. **32a** [1977] 594/7).

[11] Davies, P. B.; Russell, D. K.; Thrush, B. A. (Chem. Phys. Lett. **36** [1975] 280/2).
[12] Di Stefano, G.; Lenzi, M.; Margani, A.; Nguyen Xuan, C. (J. Chem. Phys. **68** [1978] 959/63).
[13] Droege, A. T.; Engelking, P. C. (J. Chem. Phys. **80** [1984] 5926/9).
[14] Ashfold, M. N. R.; Dixon, R. N.; Stickland, R. J. (Chem. Phys. Lett. **111** [1984] 226/33).
[15] Ashfold, M. N. R.; Stickland, R. J.; Tutcher, B. (Mol. Phys. **65** [1988] 1455/71).
[16] Ashfold, M. N. R.; Clement, S. G.; Howe, J. D.; Western, C. M. (J. Chem. Soc. Faraday Trans. **87** [1991] 2515/23).
[17] Liu, H. P. D.; Legentil, J.; Verhaegen, G. (Sel. Top. Mol. Phys. Proc. Int. Symp., Ludwigsburg, FRG, 1970 [1972], pp. 19/30).
[18] Cade, P. E. (Can. J. Phys. **46** [1968] 1989/91).

[19] Minaev, B. F.; Buketova, A. E.; Muldakhmetov, Z. M. (Izv. Akad. Nauk Kaz. SSR Ser. Khim. **1988** No. 3, pp. 26/31; C.A. **109** [1988] No. 179165).
[20] Minaev, B. F.; Buketova, A. E.; Muldakhmetov, Z. M. (Spectrosc. Lett. **22** [1989] 211/36).

[21] Bruna, P. J.; Hirsch, G.; Peyerimhoff, S. D.; Buenker, R. J. (Mol. Phys. **42** [1981] 875/98).
[22] Staemmler, V.; Jaquet, R. (Theor. Chim. Acta **59** [1981] 501/15).
[23] Nguyen, M. T.; Hegarty, A. F.; McGinn, M. A.; Ruelle, P. (J. Chem. Soc. Perkin Trans. II **1985** 1991/7).
[24] Rostas, J.; Cossart, D.; Bastien, J. R. (Can. J. Phys. **52** [1974] 1274/87).
[25] Huber, K. P.; Herzberg, G. (Molecular Spectra and Molecular Structure, Vol. 4: Constants of Diatomic Molecules, Van Nostrand Reinhold, New York 1979, pp. 534/7).
[26] Senekowitsch, J.; Rosmus, P.; Werner, H.-J.; Larsson, M. (Z. Naturforsch. **41a** [1986] 719/23).
[27] Berkowitz, J.; Cho, H. (J. Chem. Phys. **90** [1989] 1/6).

1.1.1.2.2 Ionization Potentials

The adiabatic ionization potential, $E_i(ad) = 10.149 \pm 0.008$ eV, was obtained from a photoionization (125 to 104 nm) mass-spectrometric study of PH; the PH was produced by the reaction $H + PH_3 \rightarrow PH + H_2$, and the onset of the PH^+ ion yield at 122.6 ± 0.1 nm was attributed to the process $PH(X\ ^3\Sigma^-, \ldots (4\sigma)^2 (5\sigma)^2 (2\pi)^2) \rightarrow PH^+(X\ ^2\Pi_{1/2}, \ldots (4\sigma)^2 (5\sigma)^2 (2\pi)^1)$. The extensive autoionization structure observed in the ion yield curve was interpreted to belong to three Rydberg series of PH; their mutual convergence limit at 11.852 ± 0.002 eV was attributed to the onset of $PH^+(a\ ^4\Sigma^-, \ldots (4\sigma)^2 (5\sigma)^1 (2\pi)^2)$ formation [1]. The authors improved an earlier result for the first ionization potential, $E_i(ad) = 10.1_8 \pm 0.1$ eV from a photoionization (130 to 50 nm) mass-spectrometric study of PH_3 [2].

The first vertical ionization potential, $E_i(vert) = 10.19 \pm 0.01$ eV, was derived from the He I photoelectron spectrum recorded for the $F + PH_3$ reaction [3].

Theoretical values obtained from the differences of the total energies of PH^+ and PH at their respective equilibrium bond lengths (leading to $E_i(ad)$) or at the bond length of the neutral molecule (leading to $E_i(vert)$) are as follows:

$E_i(ad)$ in eV	$E_i(vert)$ in eV	method[a)]	Ref.
10.09, 10.08		G2, G1 (MP4+QCI)	[4, 5]
10.09[b), c)]		MP4	[2, 6]
10.04[b), d)]		CEPA	[7]
	10.05, 10.06	MRD CI, "full" CI	[8]
9.83		MP4	[9]
9.5, 9.3		MCSCF-CI, MCSCF	[10]

[a)] The abbreviations are explained on pp. 3/4. – [b)] Corrected for zero-point vibrations. – [c)] The value was slightly modified into 10.14 eV by calculating the isogyric process $PH + H^+ \rightarrow PH^+ + H$ [11]. – [d)] $E_i(ad) = 10.16$ eV was calculated from the proton affinity of the P atom; a value of $E_i(ad) = 10.2$ was recommended by [7].

SCF results for the first E_i(ad) are (in eV): 9.63 (corrected for zero-point vibrations) [7], 9.6 [10], 9.8 [12], and 9.65 [13, 14]; even smaller values were calculated with the FSGO [15] and MINDO [16] methods.

In addition to the first ionization potential corresponding to the removal of the outermost 2π electron of PH, ionization from the 5σ MO has been calculated with the MRD CI method; the "full" CI estimate then gave vertical ionization potentials of 11.72, 13.44, 13.49, and 14.56 eV corresponding to the ionic states $^4\Sigma^-$, $^2\Sigma^-$, $^2\Delta$, and $^2\Sigma^+$, respectively [8].

References:

[1] Berkowitz, J.; Cho, H. (J. Chem. Phys. **90** [1989] 1/6).
[2] Berkowitz, J.; Curtiss, L. A.; Gibson, S. T.; Greene, J. P.; Hillhouse, G. L.; Pople, J. A. (J. Chem. Phys. **84** [1986] 375/84).
[3] Dyke, J. M.; Jonathan, N.; Morris, A. (Int. Rev. Phys. Chem. **2** [1982] 3/42, 7, 22).
[4] Curtiss, L. A.; Raghavachari, K.; Trucks, G. W.; Pople, J. A. (J. Chem. Phys. **94** [1991] 7221/30).
[5] Curtiss, L. A.; Jones, C.; Trucks, G. W.; Raghavachari, K.; Pople, J. A. (J. Chem. Phys. **93** [1990] 2537/45).
[6] Pople, J. A.; Curtiss, L. A. (J. Phys. Chem. **91** [1987] 155/62).
[7] Rosmus, P.; Meyer, W. (J. Chem. Phys. **66** [1977] 13/9).
[8] Bruna, P. J.; Hirsch, G.; Peyerimhoff, S. D.; Buenker, R. J. (Mol. Phys. **42** [1981] 875/98).
[9] Nguyen, M. T. (Mol. Phys. **59** [1986] 547/58).
[10] Pope, S. A.; Hillier, I. H.; Guest, M. F. (Faraday Symp. Chem. Soc. No. 19 [1984] 109/23).

[11] Pople, J. A.; Curtiss, L. A. (J. Phys. Chem. **91** [1987] 3637/9).
[12] Pope, S. A.; Hillier, I. H.; Guest, M. F.; Kendric, J. (Chem. Phys. Lett. **95** [1983] 247/9).
[13] Cade, P. E.; Huo, W. M. (J. Chem. Phys. **47** [1967] 649/72).
[14] Cade, P. E.; Huo, W. M. (At. Data Nucl. Data Tables **12** [1973] 415/66).
[15] Blustin, P. H. (J. Chem. Phys. **66** [1977] 5648/55).
[16] Goetz, H.; Frenking, G.; Marschner, F. (Phosphorus Sulfur **4** [1978] 309/16).

1.1.1.2.3 Electron Affinity A

Two experimental results are available for the adiabatic electron affinity which were obtained by photodetachment studies on PH^- ions. $A = 1.028 \pm 0.010$ eV was obtained from a laser (488 nm) photoelectron spectrum of PH^- after rotational corrections had been made to the detachment energy measured from the center of the PH(X $^3\Sigma^-$, v=0)$\leftarrow$$PH^-$ (X $^2\Pi_i$, v=0) peak [1]. A less precise result, $A = 1.00 \pm 0.06$ eV, has been obtained from the lowest threshold energy of the photodetachment cross section measured between 0.8 and 2.8 eV (1.5 to 0.4 μm; a second threshold at ~1.9 eV obviously corresponds to the PH(a $^1\Delta$)$\leftarrow$$PH^-$(X $^2\Pi_i$) transition) [2].

Ab initio calculations for ground-state PH and PH^- using the composite G1 [3] and G2 [4] methods gave total energy differences of $\Delta E_t = A = 0.91$ and 0.96 eV. MP4 calculations with large basis sets gave $A = 0.796$ and 0.812 eV [5, 6]. Applying MP4 to isogyric comparisons of XH molecules with H_2 resulted in $A = 0.96$ eV for PH [7]. Hartree-Fock wave functions for PH and PH^- and an estimate for the correlation energy difference gave $A = 0.93$ eV [8], and a semiempirical estimate is $A \approx 1.6$ eV [9].

References:

[1] Zittel, P. F.; Lineberger, W. C. (J. Chem. Phys. **65** [1976] 1236/43).
[2] Rackwitz, R.; Feldmann, D.; Kaiser, H. J.; Heinicke, E. (Z. Naturforsch. **32a** [1977] 594/7).
[3] Curtiss, L. A.; Jones, C.; Trucks, G. W.; Raghavachari, K.; Pople, J. A. (J. Chem. Phys. **93** [1990] 2537/45).
[4] Curtiss, L. A.; Raghavachari, K.; Trucks, G. W.; Pople, J. A. (J. Chem. Phys. **94** [1991] 7221/30).
[5] Nguyen, M. T. (Mol. Phys. **59** [1986] 547/58).
[6] Nguyen, M. T. (J. Mol. Struct. **180** [1988] 23/9).
[7] Pople, J. A.; Schleyer, P. von R.; Kaneti, J.; Spitznagel, G. W. (Chem. Phys. Lett. **145** [1988] 359/64).
[8] Cade, P. E. (Proc. Phys. Soc. [London] A **91** [1967] 842/54).
[9] Gaines, A. F.; Page, F. M. (Trans. Faraday Soc. **62** [1966] 3086/92).

1.1.1.2.4 Dipole Moment μ. Dipole Moment Function $\mu(r)$. Quadrupole Moment Θ

Experimental results are not available.

Ab initio calculations generally gave dipole moment values between +0.4 and +0.7 D [1 to 7] (exception $\mu = +0.052$ D [8]), where the positive sign indicates the polarization $P^- H^+$ [2, 6]. An UHF calculation resulted in $\mu = 0.70$ D, but with $P^+ H^-$ polarization [9]. Dipole moment functions calculated from highly correlated wave functions (tabulations and plots of $\mu(r)$ for PH(X $^3\Sigma^-$, A $^3\Pi_i$) at r = 2.0 to 4.5 a_0 [1] and for PH(X $^3\Sigma^-$) at r = 2.0 to 3.5 a_0 [2]) gave the following equilibrium (μ_e) and mean vibrational (μ_0) values for the dipole moment and its derivative $\mu'_e = (d\mu/dr)_e$ (μ_e, μ_0 in D, μ'_e in D/Å):

molecule	state	μ_e	μ'_e MCSCF-CI [1]	μ_0	μ_e	μ'_e CEPA [2]	μ_0
PH	X $^3\Sigma^-$	0.431	−1.30	0.403	0.481	−1.39	0.449
PH	A $^3\Pi_i$	0.584	+1.60	0.638	—	—	—
PD	X $^3\Sigma^-$	—	—	0.409	—	—	0.458

Moreover, theory predicts the sign for PH(X $^3\Sigma^-$) to change at $r \approx 3.2\ a_0$, indicating a dipole direction $P^+ H^-$ at large internuclear distances [1, 2].

The only value for the quadrupole moment, $\Theta = -6.509 \times 10^{-26}$ esu·cm² with respect to the center of mass, stems from a very simple ab initio FSGO calculation [8] (cf. the exceptional μ from [8] given above).

References:

[1] Senekowitsch, J.; Rosmus, P.; Werner, H.-J.; Larsson, M. (Z. Naturforsch. **41a** [1986] 719/23).
[2] Meyer, W.; Rosmus, P. (J. Chem. Phys. **63** [1975] 2356/75).
[3] Boyd, R. J.; Edgecombe, K. E. (J. Comput. Chem. **8** [1987] 489/98).
[4] Bader, R. F. W.; Messer, R. R. (Can. J. Chem. **52** [1974] 2268/82).
[5] Wayne, F. D.; Colbourn, E. A. (Mol. Phys. **34** [1977] 1141/55).

[6] Cade, P. E.; Bader, R. F. W.; Henneker, W. H.; Keaveny, I. (J. Chem. Phys. **50** [1969] 5313/33).
[7] Cade, P. E.; Huo, W. M. (J. Chem. Phys. **45** [1966] 1063/5).
[8] Blustin, P. H. (J. Chem. Phys. **66** [1977] 5648/55).
[9] Nguyen, M. T. (Mol. Phys. **59** [1986] 547/58).

1.1.1.2.5 Polarizability α. Magnetic Susceptibility χ

For these physical properties only theoretical results from a simple ab initio FSGO calculation are available. The average polarizability $\overline{\alpha}=4.107$ Å^3 and the molar, dia-, and paramagnetic susceptibilities (in 10^{-6} emu/mol) $\chi_m=-16.430$, $\chi_d=-21.482$, $\chi_p=5.053$ were obtained.

Reference:

Blustin, P. H. (J. Chem. Phys. **66** [1977] 5648/55).

1.1.1.2.6 Spectroscopic Constants

Hyperfine Interaction Constants

In the electronic ground state X $^3\Sigma^-$ of PH or PD, hyperfine interaction occurs between the electron spin (S=1) and the nuclear spins of ^{31}P (I=1/2), H (I=1/2) or D (I=1). Hf doublet splittings have been observed in the far-IR [1 to 3] and mid-IR [4] LMR spectra of PH and PD (cf. pp. 23/5), from which the hf constants α_P, α_H (Fermi contact term) and β_P, β_H (dipole-dipole interaction term) were derived. Small triplet splittings due to the D nucleus have been observed only on two LMR transitions of PD, insufficient to derive α_D or β_D [1]. Hf constants for ^{31}P and ^{1}H are as follows (three standard deviation in parentheses; α, β in cm^{-1}):

molecule	state	$10^3\cdot\alpha_P$	$10^3\cdot\beta_P$	$10^3\cdot\alpha_H$	$10^3\cdot\beta_H$	Ref.
PH	X $^3\Sigma^-$, v=0	4.254(77)	−5.255(96)	−1.603(53)	0.181(65)	[1]
PH	X $^3\Sigma^-$, v=0	4.18(27)	−5.07(40)	−1.63(17)	0.17(25)	[3]
PD	X $^3\Sigma^-$, v=0	4.330(39)	−5.312(32)	—	—	[1]
PD	X $^3\Sigma^-$, v=1	4.66(40)	−4.79(69)	—	—	[1]
PD	X $^3\Sigma^-$, v=0, 1	4.22(120)*)	−5.48(100)	—	—	[4]

*) Ref. [4] gives the Frosch-Foley parameters $b_P=291(21)$ MHz ($9.71(70)\times10^{-3}$ cm^{-1}) and $c_P=-493(90)$ MHz ($-16.4(3)\times10^{-3}$ cm^{-1}) which have been converted with the relations $\alpha=b+c/3$ and $\beta=c/3$; see [5, pp. 33/40].

In the first excited state a $^1\Delta$ of PH (with S=0), hf interaction occurs between the electron orbital angular momentum ($\Lambda=2$) and the nuclear spins of ^{31}P and ^{1}H. The hf constants $a_P=768(37)$ MHz (0.0256(12) cm^{-1}) and $a_H=28(15)$ MHz (0.00093(50) cm^{-1}) were obtained from the hf splittings in the far-IR LMR spectrum of PH(a $^1\Delta$, v=0) [3] (changing slightly the previous result [2]).

Fine Structure Constants

Fine structure splittings have been observed in the far-IR [1 to 3] and mid-IR [4] LMR spectra, in the IR emission [6] and absorption [7] spectra, and in the near-UV emission [8 to 12] and absorption [13] spectra of PH and PD. The fine-structure constants for the electronic ground state X $^3\Sigma^-$, i.e., λ for the spin-spin coupling, γ for the spin-rotation coupling, and λ_D, γ_D for their respective centrifugal distortions, and constants for the excited state A $^3\Pi_i$, i.e., A, A_D for the spin-orbit coupling and p, q, q_D or C_i (i=0 to 3) for the Λ-type doubling, have been derived; for definitions of the constants and relations with reference to earlier notations, see [8, table 3].

The results for the ground state obtained by diverse spectroscopic methods are in good agreement with each other yielding $\lambda \approx 2.21$ cm^{-1} for PH and PD (v=0, 1) and $\gamma \approx -0.07$ cm^{-1} for PH (v=0, 1) and $\gamma \approx -0.04$ cm^{-1} for PD (v=0, 1). The following table demonstrates the good agreement and the degree of precision of the various analyses (three standard deviation in parentheses; constants in cm^{-1}):

molecule	state	λ	$-10^2 \cdot \gamma$	$10^5 \cdot \gamma_D$	spectrum[b)]	Ref.
PH	X $^3\Sigma^-$, v=0	2.20993(24)	7.6894(24)	1.308(70)	far-IR LMR	[1]
		2.2067(15)	7.6476(48)	—	far-IR LMR	[3]
		2.2089(17)[a)]	7.668(22)[a)]	1.287(60)[a)]	IR emission	[6]
		2.212(9)	7.38(10)	—	UV emission	[8]
PH	X $^3\Sigma^-$, v=1	2.20992(176)	7.339(30)	1.308 [1]	IR absorption	[7]
		2.2092(17)[a)]	7.312(21)[a)]	1.249(61)[a)]	IR emission	[6]
		2.207(12)	7.26(19)	—	UV emission	[8]
PD	X $^3\Sigma^-$, v=0	2.20848(57)	3.9900(34)	3.23(67)	far-IR LMR	[1]
		2.2081(13)	3.96(11)	—	mid-IR LMR	[4]
		2.211(8)	3.85(6)	—	UV emission	[8]
PD	X $^3\Sigma^-$, v=1	2.2095(10)	3.882(21)	3.3(fixed)	far-IR LMR	[1]
		2.2096(26)	3.92(6)	—	mid-IR LMR	[4]
		2.202(17)	3.81(15)	—	UV emission	[8]

a) One standard deviation in parentheses; $10^5 \cdot \lambda_D = 2.09(91)$ and 1.60(88) cm^{-1} for v=0 and 1; results for v=2 to 5 are given also in [6]. – b) Rotational transitions N=2←1, 3←2, ..., 7←6 at 571 to 85 µm [1], N=5←4 at 118 µm [3]; rovibrational transitions v=1→0, 2→1, ..., 5→4 at 4.0 to 5.4 µm [6], v=1←0, N=2←3, 3←4, ..., 6←7, 8←9 and 1←0, 2←1 around 4.4 µm [7], v=1←0, N=2←3, 1←2, 0←1, 1←0, 2←1, 3←2 at 6.6 µm [4]; rovibronic transitions A $^3\Pi_i$→X $^3\Sigma^-$, v=1→0, 0→0, 0→1, 1→1 at 340 nm [8].

For the A $^3\Pi_i$ state, values for the following constants A, A_D, p, q, q_D were obtained from the highly resolved A $^3\Pi_i$→X $^3\Sigma^-$ emission spectra of PH and PD at 340 nm (three standard deviation in parentheses; constants in cm^{-1}) [8]:

molecule	state	$-A$	$10^3 \cdot A_D$	$-10^2 \cdot p$	$10^2 \cdot q$	$-10^5 \cdot q_D$
PH	A $^3\Pi_i$, v=0	115.71	1.09(10)	13.8(3)	1.3(2)	0.26(20)
PH	A $^3\Pi_i$, v=1	115.20	3.5(6)	12(1)	1.0(2)	—
PD	A $^3\Pi_i$, v=0	115.74	0.45(4)	7.2(30)	0.33(2)	—
PD	A $^3\Pi_i$, v=1	115.55	0.85(8)	7.0(2)	0.35(3)	—

The corresponding equilibrium values are $A_e = -115.8 \pm 0.1\,cm^{-1}$ for PH and PD, $A_{De} = 2.3\times10^{-3}$, $p_e = -1.47\times10^{-2}$, $q_e = 1.3\times10^{-2}\,cm^{-1}$ for PH and $A_{De} = 0.7\times10^{-3}$, $p_e = -7.2\times10^{-2}$, $q_e = 0.3\times10^{-2}$ for PD [8].

An alternative description of Λ-doubling uses the constants C_0, C_1, C_2, and C_3, where C_0 is the Λ-doubling of the $^3\Pi$ state in the absence of rotation, i.e., the $^3\Pi_{0+} - {}^3\Pi_{0-}$ splitting, and C_1, C_2, C_3 characterize the rotational terms up to $J^2(J+1)^2$; see [14, pp. 3322/7]. The above given constants p, q, and q_D have been converted into the C_i by [8].

Second-order spin-orbit constants for the interaction of A $^3\Pi_i$ with the a $^1\Delta$, b $^1\Sigma^+$, and c $^1\Pi$ states have also been deduced or estimated by [8].

An earlier fine-structure analysis of the A→X emission spectrum is to be found in [14].

Theoretical studies deal with the calculation of λ [15 to 17], C_0 [15], and A [18] by ab initio methods and of λ [14] and A [19] by semiempirical methods. A theoretical study of the triplet splitting of the A $^3\Pi_i$ state was offered by [20].

Zeeman Effect. g Factors

Zeeman splitting of rotational levels of PH and PD has been observed by laser magnetic resonance (LMR) in the far IR [1 to 3] and at 6 μm [4]. To describe the Zeeman effect in the electronic ground state X $^3\Sigma^-$, three g factors were used: g_S, the electron spin g factor; g_l^e, the orbital g factor due to the mixing of X $^3\Sigma^-$ with excited states via spin-orbit coupling; g_R, the rotational g factor. The mid-IR LMR spectra of PD(X $^3\Sigma^-$, v=0 and 1) were accounted for by the fixed values $g_S = 2.0023$ (free electron), $g_l^e = \gamma/(2B) = 0.0045$ (Curl's relation [21] with γ = spin-rotation constant and B = rotational constant), and $g_R = 0$ [4]. In a more recent analysis of the far-IR LMR spectra, on the other hand, all three g factors, more strictly speaking the parameters $g_S + g_l^e$, $\Delta g' = \Delta g - g_l^e$ (Δg = small diamagnetic correction; see [21]), and g_R were chosen as adjustable parameters; the results are (g in units of Bohr magnetons μ_B; three standard deviation in parentheses) [1]:

	PD(X $^3\Sigma^-$) v=0	PD(X $^3\Sigma^-$) v=1	PH(X $^3\Sigma^-$) v=0
$g_S + g_l^e$	2.00722(51)[a]	2.0052(11)	2.00683(39)[a]
$\Delta g'$	−0.00431(93)	fixed at values for v=0	−0.0039(10)
g_R[b]	−0.00043(13)		−0.00077(17)

[a] Using $g_S = 2.00232$ (free electron), the orbital g factors are calculated to be $g_l^e = 0.00490(51)$ and 0.00451(39) for PD, v=0, and PH, v=0, respectively. – [b] Definition of g_R as given in [22] and [5, p. 63], which is $-g_r$ of [1].

In an earlier far-IR LMR study, g_S and g_l^e were held fixed at the free-electron (2.00232) and Curl (0.00455) values, and g_R was allowed to vary in the fit which resulted in $g_R = -0.00072(39)$ [3].

The Zeeman splitting of the a $^1\Delta$ state of PH was accounted for by setting $g_L = 1.000$ (orbital g factor of the free electron) and $g_R = 0$ [3].

Rotational and Vibrational Constants. Internuclear Distance

The analysis of purely rotational, rotational-vibrational, and electronic spectra (cf. pp. 23/30) gave rotational and vibrational constants for the electronic ground state and various excited states of PH and PD. The more accurate values were obtained, of course, by IR Fourier transform and laser spectroscopy as compared to the conventional UV-visible spectroscopic methods (conventional IR spectroscopy has never been applied to PH).

The electronic ground state X $^3\Sigma^-$ has been studied by far-IR [1 to 3] and mid-IR [4] laser magnetic resonance (LMR), diode laser absorption [7], and Fourier transform emission [6] spectroscopy as well as by analysis of the electronic spectrum [8]. The rotational constants B_e or B_0, centrifugal distortion constants D_e or D_0, H_0, L_0, and the corresponding vibration-rotation interaction constants, α_e, γ_e, β_e, δ_e (where $B_v = B_e - \alpha_e(v+1/2) + \gamma_e(v+1/2)^2 + \ldots$; $D_v = D_e - \beta_e(v+1/2) + \ldots$; $H_v = H_e - \delta_e(v+1/2) + \ldots$), the vibrational constants ω_e, $\omega_e x_e$, $\omega_e y_e$, $\omega_e z_e$, and the internuclear distance r_e are given in Table 2 and the supplementary remarks a) to f), pp. 15 and 17. An analogous compilation for the excited states a $^1\Delta$, b $^1\Sigma^+$, A $^3\Pi_i$, $^3\Sigma^+$, $^3\Pi$, $^1\Phi$, and $^1\Pi$, which were determined by far-IR LMR [3], UV-visible emission [8, 23] and absorption [13, 24] as well as by multiphoton ionization (MPI) spectroscopy [25], is given in Table 3, and remarks a) to e), pp. 16/7.

Theoretical values for the spectroscopic constants of PH were derived by polynomial fits of the potential curves which had been calculated from highly correlated ab initio wave functions. The results for the ground state and the lowest excited states reproduce the experimental results, and constants for the hitherto only tentatively assigned higher valence state d $^1\Sigma^+ \ldots (4\sigma)^2 (5\sigma)^0 (2\pi)^4$ were predicted. The extent of agreement between experiment and theory is demonstrated by the following compilation and a comparison with Tables 2 and 3 (B_e, B_0, α_e, ω_e, $\omega_e x_e$ in cm^{-1}, r_e in Å):

state	B_e	B_0	α_e	ω_e	$\omega_e x_e$	r_e	wave function*)	Ref.
X $^3\Sigma^-$	8.524	—	0.253	2379	42	1.423	MCSCF-CI	[29]
	8.49	—	0.251	2365.9	44.8	1.426	CEPA	[30]
	8.55	—	0.240	2408.6	41.7	1.421	PNO-CI	[30]
	8.533	8.406	0.254	2380	43.9	1.4226	MRD CI	[31]
	8.62466	—	—	2484	—	1.415	MP2	[32]
a $^1\Delta$	8.591	8.457	0.269	2415	48.0±3	1.4178	MRD CI	[31]
b $^1\Sigma^+$	8.637	8.507	0.260	2438	48.0±3	1.4141	MRD CI	[31]
	8.64178	—	—	2506	—	1.412	MP2	[32]
A $^3\Pi_i$	8.262	—	0.450	2029	87	1.446	MCSCF-CI	[29]
	8.306	8.081	0.45	2071	80	1.4420	MRD CI	[31]
d $^1\Sigma^+(2\pi)^4$	8.244	8.114	0.26	2140±50	38.0	1.4473	MRD CI	[31]

*) The abbreviations are explained on pp. 3/4.

A number of further ab initio calculations [33 to 40] and one semiempirical (CNDO) study [41] dealt with geometry optimization for PH and predicted ground-state and, in one case [38], excited-state equilibrium internuclear distances.

Table 2

PH, PD. Ground State X $^3\Sigma^-$. Rotational and Vibrational Constants, Internuclear Distance.
B, D, α_e, ω_e, $\omega_e x_e$ in cm^{-1}, r_e in Å; three standard deviation in parentheses.

molecule	$B_e\{B_0\}$	α_e	$10^4 \cdot D_e\{D_0\}$	$\omega_e\{\Delta G_{1/2}\}$	$\omega_e x_e$	r_e	remark	Ref.
PH	8.53899(57) {8.412445(90)}	0.25343(81)	— {4.4228(84)}	2363.779(108)	43.911(81)	—	a)	[6]
	8.53864(10)	—	4.46(23)	—	—	—	b)	[7]
	8.5327(14) {8.412524(21)}	—	— {4.4367(77)}	2366.79(66)	—	1.42140(22)	c)	[1]
	{8.412018(27)}	—	{4.34(1)} [8]	—	—	—	d)	[3]
	8.5371 {8.4127(11)}	0.2514	4.36 {4.40(4)}	2365.2	44.5	1.4223_4	e)	[8, 26]
PD	4.40802(58) {4.3628675(77)}	—	— {1.1803(24)}	1700.35(12)	—	1.42140(22)	c)	[1]
	4.4100(5)	0.0939(4)	{1.161} [8]	{1653.2858(36)}	—	—	f)	[4]
	4.4081 {4.361(3)}	0.0928	1.16 {1.161(5)}	1699.2	230	1.4220	e)	[8, 26]

Table 3

PH, PD. Rotational and Vibrational Constants, Internuclear Distances for Electronically Excited States.
B, D, α_e, ω_e, $\omega_e x_e$ in cm^{-1}, r in Å; three standard deviation in parentheses except for results of [23].

molecule	state	$B_e\{B_0\}$	α_e	$10^4 \cdot D_e\{D_0\}$	$\omega_e\{\Delta G_{1/2}\}$	$\omega_e x_e$	$r_e\{r_0\}$	remark	Ref.
PH	a $^1\Delta$	{8.4392925(46)}	0.12	{4.18} [24]	—	—	$\{1.430_2\}$	a)	[3]
	b $^1\Sigma^+$	8.587(3)	0.253(3)	4.00(5)	2403.0(1)	42.0(1)	1.417	b)	[23]
	A $^3\Pi_i$	8.259 {8.0222(3)}	0.473	5.25 {5.683(8)}	2030.6 {1833.8}	98.5	1.4458 $\{1.4672_8\}$	c)	[8, 26]
	$^3\Sigma^+$	{8.404(15)}	—	{4.35(30)}	~2340	—	—	d)	[25]
	$^3\Pi$	$B_{eff}=7.3$	—	—	—	—	$\{1.5_4\}$	f)	[24]
	$^1\Phi$	$\{8.60_2\}$	0.21	$\{5.5_4\}$	—	—	$\{1.416_7\}$	f)	[24]
	$^1\Pi$	$\{8.47_8\}$	—	$\{4.1_7\}$	—	—	$\{1.427_0\}$	f)	[24]
PD	b $^1\Sigma^+$	{4.405(15)}	—	{1.20(15)}	—	—	—	d)	[25]
	A $^3\Pi_i$	4.256 {4.1720(1)}	0.167	1.44 {1.506(2)}	1458.9 {1357.3}	50.8	1.469 {1.4617}	c)	[8, 26]
	$^3\Sigma^+$	{4.364(15)}	—	{1.28(15)}	—	—	—	d)	[25]

Remarks for Table 2, p. 15:

a) Fourier transform IR emission spectrum of the 1→0 to 5→4 bands; analysis of centrifugal distortion up to octic terms: $10^8 \cdot H_0 = 1.02(33)$, $10^{11} \cdot L_0 = 0.50(12)$ cm^{-1}. Further equilibrium constants derived are (all in cm^{-1}): $10^3 \cdot \gamma_e = 0.89(57)$, $10^3 \cdot \delta_e = -0.187(36)$ (D_e and β_e are not given), and $\omega_e y_e = 0.1068(213)$, $\omega_e z_e = -0.01871(183)$; the constants B_v, D_v, and H_v for v=1 to 5 are also given [6].

b) Diode-laser absorption spectrum of the fundamental band. Using the constants of [1] for the ground vibrational level (see c)), the constants $B_1 = 8.16028(14)$, $10^4 \cdot D_1 = 4.389(22)$, $10^8 \cdot H_1 = 3.30(118)$ cm^{-1} were obtained, and the equilibrium values derived; $10^4 \cdot \beta_e = 4.8(30)$ cm^{-1} [7].

c) Far-IR LMR for PH, v=0 and PD, v=0, 1; $10^8 \cdot H_0 = 2.56(77)$ cm^{-1} for PH; analysis of the PD, v=1 spectrum by using the centrifugal distortion constant D_1 of [8]. The r_e given is the Born-Oppenheimer equilibrium distance obtained from B_0 by applying the various corrections described in [1, pp. 346/8].

d) Analysis of the far-IR LMR with D_0 held fixed at the same value as in [8] which was obtained by neglecting the sextic centrifugal distortion constant H_0 [3].

e) High-resolution UV emission spectra A $^3\Pi_i$→X $^3\Sigma^-$, 0→0, 0→1 of PH, 1→0, 0→0, 0→1, 1→1 of PD [8] and absorption spectrum A $^3\Pi_i$←X $^3\Sigma^-$, 1←0 of PH [13], all analyzed by [8] and results revised by [26]. The revised r_e values (originally 1.4221 Å for PH and 1.4218 Å for PD [8]) presumably are based on an improved value of Planck's constant [26]. $10^8 \cdot H_0 = 0.9(5)$ cm^{-1} for PH [8].

f) Analysis of the mid-IR LMR spectrum of the fundamental rotational-vibrational band [4] with D_0 and D_1 held fixed at the values derived by [8].

Remarks for Table 3:

a) Analysis of the far-IR LMR spectrum [3] with D_0 held fixed at the value of [24]; α_e and r_0 were derived by [26] from preliminary LMR results of the same team [2]. The $^1\Phi$←a $^1\Delta$ absorption spectrum had yielded $B_0 = 8.44_0$ cm^{-1}, $10^4 \cdot D_0 = 4.1_8$ cm^{-1}, and $r_0 = 1.430_3$ Å [24].

b) Analysis of the visible b $^1\Sigma^+$→X $^3\Sigma^-$, 0→0, 1→1, 2→2 emission spectrum using the ground-state constants of [8]; the values in parentheses presumably are mean deviations [23]. The MPI spectrum of PH_3 (see remark d)) gave $B_0 = 8.459(15)$ cm^{-1} and $10^4 \cdot D_0 = 4.1(3)$ cm^{-1} [25].

c) High-resolution UV emission spectrum A $^3\Pi_i$→X $^3\Sigma^-$, 0→0, 0→1 for PH and 1→0, 0→0, 0→1, 1→1 for PD [8] and absorption spectrum A $^3\Pi_i$←X $^3\Sigma^-$, 1←0 for PD [13], all analyzed by [8] and revised by [26]. Centrifugal distortion constants $10^8 \cdot H_0 = 1.6(1)$ cm^{-1} for PH, $10^4 \cdot \beta_e = 0.8$ and 0.1 cm^{-1} for PH and PD [8]. The revised r_0 values presumably are based on an improved value for Planck's constant; vibrational frequencies for the individual fine-structure substates $^3\Pi_{0^+}$, $^3\Pi_{0^-}$, $^3\Pi_1$, $^3\Pi_2$ of A $^3\Pi_i$ were derived (based also on a fine-structure analysis of earlier UV emission [9, 12] and absorption [13] spectra by [14]): v_0 ($= \Delta G_{1/2}$)=1833.78, 1833.39, 1833.74, 1834.38 cm^{-1} for PH, and 1357.40, 1357.14, 1357.21, 1357.54 cm^{-1} for PD [26].

d) Resonance enhancement due to $^3\Sigma^+$←b $^1\Sigma^+$, v=0←0 rovibronic transitions in the multiphoton ionization (MPI) spectrum of PH_3, i.e., (3+2+1)-photon process PH_3→PH b $^1\Sigma^+$→PH $^3\Sigma^+$→PH^+) [25] (a reassignment of the resonance-enhancing Rydberg state to the d $^1\Sigma^+ \ldots (4\sigma)^2 (5\sigma)^0 (2\pi)^4$ valence state was proposed, however [27], cf. p. 29).

e) The vacuum-UV absorption spectrum revealed the three Rydberg transitions $^1\Phi$←a $^1\Delta$, $^1\Pi$←a $^1\Delta$, and (tentatively) $^3\Pi$←X $^3\Sigma^-$ [24]; $B_{eff} = B_0(1 - 2B_0/A + \ldots)$ with A=spin-orbit coupling constant (see [28, p. 235]); therefrom the r_0 value for the $^3\Pi$ state was estimated by [26]; α_e for $^1\Phi$ was also derived by [26].

References:

[1] Ohashi, N.; Kawaguchi, K.; Hirota, E. (J. Mol. Spectrosc. **103** [1984] 337/49).
[2] Davies, P. B.; Russell, D. K.; Thrush, B. A. (Chem. Phys. Lett. **36** [1975] 280/2).
[3] Davies, P. B.; Russel, D. K.; Smith, D. R.; Thrush, B. A. (Can. J. Phys. **57** [1979] 522/8).
[4] Uehara, H.; Hakuta, K. (J. Chem. Phys. **74** [1981] 4326/9).
[5] Hirota, E. (Springer Ser. Chem. Phys. **40** [1985] 1/233).
[6] Ram, R. S.; Bernath, P. F. (J. Mol. Spectrosc. **122** [1987] 275/81).
[7] Anacona, J. R.; Davies, P. B.; Hamilton, P. A. (Chem. Phys. Lett. **104** [1984] 269/71).
[8] Rostas, J.; Cossart, D.; Bastien, J. R. (Can. J. Phys. **52** [1974] 1274/87).
[9] Ishaque, M.; Pearse, R. W. B. (Proc. R. Soc. [London] A **173** [1939] 265/77).
[10] Ishaque, M.; Pearse, R. W. B. (Proc. R. Soc. [London] A **156** [1936] 221/32).

[11] Pearse, R. W. B. (Proc. R. Soc. [London] A **129** [1930] 328/55).
[12] Pearse, R. W. B.; Ishaque, M. (Nature **137** [1936] 457).
[13] Legay, F. (Can. J. Phys. **38** [1960] 797/805).
[14] Horani, M.; Rostas, J.; Lefebvre-Brion, H. (Can. J. Phys. **45** [1967] 3319/31).
[15] Palmiere, P.; Sink, M. L. (J. Chem. Phys. **65** [1976] 3641/6).
[16] Wayne, F. D.; Colbourn, E. A. (Mol. Phys. **34** [1977] 1141/55).
[17] Furlani, T. R.; King, H. F. (J. Chem. Phys. **82** [1985] 5577/83).
[18] Koseki, S.; Schmidt, M. W.; Gordon, M. S. (J. Phys. Chem. **96** [1992] 10768/72).
[19] Ishiguro, E.; Kobori, M. (J. Phys. Soc. Jpn. **22** [1967] 263/70).
[20] Kovács, I. (Acta Phys. Acad. Sci. Hung. **13** [1961] 303/10).

[21] Curl, R. F., Jr. (Mol. Phys. **9** [1965] 585/97).
[22] Wayne, F. D.; Radford, H. E. (Mol. Phys. **32** [1976] 1407/22).
[23] Droege, A. T.; Engelking, P. C. (J. Chem. Phys. **80** [1984] 5926/9).
[24] Balfour, W. J.; Douglas, A. E. (Can. J. Phys. **46** [1968] 2277/80).
[25] Ashfold, M. N. R.; Dixon, R. N.; Stickland, R. J. (Chem. Phys. Lett. **111** [1984] 226/33).
[26] Huber, K. P.; Herzberg, G. (Molecular Spectra and Molecular Structure, Vol. 4: Constants of Diatomic Molecules, Van Nostrand Reinhold, New York 1979, pp. 534/7).
[27] Ashfold, M. N. R.; Clement, S. G.; Howe, J. D.; Western, C. M. (J. Chem. Soc. Faraday Trans. **87** [1991] 2515/23).
[28] Herzberg, G. (Molecular Spectra and Molecular Structure, Vol. 1: Spectra of Diatomic Molecules, Van Nostrand, Princeton, N. J., 1961).
[29] Senekowitsch, J.; Rosmus, P.; Werner, H.-J.; Larsson, M. (Z. Naturforsch. **41a** [1986] 719/23).
[30] Meyer, W.; Rosmus, P. (J. Chem. Phys. **63** [1975] 2356/75).

[31] Bruna, P. J.; Hirsch, G.; Peyerimhoff, S. D.; Buenker, R. J. (Mol. Phys. **42** [1981] 875/98).
[32] Nguyen, M. T. (Mol. Phys. **59** [1986] 547/58).
[33] Cade, P. E.; Huo, W. M. (J. Chem. Phys. **47** [1967] 649/72).
[34] Cade, P. E.; Huo, W. M. (At. Data Nucl. Data Tables **12** [1973] 415/66).
[35] Marsh, F. J.; Gordon, M. S. (Chem. Phys. Lett. **45** [1977] 255/60).
[36] Pope, S. A.; Hillier, I. H.; Guest, M. F. (Faraday Symp. Chem. Soc. No. 19 [1984] 109/23).
[37] Pople, J. A.; Luke, B. T.; Frisch, M. J.; Binkley, J. S. (J. Phys. Chem. **89** [1985] 2198/203).
[38] Yoshifuji, M.; Shibayama, K.; Inamoto, N. (J. Am. Chem. Soc. **105** [1983] 2495/7).
[39] Blustin, P. H. (J. Chem. Phys. **66** [1977] 5648/55).
[40] Trinquier, G. (J. Am. Chem. Soc. **104** [1982] 6969/77).

[41] Boyd, R. J.; Whitehead, M. A. (J. Chem. Soc. Dalton Trans. **1972** 78/81).

1.1.1.2.7 Potential Energy Functions

A Rydberg-Klein-Rees (RKR) potential function, i.e., numerical values for the energy E(v) and the corresponding classical turning points r_{min} and r_{max}, was calculated for the first six vibrational levels v=0 to 5 of the electronic ground state X $^3\Sigma^-$ (r=1.1 to 2.0 Å), using the authors' spectroscopic constants derived from the IR Fourier transform emission spectrum of PH [1]. An RKRV potential function (Vanderslice extension of the RKR method) based on the same spectroscopic data [1] was constructed for the ground-state vibrational levels v=0 to 7 [2]. Klein-Dunham potential curves (see [3]), extended by Morse potentials at large internuclear distances, have been constructed for the X $^3\Sigma^-$ and A $^3\Pi_i$ states at r=1.0 to 4.5 Å [4] using the spectroscopic constants (A↔X system in the UV) of [5]. The latter authors used their own data for the X and A states, those of [6] for the a $^1\Delta$ state, and semiempirically calculated [7] and empirically estimated term values to construct Morse potential curves for the states X $^3\Sigma^-$, a $^1\Delta$, b $^1\Sigma^+$, A $^3\Pi_i$, and $^1\Pi$ [5]. In **Fig. 1** are compiled the results of [1, 4, 5]. (Note that in [5] a somewhat lower P(^{2}D)+H(^{2}S) limit is depicted than in [4] and in the present figure.)

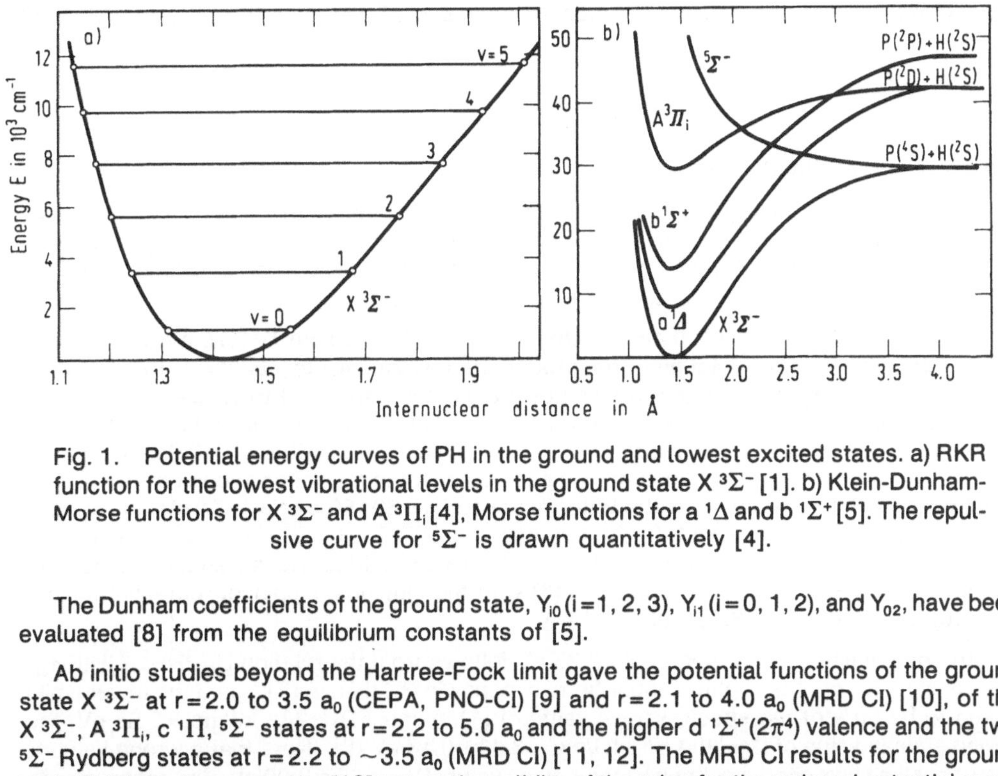

Fig. 1. Potential energy curves of PH in the ground and lowest excited states. a) RKR function for the lowest vibrational levels in the ground state X $^3\Sigma^-$ [1]. b) Klein-Dunham-Morse functions for X $^3\Sigma^-$ and A $^3\Pi_i$ [4], Morse functions for a $^1\Delta$ and b $^1\Sigma^+$ [5]. The repulsive curve for $^5\Sigma^-$ is drawn quantitatively [4].

The Dunham coefficients of the ground state, Y_{i0} (i=1, 2, 3), Y_{i1} (i=0, 1, 2), and Y_{02}, have been evaluated [8] from the equilibrium constants of [5].

Ab initio studies beyond the Hartree-Fock limit gave the potential functions of the ground state X $^3\Sigma^-$ at r=2.0 to 3.5 a_0 (CEPA, PNO-CI) [9] and r=2.1 to 4.0 a_0 (MRD CI) [10], of the X $^3\Sigma^-$, A $^3\Pi_i$, c $^1\Pi$, $^5\Sigma^-$ states at r=2.2 to 5.0 a_0 and the higher d $^1\Sigma^+$ ($2\pi^4$) valence and the two $^5\Sigma^-$ Rydberg states at r=2.2 to ~3.5 a_0 (MRD CI) [11, 12]. The MRD CI results for the ground state [11] have been used by [13] to test the validity of the rules for the reduced potential curve scheme (RPC method, see [14]). A MINDO/3+CI study dealt with the potential curves of the known states and with a number of unobserved bound and repulsive states [15].

References:

[1] Ram, R. S.; Bernath, P. F. (J. Mol. Spectrosc. **122** [1987] 275/81).
[2] Reddy, R. R.; Viswanath, R. (Astrophys. Space Sci. **155** [1989] 39/43).
[3] Jarmain, W. R. (J. Quant. Spectrosc. Radiat. Transfer **11** [1971] 421/6).

[4] Gustafsson, O.; Kindvall, G.; Larsson, M.; Senekowitsch, J.; Sigray, P. (Mol. Phys. **56** [1985] 1369/80).
[5] Rostas, J.; Cossart, D.; Bastien, J. R. (Can. J. Phys. **52** [1974] 1274/87).
[6] Balfour, W. J.; Douglas, A. E. (Can. J. Phys. **46** [1968] 2277/80).
[7] Jordan, P. C. (J. Chem. Phys. **41** [1964] 1442/9).
[8] Glushko, V. P.; Gurvich, L. V.; Bergman, G. A.; Veits, I. V.; Medvedev, V. A.; Khachkuruzov, G. A.; Yungman, V. S. (Termodinamicheskie Svoistva Individual'nykh Veshchestv, 3rd Ed., Vol. I, Book 1, Moscow 1978, p. 349).
[9] Meyer, W.; Rosmus, P. (J. Chem. Phys. **63** [1975] 2356/75).
[10] Barclay, V. J.; Wright, J. S. (Chem. Phys. **121** [1988] 381/91).

[11] Bruna, P. J.; Hirsch, G.; Peyerimhoff, S. D.; Buenker, R. J. (Mol. Phys. **42** [1981] 875/98).
[12] Bruna, P. J.; Peyerimhoff, S. D. (Bull. Soc. Chim. Belg. **92** [1983] 525/46).
[13] Jenc, F.; Brandt, B. A. (J. Chem. Soc. Faraday Trans. I **83** [1987] 2857/66).
[14] Jenc, F. (Adv. At. Mol. Phys. **19** [1983] 265/307).
[15] Minaev, B. F.; Buketova, A. E.; Muldakhmetov, Z. M. (Spectrosc. Lett. **22** [1989] 211/36).

1.1.1.2.8 Dissociation Energy

The dissociation energy of PH in the electronic ground state has not yet been accurately determined. Estimates from spectroscopic and thermochemical data indicate D_0 to lie somewhere between 3.0 and 3.5 eV; results from a few quantum-chemical calculations are also available.

The upper limits, $D_e \leq 31400\ cm^{-1} = 3.89$ eV and $D_0 \leq 30220\ cm^{-1} = 3.75$ eV, came from a linear Birge-Sponer extrapolation of the ground state vibrational levels $v'' = 0$, 1 ($A \leftrightarrow X$ emission and absorption spectra); the true values are expected to be 10 to 15% smaller [1]. The predissociation of the A $^3\Pi_i$ state by the repulsive $^5\Sigma^-$ state, which dissociates into the same products as the ground state, leads to the upper limits $D_e \leq 32150\ cm^{-1} = 3.99$ eV and $D_0 \leq 30970\ cm^{-1} = 3.84$ eV [1]. The accurately measured onset of predissociation at the $v = 0$, $J = 11$ level of the A $^3\Pi_i$ state lowered the latter result to $D_0 \leq 30150\ cm^{-1} = 3.74$ eV [2]. An older spectroscopic value, $D_0 = 3.5 \pm 0.3$ eV, recommended by [3], is based on the absorption measurements in the $A \leftarrow X$ system by [4]. Fitting of a Rydberg-Klein-Rees-Vanderslice potential function based on IR spectroscopic data [5] with the empirical Lippincott potential function (see [6]) resulted in $D_e = 3.16 \pm 0.12$ eV [7].

An upper limit of $D_0 \leq 314$ kJ/mol = 3.25 eV follows from the relation $D_0(PH) < \Delta_{at}H(PH_2)/2$ (as concluded from the slightly longer bond distance in PH than in PH_2), and the lower limit is taken to be 80% of the spectroscopic value of [1], namely $D_0 \geq 288$ kJ/mol = 2.98 eV, thus giving a mean value of $D_0 = 300 \pm 13$ kJ/mol = 3.11 ± 0.13 eV [8]. (Note that a wrong value for the upper limit was cited, and thus a lower mean value, $D_0 = 3.06 \pm 0.09$ eV, was recommended by [9].) A further value, $D_0 = 3.04 \pm 0.2$ eV, is derived [10] by employing normal standard thermodynamic values for P and H and the experimental $\Delta_f H^\circ$ value for PH from [1]. A value of $D_0 = 3.2$ eV (corresponding to $D_e = 3.34$ eV as cited by [12]) was estimated from isoelectronic series [13].

Theoretical values, obtained by quantum-chemical calculations that go beyond the SCF level (D_e(SCF) = 2.03 eV [12], 2.04 eV [14]) are given on the next page.

D_e in eV	D_0 in eV	method[a)]	Ref.
—	3.00, 2.97	G2, G1 (MP4+QCI)	[15, 16]
3.0527	—	MRD CI	[17]
3.05 ± 0.15	—	MCSCF-CI	[18]
3.07	2.93	MP4	[19]
3.03, 2.95	2.88, 2.80	"full" CI, MRD CI	[10]
3.04	3.02 ± 0.05[b)]	CEPA	[14]
2.93	—	PNO-CI	[14]
2.95 to 3.37	—	spin-density functional	[20]
3.01	—	spin-density functional	[21]
2.80	—	SCF + semiemp. E_{corr} estimate	[22]

[a)] The abbreviations are explained on pp. 3/4. – [b)] Recommended value [14] adopted by Huber and Herzberg [23], from which $D_0 = 3.06$ eV was derived for the PD molecule [23].

CNDO calculations gave values of ~3.4 to 3.5 eV [24], and simple empirical MO arguments used for the prediction of successive bond energies in PH_n (n=1 to 3) molecules resulted in $D_0 = 3.04$ eV for PH [25]. The early semiempirical estimate of [26], $D_0 = 3.05$ eV, has been adopted in the JANAF Tables [27, 28].

References:

[1] Rostas, J.; Cossart, D.; Bastien, J. R. (Can. J. Phys. **52** [1974] 1274/87).
[2] Gustafsson, O.; Kindvall, G.; Larsson, M.; Senekowitsch, J.; Sigray, P. (Mol. Phys. **56** [1985] 1369/80).
[3] Gaydon, A. G. (Dissociation Energies and Spectra of Diatomic Molecules, 3rd Ed., Chapman & Hall, London 1968, p. 279).
[4] Legay, F. (Can. J. Phys. **38** [1960] 797/805).
[5] Ram, R. S.; Bernath, P. F. (J. Mol. Spectrosc. **122** [1987] 275/81).
[6] Steele, D.; Lippincott, E. R.; Vanderslice, J. T. (Rev. Mod. Phys. **34** [1962] 239/51).
[7] Reddy, R. R.; Viswanath, R. (Astrophys. Space Sci. **155** [1989] 39/43).
[8] Glushko, V. P.; Gurvich, L. V.; Bergman, G. A.; Veits, I. V.; Medvedev, V. A.; Khachkuruzov, G. A.; Yungman, V. S. (Termodinamicheskie Svoistva Individual'nykh Veshchestv, 3rd Ed., Vol. I, Book 1, Moscow 1978, pp. 366/8).
[9] Berkowitz, J.; Curtiss, L. A.; Gibson, S. T.; Greene, J. P.; Hillhouse, G. L.; Pople, J. A. (J. Chem. Phys. **84** [1986] 375/84).
[10] Bruna, P. J.; Hirsch, G.; Peyerimhoff, S. D.; Buenker, R. J. (Mol. Phys. **42** [1981] 875/98).
[11] Di Stefano, G.; Lenzi, M.; Margani, A.; Mele, A.; Nguyen Xuan, C. (J. Photochem. **7** [1977] 335/44).
[12] Cade, P. E.; Huo, W. M. (J. Chem. Phys. **47** [1967] 649/72).
[13] Price, W. C. (private communication to [12]).
[14] Meyer, W.; Rosmus, P. (J. Chem. Phys. **63** [1975] 2356/75).
[15] Curtiss, L. A.; Raghavachari, K.; Trucks, G. W.; Pople, J. A. (J. Chem. Phys. **94** [1991] 7221/30).
[16] Curtiss, L. A.; Jones, C.; Trucks, G. W.; Raghavachari, K.; Pople, J. A. (J. Chem. Phys. **93** [1990] 2537/45).
[17] Barclay, V. J.; Wright, J. S. (Chem. Phys. **121** [1988] 381/91).

[18] Senekowitsch, J.; Rosmus, P.; Werner, H.-J.; Larsson, M. (Z. Naturforsch. **41a** [1986] 719/23).
[19] Pople, J. A.; Luke, B. T.; Frisch, M. J.; Binkley, J. S. (J. Phys. Chem. **89** [1985] 2198/203).
[20] Carroll, M. T.; Bader, R. F. W.; Vosko, S. H. (J. Phys. B **20** [1987] 3599/629).

[21] Stoll, H.; Pavlidou, C. M. E.; Preuß, H. (Theor. Chim. Acta **49** [1978] 143/9).
[22] Liu, H. P. D.; Legentil, J.; Verhaegen, G. (Sel. Top. Mol. Phys. Proc. Int. Symp., Ludwigsburg, FRG, 1970 [1972], pp. 19/30).
[23] Huber, K. P.; Herzberg, G. (Molecular Spectra and Molecular Structure, Vol. 4: Constants of Diatomic Molecules, Van Nostrand Reinhold, New York 1979, pp. 534/7).
[24] Boyd, R. J.; Whitehead, M. A. (J. Chem. Soc. Dalton Trans. **1972** 78/81).
[25] Goddard, W. A., III; Harding, L. B. (Ann. Rev. Phys. Chem. **29** [1978] 363/96).
[26] Jordan, P. C. (J. Chem. Phys. **41** [1964] 1442/9).
[27] Stull, D. R.; Prophet, H. (JANAF Thermochemical Tables, 2nd Ed., NSRDS-NBS-37 [1971]).
[28] Chase, M. W., Jr.; Davies, C. A.; Downey, J. R., Jr.; Frurip, D. J.; McDonald, R. A.; Syverud, A. N. (JANAF Thermochemical Tables, 3rd Ed., J. Phys. Chem. Ref. Data **14** Suppl. No. 1 [1985] 1254).

1.1.1.3 Heat of Formation. Thermodynamic Functions

The lack of a reliable experimental value for the dissociation energy of PH (see above) is reflected in the scarce data for the **heat of formation**. The 3rd edition of the JANAF Tables [1] recommends $\Delta_fH^\circ_0 = 253.55 \pm 33.5$ kJ/mol and $\Delta_fH^\circ_{298.15} = 254.49 \pm 33.5$ kJ/mol; these are, however, the old values [2] based on a semiempirical estimate [3] for D_0 that have been converted to a standard-state pressure of 0.1 MPa. A Russian compilation of thermodynamic data [4] gives $\Delta_fH^\circ_0 = 231.597 \pm 13$ kJ/mol and $\Delta_fH^\circ_{298.15} = 231.651 \pm 13$ kJ/mol based on estimated upper and lower limits for D_0. Ab initio calculations at the fourth-order Møller-Plesset level (MP4), using various extensions of the 6-31G basis set, showed $\Delta_fH^\circ_{298}$ to decrease with increasing basis set giving values between 261.4 and 244.0 kJ/mol [5, 6].

The **heat capacity** C°_p, **thermodynamic functions** S°, $-(G^\circ - H^\circ_{T^*})/T$, $H^\circ - H^\circ_{T^*}$, and the **equilibrium constant** K for the formation of PH as an ideal gas from the elements have been calculated for a standard-state pressure of 1 atm and tabulated for 298.15 K and between 0 and 6000 K at 100 K intervals [2, 4]. The JANAF data with $T^* = 298$ K are based on early spectroscopic data [7, 8] and semiempirical estimates for D_0 and the electronically excited states [3]. In the third edition of the JANAF Tables [1], the old values were only converted to Joule units and to a standard pressure of 0.1 MPa. In the Russian compilation [4], values for a reference temperature $T^* = 0$ K are given that are based on a partition function established by using spectroscopic data for the X $^3\Sigma^-$, a $^1\Delta$, A $^3\Pi_i$ states [9, 10] and theoretical data for the b $^1\Sigma^+$ and c $^1\Pi$ states [3, 11]. Selected values from [4] are:

T in K	C°_p	S° in $J \cdot mol^{-1} \cdot K^{-1}$	$-(G^\circ - H^\circ_0)/T$	$H^\circ - H^\circ_0$ in kJ/mol	log K
100	29.192	164.410	135.719	2.869	−153.8296
200	29.155	184.628	155.699	5.786	−75.0171
298.15	29.175	196.270	167.266	8.648	−48.9614
400	29.328	204.860	175.797	11.625	−35.3863

T in K	C_p°	S° in $J \cdot mol^{-1} \cdot K^{-1}$	$-(G^\circ - H_0^\circ)/T$	$H^\circ - H_0^\circ$ in kJ/mol	log K
600	30.297	216.907	187.617	17.574	−22.0648
800	31.733	225.816	196.098	23.775	−15.3538
1000	33.104	233.048	202.787	30.262	−11.3011
1500	35.674	246.992	215.326	47.500	−5.8522
2000	37.583	257.525	224.610	65.829	−3.1005
3000	40.693	273.378	238.367	105.033	−0.3225
4000	42.768	285.402	248.682	146.878	1.0831
5000	43.242	295.027	257.022	190.023	1.9390
6000	42.302	302.844	264.030	232.886	2.5212

Polynomial expansions of the partition function and the equilibrium constant at T=1000 to 10000 K based on the spectroscopic data compiled by Huber and Herzberg [12] are of astrophysical interest [13].

References:

[1] Chase, M. W., Jr.; Davies, C. A.; Downey, J. R., Jr.; Frurip, D. J.; McDonald, R. A.; Syverud, A. N. (JANAF Thermochemical Tables, 3rd Ed., J. Phys. Chem. Ref. Data **14** Suppl. No. 1 [1985] 1254).

[2] Stull, D. R.; Prophet, H. (JANAF Thermochemical Tables, 2nd Ed., NSRDS-NBS-37 [1971]).

[3] Jordan, P. C. (J. Chem. Phys. **41** [1964] 1442/9).

[4] Glushko, V. P.; Gurvich, L. V.; Bergman, G. A.; Veits, I. V.; Medvedev, V. A.; Khachkuruzov, G. A.; Yungman, V. S. (Termodinamicheskie Svoistva Individual'nykh Veshchestv, 3rd Ed., Vol. I, Moscow 1978, Book 1, pp. 366/8, Book 2, p. 281).

[5] Wong, M. W.; Gill, P. M. W.; Nobes, R. H.; Radom, L. (J. Phys. Chem. **92** [1988] 4875/80).

[6] Gordon, M. S.; Heitzinger, J. (J. Phys. Chem. **91** [1987] 2353/4).

[7] Ishaque, M.; Pearse, R. W. B. (Proc. R. Soc. [London] A **173** [1939] 265/77).

[8] Pearse, R. W. B. (Proc. R. Soc. [London] A **129** [1930] 328/55).

[9] Balfour, W. J.; Douglas, A. E. (Can. J. Phys. **46** [1968] 2277/80).

[10] Rostas, J.; Cossart, D.; Bastien, J. R. (Can. J. Phys. **52** [1974] 1274/87).

[11] Liu, H. P. D.; Legentil, J.; Verhaegen, G. (Sel. Top. Mol. Phys. Proc. Int. Symp., Ludwigsburg, FRG, 1970 [1972], pp. 19/30).

[12] Huber, K. P.; Herzberg, G. (Molecular Spectra and Molecular Structure, Vol. 4: Constants of Diatomic Molecules, Van Nostrand Reinhold, New York 1979, pp. 534/7).

[13] Sauval, A. J.; Tatum, J. B. (Astrophys. J. Suppl. Ser. **56** [1984] 193/209).

1.1.1.4 Spectra

1.1.1.4.1 Rotational and Vibrational Spectra

Purely rotational transitions of PH and PD, generated in microwave discharge flow systems by reacting phosphorus vapor with H or D atoms, have been observed by laser magnetic resonance (LMR) in the far IR. Using magnetic fields up to 1.4 T and eleven laser lines between

571 and 85 μm in parallel and perpendicular polarization, a great number of resonances have been assigned to $\Delta M_J=0$ and ± 1 transitions between the Zeeman levels of the lowest rotational levels with N=1 to 7 and J=N, N±1 (Hund's case (b)) for PH and PD in the ground vibronic state X $^3\Sigma^-$, v=0 and for the X $^3\Sigma^-$, v=1 state of PD [1]. Using the water vapor laser line at 118.6 μm, resonances were found for the N=5←4 transition of PH(X $^3\Sigma^-$, v=0) in magnetic fields up to 1.4 T [2] and for the J=5←4 transition of PH(a $^1\Delta$, v=0, Hund's case (a)) in fields up to 1.4 T [3] and extended up to 2.1 T [2]. Almost all resonances of PH appear as doublets of doublets due to the ^{31}P (I=1/2) and weaker ^1H (I=1/2) hyperfine interactions [1 to 3]. All resonances of PD exhibit ^{31}P hfs, whereas triplet splitting due to ^2H (I=1) was only resolved for two components of the N, J=3, 2←2, 1 transition in PD(X $^3\Sigma^-$, v=0) [1]. The following rotational transitions have been detected by LMR:

molecule	state	λ_{laser} in μm	N′←N″	J′←J″	number	Ref.
PD	X $^3\Sigma^-$, v=0	570.6	2←1	2←1, 3←2	5	[1]
		380.6	3←2	2←1, 3←2, 4←3	9	
		287.3	4←3	3←2, 4←3, 5←4	15	
		232.9	5←4	5←4	2	
		191.6	6←5	5←4, 7←6	12	
		164.6	7←6	6←5, 6←7, 8←7*)	14	
PD	X $^3\Sigma^-$, v=1	392.1	3←2	4←3, 2←3*)	4	[1]
PH	X $^3\Sigma^-$, v=0	302.3	2←1	3←2, 1←2*)	4	[1]
		294.8	2←1	1←2	2	
		287.3	2←1	3←2	1	
		118.8	5←4	4←5, 5←4, 4←3	9	
		85.3	7←6	7←6, 6←5	8	
PH	X $^3\Sigma^-$, v=0	118.6	5←4	5←4, 4←5	8	[2]
PH	a $^1\Delta$, v=0	118.6	—	5←4	9	[2, 3]

*) J and/or M_J numbers obviously misprinted in [1, Tables II and IV].

The analyses of these LMR spectra included investigations into the effects of rotation, centrifugal distortion, and fine-structure interaction. The results led to very precise values for the respective molecular constants (see pp. 12, 15/7) [1 to 3].

Rotational-vibrational transitions of PH and PD (generated by reacting phosphorus vapor with H or D atoms) in the fundamental band region were observed in the IR absorption [4] and emission [5] spectra and by mid-IR LMR [6]:

The fundamental **absorption** band of PH in its ground state X $^3\Sigma^-$ was observed at high resolution using a tunable diode laser spectrometer; eighteen lines betwen 2224.4125 and 2107.8160 cm^{-1} (4.50 to 4.74 μm) have been identified as P-branch transitions with N″=3 to 7 and 9, where each transition exhibits the fine-structure triplet splitting due to $\Delta J=\Delta N$ transitions between the three sublevels (J=N+1, N, N−1) of the respective rotational levels N″ and N′. Two further lines at 2292.1918 and 2308.1542 cm^{-1} (~4.3 μm) have been identified as the R-branch transitions with N″, J″=0,1 and 1,0. Hyperfine splitting could not be resolved. The analysis, which included rotational and centrifugal distortion effects as well as spin-spin and spin-rotation interaction, yielded the band origin at $\nu_0(1 \leftarrow 0) = 2276.2106(29)\ cm^{-1}$ (three standard deviation in parentheses) and, using the vibrational ground-state parameters of [1], spectroscopic constants for the v=1 level (see pp. 12, 17) [4].

Infrared **emission** from PH was excited in a microwave discharge tube through which a mixture of hydrogen and phosphorus vapor flowed, and the emission was observed with a high-resolution Fourier transform spectrometer. Five bands between 2500 and 1860 cm^{-1} (4.0 to 5.4 μm), which overlap each other in almost the whole region, were identified as the v=1→0, 2→1, 3→2, 4→3, and 5→4 transitions of PH(X $^3\Sigma^-$). Each band consists of three R and three P branches corresponding to the $\Delta J = \Delta N = \pm 1$ (J=N+1, N, N−1) transitions. Rotational progressions were recorded up to N=20, 17, 18, 14, and 13 for the v=1→0, 2→1, 3→2, 4→3, and 5→4 bands, respectively, leading to a total of about 380 rotational lines, whose wave numbers are given with an accuracy of ±0.001 cm^{-1}. The hyperfine structure was not resolved. The analysis yielded the molecular constants for PH(X $^3\Sigma^-$) in the lowest six vibrational states (v=0 to 5) and enabled the construction of an RKR potential curve (see p. 19) [5].

For PD in the electronic ground state (D atoms from MW discharges through D_2O or D_2 pumped over red phosphorus), rovibrational transitions have been detected by **LMR**. Measurements in fields of 0.2 to 1.9 T and using seven $^{12}C^{16}O$ laser lines around 6 μm in perpendicular polarization resulted in the following sixteen $\Delta M_J = \pm 1$ transitions within the v=1←0 band:

ν_{laser} in cm^{-1}	N′←N″	J′←J″	number
1626.1739	2←3	1←2	3
1633.5188	1←2	2←2, 2←3	6
1640.7316	0←1	1←1	1
1644.2882	0←1	1←2	1
1661.8830	1←0	2←1	2
1665.5079	2←1	1←1	2
1679.2899	3←2	2←1	1

Hyperfine splitting due to the ^{31}P nucleus was measured for six resonances, while hf splitting due to the D nucleus was not resolved. The analysis yielded precise values for the rotational, fine-structure, and hyperfine-structure constants of PD(X $^3\Sigma^-$, v=0 and 1) (see pp. 11, 12, 15) and the band origin $\nu_0(1 \leftarrow 0) = 1653.2858(36)\ cm^{-1}$ (three standard deviation) [6].

Theoretical Results. Before the fundamental rotation-vibration band of PH was observed, line positions for the main branches $P_i(N'')$ and $R_i(N'')$ (i=1, 2, 3; N″=0 to 20) in the range 5.4 to 4.0 μm had already been calculated (astrophysical interest) by [7], using the line positions of the 0−0 and 0−1 bands in the A $^3\Pi_i \leftrightarrow$ X $^3\Sigma^-$ system reported by [8] (see below).

Using the ab initio (CEPA) calculated dipole moment function for the X $^3\Sigma^-$ state of [9], the Einstein transition probabilities $A_{v',v''}$, oscillator strengths $f_{v',v''}$, and band origins $\nu(v'-v'')$ were calculated for the v′−v″=1−0, 2−1, and 2−0 transitions in PH(X $^3\Sigma^-$) [7]. Using the dipole moment functions for the X $^3\Sigma^-$ and A $^3\Pi_i$ states calculated by highly correlated MCSCF-CI wave functions, the vibrational electric dipole matrix elements $R_{v',v''}$ (v′=v″, v″+1, v″+2) and Einstein transition probabilities $A_{v',v''}$ (v′=v″+1, v″+2) have been derived for the transitions between the lowest vibrational states, v″=0, 1, 2 of X $^3\Sigma^-$ and v′=0, 1 of A $^3\Pi_i$, in PH and PD [10].

References:

[1] Ohashi, N.; Kawaguchi, K.; Hirota, E. (J. Mol. Spectrosc. **103** [1984] 337/49).
[2] Davies, P. B.; Russell, D. K.; Smith, D. R.; Thrush, B. A. (Can. J. Phys. **57** [1979] 522/8).
[3] Davies, P. B.; Russell, D. K.; Thrush, B. A. (Chem. Phys. Lett. **36** [1975] 280/2).
[4] Anacona, J. R.; Davies, P. B.; Hamilton, P. A. (Chem. Phys. Lett. **104** [1984] 269/71).
[5] Ram, R. S.; Bernath, P. F. (J. Mol. Spectrosc. **122** [1987] 275/81).
[6] Uehara, H.; Hakuta, K. (J. Chem. Phys. **74** [1981] 4326/9).
[7] De Gouveia, E. M.; Singh, P. D. (Sol. Phys. **90** [1984] 259/68).
[8] Rostas, J.; Cossart, D.; Bastien, J. R. (Can. J. Phys. **52** [1974] 1274/87).
[9] Meyer, W.; Rosmus, P. (J. Chem. Phys. **63** [1975] 2356/75).
[10] Senekowitsch, J.; Rosmus, P.; Werner, H.-J.; Larsson, M. (Z. Naturforsch. **41a** [1986] 719/23).

1.1.1.4.2 Electronic Spectra

General

The first spectroscopic identification of the PH and PD radicals was based on the emission spectrum near 340 nm which was attributed to the 0–0 band of the A $^3\Pi_i \rightarrow X\ ^3\Sigma^-$ transition; see "Phosphor" C, 1965, pp. 4/5, and [1, pp. 264, 564], [2, p. 306]. More detailed studies of the A–X transition, which could also be observed in absorption, have been carried out repeatedly, and reliable spectroscopic constants for the X and A states were derived. Furthermore, intensity and lifetime measurements enabled insight into the excitation and deexcitation processes of A $^3\Pi_i$ as well as the predissociation effects.

A second transition which has been studied in some detail is the spin-forbidden b $^1\Sigma^+ \rightarrow$ X $^3\Sigma^-$ transition identified with the visible emission near 700 nm. Attempts to observe the other forbidden transition, a $^1\Delta \rightarrow$ X $^3\Sigma^-$ expected in the near IR at 1.3 μm, were unsuccessful.

Absorption measurements in the vacuum UV resulted in the detection of two Rydberg transitions originating from the metastable a $^1\Delta$ state and of one $^3\Pi \leftarrow$ X $^3\Sigma^-$ transition which possibly is also of Rydberg type. Another Rydberg transition, $^3\Sigma^+ \leftarrow$ b $^1\Sigma^+$, was detected as a two-photon process with near-UV radiation; however, this transition was suggested to be one to a state of mainly valence character, d $^1\Sigma^+ \leftarrow$ b $^1\Sigma^+$.

Visible Emission. The b $^1\Sigma^+ \rightarrow$ X $^3\Sigma^-$ System

The spin-forbidden b $^1\Sigma^+ \rightarrow$ X $^3\Sigma^-$ transition was associated with a weak emission system in the red region centered at 697.3 nm, which was first observed in the vacuum-UV photolysis of PH_3 [3 to 6] and later in the flowing afterglow of an electrical discharge in a dilute PH_3–He mixture [7].

The spectrum recorded in the PH_3 afterglow at a resolution of 0.04 nm exhibited three vibrational bands in the range 682 to 720 nm assigned to the 0–0, 1–1, and 2–2 bands. Each of them contained in addition to small rotational branches a strong feature centered at ν_{max} (in cm^{-1}) = 14349.1 (0–0), 14391.9 (1–1), and 14439.7 (2–2) and having relative intensities of ~15:7:3. The rotational structure of the bands is the one expected from the selection rules +↔− and ΔJ = 0, ±1 for a transition between a $^1\Sigma^+$ state (J = N) and a $^3\Sigma^-$ state (Hund's case (b); J = N, N±1): The central feature consists of three Q-type (ΔN = 0) branches ^QP (N″ = 0 to 10), ^QQ and ^QR (N″ = 1 to 10) being flanked by small ^SR (N″ = 0 to 5, ΔN = +2) and ^OP (N″ = 2 to 9,

$\Delta N = -2$) branches (see the energy level diagram depicted in [3, figure 5]). Using the ground-state spectroscopic constants of [8], the rotational line positions and intensities of the b→X transition were modeled yielding the term value and the vibrational and rotational constants for the b $^1\Sigma^+$ state (see p. 16). Moreover, the intensity distribution showed that, in agreement with ab initio calculations of the $^1\Sigma^+ - {}^3\Sigma^-$ transition moments [9], the "forbidden" transition occurs because of spin-orbit mixing between the b $^1\Sigma^+$ and A $^3\Pi_i$ states at the expense of A→X intensity [7]. More recent intensity measurements in the 0–0 band and a previously determined radiative lifetime for the b $^1\Sigma^+$ state (see below) allowed the evaluation of the parallel and perpendicular transition moments $\mu_{\parallel}$ and $\mu_{\perp}$, active in the b $^1\Sigma^+$, $v'=0$→X $^3\Sigma^-$, $v''=0$ transition [10].

A spectrum recorded at lower resolution (0.5 nm) upon PH_3 photolysis showed the central "$^QP\ ^QQ\ ^QR$" peak at 697.3 nm and a few $^OP(N''=2$ to 5) and ^SR ($N''=0$ to 4) lines, all belonging to the 0–0 band [3].

Time-resolved measurements of the whole emission band gave a radiative lifetime of $\tau_{rad} = 1.25^{+0.16}_{-0.13}$ ms for the b $^1\Sigma^+$ state [5]. For the collisional quenching of the b→X emission, see p. 31.

UV Emission and Absorption. The A $^3\Pi_i \leftrightarrow$ X $^3\Sigma^-$ System

Excitation. The A→X emission at 340 nm has been observed in electric discharges through mixtures of H_2 or D_2 and phosphorus vapor [11 to 14] and discharges through PH_3 or PD_3 [8], in UV-photolyzed PH_3 or PD_3 [15 to 18], and in PH_3–H flames [19 to 21]. Furthermore, the action of shock waves on a PH_3–Ar mixture [22] and electron impact on PH_3 [23, 24] produced PH(A $^3\Pi_i$) radicals and thus A→X emission.

The A←X absorption spectrum around 340 nm has been recorded during flash photolysis of PH_3 [25 to 27].

A←X, 0←0 and 1←0 bands of PH and PD in Ar matrices (mole ratios 200 to 500) at 14 K could be observed, when $PH_3(PD_3)$–Ar mixtures were slowly deposited and simultaneously vacuum-UV-photolyzed [28]. The A←X, 0←0 band of PH in an Ar matrix at ~4 K was also observed when a PH_3–Ar mixture (volume ratio ~1:30) was exposed to a microwave discharge and subsequently condensed on a cold surface [29].

Rotational Analysis. Highly resolved emission spectra of PH and PD have been photographed in discharges through PH_3 or PD_3 in a Schüler-type discharge tube [8]: In addition to the already known 0–0 bands for PH and PD [11 to 14], the 0–1 band of PH and the 0–1, 1–1, and 1–0 bands of PD were obtained. The 1–0 band of PH could not be observed in emission (cf. below), therefore the earlier recorded absorption spectrum [26] was reanalyzed. The emission and absorption bands exhibited the rotational and fine structure expected for a $^3\Pi_i \rightarrow {}^3\Sigma^-$ transition, where the upper state approaches Hund's case (a) and turns towards (b) with increasing rotation and where the lower state belongs to Hund's case (b). In other words, for a $^3\Pi_i(a) \rightarrow {}^3\Sigma^-(b)$ transition, each of the three subbands $^3\Pi_2(F_1)$, $^3\Pi_1(F_2)$, $^3\Pi_0(F_3) \rightarrow {}^3\Sigma^-$ (with Λ-type doubling of the upper-state rotational levels into F_{ie} and F_{if} (i=1, 2, 3) components) consists of the three main branches ($\Delta N = \Delta J$) P_i, Q_i, R_i (i=1, 2, 3) and the six satellite branches exhibiting an N, O, P, Q, R, S, or T form for those $\Delta N = \Delta J \pm 1$ or $\Delta J \pm 2$ transitions that are allowed by the selection rules $+ \leftrightarrow -$ and $\Delta J = 0, \pm 1$ (see, e.g., [30], [1, pp. 264/5], and [24, p. 1371]). Lines were observed in each of the 27 branches of the 0–0 bands except in $^NP_{13}$ of PH, but some satellite branches are absent in the vibrational bands. More than 2000 lines (for PH: 190 (1–0), 417 (0–0), 222 (0–1); for PD: 321 (1–0), 544 (0–0), 357 (1–1), 362 (0–1)) were included in the so far most accurate A–X rotational analysis which considered spin-spin and second-order spin-orbit interactions as well as centrifugal distortion effects on both rotation and spin-orbit couplings. A complete list of the wave numbers and assignments is available on request; see [8, footnote on p. 1276]. The various molecular constants derived for PH and PD by [8] are given on pp. 12, 15/7.

Radiative Lifetime. Oscillator Strength. Predissociation. The fact that A→X emission in PH originates only from the v′ = 0 vibrational level, even though the excitation energy is sufficient to produce also PH(A $^3\Pi_i$, v′ = 1) and the Franck-Condon factors indicate comparable intensities for the 1−1 and 0−0 bands, was explained by a depopulation of the v′ = 1 level through a nonradiative process. This process was suggested to be a predissociation into the ground-state atoms P(^{4}S) and H(^{2}S) through the repulsive $^5\Sigma^-$ state which dissociates into the same products and crosses the A-state potential curve [8]. On the other hand, the A←X, 1←0 absorption band does not reveal any signs of predissociation effects in the form of line broadening [26], which indicates that the onset of predissociation is somewhere between the v′ = 0 and v′ = 1 levels. The onset around the J′ = 11 rotational level of the v′ = 0 level can be explained as follows [24]: Fluorescence lifetimes τ_f were measured at a high resolution of ~0.005 nm (high-frequency deflection technique) for a large number of rotational levels on twelve branches (up to J′ = 23) originating from all six fine-structure components F_{ie} and F_{if} (i = 1, 2, 3) of the A $^3\Pi_i$, v′ = 0 state. The excited PH were generated by pulsed electron impact (20 keV) on PH_3. For the low rotational quantum numbers J′ = 1 to 10, the fluorescence lifetimes are essentially independent of J′; the average value of 450 ns which was assumed to correspond to a purely radiative decay is in good agreement with τ_{rad} = 445 ± 50 ns (oscillator strength f = 0.0078 ± 0.0008) measured for the first time at low resolution (~1 nm) upon electron excitation (30 to 90 eV) of PH_3 [23]. A distinct decrease of the fluorescence lifetime starts at around J′ = 11, and values of τ_f down to 144 ns for the F_{1e}, J′ = 23 level were measured. Using the relation $\tau_f^{-1} = \tau_{rad}^{-1} + \tau_{pred}^{-1}$ and τ_{rad} = 450 ns, the lifetimes for predissociation, τ_{pred}, could be derived. A theoretical study of the (first-order spin-orbit) interaction between the A $^3\Pi_i$ (v′ = 0) and $^5\Sigma^-$ states was carried out assuming that the $^5\Sigma^-$ state belongs to Hund's case (b) and the A $^3\Pi_i$ to either Hund's case (a) or case (b): in the first case, all fine-structure levels of A $^3\Pi_i$ except F_{3e} ($^3\Pi_0$) are affected by $^5\Sigma^-$ state levels, in the second case, the F_{3e} level also interacts with $^5\Sigma^-$, i.e., it is expected to be also predissociated. Unfortunately, the P_3 and Q_3 branches originating from this level were so badly developed in the spectra that the onset of predissociation could not be monitored [24].

Earlier measurements of time-resolved luminescence at 342.2, 341.5, and 339.6 nm for the emissions from the three substates $^3\Pi_2$, $^3\Pi_1$, and $^3\Pi_0$ of PH(A $^3\Pi_i$, v′ = 0), carried out at low resolution (~0.6 nm) and following pulsed irradiation of PH_3 with an ArF excimer laser (193.3 nm), yielded the short fluorescence lifetimes τ_f = 355 ns for $^3\Pi_2$ and $^3\Pi_1$ and τ_f = 265 ns for $^3\Pi_0$; observations centered at wavelengths away from the emission maxima showed evidence for components with even shorter lifetimes down to presumably 70 ns [18]. These findings can in view of the more recent high-resolution study of [24] be explained by spectral overlap between the different substates and between predissociated and unpredissociated rotational levels leading to lifetimes shorter than τ_{rad}.

The absorption oscillator strengths for the 0−0, 1−0, and 0−1 bands of the A−X system of PH were calculated by [31] from the radiative lifetime of the A $^3\Pi_i$ state as measured by [23] and from Franck-Condon factors derived from the spectroscopic constants of [8].

Using highly correlated MCSCF-CI wave functions for the A $^3\Pi_i$ and X $^3\Sigma^-$ states, the transition moment function for the A−X transition has been calculated which in turn allowed the evaluation of Einstein coefficients of spontaneous emission $A_{v',v''}$ (v′ = 0, 1; v″ = 0, 1, 2), absorption oscillator strengths $f_{v',v''}$ (v′ = 0, 1; v″ = 0, 1), and radiative lifetimes for A $^3\Pi_i$, v′ = 0, 1 of PH and PD. The v′ = 0 lifetime τ_{rad} = 399 ns for PH (390 ns for PD) is shorter than the experimental value, probably because the large correlation energy contributions to the transition moment have not been sufficiently accounted for in the calculation [32].

Rydberg Transitions $^3\Sigma^+ \leftarrow b\ ^1\Sigma^+$; $^1\Phi$, $^1\Pi \leftarrow a\ ^1\Delta$; and $^3\Pi \leftarrow X\ ^3\Sigma^-$

A spin-forbidden $^3\Sigma^+$ {KL $(4\sigma)^2$ $(5\sigma)^2$ $(2\pi)^1$} $(4p\pi)^1 \leftarrow b\ ^1\Sigma^+$ transition has been observed to be a two-photon resonance in multiphoton ionization (MPI) spectroscopy of PH_3 (PD_3) with a tunable dye laser in the wavelength range 394 to 383 nm. The MPI spectrum of PH_3, i.e., PH^+ ion yield vs. wavelength, showed resonance enhancement (RE) in the form of a strong central Q-branch feature at 388.8 nm and O- and S-form branches with N″ up to 20, where the Q-branch feature is sensitive to the polarization state of the exciting light. Weak Q branches attributed to the 1–1 and 2–2 bands appear on the long-wavelength side of the central peak. The spectroscopic observations are consistent with the 2-photon transitions $^3\Sigma^+$ (Hund's case (b); J = N, N ± 1) $\leftarrow ^1\Sigma^+$ with the selection rules $+ \leftrightarrow +$, $- \leftrightarrow -$, $\Delta J = 0, \pm 1, \pm 2$, and $\Delta N = 0, \pm 2$ giving rise to a central ^QP ^QQ ^QR-branch feature and adjacent ^OP, ^OO and ^SR, ^SS branches (see the energy level diagram depicted in [33, figure 2]). Rotational analysis using the b $^1\Sigma^+$ state molecular constants derived by [7] resulted in term values and rotational and vibrational constants for the $^3\Sigma^+$ state (cf. pp. 5, 16). The overall REMPI process was originally interpreted to be a 3-photon excitation and dissociation, $PH_3 \rightarrow PH_3^* \rightarrow PH(b\ ^1\Sigma^+)$, followed by (2+1)-photon excitation and ionization, $PH(b\ ^1\Sigma^+) \rightarrow PH(^3\Sigma^+) \rightarrow PH^+$ [33]. A subsequent reinvestigation of the REMPI of PH_3 with mass-selective detection revealed a striking rotational-level dependence of the branching ratio in the ion yield: REMPI occurring via the lower rovibrational levels of the intermediate state yielded PH^+ ions, whereas REMPI via the higher J′ levels yielded P^+ fragment ions. These observations were rationalized in terms of a (2+2)-REMPI process of PH, in which the final 2-photon ionization presumably proceeds via a favorable near resonance (possibly a $^3\Pi$ Rydberg state with {KL $(4\sigma)^2$ $(5\sigma)^1$ $(2\pi)^2$} $(4p\pi)^1$ configuration) before terminating in the predissociated A $^2\Delta$ state of the ion [34]. More recently, striking parallels in the REMPI behavior of the PH (PD) and NH (ND) radicals (cf. "Nitrogen" Suppl. Vol. B 1) have been observed, possibly requiring a reassignment of the intermediate state: instead of the $^3\Sigma^+$ Rydberg state, the analogue of the d $^1\Sigma^+$ state of NH with predominantly valence character may be responsible for the resonance enhancement in the REMPI of PH [35].

Two Rydberg transitions originating from the metastable a $^1\Delta$ state were observed in the vacuum-UV absorption spectrum of PH generated in a discharge flow system in diluted PH_3–Ar mixtures. The $^1\Phi \leftarrow a\ ^1\Delta$ system near 162.5 nm consists of single P (J = 4 to 14), Q (J = 3 to 20), and R (J = 2 to 18) branches attributed to the 0–0 transition. Weaker branches accompanying the R branch were assigned to the 1–1 (J = 2 to 12) and 2–2 (J = 2 to 8) transitions. The $^1\Pi \leftarrow a\ ^1\Delta$ transition was identified with a weak (about half the intensity of the $^1\Phi \leftarrow a\ ^1\Delta$, 0–0 band) and isolated band at 159.5 nm which because of Λ-type doubling in the $^1\Pi$ state exhibits two P (J=1 to 12), two Q (J=1 to 12), and two R (J=3 to 10) branches. A third, extremely weak absorption band of complex rotational structure occurring at 143.5 nm was tentatively assigned to a $^3\Pi \leftarrow X\ ^3\Sigma^-$ transition. A number of lines could be identified as the main branches P_1, Q_1, R_1 (N″ up to 11) and the corresponding satellite branches $^OP_{12}$, $^PQ_{12}$, and $^QR_{12}$ (N″ up to 12) of the $^3\Pi_0(F_1) \leftarrow X\ ^3\Sigma^-_{0,1}(F_1, F_2)$ transitions between the substates of $^3\Pi$ and X $^3\Sigma^-$. For the metastable a $^1\Delta$ and the three Rydberg states $^1\Phi$, $^1\Pi$, and $^3\Pi$, which possibly arise from the electron configuration {KL $(4\sigma)^2$ $(5\sigma)^2$ $(2\pi)^1$} $(3d\delta)^1$, molecular constants (cf. p. 16) have been derived [36].

References:

[1] Herzberg, G. (Molecular Spectra and Molecular Structure, Vol. 1: Spectra of Diatomic Molecules, Van Nostrand, Princeton, N. J., 1961).

[2] Rosen, B. (International Tables of Selected Constants **17**, Spectroscopic Data Relative to Diatomic Molecules, Pergamon Press, Oxford 1970).

[3] Di Stefano, G.; Lenzi, M.; Margani, A.; Nguyen Xuan, C. (J. Chem. Phys. **68** [1978] 959/63).
[4] Lenzi, M.; Margani, A.; Nguyen Xuan, C.; Di Stefano, G. (Chem. Phys. Lett. **63** [1979] 86/9).
[5] Nguyen Xuan, C.; Di Stefano, G.; Lenzi, M.; Margani, A.; Mele, A. (Chem. Phys. Lett. **57** [1978] 207/10).
[6] Margani, A.; Nguyen Xuan, C.; Di Stefano, G.; Lenzi, M. (J. Chem. Phys. **75** [1981] 4912/20).
[7] Droege, A. T.; Engelking, P. C. (J. Chem. Phys. **80** [1984] 5926/9).
[8] Rostas, J.; Cossart, D.; Bastien, J. R. (Can. J. Phys. **52** [1974] 1274/87).
[9] Wayne, F. D.; Colbourn, E. A. (Mol. Phys. **34** [1977] 1141/55).
[10] Di Stefano, G.; Lenzi, M.; Piciacchia, G.; Ricci, A. (Chem. Phys. Lett. **185** [1991] 212/4).

[11] Ishaque, M.; Pearse, R. W. B. (Proc. R. Soc. [London] A **156** [1936] 221/32).
[12] Ishaque, M.; Pearse, R. W. B. (Proc. R. Soc. [London] A **173** [1939] 265/77).
[13] Pearse, R. W. B. (Proc. R. Soc. [London] A **129** [1930] 328/55).
[14] Pearse, R. W. B.; Ishaque, M. (Nature **137** [1936] 457).
[15] Norrish, R. G. W.; Oldershaw, G. A. (Proc. R. Soc. [London] A **262** [1961] 1/9).
[16] Becker, K. H.; Welge, K. H. (Z. Naturforsch. **19a** [1964] 1006/15).
[17] Di Stefano, G.; Lenzi, M.; Margani, A.; Mele, A.; Nguyen Xuan, C. (J. Photochem. **7** [1977] 335/44).
[18] Sam, C. L.; Yardley, J. T. (J. Chem. Phys. **69** [1978] 4621/7).
[19] Guenebaut, H.; Pascat, B. (C. R. Hebd. Seances Acad. Sci. **256** [1963] 677/80).
[20] Guenebaut, H.; Pascat, B. (J. Chim. Phys. Phys. Chim. Biol. **61** [1964] 592/5).

[21] Pascat, B.; Guenebaut, H. (Ann. Univ. ARERS **2** [1963] 100/7).
[22] Guenebaut, H.; Pascat, B. (C. R. Hebd. Seances Acad. Sci. **255** [1962] 1741/3).
[23] Fink, E.; Welge, K. H. (Z. Naturforsch. **19a** [1964] 1193/201).
[24] Gustafsson, O.; Kindvall, G.; Larsson, M.; Senekowitsch, J.; Sigray, P. (Mol. Phys. **56** [1985] 1369/80).
[25] Nelson, L. S.; Ramsay, D. A. (J. Chem. Phys. **25** [1956] 372/3).
[26] Legay, F. (Can. J. Phys. **38** [1960] 797/805).
[27] Kley, D.; Welge, K. H. (Z. Naturforsch. **20a** [1965] 124/31).
[28] Larzillière, M.; Jacox, M. E. (NBS-SP-561-1 [1979] 529/43).
[29] McCarty, M., Jr.; Robinson, G. W. (J. Chim. Phys. Phys. Chim. Biol. **56** [1959] 723/30).
[30] Dixon, R. N. (Can. J. Phys. **37** [1959] 1171/86).

[31] De Gouveia, E. M.; Singh, P. D. (Sol. Phys. **90** [1984] 259/68).
[32] Senekowitsch, J.; Rosmus, P.; Werner, H.-J.; Larsson, M. (Z. Naturforsch. **41a** [1986] 719/23).
[33] Ashfold, M. N. R.; Dixon, R. N.; Stickland, R. J. (Chem. Phys. Lett. **111** [1984] 226/33).
[34] Ashfold, M. N. R.; Stickland, R. J.; Tutcher, B. (Mol. Phys. **65** [1988] 1455/71).
[35] Ashfold, M. N. R.; Clement, S. G.; Howe, J. D.; Western, C. M. (J. Chem. Soc. Faraday Trans. **87** [1991] 2515/23).
[36] Balfour, W. J.; Douglas, A. E. (Can. J. Phys. **46** [1968] 2277/80).

1.1.1.5 Collisional Quenching of PH(b $^1\Sigma^+$) and PH(A $^3\Pi_i$)

Time-resolved fluorescence of PH in the b $^1\Sigma^+ \rightarrow X\ ^3\Sigma^-$ and A $^3\Pi_i \rightarrow X\ ^3\Sigma^-$ systems has been measured in the presence of the parent molecules PH_3 only or in the presence of an added foreign gas species. Stern-Volmer plots of the fluorescence decay rates as a function of the concentration (or pressure) of the collision partner Q, i.e., τ^{-1} vs. [Q], allowed the determination of the radiative lifetimes (cf. pp. 27 and 28) and the quenching rate constants k_Q or quenching cross sections σ_Q for the two excited states. In the following paragraphs, quenching rate constants k_Q are given in $cm^3 \cdot molecule^{-1} \cdot s^{-1}$ and quenching cross sections σ_Q in Å^2; the latter were obtained by the relation $\sigma_Q = k_Q \cdot (\pi\mu/8\ kT)^{1/2}$, where k = Boltzmann constant, T = absolute temperature, μ = reduced mass of the colliding pair [4].

Quenching of the b $^1\Sigma^+$ State. Rare-gas atoms, four diatomic molecules, and the parent molecule PH_3 were chosen as collision partners Q. The following rate constants and cross sections were measured [1, 2]:

Q	$10^{15} \cdot k_Q$	$10^6 \cdot \sigma_Q$	Q	$10^{15} \cdot k_Q$	$10^6 \cdot \sigma_Q$
He	0.319±0.005 a)	23.8 a)	H_2	23.5±0.2 c)	1275 c)
Ne	0.127±0.011	17.9	N_2	1.71±0.02	265
Ar	0.502±0.033	84.0	CO	1.81±0.03	278
Kr	0.823±0.074	156	NO	5.66±0.2	879
Xe	(0.795, 1.34) b)	(160, 270.7) b)	PH_3	320±10 d)	—

a) Results possibly erroneous due to a contaminant; extrapolation of the data for Ne, Ar, and Kr gave $10^{15} \cdot k_Q = 0.0586$, 0.106 and $10^6 \cdot \sigma_Q = 4.38$, 7.92. – b) Extrapolated. – c) Several reaction channels are thermodynamically allowed for the formation of PH_2 and PH_3. – d) Previous result (275±7) of [3] improved by [1].

Quenching of the A $^3\Pi_i$ State. Electronic quenching of PH(A $^3\Pi_i$, v′ = 0) with a thermally equilibrated rovibrational population was studied at 243, 296, and 415 K for a number of molecular collision partners; in all cases, the rate constants decreased with increasing temperature as shown in the following table [4, 5]:

Q	T = 243±3 K [5] $10^{11} \cdot k_Q$	T = 296±3 K [4] $10^{11} \cdot k_Q$	σ_Q	T = 415±5 K [4] $10^{11} \cdot k_Q$	σ_Q
H_2	9.4±0.3	8.8±0.5 8.9±0.6 [5]*) 8.7±0.3 [5]*)	4.8±0.3	8.1±0.3	3.7±0.1
D_2	—	5.9±0.4 [5]	—	—	—
N_2	0.49±0.02	0.39±0.03	0.6±0.04	0.27±0.01	0.36±0.01
O_2	8.95±0.4*) 9.0±0.6*)	8.3±0.4	13.2±0.6	7.9±0.5	10.7±0.7
CO	22.5±0.6	21±1	32±1	17.8±0.8	23±1
H_2O	43.5±1.3 (253 K)	40.6±1.9	54.8±2.6	34.4±0.7	39.5±0.8
CO_2	0.38±0.02	0.34±0.014	0.58±0.02	0.31±0.01	0.45±0.02

Q	T=243±3 K [5]	T=296±3 K [4]		T=415±5 K [4]	
	$10^{11}\cdot k_Q$	$10^{11}\cdot k_Q$	σ_Q	$10^{11}\cdot k_Q$	σ_Q
N_2O	0.87±0.04	0.80±0.04	1.36±0.06	0.73±0.04	1.06±0.06
NH_3	78±4	76±7	101±10	72±7	81±8
PH_3	—	53±17	80±24	—	—
C_2H_6	17.1±0.4	15.3±0.5	24.0±0.8	12.6±0.4	16.7±0.5

*) Two separate measurements.

Earlier results for the quenching by the parent molecule PH_3 are $10^{10}\cdot k_Q$=7.9, 5.5, 6.7 (all ±0.6) and σ_Q=130, 91, 110 for the A-state spin components $^3\Pi_2$, $^3\Pi_1$, $^3\Pi_0$ (original values $10^{-6}\cdot k_Q$=22, 18, 22 (all ±2) $Torr^{-1}\cdot s^{-1}$ converted assuming a temperature of 295 K) [6] and $10^{10}\cdot k_Q$=9.2 for the three substates (converted from $10^{-7}\cdot k_Q$=3 $Torr^{-1}\cdot s^{-1}$ assuming a temperature of 295 K) [7]. Rate constants $k_Q=4.2\times10^{-10}$, $\sim2.2\times10^{-11}$, and $\sim8.2\times10^{-12}$ (σ_Q=63, 1.2, and 0.6) for collisions with PH_3, H_2, and D_2 were reported by [8], $\sigma_Q\approx100$ and 0.2 for collisions with PH_3 and N_2 by [9].

References:

[1] Lenzi, M.; Margani, A.; Nguyen Xuan, C.; Di Stefano, G. (Chem. Phys. Lett. **63** [1979] 86/9).
[2] Margani, A.; Nguyen Xuan, C.; Di Stefano, G.; Lenzi, M. (J. Chem. Phys. **75** [1981] 4912/20).
[3] Nguyen Xuan, C.; Di Stefano, G.; Lenzi, M.; Margani, A.; Mele, A. (Chem. Phys. Lett. **57** [1978] 207/10).
[4] Kenner, R. D.; Pfannenberg, S.; Stuhl, F. (Chem. Phys. Lett. **156** [1989] 305/11).
[5] Kenner, R. D.; Pfannenberg, S.; Heinrich, P.; Stuhl, F. (J. Phys. Chem. **95** [1991] 6585/93).
[6] Sam, C. L.; Yardley, J. T. (J. Chem. Phys. **69** [1978] 4621/7).
[7] Gustafsson, O.; Kindvall, G.; Larsson, M.; Senekowitsch, J.; Sigray, P. (Mol. Phys. **56** [1985] 1369/80).
[8] Allen, J. E., Jr.; Cody, R. J. (NBS-SP-716 [1986] 3/10).
[9] Becker, K. H.; Welge, K. H. (Z. Naturforsch. **19a** [1964] 1006/15).

1.1.1.6 Chemical Reactions

The intermediate formation and decomposition of PH upon sorption of PH_3 on metal surfaces is discussed in Section 1.3.1.5.11, p. 287. The reaction of the spectroscopically observed intermediate PH (from the reaction of red phosphorus with atomic H) with atomic oxygen yielded PO and PO_2 [1, 2]; the same mechanism was also formulated as a step in the reactions of PH_3 with O atoms [3]. Cocondensation of PH (from UV photolysis of PH_3) with CO in an Ar matrix at 12 K gives HPCO [4].

The interaction of singlet PH with some hydrogen compounds XH_n of main-group elements was studied by ab initio MO calculations at the MP4/6-31G* [5] and MP4/6-311G** levels [6]. The bonding in the initial donor-acceptor complexes $HPXH_n$ was predicted to originate mainly from van der Waals interactions of the reactants, except for the products with H_2S and PH_3

where some ylide-type stabilization was discussed. The rearrangement of these initial products via a hydrogen shift gives the energetically more favorable tautomers H_2PXH_{n-1}. Their formation requires the crossing of an energy barrier (data are given in the original papers) and corresponds to the insertion of PH into the H–X bonds [5, 6]. Energies in kcal/mol (corrected for zero point energies) of the interaction of the reactants in the donor-acceptor complexes and the energies set free upon rearrangement to the insertion products were calculated to be as follows:

XH_n	FH	ClH	OH_2	SH_2	NH_3	PH_3
interaction energy in $HPXH_n$	7.0	6.0	21.0	25.9	42.1	53.8
energy gained by hydrogen shift	67.1	66.9	51.2	48.3	32.3	27.6
Ref.	[6]	[6]	[6]	[6]	[5]	[5]

The potential energy surfaces for the addition and insertion reactions of PH with C_2H_4 were studied by an ab initio SCF-CI calculation with a 4-31G* basis set. The cycloaddition of PH to the double bond yields phosphirane and was predicted to proceed without an activation energy. The insertion of PH into a C–H bond with formation of vinylphosphane was predicted to have an activation energy of about 6 kcal/mol. Both reaction channels were computed to be exothermic by about 75 kcal/mol [7].

Reactions of PH were discussed in the hypothetical mechanism of the UV photolysis of PH_3 (see Section 1.3.1.5.1.3.1, p. 206) and its mixture with CO or SiH_4 (see Section 1.3.1.5.6, p. 249). Reactions of PH were also thought to be involved in the photosensitized reaction upon IR laser irradiation of PH_3 with SiH_4 in the presence of SiF_4 (see Section 1.3.1.5.6, p. 250). The insertion of intermediately formed PH into an Si–D bond was thought to be responsible for the formation of $(CH_3)_2Si(D)PHD$ during the pyrolysis of $SiH_3PH_2/(CH_3)_2SiD_2$ mixtures at 300 K [8].

References:

[1] Kawaguchi, K.; Seito, S.; Hirota, E. (J. Chem. Phys. **79** [1983] 629/34).
[2] Hirota, E. (Springer Ser. Chem. Phys. **40** [1985] 1/233, 201/2).
[3] Davies, P. B.; Thrush, B. A. (Proc. R. Soc. [London] A **302** [1968] 243/52).
[4] Mielke, Z.; Andrews, L. (Chem. Phys. Lett. **181** [1991] 355/60).
[5] Sudhakar, P. V.; Lammertsma, K. (J. Am. Chem. Soc. **113** [1991] 1899/906).
[6] Sudhakar, P. V.; Lammertsma, K. (J. Org. Chem. **56** [1991] 6067/71).
[7] Gonbeau, D.; Pfister-Guillouzo, G. (Inorg. Chem. **26** [1987] 1799/805).
[8] Elliot, L. E.; Estacio, P.; Ring, M. A. (Inorg. Chem. **12** [1973] 2193/4).

1.1.2 PH^+

CAS Registry Numbers: PH^+ *[12339-19-4]* Phosphorus(1+), hydro-; PD^+ *[12191-23-0]* Phosphoniumylidyne-*d*

Formation

Free PH^+ ions were observed in the mass spectrum of PH_3 (cf. Section 1.3.1.5.1.4, p. 212). The appearance potential for the (energetically lowest) fragmentation process $PH_3 \rightarrow PH^+ + H_2 + e^-$ was measured to be AP(in eV)=12.4±0.2 [1], 12.6±0.2 [2], or 12.9 [3]. This was confirmed by a photoionization mass spectrometric study of PH_3 between 130 and 60 nm: the PH^+

ion yield curve gave a threshold energy of 12.451±0.005 eV (corrected to 12.492±0.005 eV at 0 K) which is about 1 eV lower than that for the PH_2^+ ion [4].

Photoionization of a PH molecular beam via $PH \rightarrow PH^+ + e^-$ sets in at 10.149±0.008 eV [5].

PH^+ ions were obtained by the gas-phase reactions of PH_3 with rare gas cations and CO_2^+ ions [6].

PH^+ and PD^+ ions for UV-visible spectroscopic studies (cf. pp. 40/2) were obtained by hollow-cathode [7, 8] or microwave [9, 10] discharges through helium that contained traces of H_2 or D_2 and small amounts of phosphorus vapor, by collisions of P^+ ions with H_2 or D_2 molecules [11], and also by electron impact on PH_3 [8, 12 to 14].

Resonance-enhanced multiphoton ionization (REMPI) of PH_3 (PD_3) terminated in the production of PH^+ (PD^+) ions resulting from the (3+2+1)- or (3+2+2)-photon process $PH_3 \rightarrow PH^* \rightarrow PH^{**} \rightarrow PH^+$ (PH^* and PH^{**} = PH radicals in different excited states) [15, 16].

Electronic Structure

The ground state X $^2\Pi_r$ of the PH^+ ion results from ionization of the 2π MO of the ground-state PH X $^3\Sigma^-$ radical (cf. p. 4), and the lowest excited states arise from $2\pi \leftarrow 5\sigma$, $6\sigma^* \leftarrow 5\sigma$, $(2\pi)^2 \leftarrow (5\sigma)^2$, and $6\sigma^* \leftarrow 2\pi$ excitations. Electron configurations representing the PH^+ states and the corresponding dissociation limits are as follows [17, pp. 335/7, 341], [7, 18, 19]:

electron configuration	state	dissociation limit
KL $(4\sigma)^2 (5\sigma)^2 (2\pi)^1$	$^2\Pi_r$	$P^+(^3P) + H(^2S)$
KL $(4\sigma)^2 (5\sigma)^1 (2\pi)^2$	$^4\Sigma^-$, $^2\Sigma^-$	$P^+(^3P) + H(^2S)$
	$^2\Delta$, $^2\Sigma^+$	$P^+(^1D) + H(^2S)$
KL $(4\sigma)^2 (5\sigma)^1 (2\pi)^1 (6\sigma^*)^1$	$^4\Pi_r$, $^2\Pi$, $^2\Pi$	$P^+(^3P) + H(^2S)$
KL $(4\sigma)^2 (5\sigma)^0 (2\pi)^3$	$^2\Pi_i$	$P^+(^1D) + H(^2S)$
KL $(4\sigma)^2 (5\sigma)^2 (2\pi)^0 (6\sigma^*)^1$	$^2\Sigma^+$	$P^+(^1S) + H(^2S)$

Higher excited states that correlate with H^+ and the ground and first excited states of the neutral P atom, i.e., $^4\Sigma^-$ ($P(^4S) + H^+$) and $^2\Pi$, $^2\Delta$, $^2\Sigma^-$ ($P(^2D) + H^+$), have been mentioned (cf. p. 38) [18].

Experimentally observed electronic states of PH^+ are the ground state X $^2\Pi_r$ and the excited states A $^2\Delta$ and a $^4\Sigma^-$. The X and A states have been identified with the lower and upper states of the UV-visible A–X transition detected by emission [7 to 11, 14], and laser photofragment [2, 8, 13] spectroscopy. The term value for the A $^2\Delta$ state, T_e = 26221.1 cm^{-1}, was derived [8] by simultaneously fitting the combined early emission data of [7], improved emission data for the 0→1 band of the A→X system [8], and laser photofragment data [8]. From the early emission data [7], Huber and Herzberg [20] derived $T_e \approx 26211$ cm^{-1}. Some spectroscopic information concerning orbital population in the X $^2\Pi_r$ and A $^2\Delta$ states was provided by an analysis of the hyperfine splittings in the A←X system [13] (see pp. 35/6). The a $^4\Sigma^-$ state has been identified with the convergence limit (11.852±0.002 eV) of three Rydberg series of PH, observed as autoionization structure in the PH^+ ion yield curve resulting from photoionization of PH [5].

A number of quantum-chemical ab initio calculations, mostly at levels beyond the Hartree-Fock approximation, namely MP4 [4, 21 to 23], MP4+CI [24, 25], MCSCF and MCSCF-CI [14, 26 to 28], MRD CI [18, 19, 29], CEPA, PNO-CI [30], GVB [31], and SCF [32 to 34] and

FSGO [35] (the abbreviations are explained on pp. 3/4), deal with various ground-state properties of PH^+ and, in some cases [14, 18, 19, 29, 31], with the lowest excited states. The so far best theoretical value for the ground-state total molecular energy at the equilibrium internuclear distance, $E_t = -341.05765\ E_h$, was obtained by a recent G2 calculation using a method that treats the electron correlation by Møller-Plesset perturbation theory (MP4) and quadratic configuration interaction (QCI) [25].

For the six low-lying excited states $^4\Sigma^-$, $^2\Sigma^-$, $^4\Pi$ ($P^+(^3P) + H(^2S)$) and $^2\Delta$, $^2\Sigma^+$, $^2\Pi$ ($P^+(^1D) + H(^2S)$), MRD CI calculations and "full" CI extrapolations have been carried out. Only the $^4\Sigma^-$ and $^2\Delta$ states are stable, their term values being $T_e(^4\Sigma^-) = 1.64$ eV and (in good agreement with the experiment) $T_e(A\ ^2\Delta) = 3.29$ eV; the remaining states turned out to be either repulsive ($^4\Pi$, $^2\Pi$) or quasidissociative ($^2\Sigma^-$, $^2\Sigma^+$), which may explain the failure [7] to observe transitions other than the A–X system within the 200- to 800-nm region ($^4\Sigma^- - X\ ^2\Pi_r$ is spin-forbidden). Vertical transition energies at r = 1.43 Å with respect to the ground state are ("full" CI estimate): 3.38 eV ($^2\Sigma^-$), 4.50 eV ($^2\Sigma^+$), and 8.21 eV ($^4\Pi$ and $^2\Pi$) [18, 19, 29]. A GVB calculation also predicted the $^4\Sigma^-$ state to lie 1.63 eV above the ground state and to be stable by at least 122 kJ/mol [31].

Ionization Potential

Experimental results are not available. Ab initio calculations give for the first adiabatic ionization potential $E_i = 17.7$ eV (MCSCF), 18.7 eV (MCSCF-CI) [26], 18.8 eV (MCSCF-CI) [27].

Polarizability

A theoretical value for the average polarizability, $\alpha_{av} = 2.807\ Å^3$, was obtained from a simple ab initio FSGO calculation [35].

Spectroscopic Constants

Hyperfine Interaction Constants. In the doublet states $X\ ^2\Pi_r$ and $A\ ^2\Delta$, where the unpaired electron occupies the 2π or the 5σ MO, respectively, hyperfine interaction occurs between the electron spin (S = 1/2) and the nuclear spin of ^{31}P (I = 1/2) or ^1H (I = 1/2). This results in a splitting of the rotational levels into four sublevels. Hf splitting of the rotational lines have been observed in the v = 1←2 band of the $A\ ^2\Delta \leftarrow X\ ^2\Pi_r$ system applying high-resolution laser photofragment spectroscopy [12, 13]. The following Frosch-Foley hf interaction constants a, b_F (= b + c/3), c, d (see [36, pp. 34/8]) for the ^{31}P and ^1H nuclei have been derived (constants in MHz; three standard deviation in parentheses) [13]:

	^{31}P		^1H	
	$X\ ^2\Pi_r$, v″ = 2	$A\ ^2\Delta$, v′ = 1	$X\ ^2\Pi_r$, v″ = 2	$A\ ^2\Delta$, v′ = 1
a	915(150)	894(24)	(24.9)[a)]	23(6)
b_F	195(36)	1315(12)	−67(15)	903(12)
c	−572(117)	710(49)	(24.7)[a)]	−38(45)
d	1184(8)	0[b)]	27(8)	0[b)]

[a)] Estimates using experimental a(^1H) and c(^1H) values for the $SH(^2\Pi_i)$ radical. – [b)] Assumed to be negligibly small.

Some conclusions concerning the electronic structure of PH^+ in the X and A states have been drawn by considering the magnitudes and signs of the hf constants: For the ground state, the experimental dipole-dipole interaction constants a(^{31}P) and c(^{31}P) are close to the calculated (ab initio SCF) values for the free P atom or P^+ ion, and the Fermi contact term $b_F(^{31}P)$ is very small as compared to that of a 3s electron in the free P atom. This indicates that the simple picture of P3p occupancy of the unpaired electron holds. The negative value of $b_F(^1H)$ arises from a negative spin density on the proton due to electron spin polarization. For the excited state, on the other hand, the P3pσ and substantial H1s contributions to the unpaired electron spin density are indicated by the large values of the parameters $c(^{31}P)$ and $b_F(^1H)$. Moreover, approximately 10% of the spin density is in the P3s orbital as concluded from the Fermi contact term $b_F(^{31}P)$ [13].

Fine-Structure Constants. Fine-structure splittings have been observed in the $A\,^2\Delta-X\,^2\Pi_r$ system by emission [7, 9, 10] and laser photofragment [8] spectroscopy. Therefrom, the constants A for spin-orbit coupling, p and q for Λ-doubling, and γ for spin-rotation coupling have been derived. The following constants for the $X\,^2\Pi_r$ and $A\,^2\Delta$ states of PH^+ were obtained by simultaneously fitting the early emission data for the 0→0, 0→1, and 1→0 bands of the A→X system [7], the improved emission data for the 0→1 band [8], and laser photofragment data for the 0←1 and 1←2 bands of the A←X transition [8] (constants in cm^{-1}; three standard deviation in parentheses) [8]:

state		A	p	q	γ
$X\,^2\Pi_r$	v=0	296.040(24)	0.2061(19)	0.01170(27)	−0.0905(58)
	v=1	296.327(27)	0.2017(20)	0.01149(15)	−0.0813(40)
	v=2	296.480(42)	0.2073(30)	0.01196(47)	−0.0802(80)
$A\,^2\Delta$	v=0	1.366(20)	*)	*)	0.1723(13)
	v=1	0.850(27)	*)	*)	0.1680(24)

*) The Λ-doubling in the $A\,^2\Delta$ state was assumed to be negligibly small [8].

The early results for PH^+ by [7] (in cm^{-1}: $A_0=295.94$, $A_1=296.2$, $|p_0|=0.23$, $|q_0|=0.011$ for $X\,^2\Pi_r$, $A_0=1.38$, $A_1=0.82$, $\gamma_0=\gamma_1=0.175$ for $A\,^2\Delta$) and for PD^+ from [9] ($A_0=295.83$, $|p_0|=0.08$ for $X\,^2\Pi_r$ and $A_0=1.35$, $\gamma_0=0.096$ for $A\,^2\Delta$) are quoted by Huber and Herzberg [20]. Reanalysis of the PH^+ spectrum and remeasurement and analysis including centrifugal distortion in the spin-orbit coupling ($A=A_v+A_J(J+1/2)^2+a_J(J+1/2)^4$), was reported by [10] to give the following values for A, A_J, and a_J (in cm^{-1}):

ion	state		A_v	$10^2\cdot A_J$	$10^5\cdot a_J$
PH^+	$X\,^2\Pi_r$	v=0	295.95	0.720	—
		v=1	296.29	0.569	—
	$A\,^2\Delta$	v=0	1.022	6.023	5.112
		v=1	1.115	3.572	2.017
PD^+	$X\,^2\Pi_r$	v=0	295.83	0.235	—
	$A\,^2\Delta$	v=0	1.184	3.973	7.5646

Theoretical results for $PH^+(X\,^2\Pi_r)$, A=343.53 or 296.39 cm^{-1}, were obtained by an ab initio MCSCF calculation and choosing two different approximations for the effective nuclear charges Z_{eff} of P and H [28].

Rotational and Vibrational Constants. Internuclear Distance. Experimental results are available for the ground state $X\,^2\Pi_r$ and the excited state $A\,^2\Delta$, which have been obtained from rotational analyses of the UV-visible A–X system. Rotational constants B_e or B_0, α_e, γ_e, D_0, vibrational constants ω_e, $\omega_e x_e$, or $\Delta G_{1/2}$, and internuclear distances r_e or r_0 are given in Table 4.

Table 4

PH^+, PD^+. Experimental Rotational and Vibrational Constants and Internuclear Distances. B_e, B_0, α_e, D_0, ω_e, $\omega_e x_e$, $\Delta G_{1/2}$ in cm^{-1}, r_e, r_0 in Å.

ion	state	B_e	B_0	α_e	$10^4 \cdot D_0$	ω_e	$\Delta G_{1/2}$	$\omega_e x_e$	r_e	r_0	remark	Ref.
PH^+	$X\ ^2\Pi_r$	8.50853	8.38677	0.24399	4.260	2382.75	2299.41	41.67	—	—	a)	[8]
		8.5051	—	0.2401	4.16	—	2299.60	—	1.4251	1.4352	b)	[7, 20]
		—	—	—	—	2373.1	—	36.8	—	—	c)	[11]
	$A\ ^2\Delta$	7.19635	6.98515	0.42240	6.472	1534.6	1397.01	68.8	—	—	a)	[8]
		7.1955	6.9833	0.4245	6.28	—	1398.76	—	1.5483	1.5726	b)	[7, 20]
		—	—	—	—	1461.4	—	31.3	—	—	c)	[11]
PD^+	$X\ ^2\Pi_r$	—	4.3505	—	1.16	—	1666	—	—	1.4314	b)	[9, 20]
		—	—	—	—	1703.9	—	19.0	—	—	c)	[11]
	$A\ ^2\Delta$	—	3.635	—	1.71	—	1017	—	—	1.5660	b)	[9, 20]
		—	—	—	—	1049.3	—	16.2	—	—	c)	[11]

a) Simultaneous fit of the early A→X emission data of [7], the emission data from a remeasurement of the A→X, v=0→1 band [8], and the laser photofragment data of the A←X, v=0←1, 1←2 transitions [8]. $10^3 \cdot \gamma_e = 0.94\ cm^{-1}$. B_1, B_2, D_1, D_2 values for $X\ ^2\Pi_r$ and B_1, D_1 values for $A\ ^2\Delta$ are also given in [8].

b) Analysis of the A→X, v=0→0, 0→1, 1→0 bands of PH^+ [7] and the A→X, v=0→0 band of PD^+ [9]. Huber and Herzberg [20] quoted the B_0, D_0, and $\Delta G_{1/2}$ values of [7, 9] and derived slightly improved r_0 values. – A reanalysis of the emission data for PH^+ and a remeasurement and analysis of the A→X, v=0→0 band of PD^+ gave the same B_v and D_v values [10] as those given in [7, 9].

c) Derived from $\Delta G_{1/2}$ of [7, 9] by using the isotopic rules (see [17, pp. 141/2]) and $\rho = 0.718$ [11].

Theoretical values have been derived by various quantum-chemical ab initio calculations for the two observed states X $^2\Pi_r$ and A $^2\Delta$ of PH^+ and for the unobserved state a $^4\Sigma^-$ predicted to be stable. The results from calculations including electron correlation are compiled in Table 5. SCF calculations of r_e and ω_e are in [24, 25].

Potential Energy Functions

Rydberg-Klein-Dunham potential curves have been constructed for the states X $^2\Pi_r$ and A $^2\Delta$ using the molecular constants of [7] and [37] (publication in [8]); these have been combined with theoretical (ab initio MCSCF-CI) curves for the X and A states to obtain also the short- and long-range parts of the curves; MCSCF-CI potential curves have also been calculated for the unobserved excited states $^4\Pi$ (repulsive) and $^2\Sigma^-$ (quasidissociative) [14]. Potential curves have been calculated for the states X $^2\Pi_r$, a $^4\Sigma^-$, $^2\Sigma^-$ (quasidissociative), and $^4\Pi$ (repulsive) with the dissociation limit $P^+(^3P) + H(^2S)$ and for the states A $^2\Delta$, $^2\Sigma^+$ (quasidissociative), and $^2\Pi$ (repulsive) with the dissociation limit $P^+(^1D) + H(^2S)$ using the MRD CI method; the results are shown in **Fig. 2** along with further possible dissociation limits and corresponding excited states of PH^+ [18, 19].

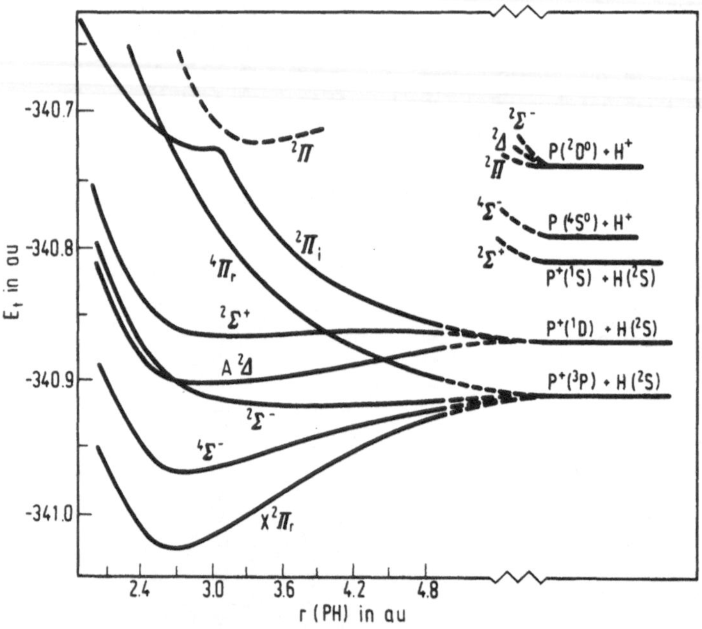

Fig. 2. Calculated MRD CI potential curves for the low-lying states of PH^+ and the corresponding dissociation limits [18].

Crossings of the A $^2\Delta$ curve in its long-range region with the repulsive $^4\Pi$ curve and near its potential minimum or in the short-range region with the quasidissociative $^2\Sigma^-$ curve were obtained from two theoretical studies [14, 18]; this is in agreement with the spectroscopically observed predissociation of the A $^2\Delta$, v′ = 0 and 1 levels, which, however, has been interpreted differently by [14] and [18] (see p. 41).

Table 5

PH^+. Theoretical Rotational and Vibrational Constants and Internuclear Distances. B_e, B_0, α_e, γ_e, ω_e, $\omega_e x_e$ in cm^{-1}, r_e in Å.

state	B_e	B_0	α_e	$10^3 \cdot \gamma_e$	ω_e	$\omega_e x_e$	r_e	method[a)]	Ref.
X $^2\Pi_r$	8.42	—	0.235	—	2375.8	44.9	1.431	CEPA[b)]	[30]
	8.509	8.3735	0.271	—	2354	47.2	1.4246	MRD CI	[18]
	8.597	—	0.280	3	2424.82	46.86	1.4205	MCSCF-CI[c)]	[14]
	8.552	—	0.256	1	2423.12	46.98	1.4202		
	8.6613[d)]	—	—	—	2564	—	1.412	MP2	[22]
	—	—	—	—	—	—	1.437	GVB	[31]
a $^4\Sigma^-$	7.7803	7.5843	0.392	—	1781	56	1.4899	MRD CI	[18]
	—	—	—	—	—	—	1.431	GVB	[31]
A $^2\Delta$	7.2224	7.0149	0.415	—	1458	60	1.5465	MRD CI	[18]
	7.1979	—	0.4542	—	1539.10	69.77	1.5402	MCSCF-CI[c)]	[14]
	7.2152	—	0.436	—	1534.78	71.03	1.5407		

[a)] Abbreviations are explained on pp. 3/4. – [b)] PNO-CI and SCF results are also given in [30]. – [c)] Two different fits of rotational and vibrational levels to the MCSCF-CI potential curves [14]. – [d)] B_e=259660 MHz is given in [22].

Dissociation Energy

$D_0(P^+-H) = 325 \pm 2$ kJ/mol $= 3.37 \pm 0.02$ eV for the ground state X $^2\Pi_r$ was derived from the photoionization threshold of PH^+ from PH_3 and using literature values for the heat of formation and ionization potential of the P atom [4].

Lower and upper limits, $D_0 = 3.348$ to 3.369 eV for the X $^2\Pi_r$ state and $D_0 = 1.230$ to 1.251 eV for the A $^2\Delta$ state, have been determined from the predissociation of the A $^2\Delta$, v'=0 and 1 levels observed in the laser photofragment spectrum of PH^+ [8]. This confirms the result $D_0 \leq 3.369$ eV for X $^2\Pi_r$ derived from the A $^2\Delta$ predissociation observed by lifetime measurements on various rotational levels of A $^2\Delta$, v'=0, 1 [14]. It also confirms the earlier estimates $D_0 \leq 3.36$ eV [20], ≤ 3.5 eV [39], and ≤ 3.35 eV [7] which are based on the A $^2\Delta$, v'=0, 1 predissociation observed in the A→X emission spectrum [7].

A rough Birge-Sponer extrapolation for the A $^2\Delta$ state (estimated $\omega_e x_e$ value) to the $P^+(^1D) + H(^2S)$ dissociation limit gave only $D_0 \approx 3.06$ eV [7, 20, 38].

Using the experimental dissociation energy and the calculated ionization potential of PH (X $^3\Sigma^-$), and the ionization potential of the P atom, a value of $D_0 = 3.34 \pm 0.20$ eV for X $^2\Pi_r$ was estimated [18]. Another estimate, $D_0 = 3.31$ eV, is based on the calculated (CEPA) proton affinity of the P atom and literature values for the ionization potentials of P and H [30].

Theoretical values obtained by a "full" CI extrapolation following an MRD CI calculation are: $D_e = 3.41$ eV and $D_0 = 3.26$ eV (SCF value $D_e = 2.56$ eV) for X $^2\Pi_r$ [18]. A GVB calculation gave $D_e = 2.90$ eV [31].

Heat of Formation. Partition Function. Equilibrium Constant

$\Delta_f H^\circ_0 = 1219 \pm 2$ kJ/mol (converted to $\Delta_f H^\circ_{298} = 1225 \pm 2$ kJ/mol by [21, 39]) follows from the threshold energy of PH^+ ions measured upon photoionization of PH_3 [4]. This confirms the earliest mass-spectrometric result, $\Delta_f H^\circ = 1218$ kJ/mol [2]. An estimate, $\Delta_f H^\circ_0 = 1221 \pm 20$ kJ/mol, is based on the ionization potential and the dissociation energy of the PH molecule [12].

Theoretical results obtained by ab initio calculations are (in kJ/mol):

$\Delta_f H^\circ_0$	1222	1222	1227	1237	1229
$\Delta_f H^\circ_{298}$	—	1228	1233	—	—
method (cf. pp. 3/4)	MP4	MP4	MP4	MRD CI	"full" CI
Ref.	[4]	[21]	[39]	[18]	[18]

Polynomial expansions of the partition function and the equilibrium constant between 1000 and 10000 K, based on the spectroscopic data compiled by Huber and Herzberg [20], are of astrophysical interest [40].

Spectra

The only observed electronic transition of PH^+ and PD^+ is the A $^2\Delta$ – X $^2\Pi_r$ transition lying in the violet and near UV between ~450 and 360 nm.

The A $^2\Delta$→X $^2\Pi_r$ Emission Spectrum. The first time, the A→X emission spectrum was excited in a hollow-cathode discharge through helium that contained small amounts of hydrogen and phosphorus vapor. A system of three red-degraded bands at 422.8, 385.4, and 356.7 nm was identified with the v=0→1, 0→0, and 1→0 bands. Their rotational and fine structures are those expected for a $^2\Delta$→$^2\Pi_r$ transition, where the upper state approaches Hund's case (b) (small spin-orbit splitting) and the lower state approaches Hund's case (a) (large spin-orbit splitting): The six main branches R_1, Q_1, P_1 ($^2\Delta_{3/2}(F_1) \rightarrow {}^2\Pi_{1/2}(F_1)$) and R_2, Q_2, P_2 ($^2\Delta_{5/2}(F_2) \rightarrow {}^2\Pi_{3/2}(F_2)$) are accompanied by the satellite branches $^SR_{21}$, $^RQ_{21}$, $^QP_{21}$ ($\Delta N = 2, 1, 0$) and $^QR_{12}$, $^PQ_{12}$, $^OP_{12}$ ($\Delta N = 0, -1, -2$), respectively; all lines form close pairs of Λ-doublets (due to Λ-doubling of the $^2\Pi$ levels). The band origins were determined to be at 25770.59 cm^{-1} (0→0), 27169.35 cm^{-1} (1→0), and 23470.99 cm^{-1} (0→1). Predissociation was observed in the A $^2\Delta$, v=0, 1 rotational levels; for v=0, the F_1 levels with $N \geq 13$ and the F_2 levels with $N \geq 15$ are predissociated, i.e., a sudden break-off of the branches was observed. The lines of the F_1 levels with v=1 are extremely weak. The repulsive $^4\Pi$ ($P^+(^3P) + H(^2S)$) state, even though forbidden according to the selection rules for predissociation, was suggested to be responsible for the A-state predissociation [7]. Remeasurements of the 0→1 band in a hollow-cathode discharge through helium with hydrogen and phosphorus vapor have been reported [8], and a list of all measured lines is available on request [8, footnote 13].

The A→X, v=0→1, 0→0, 1→0 bands of PD^+ were observed at 408.6, 382.6, and 368.3 nm upon microwave excitation of He gas containing small amounts of D_2 and phosphorus vapor. Rotational analysis has been performed for the 0→0 band which shows the same $^2\Delta(b) \rightarrow {}^2\Pi_r(a)$ characteristics as the PH^+ bands [9, 10]. Predissociation occurs for the A $^2\Delta$, v=0 rotational levels with $N \geq 17$ [10].

Chemiluminescence in the 450- to 360-nm range resulting from collisions of P^+ ions with H_2 or D_2 molecules (center-of-mass kinetic energies $E_{c.m.}$ = 2.3 to 12.2 eV) was identified with the A→X emission of PH^+ and PD^+ ions. A spectral analysis has been performed at the maximum cross section ($E_{c.m.}$ = 6.4 eV) by comparison with computer-simulated spectra which were based on the spectroscopic constants of [7, 9, 10]; besides the already known 0→1, 0→0, and 1→0 bands, some new bands, 1→1 for PH^+ and 2→0, 1→2, 0→2 for PD^+, could be localized. Additional experiments were carried out to verify that the observed emission is indeed due to the chemiluminescent exchange reaction $P^+(^3P, {}^1D) + H_2 \rightarrow PH^+(A\ {}^2\Delta) + H$ and not to collisionally excited PH^+ impurities in the P^+ beam [11].

For a more detailed study of the A-state predissociation, the lifetimes of twenty-seven A $^2\Delta$, v=0, 1 rotational levels have been measured using the high-frequency-deflection (HFD) technique, whereby the A→X, 0→0, 1→0 emission was excited by 20-keV electron impact on PH_3 gas. The lifetimes τ of the v′=0, N′≤12, F_1 and F_2 levels are constant between 1300 and 1600 ns (error limits: 150 ns); at N′=13 a drastic shortening in the F_2 component to $\tau = 280 \pm 50$ ns was observed, while (unfortunately) the F_1 component was absent. The lifetimes of all v′=1 levels were considerably shorter, τ = 250 to 600 (±50) ns, which shows predissociation of the whole sequence. A discussion on the origin of the predissociation was based on calculated predissociation rates that have been obtained by combining the experimental results and theoretically (MCSCF-CI) derived potential energy functions for the X $^2\Pi_r$, A $^2\Delta$, $^4\Pi$, and $^2\Sigma^-$ states and A→X transition moments. These calculations are in conflict with the suggestions of [7] (see above) and the theoretical results (curve crossing of A $^2\Delta$ and $^4\Pi$) of [18]: Not a direct interaction with the repulsive $^4\Pi$ state, but a second-order interaction with the $^2\Sigma^-$ state via an excited $^2\Pi$ state is a more plausible mechanism for the A $^2\Delta$ predissociation [14].

The A $^2\Delta$←X $^2\Pi_r$ Laser Photofragment Spectrum. A high-resolution spectrum of the A $^2\Delta$←X $^2\Pi_r$, v=1←2 and 0←1 bands has been recorded by laser photofragment spectroscopy: A PH^+ ion beam generated by 70-eV electron impact on PH_3 was irradiated coaxially with a tunable laser at 420 to 460 nm; P^+ ions arising from predissociation of the A $^2\Delta$ state are selected and detected with an electron multiplier. Thus, a sharp onset of the rotational branches of the A $^2\Delta$, v′=0 levels was observed at N′=13, just where disappearance or weakening of the rotational structure was observed in the emission spectrum (see above). In the 1←2 band, on the other hand, the rotational branches extend to the lowest J′ values, and branch heads could be localized. 100 rotational lines for the 0←1 band and 117 lines for the

1←2 band have been assigned and used together with emission data [7, 8] to derive the so far most precise spectroscopic constants (term value of A $^2\Delta$, rotational, vibrational, and fine-structure constants for A $^2\Delta$ and X $^2\Pi_r$) for PH^+ [8]. An even higher resolution of the spectrum (to observe the hyperfine structure) was achieved by Doppler tuning of rotational lines into resonance, i.e., by changing the PH^+-beam velocity with the laser wavelength held fixed at 430 nm [12, 13]. Hyperfine measurements (^{31}P and ^1H hfs) were made for 58 rotational lines of the 1←2 band and for a few high-J′ lines of the 0←1 band [13].

Chemical Reactions

Reactions of PH^+ with a series of neutral molecules were investigated by the selected ion flow tube (SIFT) [41] and the ion cyclotron resonance (ICR) techniques [42, 43]. The formation of condensation products via binary reactions predominates for reactant molecules of moderate ionization energy [41, 43]. Reactions via charge or proton transfer are less important except for HCN, where proton transfer is the sole exothermal binary reaction channel. For reactant molecules of high ionization energy, exothermal binary reaction channels are not available and ternary association was observed; this is a much slower process [41]. The following table lists the cationic products identified and the rate constants of the reactions from the more recent investigations [41, 43]; the reactant molecules are given in order of increasing ionization energy:

reactant	cationic products (relative amount in %)	rate constant[a)] in 10^{-10} $cm^3 \cdot molecule^{-1} \cdot s^{-1}$
CH_3NH_2	PNH_3^+ (46), PNH_2^+ (16), $CH_2NH_2^+$ (38)	18 [41]
PH_3	$P_2H_2^+$ (53), P_2^+ (12), $P_2H_3^+$ (8), PH_3^+ (18), PH_4^+ (9)	13 [41]
	$P_2H_2^+$ (29), P_2^+ (35), $P_2H_3^+$ (14), PH_3^+ (15), P_2H^+ (7)	15±5 [43]
NH_3	PNH_3^+ (53), PNH_2^+ (28), NH_4^+ (19)	21 [41]
	PNH_3^+ (58), PNH_2^+ (42)	21±2 [43]
CH_3CCH	$PC_2H_2^+$ (64), $C_2H_4^+$ (19), $C_3H_3^+$ (17)	17 [41]
H_2S	HPS^+ (64), H_2PS^+ (27), H_3S^+ (9)	15 [41]
C_2H_4	$PC_2H_3^+$ (70), PCH_2^+ (30)	12 [41]
CH_3OH	H_2PO^+ (100)	19 [41]
COS	HPS^+ (100)	13 [41]
C_2H_2	$PC_2H_2^+$ (100)	13 [41]
O_2	PO^+ (100)	5.4 [41]; 4.9±0.5 [43]
H_2O	HPO^+ (62), H_2PO^+ (17), H_3O^+ (21)	12 [41]
	HPO^+ (70), H_2PO^+ (30)	10.8±1.1 [43]
CH_4	PCH_3^+ (95), PCH_4^+ (5)	11 [41]
	PCH_3^+ (83), PCH_4^+ (17)	8.4±0.8 [43]
HCN	H_2CN^+ (95), $HCN \cdot PH^+$ (5)	17[b)] [41]
	H_2CN^+ (65), $PCNH^+$ (35)	4.7±0.5 [43]

reactant	cationic products (relative amount in %)	rate constant[a)] in 10^{-10} $cm^3 \cdot molecule^{-1} \cdot s^{-1}$
CO_2	$CO_2 \cdot PH^+$ (100)	0.086[c)] [41]; no reaction [43]
CO	$CO \cdot PH^+$ (100)	0.01[d)] [41]; no reaction [43]
H_2	PH_3^+ (100)	0.0043[e)] [41]; no reaction [43]
N_2	—	no reaction [41, 43]

[a)] At 300 K in [41]; no temperature given in [43]. – [b)] $>7.3\times10^{-28}$, [c)] 5.9×10^{-28}, [d)] 7.0×10^{-29}, [e)] 2.4×10^{-29} $cm^6 \cdot molecule^{-2} \cdot s^{-1}$ were obtained as ternary association rate constants at a fixed He pressure of 0.48 Torr; the tabulated effective binary rate constants were calculated from these values [41].

The results of early ICR investigations of the reactions of PH^+ with PH_3, NH_3, H_2O, and CH_4 [42] agree with the more recent ICR investigation only in the case of CH_4, but reactions of the other molecules gave fewer products and lower rate constants [43]. In the reaction of PH_3 with H_2O upon exposure to an electron beam, the intermediately formed PH^+ was identified to be the precursor of HPO^+ and H_2PO^+. Ions containing more oxygen atoms did not form [42]; the formation of the species $H_mPO_n^+$ with m = 0 to 3 and n = 2 to 4 [44] could not be confirmed [42].

A rate constant of $(3.1\pm0.8)\times10^{-10}$ $cm^3 \cdot molecule^{-1} \cdot s^{-1}$ (probably at ambient temperature) was measured for the isotope exchange reaction $PH^+ + D_2 \rightarrow PD^+ + HD$ [43].

The formation of molecules containing phosphorus in interstellar clouds from reactions of PH^+ with gaseous molecules via ions of phosphorus compounds was discussed in [45, 46] and less extensively in [43].

References:

[1] Saalfeld, F. E.; Svec, H. J. (Inorg. Chem. **2** [1963] 46/50).
[2] Fehlner, T. P.; Callen, R. B. (Adv. Chem. Ser. **72** [1968] 181/90).
[3] Morrison, J. D.; Traeger, J. C. (Int. J. Mass Spectrom. Ion Phys. **11** [1973] 277/88).
[4] Berkowitz, J.; Curtiss, L. A.; Gibson, S. T.; Greene, J. P.; Hillhouse, G. L.; Pople, J. A. (J. Chem. Phys. **84** [1986] 375/84).
[5] Berkowitz, J.; Cho, H. (J. Chem. Phys. **90** [1989] 1/6).
[6] Chau, M.; Bowers, M. T. (Chem. Phys. Lett. **44** [1976] 490/4).
[7] Narasimham, N. A. (Can. J. Phys. **35** [1957] 901/11).
[8] Edwards, C. P.; Jackson, P. A.; Sarre, P. J.; Milton, D. J. (Mol. Phys. **57** [1986] 595/604).
[9] Narasimham, N. A.; Dixit, M. N. (Curr. Sci. **36** [1967] 1/3).
[10] Rao, P. M. R.; Dixit, M. N.; Balasubramanian, T. K.; Narasimham, N. A. (Indian J. Pure Appl. Phys. **18** [1980] 276/80).

[11] Müller, B.; Ottinger, C. (J. Chem. Phys. **85** [1986] 243/8).
[12] Edwards, C. P.; Maclean, C. S.; Sarre, P. J. (J. Chem. Phys. **76** [1982] 3829/31).
[13] Edwards, C. P.; Sarre, P. J.; Milton, D. J. (Mol. Phys. **58** [1986] 53/63).
[14] Elander, N.; Erman, P.; Gustafsson, O.; Larsson, M.; Rittby, M.; Rurarz, E. (Phys. Scr. **31** [1985] 37/44).
[15] Ashfold, M. N. R.; Dixon, R. N.; Stickland, R. J. (Chem. Phys. Lett. **111** [1984] 226/33).
[16] Ashfold, M. N. R.; Stickland, R. J.; Tutcher, B. (Mol. Phys. **65** [1988] 1455/71).
[17] Herzberg, G. (Molecular Spectra and Molecular Structure, Vol. 1: Spectra of Diatomic Molecules, Van Nostrand, Princeton, N. J., 1961).

[18] Bruna, P. J.; Hirsch, G.; Peyerimhoff, S. D.; Buenker, R. J. (Mol. Phys. **42** [1981] 875/98).
[19] Bruna, P. J.; Peyerimhoff, S. D. (Bull. Soc. Chim. Belg. **92** [1983] 525/46).
[20] Huber, K. P.; Herzberg, G. (Molecular Spectra and Molecular Structure, Vol. 4: Constants of Diatomic Molecules, Van Nostrand Reinhold, New York 1979, pp. 534/7).

[21] Pople, J. A.; Curtiss, L. A. (J. Phys. Chem. **91** [1987] 155/62).
[22] Nguyen, M. T. (Mol. Phys. **59** [1986] 547/58).
[23] Maclagan, R. G. A. R. (Chem. Phys. Lett. **163** [1989] 349/53).
[24] Curtiss, L. A.; Jones, C.; Trucks, G. W.; Raghavachari, K.; Pople, J. A. (J. Chem. Phys. **93** [1990] 2537/45).
[25] Curtiss, L. A.; Raghavachari, K.; Trucks, G. W.; Pople, J. A. (J. Chem. Phys. **94** [1991] 7221/30).
[26] Pope, S. A.; Hillier, I. H.; Guest, M. F. (Faraday Symp. Chem. Soc. No. 19 [1984] 109/23).
[27] Pope, S. A.; Hillier, I. H.; Guest, M. F.; Kendric, J. (Chem. Phys. Lett. **95** [1983] 247/9).
[28] Koseki, S.; Schmidt, M. W.; Gordon, M. S. (J. Phys. Chem. **96** [1992] 10768/72).
[29] Buenker, R. (Int. J. Quantum Chem. **29** [1986] 435/60).
[30] Rosmus, P.; Meyer, W. (J. Chem. Phys. **66** [1977] 13/9).

[31] Harrison, J. F. (J. Am. Chem. Soc. **103** [1981] 7406/13).
[32] Cade, P. E.; Huo, W. M. (At. Data Nucl. Data Tables **12** [1973] 415/66).
[33] Bader, R. F. W.; Messer, R. R. (Can. J. Chem. **52** [1974] 2268/82).
[34] Thakkar, A. J.; Pedersen, W. A. (Int. J. Quantum Chem. Symp. **23** [1989] 245/53).
[35] Blustin, P. H. (J. Chem. Phys. **66** [1977] 5648/55).
[36] Hirota, E. (Springer Ser. Chem. Phys. **40** [1985] 1/233).
[37] Edwards, C. P. (Diss. Univ. of Nottingham 1984 from [8]).
[38] Gaydon, A. G. (Dissociation Energies and Spectra of Diatomic Molecules, 3rd Ed., Chapman & Hall, London 1968, p. 279).
[39] Pople, J. A.; Curtiss, L. A. (J. Phys. Chem. **91** [1987] 3637/9).
[40] Sauval, A. J.; Tatum, J. B. (Astrophys. J. Suppl. Ser. **56** [1984] 193/209, 201).

[41] Smith, D.; McIntosh, B. J.; Adams, N. G. (J. Chem. Phys. **90** [1989] 6213/9).
[42] Holtz, D.; Beauchamp, J. L.; Eyler, J. R. (J. Am. Chem. Soc. **92** [1970] 7045/55).
[43] Thorne, L. R.; Anicich, V. G.; Huntress, W. T. (Chem. Phys. Lett. **98** [1983] 162/6).
[44] Platzner, I. (Isr. J. Chem. **6** [1968] 34p).
[45] Adams, N. G.; McIntosh, B. J.; Smith, D. (Astron. Astrophys. **232** [1990] 443/6; C. A. **113** [1990] No. 139646).
[46] Thorne, L. R.; Anicich, V. G.; Prasad, S. S.; Huntress, W. T., Jr. (Astrophys. J. **280** [1984] 139/43).

1.1.3 PH^{2+}

CAS Registry Numbers: PH^{2+} *[65756-17-4]* Phosphorus(2+), hydro-; PD^{2+} *[65756-19-6]*

No experimental results on the PH^{2+} ion have been published so far. The species has been the subject of a few quantum-chemical ab initio calculations. CASSCF and MRD CI calculations of all ten electronic states correlating with the four lowest asymptotes, $P^+(^3P, {}^1D, {}^1S) + H^+$ and $P^{2+}(^2P^{\circ}) + H(^2S)$, reveal three quasi-bound and seven repulsive states. The quasi-bound states are the previously known [1 to 3] ground state X $^1\Sigma^+$ (KL $(4\sigma)^2$ $(5\sigma)^2$) correlating with the third asymptote and the excited states a $^3\Pi$ and A $^1\Pi$ (KL $(4\sigma)^2$ $(5\sigma)^1$ $(2\pi)^1$) correlating with the first and second asymptotes. The MRD CI results for the three quasi-bound states, i.e., the internuclear distances r_e and r_b at the potential minima and maxima, the term values T_e, the activation barriers D_b, dissociation (deprotonation) energies D_e, and the vibrational and rotational constants ω_e, $\omega_e x_e$, B_e, and α_e are as follows (values in parentheses were obtained from fits to 16, 4, and 5 variationally calculated rovibrational levels of the X, a, and A states, respectively) [4]:

constant	X $^1\Sigma^+$ ($P^+(^1S)+H^+$)[a]	a $^3\Pi$ ($P^+(^3P)+H^+$)	A $^1\Pi$ ($P^+(^1D)+H^+$)
r_e in Å	1.472	1.770	1.988
r_b in Å	6.401	5.399	6.054
T_e in eV[b]	0.00	1.74	2.49
D_b in eV	2.77	0.44	0.48
D_e in eV	−0.92	−3.75	−3.38
ω_e in cm^{-1}	2105.1 (2121.7)	966.9 (967.5)	840.6 (846.9)
$\omega_e x_e$ in cm^{-1}	44.48 (48.21)	39.18 (31.40)	22.45 (17.00)
B_e in cm^{-1}	7.954 (7.966)	5.565 (5.518)	4.427 (4.391)
α_e in cm^{-1}	0.268 (0.269)	0.375 (0.315)	0.185 (0.172)

[a] Previous MRD CI results: r_e=1.474 Å, D_b=2.91 eV, D_e=−0.62 eV [1, 2]; previous SCF results: r_e=1.431 Å, ω_e=2502 cm^{-1} [3]. – [b] Derived from the total molecular energies at the potential minima given in the original paper [4].

References:

[1] Pope, S. A.; Hillier, I. H.; Guest, M. F. (Faraday Symp. Chem. Soc. No. 19 [1984] 109/23).
[2] Pope, S. A.; Hillier, I. H.; Guest, M. F.; Kendric, J. (Chem. Phys. Lett. **95** [1983] 247/9).
[3] Pyykkö, P. (Mol. Phys. **67** [1989] 871/8).
[4] Senekowitsch, J.; O'Neil, S. V.; Werner, H.-J.; Knowles, P. J. (Chem. Phys. Lett. **175** [1990] 548/54).

1.1.4 PH^-

CAS Registry Number *[23841-34-1]* Phosphate(1−), hydro-

CAS assigned to the ion $\mathbf{PH^{2-}}$ the registry number *[31427-78-8]* Phosphide (PH^{2-}), however, without citing any references.

Dissociative electron capture by PH_3 molecules resulted in the formation of PH^- (and PH_2^-, P^-, H^-) ions [1 to 4]. The appearance potentials measured mass-spectrometrically in the resonance electron capture region by the retarding potential difference method, AP = 2.2±0.2, 6.3±0.1, and 8.1±0.3 eV, were attributed to the processes $PH_3+e^- \rightarrow PH^-+H_2$, PH^-+2H, and $(PH^-)^*+2H$, respectively, where $(PH^-)^*$ is possibly a long-lived (>10 μs) excited ion; the ion-pair process $PH_3 \rightarrow PH^-+H^++H$ was found at AP=19.8±0.1 eV [4].

A beam of PH^- ions was produced by burning PH_3 with N_2O in a low pressure discharge source. Upon crossing the PH^- beam with the beam of a 488-nm Ar^+ ion laser, the photoelectron spectrum was recorded, exhibiting two strong detachment peaks at ~0.5 and ~1.5 eV (kinetic energies; intensity ratio ~1:3); the latter peak is accompanied by small peaks on both wings at ~1.8 and 1.260±0.015 eV. The main peaks were assigned to transitions of the ground-state ion $PH^-(X\ ^2\Pi_i,\ v''=0)$ into the lowest molecular states $PH(X\ ^3\Sigma^-,\ v'=0)$ and $PH(a\ ^1\Delta,\ v'=0)$, and an ionization potential of PH^- (or electron affinity of PH) of 1.028±0.010 eV was derived. The hot band ($v' \leftarrow v''=0 \leftarrow 1$) at ~1.8 eV led to the fundamental vibrational frequency $\nu=2230\pm100\ cm^{-1}$ for the ground-state ion. The second small peak presumably was partly due to the $v' \leftarrow v''=1 \leftarrow 0$ transition. A Franck-Condon factor analysis gave an estimate for the equilibrium internuclear distance: $1.414\ Å \geq r_e \geq 1.400\ Å$ [5]. These are the only experimental results for molecular properties of PH^- and were quoted by Huber and Herzberg [6].

A few quantum-chemical ab initio calculations dealt with the electronic structure and some molecular parameters of PH^-. Accordingly, the ground state, derived from that of the neutral molecule by adding a 2π (P3pπ) electron, is KL $(4\sigma)^2\ (5\sigma)^2\ (2\pi)^3$, $X\ ^2\Pi_i$ [7]. The best theoretical value so far for the total molecular energy at the equilibrium internuclear distance, $E_t = -341.46363\ E_h$, was obtained from a recent G2 calculation by using a method that treats the electron correlation by Møller-Plesset perturbation theory (MP4) and quadratic configuration interaction (QCI) [8].

The spin-orbit coupling constant of the inverted ground state $A=-223.16$ or $-192.55\ cm^{-1}$ was obtained by an ab initio MCSCF calculation by choosing different approximations for the effective nuclear charges Z_{eff} of P and H [9]. $A=-211.7\ cm^{-1}$ was obtained by an ab initio calculation that took a small amount of correlation energy into account ("core polarization") [10], and values of $A=-173.1$ to $-178.5\ cm^{-1}$ [10 to 12] resulted from an SCF wave function of near-Hartree-Fock quality [7].

Spectroscopic constants for the ground state were calculated by the CEPA method: the rotational and vibrational constants $B_e=8.31\ cm^{-1}$, $\alpha_e=0.269\ cm^{-1}$, $\omega_e=2248\ cm^{-1}$, and $\omega_e x_e = 50\ cm^{-1}$, the internuclear distance $r_e=1.441\ Å$, and the dissociation energy $D_0(P^- - H)=3.25$ eV (values obtained by PNO-CI and SCF calculations were also given) [13]. For other r_e values, see also [14 to 16]. Franck-Condon factors for the $PH \leftarrow PH^-$, $v=0 \leftarrow 0$, $0 \leftarrow 1$, and $0 \leftarrow 2$ transitions were also calculated by the CEPA method [13].

A proton affinity of 1531 kJ/mol was obtained from MP4 calculations for PH^- and PH_2 [14].

References:

[1] Rosenbaum, O.; Neuert, H. (Z. Naturforsch. **9a** [1954] 990/1).
[2] Ebinghaus, H.; Kraus, K.; Müller-Duysing, W.; Neuert, H. (Z. Naturforsch. **19a** [1964] 732/6).
[3] Holtz, D.; Beauchamp, J. L.; Eyler, J. R. (J. Am. Chem. Soc. **92** [1970] 7045/55).
[4] Halmann, M.; Platzner, I. (J. Phys. Chem. **73** [1969] 4376/8).
[5] Zittel, P. F.; Lineberger, W. C. (J. Chem. Phys. **65** [1976] 1236/43).
[6] Huber, K. P.; Herzberg, G. (Molecular Spectra and Molecular Structure, Vol. 4: Constants of Diatomic Molecules, Van Nostrand Reinhold, New York 1979, pp. 536/7).
[7] Cade, P. E.; Huo, W. M. (At. Data Nucl. Data Tables **12** [1973] 415/66).
[8] Curtiss, L. A.; Raghavachari, K.; Trucks, G. W.; Pople, J. A. (J. Chem. Phys. **94** [1991] 7221/30).
[9] Koseki, S.; Schmidt, M. W.; Gordon, M. S. (J. Phys. Chem. **96** [1992] 10768/72).
[10] Walker, T. E. H.; Richards, W. G. (J. Chem. Phys. **52** [1970] 1311/4).

[11] Trivedi, H. P.; Richards, W. G. (J. Chem. Phys. **72** [1980] 3438/9).
[12] Walker, T. E. H.; Richards, W. G. (Symp. Faraday Soc. No. 2 [1968] 64/8).

[13] Rosmus, P.; Meyer, W. (J. Chem. Phys. **69** [1978] 2745/51).
[14] Pople, J. A.; Schleyer, P. von R.; Kaneti, J.; Spitznagel, G. W. (Chem. Phys. Lett. **145** [1988] 359/64).
[15] Spitznagel, G. W.; Clark, T.; Schleyer, P. von R.; Hehre, W. J. (J. Comput. Chem. **8** [1987] 1109/16).
[16] Nguyen, M. T. (Mol. Phys. **59** [1986] 547/58).

1.2 PH_2 and Ions

1.2.1 PH_2 Phosphanyl

CAS Registry Numbers: PH_2 *[13765-43-0]* Phosphino; PHD *[15117-77-8]*; PD_2 *[15117-85-8]*

1.2.1.1 Formation and Detection

In the Vapor Phase

The $PH_2(PD_2)$ radical was first detected by its electronic absorption bands between 3600 (3800) and 5500 Å during flash photolysis of **phosphane**, $PH_3(PD_3)$, with quartz UV [1, 2] (the bands represent a progression in the upper-state ($\tilde{A}\ ^2A_1$) bending frequency ν_2' and were erroneously ascribed to emission in "Phosphor" C, 1965, p. 6). Many physical and chemical methods of H abstraction from PH_3 have been applied since then together with various PH_2 detection methods. These investigations are gathered in a table below. For PHD and PD_2 formation, see the remarks below the table:

PH_3 treatment	PH_2 detection	Ref.	remark
flash photolysis in the quartz UV	electronic absorption	[1 to 5]	a)
flash photolysis in the vacuum UV	laser-induced fluorescence	[6]	b)
pulsed laser photolysis at 193 nm	electronic emission	[7]	c)
	laser-induced fluorescence (LIF)	[8, 9]	d)
photolysis in the vacuum UV	electronic emission	[10, 11]	e)
pyrolysis of 2% PH_3 in H_2	Raman scattering	[12]	f)
shock waves through PH_3/Ar mixture	electronic emission	[13]	g)
dc discharge in PH_3/O_2 or PH_3/CF_4	microwave (MW) absorption	[14]	h)
glow discharge in 1% PH_3 in SiH_4	electronic emission	[15]	
high-voltage discharge in PH_3/N_2O	electronic emission	[16]	i)
high-voltage discharge in flowing $PH_3(PD_3)$/He(Ar, H_2)	electronic emission	[17, 18]	j)

PH_3 treatment	PH_2 detection	Ref.	remark
Ar afterglow	electronic emission	[19]	k)
reaction with H from discharged H_2	electronic emission	[20 to 22]	l)
	IR laser magnetic resonance	[23]	m)
	photoionization mass spectrometry	[24]	n)
reaction with F from discharged CF_4	far-IR laser magnetic resonance	[25, 26]	o)
	IR laser magnetic resonance	[23]	m)
	intermodulated fluorescence	[27]	p)
	photoelectron spectroscopy	[28]	
	microwave absorption	[14]	q)
reaction with N	electronic emission	[29]	r)

a) Analogous treatment of PD_3 yielded PD_2 [1, 2]. Intracavity laser spectroscopy at 455 nm (most intense absorption band) was used in [5]. PH_2 was indirectly detected by P_2H_4 formation in [30]. – b) Time-resolved determination of [H] yielded an absolute rate constant for PH_3+H (see p. 233) [31]. – c) ArF excimer laser for PH_3 irradiation. A large population of the lowest vibrational level ($v_2'=0$) of the upper electronic state was detected. – d) Appreciable internal excitation of PH_2 was detected. Quantitative results were obtained from sub-Doppler spectroscopy of the H fragment [8, 9]. For LIF spectroscopic work, see also [32]. – e) A threshold of 208.0 ± 1 nm for the formation of PH_2 ($\tilde{A}\ ^2A_1$) is attributed to dissociation via PH_3 ($\tilde{X}\ ^1A_1$)$\rightarrow PH_2$ ($\tilde{X}\ ^2B_1$)$+H(^2S)$ and to excitation via $\tilde{A}\ ^2A_1 \leftarrow \tilde{X}\ ^2B_1$ of PH_2 [10]. Vibrationally excited levels of PH_2 ($\tilde{A}$) are appreciably populated at lower wavelengths [10, 11]. For 147 nm irradiation, formation of $PH_2(\tilde{X})$ and $PH_2(\tilde{A})$ from two different excited electronic states of PH_3 was proposed [33]. See also work on PH_3/SiH_4 mixtures in [34]. – f) Heating up to 632°C. A new Raman band appeared at ~2310 cm^{-1} at elevated temperatures and was ascribed to a P–H stretching vibration of PH_2. But the magnitude of the H_2 partial pressures contradicted a PH_2 formation at elevated temperatures. – g) The background emission is ascribed to PH_2. – h) PH_3/O_2 mixtures at optimum PH_3 and O_2 partial pressures of 25 mTorr each gave stronger MW signals than PH_3/CF_4 mixtures. – i) 30 kV discharge with a pulse duration of ~1 μs. – j) PHD and PD_2 were also detected. Flowing He yielded the best emission spectra. – k) PH_3 and MW-discharged Ar flowing in opposite directions. – l) PD_2 was formed from PD_3 [20 to 22]. $PH_3(PD_3)$ was pure [22] or diluted in Ar [21]. – m) Conditions: slow flow, total pressures of 0.6 to 2.1 Torr, and $PH_3:H_2:Ar=1:2:20$. Using a PH_3+H source introduced less noise from the discharge into the laser output than the PH_3+F source. – n) The relative flow rates of PH_3 and H were controlled to optimize the PH_2 production rate. – o) Fast flow system at a total pressure of 0.5 [25] or 1 Torr [26]. – p) This PH_2 source yielded a stronger fluorescence than the reaction of red phosphorus with atomic hydrogen (see below), but produced a larger deposit. – q) Only used during the early stage of the work [14]. – r) PH_3 was diluted in 75% Ar [29]. The PH_2 radical was not detected by molecular-beam sampling mass spectrometry of that reaction [35]; see also earlier work in [38].

A few examples are given in the table below of reactions of PH_3 with other atoms or radicals and of **ion-molecule** reactions, where PH_2 was not detected but was assumed to have been formed (abbreviations: k = rate constant, † = vibrational excitation, MW = microwave, UHF = ultrahigh frequency, FP = flash photolysis, LIF = laser-induced fluorescence, MS = mass spectrometry, ICR = ion cyclotron resonance, FA = flowing afterglow).

reaction	comment	Ref.
$PH_3+Cl \rightarrow PH_2+HCl$	Cl from MW-discharged Cl_2; IR emission of HCl	[36]
	step of PH_3+Cl_2 mechanism; Cl possibly from $HCl^†$, $PH_2Cl^†$	[37]
$PH_3+O \rightarrow PH_2+OH$	O from discharged O_2 or from NO + N titration	[35, 38, 39]
	O from UHF-discharged N_2O; k determined from [O] and [OH] concentrations	[40]
$PH_3+{}^{32}P \rightarrow PH_2+{}^{32}PH$	^{32}P from ^{31}P (n, γ) reaction	[41]
$PH_3+OH \rightarrow PH_2+H_2O$	OH from HNO_3; k determined by [OH]	[42]
$PH_3+NH_2 \rightarrow PH_2+NH_3$	NH_2 by FP of NH_3; k by LIF	[43]
$PH_3+(CF_3)_2C^- \rightarrow PH_2+(CF_3)_2CH^-$	dissociative electron attachment to $(CF_3)_2C{=}N_2$	[44]
$PH_3+PH_3^+ \rightarrow PH_2+PH_4^+$	PH_4^+ detected by MS	[45]
	PH_4^+ detected by ICR	[46 to 48]
$PH_3+PH_2^+ \rightarrow PH_2+PH_3^+$	PH_3^+ detected by ICR	[46]
$PH_3+PH^+ \rightarrow PH_2+PH_2^+$	PH_2^+ detected by ICR	[46]
$PH_3^++NH_3 \rightarrow PH_2+NH_4^+$	NH_4^+ detected by ICR	[46]
$PH_2^-+NO_2 \rightarrow PH_2+NO_2^-$	NO_2^- detected by FA	[70]
$PH_2^-+SO_2 \rightarrow PH_2+SO_2^-$	SO_2^- detected by FA	[70]

Another PH_2 source is **benzylphosphane** $C_6H_5CH_2PH_2$. Its thermal decomposition starts at 650°C [49] or 600°C [50] and was approximately complete at 850°C [49] or 800°C [50]. The PH_2 fragment was mass-spectrometrically detected following either electron impact [49] or photoionization with the Lyman-α line [50]. The formation of PH_2 was confirmed by its reaction with CH_3 (from the simultaneous pyrolysis of ethyl nitrite) to CH_3PH_2 which was mass-spectrometrically detected [49].

The reaction of **elemental phosphorus** with hydrogen was also used for PH_2 production. H atoms from MW-discharged H_2 were passed over the red modification, and the resulting PH_2 was detected by LIF [51, 52], IMF (intermodulated fluorescence) [27], or MW spectroscopy [14]. The reaction of phosphorus vapor with atomic or molecular hydrogen did not lead to any PH_2 electronic emission [18]. Emission bands, which had been observed during the reaction of elemental P with atomic H, were first attributed to PH_2 [53, 54], but they were later shown to be due to HPO [55].

In Solid Matrices

PH_3 was used as a precursor and generally isolated in a low-temperature matrix. An exceptional case is shown in the first row of the table below, which lists the PH_3 treatment and PH_2 detection methods:

matrix	T in K	PH_3 treatment	PH_2 detection	Ref.	remark
Ar	4.2	condensation of MW-discharged PH_3/Ar mixture	electronic absorption	[56]	a)
	14	photolysis at 121.6 nm	IR and electronic absorption	[57]	b)
Kr	4.2	photolysis at 106.7 and 104.8 nm	ESR at ~10 K	[58]	c)
	4.2	^{60}Co γ irradiation	ESR	[59]	d)
Xe	4.2	^{60}Co γ irradiation	ESR	[60]	e)
	77	^{60}Co γ irradiation	ESR at 4.2, 77 K	[61]	f)
	4.2	reaction with H from HI photolysis	ESR at 4.2, 30 K	[61]	
cancrinite	77	photolysis at 106.7 and 104.8 nm	ESR	[62]	c)

[a] Ar:PH_3=25:1 according to [63]. A CH_4/Ar mixture in a P-coated tube also yielded PH_2 absorption bands [56]. – [b] The mole ratios Ar:PH_3(PD_3) were between 200 and 500 [57]. For detecting PH_2 by IR absorption, see [64]. – [c] Ar resonance lamp (for wavelengths, see [65]). – [d] Dissociation via $PH_3 \rightarrow P+3H$ is more probable than a dissociation via $PH_3 \rightarrow PH_2+H$. The reverse was true for PD_3 [59, 60]. – [e] 2% PH_3 in Xe. The PH_2 signals dominated over those of P. – [f] 1% PH_3 in Xe.

At Solid Surfaces

Adsorbed PH_2 radicals from the dissociative adsorption of PH_3 on Al_2O_3-supported Rh surfaces at 190 to 400 K [66] or on an Si(111) surface at 80 K [67] were detected by IR transmission spectroscopy in the range of the P–H stretching vibration [66] or by high-resolution electron energy loss spectroscopy (EELS) in the range of both P–H stretching and HPH bending vibrations [67]. For earlier Auger spectroscopic work on the Si(111) surface, see [68, 69].

References:

[1] Ramsay, D. A. (Nature **178** [1956] 374/5).
[2] Ramsay, D. A. (Ann. N.Y. Acad. Sci. **67** [1956/57] 485/98, 492/5).
[3] Norrish, R. G. W.; Oldershaw, G. A. (Proc. R. Soc. [London] A **262** [1961] 1/9, 10/8).
[4] Kley, D.; Welge, K. H. (Z. Naturforsch. **20a** [1965] 124/31).
[5] Zakhar'in, V. I.; Nadtochenko, V. A.; Sarkisov, O. M.; Teitel'boim, M. A. (Dokl. Akad. Nauk SSSR **263** [1982] 127/30; Dokl. Phys. Chem. [Engl. Transl.] **262/267** [1982] 168/70; Khim. Fiz. **1982** 1068/74; Oxid. Commun. **4** [1983] 443/9).
[6] Nguyen Xuan, C.; Margani, A. (J. Chem. Phys. **93** [1990] 136/46).
[7] Sam, C. L.; Yardley, J. T. (J. Chem. Phys. **69** [1978] 4621/7).

[8] Koplitz, B.; Xu, Z.; Baugh, D.; Buelow, S.; Häusler, D.; Rice, J.; Reisler, H.; Qian, C. X. W.; Noble, M.; Wittig, C. (Discuss. Faraday Soc. No. 82 [1986] 125/48, 131/5).
[9] Baugh, D.; Koplitz, B.; Xu, Z.; Wittig, C. (J. Chem. Phys. **88** [1988] 879/87).
[10] Di Stefano, G.; Lenzi, M.; Margani, A.; Mele, A.; Nguyen Xuan, C. (J. Photochem. **7** [1977] 335/44).

[11] Di Stefano, G.; Lenzi, M.; Margani, A.; Nguyen Xuan, C. (J. Chem. Phys. **68** [1978] 959/63).
[12] Abraham, P.; Bekkaoui, A.; Soulière, V.; Bouix, J.; Monteil, Y. (J. Cryst. Growth **107** [1991] 26/31).
[13] Guenebaut, H.; Pascat, B. (C. R. Hebd. Seances Acad. Sci. **255** [1962] 1741/3).
[14] Endo, Y.; Saito, S.; Hirota, E. (J. Mol. Spectrosc. **97** [1983] 204/12).
[15] Kampas, F. J.; Griffith, R. W. (AIP Conf. Proc. No. 73 [1981] 1/5; C.A. **95** [1981] No. 52147).
[16] Harris, D. G.; Chou, M. S.; Cool, T. A. (J. Chem. Phys. **82** [1985] 3502/15, 3512).
[17] Guenebaut, H.; Pascat, B. (C. R. Hebd. Seances Acad. Sci. **259** [1964] 2412/5).
[18] Guenebaut, H.; Pascat, B.; Berthou, J.-M. (J. Chim. Phys. Phys. Chim. Biol. **62** [1965] 867/77).
[19] Nguyen Xuan, C.; Tamanini, M.; Di Stefano, G.; Margani, A. (Gazz. Chim. Ital. **116** [1986] 243/53).
[20] Pascat, B.; Guenebaut, H. (Bull. Univ. ARERS **2** [1963] 100/7).

[21] Guenebaut, H.; Pascat, B. (C. R. Hebd. Seances Acad. Sci. **256** [1963] 677/80).
[22] Guenebaut, H.; Pascat, B. (J. Chim. Phys. Phys. Chim. Biol. **61** [1964] 592/5).
[23] Hills, G. W.; McKellar, A. R. W. (J. Chem. Phys. **71** [1979] 1141/9).
[24] Berkowitz, J.; Cho, H. (J. Chem. Phys. **90** [1989] 1/6).
[25] Davies, P. B.; Russell, D. K.; Thrush, B. A. (Chem. Phys. Lett. **37** [1976] 43/6).
[26] Davies, P. B.; Russell, D. K.; Thrush, B. A.; Radford, H. E. (Chem. Phys. **44** [1979] 421/6).
[27] Kakimoto, M.; Hirota, E. (J. Mol. Spectrosc. **94** [1982] 173/91, 175/6).
[28] Dyke, J. M.; Jonathan, N.; Morris, A. (Int. Rev. Phys. Chem. **2** [1982] 3/42, 22).
[29] Guenebaut, H.; Pascat, B. (C. R. Hebd. Seances Acad. Sci. **256** [1963] 2850/3).
[30] Ferris, J. P.; Benson, R. (J. Am. Chem. Soc. **103** [1981] 1922/7).

[31] Lee, J. H.; Michael, J. V.; Payne, W. A.; Whytock, D. A.; Stief, L. J. (J. Chem. Phys. **65** [1976] 3280/3).
[32] Stephens, K. M. (Diss. Univ. Alabama 1990 from Diss. Abstr. Int. B **52** [1991] 865).
[33] Blazejowski, J.; Lampe, F. W. (J. Phys. Chem. **85** [1981] 1856/64).
[34] Blazejowski, J.; Lampe, F. W. (J. Photochem. **16** [1981] 105/20).
[35] Hamilton, P. A.; Murrells, T. P. (J. Chem. Soc. Faraday Trans. II **81** [1985] 1531/41, 1533, 1539).
[36] Wickramaaratchi, M. A.; Setser, D. W. (J. Phys. Chem. **87** [1983] 64/72).
[37] Azatyan, V. V.; Gagarin, S. G.; Zakhar'in, V. I.; Kalkanov, V. A.; Kolbanovskii, Yu. A. (Kinet. Katal. **26** [1985] 222/6; Kinet. Catal. [Engl. Transl.] **26** [1985] 191/4).
[38] Clyne, M. A. A.; Heaven, M. C. (Chem. Phys. **58** [1981] 145/50).
[39] Davies, P. B.; Thrush, B. A. (Proc. R. Soc. [London] A **302** [1968] 243/52).
[40] Aleksandrov, E. N.; Arutyunov, V. S.; Dubrovina, I. V.; Kozlov, S. N. (Fiz. Goreniya Vzryva **18** No. 4 [1982] 73/8; Combust. Explos. Shock Waves [Engl. Transl.] **18** [1982] 451/5).

[41] Stewart, G. W.; Hower, C. O. (J. Inorg. Nucl. Chem. **34** [1972] 39/45).
[42] Fritz, B.; Lorenz, K.; Steinert, W.; Zellner, R. (EUR-7624 [1982] 192/202, 197/9; C.A. **96** [1982] No. 223993).

[43] Bosco, S. R.; Brobst, W. D.; Nava, D. F.; Stief, L. J. (J. Geophys. Res. C **88** [1983] 8543/9).
[44] McDonald, R. N.; Chowdhury, A. K.; McGhee, W. D. (J. Am. Chem. Soc. **106** [1984] 4112/6).
[45] Halmann, M.; Platzner, I. (J. Phys. Chem. **71** [1967] 4522/6).
[46] Eyler, J. R. (Inorg. Chem. **9** [1970] 981/2).
[47] Holtz, D.; Beauchamp, J. L.; Eyler, J. R. (J. Am. Chem. Soc. **92** [1970] 7045/55, 7046).
[48] Thorne, L. R.; Anicich, V. G.; Huntress, W. T. (Chem. Phys. Lett. **98** [1983] 162/6).
[49] McAllister, T.; Lossing, F. P. (J. Phys. Chem. **73** [1969] 2996/8).
[50] Berkowitz, J.; Curtiss, L. A.; Gibson, S. T.; Greene, J. P.; Hillhouse, G. L.; Pople, J. A. (J. Chem. Phys. **84** [1986] 375/84, 378).

[51] Huie, R. E.; Long, N. J. T.; Thrush, B. A. (J. Chem. Soc. Faraday Trans. II **74** [1978] 1253/62, 1254).
[52] Curl, R. F.; Endo, Y.; Kakimoto, M.; Saito, S.; Hirota, E. (Chem. Phys. Lett. **53** [1978] 536/8).
[53] Peyron, M. (Proc. Int. Symp. Mol. Struct. Spectrosc., Tokyo 1962, pp. B403-1/B403-4).
[54] Lam Thanh My; Peyron, M. (J. Chim. Phys. Phys. Chim. Biol. **59** [1962] 688/95).
[55] Lam Thanh My; Peyron, M. (J. Chim. Phys. Phys. Chim. Biol. **60** [1963] 1289/93).
[56] McCarty, M., Jr.; Robinson, G. W. (J. Chim. Phys. Phys. Chim. Biol. **56** [1959] 723/30).
[57] Larzillière, M.; Jacox, M. E. (NBS Spec. Publ. [U.S.] 561-1 [1979] 529/43, 533, 537).
[58] McDowell, C. A.; Mitchell, K. A. R.; Raghunathan, P. (J. Chem. Phys. **57** [1972] 1699/703).
[59] Morehouse, R. L.; Christiansen, J. J.; Gordy, W. (J. Chem. Phys. **45** [1966] 1747/51).
[60] Jackel, G. S.; Gordy, W. (Phys. Rev. [2] **176** [1968] 443/52).

[61] Shimokoshi, K.; Nakamura, K.; Sato, S. (Mol. Phys. **53** [1984] 1239/49).
[62] Raghunathan, P.; Sur, S. K. (Proc. Indian Acad. Sci. Chem. Sci. **92** [1983] 597/604).
[63] Robinson, G. W. (Adv. Chem. Ser. No. 36 [1962] 10/25, 11).
[64] Larzillière, M.; Jacox, M. E. (J. Mol. Spectrosc. **79** [1980] 132/50, 134).
[65] McNesby, J. R.; Okabe, H. (Adv. Photochem. **3** [1964] 157/240, 161).
[66] Lu, G.; Crowell, J. E. (J. Phys. Chem. **94** [1990] 5644/6).
[67] Chen, P.-J.; Colaianni, M. L.; Wallace, R. M.; Yates, J. T., Jr. (Surf. Sci. **244** [1991] 177/84).
[68] Wallace, R. M.; Taylor, P. A.; Choyke, W. J.; Yates, J. T., Jr. (J. Appl. Phys. **68** [1990] 3669/78).
[69] Taylor, P. A.; Wallace, R. M.; Choyke, W. J.; Yates, J. T., Jr. (Surf. Sci. **238** [1990] 1/12, 10).
[70] Anderson, D. R.; Bierbaum, V. M.; De Puy, C. H. (J. Am. Chem. Soc. **105** [1983] 4244/8).

1.2.1.2 Molecular Properties

1.2.1.2.1 Point Group. Electronic States

Only two valence states have been observed up to now; some further valence states were predicted or theoretically calculated (p. 56). More recent photoionization work on PH_2 has led to preliminary data of several Rydberg states (p. 57).

Observed Valence States

Two states, but for the time being without any electronic characterization, were first concluded from the respective gas-phase [1, 2] and matrix [3] optical absorption spectra, and somewhat later also from gas-phase emission spectra [4 to 7]. The transition between both states has since then been studied many times; for the complete literature, see the chapter on "Electronic Absorption and Emission", p. 84; compare also "Formation and Detection", p. 47 (the radical detected by electronic absorption or emission and fluorescence methods). The PH_2 radical shows a bent (symmetrical) equilibrium structure in both states (point group C_{2v}) with the upper state being less bent (123°) than the lower state (92°) [8 to 10]. Accurate numerical values for the interbond angle α were derived from rotational analyses of the electronic spectra; see "Geometrical Structure", p. 72. Many approximate or theoretical methods have been applied, however, to estimate the degree of bending in both electronic states; the resulting data for α are dealt with in the present chapter.

The respective **term symbols** $\tilde{X}\ ^2B_1$ (electronic ground state) and $\tilde{A}\ ^2A_1$ were first proposed [11, 12] in view of the Walsh [13] correlation diagram for molecular orbitals of the valence electrons of AH_2 molecules. The orbitals $1a_1$ (or $1\sigma_g$ for the linear arrangement), $1b_2$ (or $1\sigma_u$), $2a_1$ (or $1\pi_u$), and $1b_1$ (or $1\pi_u$; all for valence-electron numbering) then yielded the **electronic configurations** of the seven valence electrons of PH_2 as follows:

$$1a_1^2\ 1b_2^2\ 2a_1^2\ 1b_1,\ ^2B_1, \text{ and } 1a_1^2\ 1b_2^2\ 2a_1\ 1b_1^2,\ ^2A_1$$

Thus, both states are correlated with $1\sigma_g^2\,1\sigma_u^2\,1\pi_u^3,\ ^2\Pi_u$ of the hypothetical linear molecule [12]. The complete electronic configuration for the $\tilde{X}\ ^2B_1$ ground state [14, 15] is based on an ab initio MO-SCF calculation of orbital energies [14]:

$$\underset{P1s}{1a_1^2}\ \underset{P2s}{2a_1^2}\ \underbrace{1b_2^2\ 1b_1^2\ 3a_1^2}_{P2p}\ \ \underbrace{4a_1^2\ 2b_2^2\ 5a_1^2\ 2b_1,}_{valence}\ \ ^2B_1$$

The positions of $1b_1$ and $3a_1$, both correlated with $1\pi_u$ (core; see below) were interchanged in another SCF calculation [16]. The corresponding expression for the linear case will be [15]

$$1\sigma_g^2\ 2\sigma_g^2\ 1\sigma_u^2\ 1\pi_u^4\ 3\sigma_g^2\ 2\sigma_u^2\ 2\pi_u^3,\ ^2\Pi_u$$

The term symbols $^2B_1$ and $^2A_1$ were confirmed by rotational analyses, which yielded a type-C character of the observed electronic transition [8, 17, 18], see also a preliminary statement in [11].

The case of an asymmetric distortion (point group C_s) was considered in [19, 20]; the term designations are then $\tilde{X}\ ^2A''$ and $\tilde{A}\ ^2A'$.

In view of the bent-linear correlation shown above the two states $^2B_1$ and $^2A_1$ may also be regarded as arising from a $^2\Pi_u$ state of a linear PH_2 modification by an extreme **Renner-Teller effect** [21]: the respective bending potential energy curves will then behave like those in **Fig. 3** (b), while the original weak Renner-Teller case (describing the splitting caused by the coupling

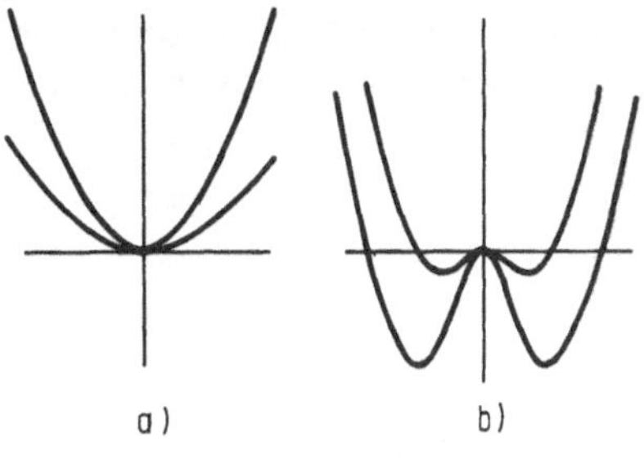

Fig. 3. Bending potential energy curves for weak (a) and strong (b) Renner-Teller coupling.

of the electronic and vibrational angular momenta around the axis of a linear molecule) is shown in Fig. 3 (a) (Fig. 3, p. 53, due to [22]). Compare also the qualitative approaches to the Renner-Teller effect using Walsh diagrams in [23, 24]. According to [25], the theoretical treatment [26] of the observed vibrational intervals in the upper state [1, 2] showed that this state was a component of a Π state (see also p. 75).

Spectroscopic **term energies** T_0 of $\tilde{A}\ ^2A_1$ are shown in the table below. They were obtained from rotational or vibrational analyses of absorption and emission spectra in the vapor phase (in the left part) and of matrix absorption spectra (in the right part):

T_0 in cm^{-1}	18276.59(3)	18276.62	18271.8	18285.3	18215(4)
spectrum	absorption	absorption	absorption	emission	absorption
Ref.	[10]	[8]	[1]	[6]	[27]
remark	a)	b)	c)	d)	e)

a) Rotational analysis of the $0 \leftarrow 0$ band [10]. Also reported in a 1988-table of electronic energy levels of small polyatomic transient molecules [29]. – b) Rotational analysis of the $0 \leftarrow 0$ band [8]. Reported to be 18276.6 cm^{-1} in [23] and apparently misprinted as 18272.62 cm^{-1} in [9, 30]. – c) Long wavelength edge of the $0 \leftarrow 0$ band. Given for PD_2: 18271.6 cm^{-1} [1]. The similar value of 18270 cm^{-1} was erroneously ascribed [31] to a matrix study [3]. – d) $0 \rightarrow 0$ band. Given for PD_2: 18282.2 cm^{-1} [6]; 18282.1 cm^{-1} reported in [29]. – e) Also reported in [32]. An older value was T_0=18188(10) cm^{-1} [28, 29].

Equilibrium term energies T_e of $\tilde{A}\ ^2A_1$ were derived from theoretically calculated bending potential energy curves for both $\tilde{X}\ ^2B_1$ and $\tilde{A}\ ^2A_1$, usually at experimental data for the P–H internuclear distance (see the section on potential curves below). T_e may also be derived from the total energies E_t calculated for both states, usually at the optimized geometries (a few explicit E_t values for $\tilde{X}\ ^2B_1$ are given at the end of the present chapter). T_e=18760 cm^{-1} (≙2.32 eV) was obtained from two potential curves shown in Fig. 4 (p. 56) [19, 50], which are based on ab initio CI calculations [30]. T_e=19650 cm^{-1} (≙2.44 eV) comes from potential curves based on semiempirical calculations (VB theory, combined with the "atoms in molecules" method, and an appropriate d-orbital contribution) [31]. T_e=2.8 eV results from potential curves calculated by an ab initio SCF method [33]. Two values, based on total energies from (unrestricted) Hartree-Fock calculations, are T_e=2.19 eV [15] and 2.41 eV (not explicitly reported) [34].

Vertical excitation energies of $\tilde{A}\ ^2A_1$ from the ground state which are based on available absorption measurements are reported to be 22500 cm^{-1} at the lower state bond angle 92° and 14000 cm^{-1} at the upper state bond angle 123° [12]. A vertical value of 21600 cm^{-1} is given for PD_2 [35].

The **degree of bending** of the PH_2 radical in its two observed valence states $\tilde{X}\ ^2B_1$ and $\tilde{A}\ ^2A_1$, specifically the change $\Delta\alpha = \alpha(\tilde{A}) - \alpha(\tilde{X})$ of the interbond angle, has been discussed ever since the corresponding electronic transition was first detected. Approximate data, extracted from early experimental work (optical absorption or emission and ESR spectra) or based on more qualitative theoretical concepts (Walsh, point charge, and generalized valence bond (GVB) models) are listed below in a chronological order:

$\alpha(\tilde{A})$	~180°	~125°	~120°	~115°	–
$\alpha(\tilde{X})$	~ 90°	~ 97°	–	~ 91°	105°
source	vapor absorption	matrix absorption	vapor absorption	vapor emission	matrix ESR
Ref.	[2]	[3]	[26]	[11]	[36]
remark	a)	b)	c)	d)	e)

$\alpha(\tilde{A})$	118°	~118°	~135°	~120°
$\alpha(\tilde{X})$	99°	~ 90°	~100°	~90°
source	point charge	GVB model	Walsh rules	Walsh rules
Ref.	[37]	[38]	[39]	[40]
remark	f)	g)	h)	i)

a) The long progression of bending vibrational bands ($v_2' \leftarrow v_2''=0$) [1] points to a large $\Delta\alpha$ value. $\alpha(\tilde{X})$ is assumed to be similar to the interbond angle of ground state PH_3; $\alpha(\tilde{A})$ was taken from the analogous excited state of NH_2 [2], see also the review [41]. – b) $\alpha(\tilde{A})$ was obtained from approximate rotational constants. $\Delta\alpha \approx 28°$ was derived from the Franck-Condon principle and led to $\alpha(\tilde{X})$. – c) From an interpretation of the dependence of the vibrational intervals in the upper state on v_2' (exhibiting a minimum) [1] by a harmonic potential function perturbed by an exponential hump [26]. – d) $\alpha(\tilde{X})$ is based [11] on a comparison with the related H_2S molecule and on semiempirical calculations [31]. $\Delta\alpha \approx 24°$ was derived from the Franck-Condon principle and led to $\alpha(\tilde{A})$ [11]. – e) From the s-character (x = 20.6%) of the PH_2 bonding orbitals, based in turn on measurements of ^{31}P and ^{1}H coupling constants in an Xe matrix [36]. ESR spectra in a Kr matrix had yielded x = 19 or 27% and thus $\alpha(\tilde{X}) = 104°$ or 112° [42]. – f) Discussion of the Walsh rules in terms of the valence bond theory. – g) Bond-bond repulsions should increase $\alpha(\tilde{X})$ slightly. – h) Average values for AH_2 molecules (including cations and anions) with A = 1st and 2nd row atoms. – i) "Standard" angles [40]; given earlier in [19, 20].

The observed interbond angles of PH_2 in its two valence states $\tilde{X}\ ^2B_1$ (92°) and $\tilde{A}\ ^2A_1$ (123°) were in turn discussed in terms of the Walsh model [43 to 45], the electrostatic force theory [46], and the valence shell electron pair repulsion (VSEPR) theory [47].

Potential Energy Curves. Barriers to Linearity. An analytical expression for a bending potential function, which considers the barrier to linearity H, the equilibrium interbond angle α_e, the bending force constant at the minimum, and an additional "shape" parameter, was fitted [48] either to rotational and vibrational constants of the ground state $\tilde{X}\ ^2B_1$ [8, 18] or to rotational constants and vibronic energy levels of the excited state $\tilde{A}\ ^2A_1$ [10]. Hence, $H(\tilde{A})$ = 6840 cm^{-1} (≙0.85 eV), leading in turn to $H(\tilde{X}) \approx 25100$ cm^{-1} (≙3.12 eV; using the observed term value $T_0(\tilde{A})$ = 2.27 eV [8]), $\alpha_e(\tilde{A})$ = 121.7°, and $\alpha_e(\tilde{X})$ = 91.27° (for the bending force constants, see p. 76) [48]. Both potential functions [48] are represented as circles in Fig. 4, p. 56. $H(\tilde{A})$ = 6700 cm^{-1} was derived earlier [49]. $H(\tilde{A}) \approx 6000$ cm^{-1} had been obtained [26] for a harmonic potential perturbed by an exponential hump, when the upper state vibrational intervals due to [1] were used (for the angle, see the table above).

Bending potential energy curves for $\tilde{X}$ and $\tilde{A}$ have also been obtained from calculations of the total energies E_t (see below) by ab initio SCF and CI methods. The SCF and the MRD (multi-reference double excitation)-CI curves are shown in the original paper [30]; the curves [30] reproduced in **Fig. 4**, p. 56 (from [19, 50]), were obtained by estimating the energy contribution of higher than double-excitation configurations ("full-CI limit"). Curves for both states were also calculated by the (unrestricted) HF [15, 33] and VB (plus "atoms in molecules") [31] methods or they are qualitatively based on the GVB theory [38].

Symmetric and antisymmetric stretching potential energy curves for both $\tilde{X}\ ^2B_1$ at α = 91.7° and $\tilde{A}\ ^2A_1$ at α = 123° (additionally for symmetric stretching at α = 91.7° and 180°) have also been calculated; graphs in the original paper are for the "full-CI limit" [30].

Total Energy E_t (in au). The lowest value for $\tilde{X}\ ^2B_1$, E_t = –342.04366, was calculated by means of Møller-Plesset 4th order (MP4) perturbation theory for an optimized geometry [51]. Somewhat higher MP4 energies are given in [52 to 54]. $E_t \approx -342.01$ was obtained when an

MRD-CI calculation for the experimental bond length (yielding $E_t = -342.0016$) was corrected by an estimate of 0.01 au (approaching the "full-CI limit"; see Fig. 4, where also an energy for $\tilde{A}\ ^2A_1$ may be read off) [30]. The lowest value from SCF (unrestricted Hartree-Fock) calculations appears to be $E_t = -341.879403$ [16]. Somewhat higher energies were calculated in [14, 15, 30, 34, 47, 54]; total energies for $\tilde{A}$ are given in [15, 34] (see also the excitation energies for $\tilde{A}\ ^2A_1$ given above).

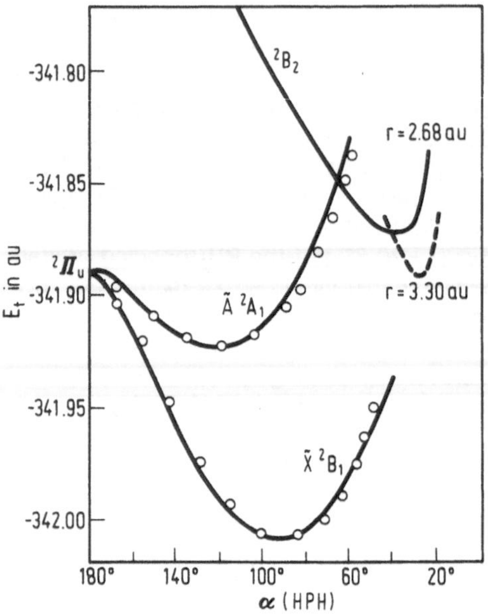

Fig. 4. Bending potential energy curves for PH_2. —— CI calculation, ○ fit to experiment (see text)

Predicted Valence States

Excitation of an electron from the $2b_2(2\sigma_u)$ to the $2b_1(2\pi_u)$ MO leads to another excited state, ...$4a_1^2\ 2b_2\ 5a_1^2\ 2b_1^2$, $^2B_2$ correlated with ...$3\sigma_g^2\ 2\sigma_u\ 2\pi_u^4$, $^2\Sigma_u^+$ [11, 15, 19, 20, 23, 50]. A slightly asymmetric distortion (point group C_s) would give the term symbol $^2A'$ [19, 20]. Both the depletion of $2b_2$ and the double occupation of $5a_1$ leads to a strongly bent character of $^2B_2$ according to the Walsh [13] correlation diagram [15, 19, 20, 50], see also [39] (AH_2 species with 1st row atoms A). A "standard" interbond angle of 30° is given in [19, 20]. An angle of $<90°$ is also indicated by the electronic configuration ...$5a_1^2\ 2b_2\ 2b_1^2$ as written in [13] ($5a_1$ becoming lower than $2b_2$ at small angles). The transition between $^2B_2$ and the $\tilde{X}\ ^2B_1$ ground state is forbidden [11, 13]. On the other hand, considerable interaction between $^2B_2$ and the $\tilde{A}\ ^2A_1$ excited state is expected for asymmetric perturbations, since both states then assume $^2A'$ character [50].

The following equilibrium term energies T_e and angles α_e have been derived from two theoretical calculations:

T_e in eV	α_e	method	Ref.	remark
3.23	27°	MRD-CI	[50]	a)
3.53	35.0°	UHF	[15]	b)

a) Apparently from a bending potential energy curve for an optimized P–H bond length. – b) From the total energy at optimized geometry.

The vertical excitation energy was calculated by the MRD-CI method to be $\Delta E = 5.50$ eV [50]. $\Delta E = 9.3$ eV (75000 ± 10000 cm^{-1}) was estimated from the spin-rotation interaction constant ε_{bb} of ground-state PH_2 and PD_2 [35].

Bending potential energy curves have been theoretically calculated by the UHF method at the experimental bond length of the $\tilde{X}\ ^2B_1$ ground state [15] and by the MRD-CI method for this bond length and an optimized bond length (over a smaller range of angles) [50]. These latter two curves [50] are also shown in Fig. 4.

Two further states, 2 $^2A_1$ and 2 $^2B_2$, result from the excitation of an electron from the $2b_1(2\pi_u)$ MO into the unoccupied $6a_1(4\sigma_g)$ and $3b_2(3\sigma_u)$ MO's. Electronic configurations for bent (C_{2v}) and linear ($D_{\infty h}$) shapes, equilibrium term energies T_e, and angles α_e from UHF calculations [15] are shown below:

state	C_{2v}	$D_{\infty h}$	T_e in eV	α_e
2 $^2A_1$	$\ldots 4a_1^2\ 2b_2^2\ 5a_1^2\ 6a_1$	$\ldots 3\sigma_g^2\ 2\sigma_u^2\ 2\pi_u^2\ 4\sigma_g$	4.77	180°
2 $^2B_2$	$\ldots 4a_1^2\ 2b_2^2\ 5a_1^2\ 3b_2$	$\ldots 3\sigma_g^2\ 2\sigma_u^2\ 2\pi_u^2\ 3\sigma_u$	5.77	80.8°

2 $^2A_1$ and 2 $^2B_2$ dot not appear to be Rydberg states [15].

Rydberg States

Several autoionization features observed in a plot of the photoionization yield of PH_2^+ from PH_2 (obtained from the $PH_3 + H$ reaction) between 1259 and 1113 Å (9.85 to 11.1 eV) probably are members of Rydberg series converging on the excited states $\tilde{a}\ ^3B_1$ and $\tilde{A}\ ^1B_1$ of the PH_2^+ ion (see p. 95). The bending vibration apparently leads to a vibrational structure (for a more detailed discussion, see the original paper) [55].

References:

[1] Ramsay, D. A. (Nature **178** [1956] 374/5).
[2] Ramsay, D. A. (Ann. N.Y. Acad. Sci. **67** [1956/57] 485/98, 492/5).
[3] McCarty, M., Jr.; Robinson, G. W. (J. Chim. Phys. Phys. Chim. Biol. **56** [1959] 723/30).
[4] Pascat, B.; Guenebaut, H. (Bull. Univ. ARERS **2** [1963] 100/7).
[5] Guenebaut, H.; Pascat, B. (C. R. Hebd. Seances Acad. Sci. **256** [1963] 677/80, 2850/3).
[6] Guenebaut, H.; Pascat, B. (J. Chim. Phys. Phys. Chim. Biol. **61** [1964] 592/5).
[7] Guenebaut, H.; Pascat, B. (C. R. Hebd. Seances Acad. Sci. **259** [1964] 2412/5).
[8] Dixon, R. N.; Duxbury, G.; Ramsay, D. A. (Proc. R. Soc. [London] A **296** [1967] 137/60, 145).
[9] Dixon, R. N.; Duxbury, G. (Chem. Phys. Lett. **1** [1967/68] 330/2).
[10] Berthou, J.-M.; Pascat, B.; Guenebaut, H.; Ramsay, D. A. (Can. J. Phys. **50** [1972] 2265/76, 2269).

[11] Guenebaut, H.; Pascat, B.; Berthou, J.-M. (J. Chim. Phys. Phys. Chim. Biol. **62** [1965] 867/77, 875/6).
[12] Dixon, R. N. (Mol. Phys. **10** [1965/66] 1/6).
[13] Walsh, A. D. (J. Chem. Soc. **1953** 2260/6).
[14] Ball, J. R.; Thomson, C. (Int. J. Quantum Chem. **14** [1978] 39/53, 43).
[15] Gu, J.-P.; Huang, M.-B.; Kong, F.; Liu, S.-H. (J. Mol. Struct. **201** [1989] 39/47 [THEOCHEM **60**]).
[16] Hinchliffe, A.; Bounds, D. G. (J. Mol. Struct. **54** [1979] 231/8).
[17] Pascat, B.; Berthou, J.-M.; Guenebaut, H.; Ramsay, D. A. (C. R. Seances Acad. Sci. B **263** [1966] 1397/9).

[18] Pascat, B.; Berthou, J.-M.; Prudhomme, J.-C.; Guenebaut, H.; Ramsay, D. A. (J. Chim. Phys. Phys. Chim. Biol. **65** [1968] 2022/9).
[19] Bruna, P. J.; Hirsch, G.; Buenker, R. J.; Peyerimhoff, S. D. (NATO-ASI Ser. B **90** [1983] 309/54, 317).
[20] Bruna, P. J.; Peyerimhoff, S. D. (Bull. Soc. Chim. Belg. **92** [1983] 525/46, 534, 539).

[21] Brus, L. E. (J. Chem. Phys. **57** [1972] 3167/74).
[22] Pople, J. A.; Longuet-Higgins, H. C. (Mol. Phys. **1** [1958] 372/83).
[23] Herzberg, G. (Molecular Spectra and Molecular Structure, Vol. 3, van Nostrand, New York 1966, pp. 316/7).
[24] Duxbury, G. (Mol. Spectrosc. Chem. Soc. [London] **3** [1975] 497/573, 512/3).
[25] Dixon, R. N.; Duxbury, G.; Horani, M.; Rostas, J. (Mol. Phys. **22** [1971] 977/92, 985/6).
[26] Dixon, R. N. (Trans. Faraday Soc. **60** [1964] 1363/8).
[27] Withnall, R.; McCluskey, M.; Andrews, L. (J. Phys. Chem. **93** [1989] 126/9).
[28] Larzillière, M.; Jacox, M. E. (NBS Spec. Publ. [U.S.] 561-1 [1979] 529/43, 533).
[29] Jacox, M. E. (J. Phys. Chem. Ref. Data **17** [1988] 269/511, 283/4).
[30] Perić, M.; Buenker, R. J.; Peyerimhoff, S. D. (Can. J. Chem. **57** [1979] 2491/7).

[31] Jordan, P. C. (J. Chem. Phys. **41** [1964] 1442/9).
[32] Jacox, M. E. (J. Phys. Chem. Ref. Data **19** [1990] 1387/546, 1397).
[33] So, S. P.; Richards, W. G. (Int. J. Quantum Chem. **11** [1977] 73/9).
[34] Müller, J.; Ågren, H.; Canuto, S. (J. Chem. Phys. **76** [1982] 5060/8).
[35] Vervloet, M.; Berthou, J.-M. (Can. J. Phys. **54** [1976] 1375/82).
[36] Jackel, G. S.; Gordy, W. (Phys. Rev. [2] **176** [1968] 443/52, 446).
[37] Takahata, Y.; Schnuelle, G. W.; Parr, R. G. (J. Am. Chem. Soc. **93** [1971] 784/5).
[38] Goddard, W. A., III; Harding, L. B. (Annu. Rev. Phys. Chem. **29** [1978] 363/96, 382/5).
[39] Casida, M. E.; Chen, M. M. L.; MacGregor, R. D.; Schaefer, H. F., III (Isr. J. Chem. **19** [1980] 127/31).
[40] Pope, S. A.; Hillier, I. H.; Guest, M. F. (Faraday Symp. Chem. Soc. No. 19 [1984] 109/23, 112).

[41] Herzberg, G. (Annu. Rev. Phys. Chem. **9** [1958] 315/38, 326).
[42] Morehouse, R. L.; Christiansen, J. J.; Gordy, W. (J. Chem. Phys. **45** [1966] 1747/51).
[43] Gimarc, B. M. (J. Am. Chem. Soc. **93** [1971] 593/9; Acc. Chem. Res. **7** [1974] 384/92).
[44] Buenker, R. J.; Peyerimhoff, S. D. (Chem. Rev. **74** [1974] 127/88, 149).
[45] Baird, N. C. (J. Chem. Educ. **55** [1978] 412/7).
[46] Nakatsuji, H. (J. Am. Chem. Soc. **95** [1973] 354/61).
[47] Baird, N. C.; Kuhn, M.; Lauriston, T. M. (Can. J. Chem. **67** [1989] 1952/8).
[48] Barrow, T.; Dixon, R. N.; Duxbury, G. (Mol. Phys. **27** [1974] 1217/34, 1225, 1229).
[49] Dixon, R. N. (J. Phys. Colloq. [Paris] **32** [1971] C5a-147/C5a-154).
[50] Bruna, P. J.; Hirsch, G.; Perić, M.; Peyerimhoff, S. D.; Buenker, R. J. (Mol. Phys. **40** [1980] 521/37, 524/6, 530).

[51] Wong, M. W.; Gill, P. M. W.; Nobes, R. H.; Radom, L. (J. Phys. Chem. **92** [1988] 4875/80).
[52] Pople, J. A.; Luke, B. T.; Frisch, M. J.; Binkley, J. S. (J. Phys. Chem. **89** [1985] 2198/203).
[53] Cattani-Lorente, M.; Geoffroy, M.; Mishra, S. P.; Weber, J.; Bernardinelli, G. (J. Am. Chem. Soc. **108** [1986] 7148/53).
[54] Gill, P. M. W.; Radom, L. (J. Am. Chem. Soc. **110** [1988] 4931/41, 4934).
[55] Berkowitz, J.; Cho, H. (J. Chem. Phys. **90** [1989] 1/6).

1.2.1.2.2 Electronic Structure

The inertial axes a, b (∥C_2 symmetry axis), and c (⊥ molecular plane) are parallel to the axes y, z, and x, respectively, of the C_{2v} point group of the PH_2 radical in its two observed valence states $\tilde{X}\ ^2B_1$ and $\tilde{A}\ ^2A_1$ [1].

Ground State $\tilde{X}\ ^2B_1$

The π-radical nature of ground-state PH_2 is indicated by the small value of the isotropic ^{31}P magnetic hyperfine (hf) coupling constant $a_F(^{31}P)$ (see p. 65) as compared to its value for the free P atom or for PH_2 in its excited state $\tilde{A}\ ^2A_1$ [2]. An essential location of the unpaired electron ($2b_1$) in a $P3p_x$ orbital is in accordance with an approximate relation found for the components T_{qq} (q = a, b, c) of the anisotropic (dipolar) ^{31}P magnetic hf coupling (see p. 65), i.e., $T_{aa} \approx T_{bb} \approx -T_{cc}/2$ [2, 3]. A small P3s character of <2.2% of the unpaired electron was first derived from $a_F(^{31}P) = 225$ MHz (≙80 G) on account of an ESR spectrum of PH_2 isolated in a Kr matrix and from an atomic value [4] of 10178 MHz (≙3630 G) [5]. A 3s contribution of ~2% was reported [3] in view of this matrix value for $a_F(^{31}P)$ [5] and of a similar value from a far-IR laser magnetic resonance (LMR) spectrum [3]. A later ESR spectrum of PH_2 in a cancrinite matrix yielded 2.3% [6].

Unpaired electron **spin density** ρ. Negative values of ρ for each H nucleus of PH_2 (and thus for the related isotropic ^1H magnetic hf coupling constant $a_F(^1H)$; see p. 65) are to be expected in view of a spin-polarization effect in the P–H σ bond [5 to 7]. Numerical data were derived from $a_F(^1H)$ from two ESR spectra: ρ = (−)0.035 (Kr matrix) [5] and −0.046 (cancrinite matrix) [6]. The following spin densities were theoretically calculated by ab initio (unrestricted Hartree-Fock, UHF) and semiempirical (INDO) methods [8]:

method	$\rho(P3p_x)$	$\rho(P3p_y)$	$\rho(P3p_z)$	ρ(P3s)	ρ(H1s)
UHF	0.99	0.04	−0.03	0.09	−0.06
INDO	1.00	0.07	0.04	0.01	−0.06

The unit spin density in the $P3p_x$ orbital was also obtained from another semiempirical calculation (INDO/2) [9].

The $P3p_x$ character of the $2b_1$ MO was also stressed in an ab initio MO-SCF calculation [10] and in the qualitative MO picture of Walsh [11]; see also [1].

The $5a_1$ MO probably contains a large contribution of $P3p_z$ (contrary to its essential P3s character in the Walsh [11] picture), as may be inferred from the spin-rotation interaction constant ε_{aa} (see p. 67). The observed ε_{aa} agrees reasonably well with a value based on an approximate model for the electronic structure of PH_2, which was explained in [1] and which assumed $5a_1$ to be a pure $P3p_z$ orbital (model of "pure precession") [12]. A preliminary, almost vanishing value of ε_{aa} was, however, used in [1], and thus very little p character was ascribed to $5a_1$. This agrees with another approximate model, i.e., one of "directive hybridization" (with $5a_1$ made orthogonal to hybridized P orbitals) [1]. Ab initio MO-SCF calculations showed $5a_1$ to be composed of $P3p_z$, P3s, and H1s orbitals with a small antibonding P–H overlap, a small bonding H–H overlap [10], and a larger contribution to $5a_1$ of P3p than that of P3s [13].

The $2b_2$ orbital is shown by ab initio MO-SCF calculations to be composed of $P3p_y$ and H1s and to be P–H bonding with a small H–H antibonding contribution [10]. For its bonding character, see also [1].

The $4a_1$ orbital was similarly shown to be a bonding orbital between P3s and H1s with a noticeable contribution from $P3p_z$ (and from polarization functions) [10]; see also [1].

The fractional s-character of the P–H σ bond was derived from spin densities on H, obtained by ESR spectra of PH_2 in three matrices, and is shown below, correcting a table given in [6]:

matrix	Kr	Xe	cancrinite
s-character	0.19	0.206[1)]	0.162[2)]
Ref.	[5]	[14]	[6]

1) Erroneously 0.026 in [6]. – 2) Erroneously 0.126 in [6].

Net atomic **charges** at P and H and the **dipole moment** μ from an ab initio MO-SCF calculation (unrestricted Hartree-Fock method) are shown below [15]:

P	H	$\lvert\mu\rvert$ in D
−0.1319	+0.0659	0.9072 (0.3569 au)

Reversed signs were obtained for the charges in another ab initio MO-SCF calculation, i.e., +0.219 and −0.109 (at H; $\lvert\mu\rvert = 0.816$ D) [10] and in a semiempirical (CNDO/2) calculation, i.e., −0.0894 (at H) [16].

The bonding in ground state PH_2 was also described in the framework of a generalized valence bond (GVB) theory [17]. A VB method, which was combined with the method of "atoms in molecules" [18], yielded a small d contribution (4%) in the P valence state [19]. For an application of the valence shell electron pair repulsion (VSEPR) theory, see [20].

Excited State $\tilde{A}\ ^2A_1$

The σ-radical nature of PH_2 in its excited state $\tilde{A}\ ^2A_1$ is indicated by the magnitude of the isotropic ^{31}P magnetic hf coupling constant $a_F(^{31}P)$ (see p. 66). A $P3p_z$ character of the unpaired electron ($5a_1$) is consistent with an approximate relation between the anisotropic ^{31}P coupling constants T_{qq} (see above), i.e., $T_{aa} \approx T_{cc} \approx -T_{bb}/2$ [2]. A large contribution of $P3p_z$ to $5a_1$ was also inferred from the spin-rotation interaction constant ε_{aa} in the excited state (see p. 70), lying between two values, which were derived either from the "directive hybridization" model ($5a_1 \perp$ P hybridization orbitals) or from the "pure precession" model ($5a_1 = P3p_z$) [1]; see also [12]. A P3s character of 18% was derived for the odd electron from $a_F(^{31}P)$. A value of 33% is expected from sp^2 hybridization, which is in turn consistent with a bond angle of 123° [21].

Ab initio MO-SCF calculations yielded a higher P3p than P3s contribution to $5a_1$ and a loss of 0.56 of an electron by $P3p_z$ on excitation from the ground state to $\tilde{A}\ ^2A_1$ [13]. The bonding in $\tilde{A}\ ^2A_1$ can also be described by a GVB picture [17] and by the VB method above (yielding a 6% d-character in the P valence state) [19].

References:

[1] Dixon, R. N. (Mol. Phys. **10** [1965/66] 1/6).
[2] Kakimoto, M.; Hirota, E. (J. Mol. Spectrosc. **94** [1982] 173/91, 186).
[3] Davies, P. B.; Russell, D. K.; Thrush, B. A.; Radford, H. E. (Chem. Phys. **44** [1979] 421/6).
[4] Morton, J. R. (Chem. Rev. **64** [1964] 453/71, 456).
[5] Morehouse, R. L.; Christiansen, J. J.; Gordy, W. (J. Chem. Phys. **45** [1966] 1747/51).
[6] Raghunathan, P.; Sur, S. K. (Proc. Indian Acad. Sci. Chem. Sci. **92** [1983] 597/604).
[7] Endo, Y.; Saito, S.; Hirota, E. (J. Mol. Spectrosc. **97** [1983] 204/12, 211).
[8] Cattani-Lorente, M.; Geoffroy, M.; Mishra, S. P.; Weber, J.; Bernardinelli, G. (J. Am. Chem. Soc. **108** [1986] 7148/53).
[9] Kilcast, D.; Thomson, C. (J. Chem. Soc. Faraday Trans. II **68** [1972] 435/43).
[10] Ball, J. R.; Thomson, C. (Int. J. Quantum Chem. **14** [1978] 39/53, 44/7).

[11] Walsh, A. D. (J. Chem. Soc. **1953** 2260/6).
[12] Dixon, R. N.; Duxbury, G.; Ramsay, D. A. (Proc. R. Soc. [London] A **296** [1967] 137/60, 145).
[13] So, S. P.; Richards, W. G. (Int. J. Quantum Chem. **11** [1977] 73/9).
[14] Jackel, G. S.; Gordy, W. (Phys. Rev. [2] **176** [1968] 443/52).
[15] Hinchliffe, A.; Bounds, D. G. (J. Mol. Struct. **54** [1979] 231/8).
[16] Takahata, Y. (Chem. Phys. Lett. **59** [1978] 472/7).
[17] Goddard, W. A., III; Harding, L. B. (Annu. Rev. Phys. Chem. **29** [1978] 363/96, 382, 384).
[18] Jordan, P. C. H.; Longuet-Higgins, H. C. (Mol. Phys. **5** [1962] 121/38).
[19] Jordan, P. C. (J. Chem. Phys. **41** [1964] 1442/9).
[20] Baird, N. C.; Kuhn, M.; Lauriston, T. M. (Can. J. Chem. **67** [1989] 1952/8).

[21] Curl, R. F.; Endo, Y.; Kakimoto, M.; Saito, S.; Hirota, E. (Chem. Phys. Lett. **53** [1978] 536/8).

1.2.1.2.3 **Ionization Potential** E_i, **Electron Affinity** A, both in eV

The following adiabatic **ionization potentials** from three measurements employing mass spectrometric (MS) and photoelectron spectroscopic (PES) methods agree within the error limits given. The values refer to ionization of the outermost ($2b_1$) electron, yielding the $\tilde{X}\ ^1A_1$ ground state of the PH_2^+ ion. For each measurement, details and additional data, also for ionization of the $5a_1$ electron (yielding the $\tilde{a}\ ^3B_1$ excited state of PH_2^+), are given below the table (PI = photoionization):

E_i(ad) in eV	9.824 ± 0.002	9.83 ± 0.02	9.84 ± 0.01
measurement	PIMS	MS	PES
Ref.	[1]	[2]	[3]

PIMS: A photoion yield curve of PH_2^+ from PH_2 (generated by pyrolysis of benzylphosphane at 800°C) showed deviations in the threshold region from idealized step function behavior, which were ascribed to rotational effects. Simulating this behavior on the basis of experimental and calculated rotational levels for the ground states of PH_2 and PH_2^+, respectively, enabled the selection of the rotationally adiabatic ionization threshold as $1262.0_5 \pm 0.3$ Å or 79236 ± 20 cm^{-1}. A further increase in the ion yield began at ~1177 Å; thus the second ionization potential (to $\tilde{a}\ ^3B_1$ of PH_2^+) of $E_i \geqq 10.534$ eV was given [1]; see also a discussion of the curve in [4]. Another PIMS study, where PH_2 was generated by the PH_3 + H abstraction reaction, showed the first threshold more clearly (at ~1261.5 Å). A step-like structure at ~1178 Å ≙ ~10.525 eV is attributed to direct ionization (of $5a_1$) forming $PH_2^+(\tilde{a}\ ^3B_1)$ in its lowest vibrational level [5].

MS: Monoenergetic electron impact on PH_2 (from benzylphosphane) was used, and the ionization efficiency curve was extrapolated to zero ion current. The sharp onset pointed to the adiabatic character of E_i. A measurement with conventional electron energy spread and the usual semilogarithmic method due to [6] yielded $E_i = 9.96$ eV [2].

PES: A preliminary He I spectrum of PH_2 from the PH_3 + F reaction showed a sharp band at the position tabulated above and a broad band in the region 11.0 to 11.5 eV, which was assigned to ionization to $PH_2^+(\tilde{a}\ ^3B_1)$ [3]. According to [5], however, energetically lower vibrational members of that latter band were probably not seen in the PE spectrum [3], because they were masked by ionization bands from P, P_2, and PH_3.

Correcting E_i by the use of updated thermochemical data [1] yields $E_i = 9.82$ eV (≙226.6 kcal/mol) instead of the original value $E_i = 9.28$ eV (214.0 kcal/mol; based on the formation enthalpies of PH_2 and PH_2^+) [7] and $E_i = 9.83$ eV instead of 10.2 eV (derived from $D(H_2P\text{-}H)$ and the appearance potential of PH_2^+ from PH_3) [8].

Purely theoretical data with an accuracy of ~0.1 eV have been calculated more recently. Ab initio MO theory corrected by 4th order Møller-Plesset (MP4) perturbation calculations [9], gave $E_i = 9.72$ and 10.64 (for $\tilde{X}\ ^1A_1$ and $\tilde{a}\ ^3B_1$ of PH_2^+) [1, 10]. A modification of that method (to give also singlet-triplet separations) yielded $E_i = 9.77$ and 10.64 [11]. A treatment of the electron correlation not only by MP4 but also by quadratic configuration interaction (CI) [12, 13] led to $E_i = 9.71$ [14]. Older data were obtained by CI [15, 16] and SCF [16, 17] calculations. – For orbital energies from SCF calculations, see [17, 18].

The **electron affinities** listed in the table below agree within the error limits given. They were obtained by laser photoelectron spectroscopy (LPES) on the PH_2^- ion [19] and PH_2^- photodetachment using a tunable laser (LPD) [20, 21] or an Xe arc lamp (with a grating monochromator) [21] and ion cyclotron resonance (ICR) spectrometry [20, 21]. All values are reported in two reviews on electron affinities [22] and electron photodetachment [23], and may be considered as adiabatic (see the remarks below the table):

A in eV	1.271 ± 0.010	1.26 ± 0.03	1.25 ± 0.03
measurement	LPES	LPD + ICR	PD + ICR
Ref.	[19]	[20, 21]	[21]
remark	a)	b)	c)

a) 4880 Å Ar ion laser. The absence of a significant vibrational structure is in agreement with the detachment of a nonbonding $2b_1$ electron and points to nearly identical geometries of the ground states $\tilde{X}\ ^1A_1$ and $\tilde{X}\ ^2B_1$ of PH_2^- and PH_2. – b) Continuously tunable laser. A sharp increase of the PD cross section was found at the threshold wavelength of 987 nm, pointing to similar structures of PH_2^- and PH_2 [20]. A threshold of 987.5 ± 7.2 nm was given [21]. – c) Xe arc lamp. The threshold wavelength was 994.3 ± 6.7 nm. A very similar energy dependence of the photodetachment cross section was found for PD_2^- [21].

Correcting A by the use of an updated value for $D(H_2P\text{-}H)$ [1] yields A = 1.27 instead of A = 1.60 (≙36.8 kcal/mol), which had been inferred from magnetron measurements with PH_3 using a W filament (or with P_2H_4 and a Pt filament) [24]. That value had earlier been modified into A = 1.34 [21]. A rough value $A = 1.5 \pm 0.5$ is based on $D(H_2P\text{-}H)$ and the proton affinity of PH_2^- (see p. 104) [25] (the original paper gives $A = 35 \pm 11$ kcal/mol, erroneously converted to 1.4 ± 0.5 eV). Two lower limits, $A \geqq 1.6$ [26] and $A \geqq 1$ [27], were derived from appearance potentials of PH_2^- from PH_3 and older data for $D(H_2P\text{-}H)$. An upper limit of $A < A(SO_2) < 1.1$ eV was based on the observation of a charge transfer reaction between PH_2^- and SO_2 [27].

A numerical relation between the electron affinity A and the ionization potential E_i of a radical X and the bond dissociation energy D(X–H) was for a certain class of hydrides D (in kcal/mol) = 11.1 [E_i (in eV) – A (in eV)] – 6.8. Then, $A \approx 1.8$ is obtained from D = 85 kcal/mol (3.7 eV) and $E_i = 10.2$ eV [8]. A = 0.5 follows from D = 90 kcal/mol (3.9 eV) and $E_i = 9.3$ eV [7].

Electron propagator theory (EPT) yielded a vertical value of 1.160 eV [28] (see also [29 to 31]), MP2 theory gave 1.162 eV (zero point correction included) [32], MP4 theory (using isogyric comparisons with the H_2 molecule) 1.23 eV [33], and MP4 + quadratic CI calculations 1.20 eV [14]. Compare also a 1985-review on theoretical calculations of A [34]. HF calculations on PH_2^- yielded Koopmans' Theorem (KT) electron detachment energies, which depend strongly on the basis sets used [35]. Older ab initio MO-SCF calculations gave unsatisfactory results [17].

References:

[1] Berkowitz, J.; Curtiss, L. A.; Gibson, S. T.; Greene, J. P.; Hillhouse, G. L.; Pople, J. A. (J. Chem. Phys. **84** [1986] 375/84, 378/81).
[2] McAllister, T.; Lossing, F. P. (J. Phys. Chem. **73** [1969] 2996/8).
[3] Dyke, J. M.; Jonathan, N.; Morris, A. (Int. Rev. Phys. Chem. **2** [1982] 3/42, 24).
[4] Berkowitz, J. (J. Chem. Phys. **89** [1988] 7065/76, 7066, 7074).
[5] Berkowitz, J.; Cho, H. (J. Chem. Phys. **90** [1989] 1/6).
[6] Fisher, I. P.; Homer, J. P.; Lossing, F. P. (J. Am. Chem. Soc. **87** [1965] 957/60).
[7] Saalfeld, F. E.; Svec, H. J. (Inorg. Chem. **3** [1964] 1442/3).
[8] Neale, R. S. (J. Phys. Chem. **68** [1964] 143/6).
[9] Pople, J. A.; Luke, B. T.; Frisch, M. J.; Binkley, J. S. (J. Phys. Chem. **89** [1985] 2198/203).
[10] Pople, J. A.; Curtiss, L. A. (J. Phys. Chem. **91** [1987] 155/62).

[11] Pople, J. A.; Curtiss, L. A. (J. Phys. Chem. **91** [1987] 3637/9).
[12] Pople, J. A.; Head-Gordon, M.; Raghavachari, K. (J. Chem. Phys. **87** [1987] 5968/75).
[13] Pople, J. A.; Head-Gordon, M.; Fox, D. J.; Raghavachari, K.; Curtiss, L. A. (J. Chem. Phys. **90** [1989] 5622/9).
[14] Curtiss, L. A.; Jones, C.; Trucks, G. W.; Raghavachari, K.; Pople, J. A. (J. Chem. Phys. **93** [1990] 2537/45).
[15] Pope, S. A.; Hillier, I. H.; Guest, M. F.; Kendric, J. (Chem. Phys. Lett. **95** [1983] 247/9).
[16] Pope, S. A.; Hillier, I. H.; Guest, M. F. (Faraday Symp. Chem. Soc. No. 19 [1984] 109/23, 118).
[17] Ball, J. R.; Thomson, C. (Int. J. Quantum Chem. **14** [1978] 39/53, 46).
[18] Hinchliffe, A.; Bounds, D. G. (J. Mol. Struct. **54** [1979] 231/8).
[19] Zittel, P. F.; Lineberger, W. C. (J. Chem. Phys. **65** [1976] 1236/43).
[20] Smyth, K. C.; McIver, R. T., Jr.; Brauman, J. I.; Wallace, R. W. (J. Chem. Phys. **54** [1971] 2758/9).

[21] Smyth, K. C.; Brauman, J. I. (J. Chem. Phys. **56** [1972] 1132/42, 1140, 1142).
[22] Janousek, B. K.; Brauman, J. I. (in: Bowers, M. T.; Gas Phase Ion Chemistry, Vol. 2, Academic, New York 1979, pp. 53/86, 80).
[23] Drzaic, P. S.; Marks, J.; Brauman, J. I. (in: Bowers, M. T.; Gas Phase Ion Chemistry, Vol. 3, Academic, New York 1984, pp. 167/211, 202).
[24] Page, F. M.; Goode, G. C. (Negative Ions and the Magnetron, Wiley-Interscience, New York 1969, pp. 130, 138).
[25] Holtz, D.; Beauchamp, J. L.; Eyler, J. R. (J. Am. Chem. Soc. **92** [1970] 7045/55, 7054/5).
[26] Halmann, M.; Platzner, I. (J. Phys. Chem. **73** [1969] 4376/8).
[27] Ebinghaus, H.; Kraus, K.; Müller-Duysing, W.; Neuert, H. (Z. Naturforsch. **19a** [1964] 732/6).
[28] Ortiz, J. V. (J. Chem. Phys. **86** [1987] 308/12).
[29] Ortiz B., J. V.; Öhrn, Y. (Chem. Phys. Lett. **77** [1981] 548/54).
[30] Ortiz, J. V. (Chem. Phys. Lett. **136** [1987] 387/91).

[31] Ortiz, J. V. (Int. J. Quantum Chem. Quantum Chem. Symp. No. 22 [1988] 431/6).
[32] Nguyen, M. T. (J. Mol. Struct. **189** [1988] 23/9 [THEOCHEM **49**]).
[33] Pople, J. A.; Schleyer, P. von R.; Kaneti, J.; Spitznagel, G. W. (Chem. Phys. Lett. **145** [1988] 359/64).
[34] Gutsev, G. L.; Boldyrev, A. I. (Adv. Chem. Phys. **61** [1985] 169/221, 176, 189).
[35] Lohr, L. L.; Ponas, S. H. (J. Phys. Chem. **88** [1984] 2992/7).

1.2.1.2.4 g Factors. Spin-Orbit Coupling

The following **electron spin** g factors were obtained from ESR spectra of the PH_2 radical isolated in various solid matrices:

matrix	Kr	Kr	Xe	Xe	cancrinite
T in K	4.2	~10	4.2	4.2	77
g	2.0087	2.0090(2)	2.0050(10)	2.0048(10)	2.0048(10)
Ref.	[1]	[2]	[3]	[4]	[5]

g = 2.0090 was derived as an average of the principal components g^{qq} of the g tensor (q = inertial axes a, b, c), which were approximated by an expression due to [6], i.e., $g^{qq} = 2.0023 - \varepsilon_{qq}/2X$; gas-phase values for the spin-rotation interaction constants ε_{qq} and the rotational constants X = A, B, C were used [7]. An anisotropic ESR spectrum of PH_2 in a frozen concentrated aqueous solution of sulfuric acid (see p. 82) yielded a g-tensor component of $g^{xx} = 2.002$ [8] which may be identified (in analogy to the PF_2 case [9]) with g^{cc}.

Rotational g factors g_r^{qq} have been derived from an expression due to [10] for the electronic contribution to these factors, i.e., $g_r^{qq} = -|\varepsilon_{qq}|/A_{so}$ with A_{so} = **spin-orbit coupling** constant. Using available data for ε_{qq} and an atomic P value for A_{so} (299 cm^{-1} given in [11]), the following factors were obtained [12]:

g_r^{aa}	g_r^{bb}	g_r^{cc}
−0.00094	−0.00027	0

$g_r^{aa} = -0.00123$ was apparently derived using $A_{so} = 228\ cm^{-1}$ (while in the text $A_{so} = 300\ cm^{-1}$ was given) [13]. $A_{so} = 232\ cm^{-1}$ was based on data [14] for the A $^3\Pi_i$ state of the PH radical (see p. 12) and was in turn used to determine the spin-rotation interaction constant ε_{aa} from two approximate models for the electronic structure of PH_2 ("directive hybridization" and "pure precession"; see pp. 59/60) [15]; for the second model, see also [16].

A single spin-orbit coupling constant of 170 cm^{-1} was chosen for both the ground state $\tilde{X}\ ^2B_1$ and the excited state $\tilde{A}\ ^2A_1$ of PH_2 to simulate the dependence of ε_{aa} on the vibrational and "rotational" quantum numbers v_2 and K (in the excited state) within a theory of the Renner-Teller effect for a bent triatomic molecule executing large-amplitude bending vibrations. The theory describes the vibronic coupling and the spin-orbit coupling between the states $^2B_1$ and $^2A_1$ in terms of the correlated $^2\Pi_u$ state of the linear radical [17]. This theory was discussed and extended in [18].

References:

[1] Morehouse, R. L.; Christiansen, J. J.; Gordy, W. (J. Chem. Phys. **45** [1966] 1747/51).
[2] McDowell, C. A.; Mitchell, K. A. R.; Raghunathan, P. (J. Chem. Phys. **57** [1972] 1699/703).
[3] Jackel, G. S.; Gordy, W. (Phys. Rev. [2] **176** [1968] 443/52, 446).
[4] Shimokoshi, K.; Nakamura, K.; Sato, S. (Mol. Phys. **53** [1984] 1239/49, 1243).
[5] Raghunathan, P.; Sur, S. K. (Proc. Indian Acad. Sci. Chem. Sci. **92** [1983] 597/604).
[6] Curl, R. F., Jr. (J. Chem. Phys. **37** [1962] 779/84; Mol. Phys. **9** [1965] 585/97, 587).
[7] Davies, P. B.; Russell, D. K.; Thrush, B. A. (Chem. Phys. Lett. **37** [1976] 43/6).
[8] Fullam, B. W.; Mishra, S. P.; Symons, M. C. R. (J. Chem. Soc. Dalton Trans. **1974** 2145/8).
[9] Wei, M. S.; Current, J. H.; Gendell, J. (J. Chem. Phys. **52** [1970] 1592/602), Nelson, W.; Jackel, G.; Gordy, W. (J. Chem. Phys. **52** [1970] 4572/8).
[10] Barnes, C. E.; Brown, J. M.; Carrington, A.; Pinkstone, J.; Sears, T. J. (J. Mol. Spectrosc. **72** [1978] 86/101, 98/9).

[11] Bower, H. J.; Symons, M. C. R.; Tinling, D. J. A. (in: Kaiser, E. T.; Kevan, L.; Radical Ions, Interscience, New York 1968, pp. 417/73, 424).
[12] McKellar, A. R. W. (Faraday Discuss. Chem. Soc. No. 71 [1981] 63/76, 72/4).
[13] Endo, Y.; Saito, S.; Hirota, E. (J. Mol. Spectrosc. **97** [1983] 204/12).
[14] Legay, F. (Can. J. Phys. **38** [1960] 797/805).
[15] Dixon, R. N. (Mol. Phys. **10** [1965/66] 1/6).
[16] Dixon, R. N.; Duxbury, G.; Ramsay, D. A. (Proc. R. Soc. [London] A **296** [1967] 137/60, 145).
[17] Barrow, T.; Dixon, R. N.; Duxbury, G. (Mol. Phys. **27** [1974] 1217/34, 1227/30).
[18] Jungen, C.; Hallin, K.-E. J.; Merer, A. J. (Mol. Phys. **40** [1980] 65/94, 69).

1.2.1.2.5 Magnetic Hyperfine Coupling Constants a_F, T_{qq}. Nuclear Spin-Rotation Interaction Constants C_{qq}

The isotropic or Fermi contact hyperfine (hf) coupling constant a_F was first obtained for both nuclei ^{31}P and ^1H from ESR spectra of ground-state PH_2 isolated at low temperature in rare-gas matrices (see the second table below). Anisotropic or dipolar features were first detected for ^{31}P with PH_2 anchored by H bonding in a frozen aqueous solution of sulfuric acid [1]. Complete sets of a_F and the anisotropic components T_{qq} (q = inertial axes a, b, c; $\Sigma T_{qq} = 0$) for both nuclei of ground-state PH_2 were later determined by far-IR laser magnetic resonance (FIR LMR) [2] and microwave (MW) [3, 4] spectra. The latter [3, 4] also yielded data for the interaction constants $C_{qq}(^{31}P)$. Interaction constants were also obtained for the electronically excited state [5, 6].

Ground State $\tilde{X}$ $^2B_1$

Two sets of hf coupling constants (all in MHz) measured in the **vapor phase** are given below:

^{31}P

a_F	T_{aa}	T_{bb}	T_{cc}	C_{aa}	C_{bb}	C_{cc}
207.316(99)	−300.25(17)	−322.01(14)	622.26	0.938(56)	0.507(49)	0.118(48)
217(10)	−303(16)	−318(12)	621(12)	—	—	—

^1H

a_F	T_{aa}	T_{bb}	T_{cc}	method	Ref.	remark
−48.866(70)	−1.04(14)	−4.42(11)	5.46	MW	[4]	a)
−49(2)	−3(7)	−3(7)	5(7)	FIR LMR	[2]	b)

a) Five rotational transitions were measured. The analysis [4] included earlier hf data from MW [3] and FIR LMR [2, 7] spectra. T_{cc} was derived from the sum rule $\Sigma T_{qq} = 0$. Very similar values were given in [3] (see also [6]). – b) Eight rotational transitions were measured. $T_{cc}/2$ for ^{31}P (and the absolute values of T_{aa} and T_{bb}) were in the order of magnitude of $B(^{31}P) = 287$ MHz calculated [8] for the P3p orbital in the ground state of the free atom [2]. In a preliminary FIR LMR study [7] $a_F(^{31}P) = 224$ MHz from a matrix ESR spectrum (see the next table) and $T_{cc}/2(^{31}P) = 287$ MHz [8] were used to interpret the observed ^{31}P hf structure.

A set of ^{31}P hf coupling constants were also obtained in an ESR study of PH_2 in a **frozen aqueous solution** of sulfuric acid [1]: a_F (or A_{iso}) = 224 MHz (≙8 mT) was taken from a Kr matrix ESR spectrum (see below); T_{cc} (or 2B) = 546 MHz (≙19.5 mT) was derived from a measured value $A_{\parallel}$ = 27.5 mT and A_{iso} = 8 mT [1]; $T_{aa} = T_{bb}$ = −273 MHz (≙9.75 mT) may be added in view of an axial symmetry and of the sum rule.

The following table lists the isotropic constant a_F for both ^{31}P and ^{1}H measured on ESR spectra of PH_2 in solid **rare-gas matrices** (a positive sign had earlier been ascribed to $a_F(^1H)$; the original data in G or cm^{-1} are also given):

matrix	T in K	$a_F(^{31}P)$ in MHz	$a_F(^{31}P)$ in G	$-a_F(^1H)$ in MHz	$-a_F(^1H)$ in G	Ref.
Kr	4.2	224	80	51	18	[9]
	~10	230	82.3(5)	50.4	18.0(5)	[10]
Xe	4.2	229	81.79(50)	47.6	17.0(10)	[11]
	4.2	224	79.75(100)	55.3	19.74(100)	[12]
cancrinite	77	234	83.4(5)[1)]	64.6	23.1(5)[1)]	[13]

[1)] $(77.95 \pm 0.5) \times 10^{-4}$ and $(21.56 \pm 0.5) \times 10^{-4}$ cm^{-1} in the original paper [13].

Purely theoretically derived values for a_F of ^{31}P and ^{1}H were obtained by ab initio (UHF) [14 to 16] and semiempirical (INDO) [14] MO methods. Values without spin annihilation seem to agree better with the experimental data [14, 15].

Excited State $\tilde{A}\ ^2A_1$

A complete set of magnetic hf coupling constants and the ^{31}P nuclear spin-rotation interaction constants (all in MHz) were obtained from an analysis of the hf structure of twelve I_H(total ^1H nuclear spin) = 0 and fourteen I_H = 1 rovibronic lines of the $\tilde{A}\ ^2A_1$ (0, 0, 0) ← $\tilde{X}\ ^2B_1$ (0, 0, 0) band. An intermodulated fluorescence (IMF) method was applied, which was first shown in [17] to be able to eliminate the Doppler width. A dye laser was employed as a source. Ground state constants were taken from a preliminary publication of [3] (T_{cc} was obtained with the sum rule) [6]:

nucleus	a_F	T_{aa}	T_{bb}	T_{cc}	C_{aa}	C_{bb}	C_{cc}
^{31}P	1747.2(14)	−259.5(38)	477.4(34)	−217.9	−1.58(66)	−1.9(10)	1.4(13)
^{1}H	190.06(69)	12.3(16)	−4.0(32)	−8.3	—	—	—

$a_F(^{31}P) = 1.8 \pm 0.1$ GHz had been derived from a Doppler-limited, laser-excited photoluminescence spectrum [5].

References:

[1] Fullam, B. W.; Mishra, S. P.; Symons, M. C. R. (J. Chem. Soc. Dalton Trans. **1974** 2145/8).
[2] Davies, P. B.; Russell, D. K.; Thrush, B. A.; Radford, H. E. (Chem. Phys. **44** [1979] 421/6).
[3] Endo, Y.; Saito, S.; Hirota, E. (J. Mol. Spectrosc. **97** [1983] 204/12).
[4] Kajita, M.; Endo, Y.; Hirota, E. (J. Mol. Spectrosc. **124** [1987] 66/71).
[5] Curl, R. F.; Endo, Y.; Kakimoto, M.; Saito, S.; Hirota, E. (Chem. Phys. Lett. **53** [1978] 536/8).

[6] Kakimoto, M.; Hirota, E. (J. Mol. Spectrosc. **94** [1982] 173/91, 185).
[7] Davies, P. B.; Russell, D. K.; Thrush, B. A. (Chem. Phys. Lett. **37** [1976] 43/6).
[8] Morton, J. R. (Chem. Rev. **64** [1964] 453/71, 456).
[9] Morehouse, R. L.; Christiansen, J. J.; Gordy, W. (J. Chem. Phys. **45** [1966] 1747/51).
[10] McDowell, C. A.; Mitchell, K. A. R.; Raghunathan, P. (J. Chem. Phys. **57** [1972] 1699/703).

[11] Jackel, G. S.; Gordy, W. (Phys. Rev. [2] **176** [1968] 443/52, 446).
[12] Shimokoshi, K.; Nakamura, K.; Sato, S. (Mol. Phys. **53** [1984] 1239/49, 1243).
[13] Raghunathan, P.; Sur, S. K. (Proc. Indian Acad. Sci. Chem. Sci. **92** [1983] 597/604).
[14] Hudson, A.; Wiffen, J. T. (Chem. Phys. Lett. **29** [1974] 113/5).
[15] Hudson, A.; Treweek, R. F. (Chem. Phys. Lett. **39** [1976] 248/9).
[16] Hinchliffe, A.; Bounds, D. G. (J. Mol. Struct. **54** [1979] 231/8).
[17] Sorem, M. S.; Schawlow, A. L. (Opt. Commun. **5** [1972] 148/51).

1.2.1.2.6 Spin-Rotation Interaction Constants, Rotational Constants, all in MHz (if not otherwise stated)

Tables in the present chapter contain in their upper part data on the spin-rotation interaction constants ε_{qq} (q = inertial axes a, b, c) and in their lower part data on the rotational constants A, B, and C. Methods of measurement, references, and remarks are the same for either constant. The quartic (Δ^S, δ^S and Δ, δ) and sextic (Φ, φ) centrifugal distortion constants listed in the first table (for the electronic ground state of PH_2) are defined by the "A-reduced" form (asymmetric reduction) of the respective Hamiltonian for spin-rotation interaction and rotation [1 to 3]. Abbreviations used in the tables are MW for microwave absorption, EL AB for electronic absorption, LMR for laser magnetic resonance, FIR for far IR, IMF for intermodulated fluorescence.

Rotation-vibration interaction constants α_i are given for the bending vibration (i = 2; see the short section on vibrationally excited states of ground-state PH_2).

Ground State $\tilde{X}\ ^2B_1$

Sets of constants from three sources are listed below for the **vibrational ground state** (0, 0, 0). Error limits in parentheses denote 2.5 times the standard deviation and are given in units of the last digit.

ε_{aa}	ε_{bb}	ε_{cc}	source	remark	Ref.
−8428.67(24)	−2458.13(24)	−7.80(24)	MW + LMR + IMF	a)	[4]
−8438(79)	−2454(69)	−2(63)	EL AB + LMR	b), d)	[5]
−8436(28)	−2434(18)	−22(13)	LMR	c), d)	[6]

Δ^S_N	$\Delta^S_{NK}+\Delta^S_{KN}$	Δ^S_K	δ^S_N	δ^S_K	Ref.
0.45(12)	−2.284(67)	7.117(61)	0.2475(71)	0.354(21)	[4]
0.36(80)	−0.7(30)	4.7(28)	0.18(40)	1.80(145)	[5]
—	—	4.9(15)	—	—	[6]

A	B	C	Ref.
273783.58(17)	242347.70(13)	126343.89(13)	[4]
273766.0(150)	242342.3(115)	126351.7(70)	[5]
273784.0(60)	242363.3(82)	126344.5(60)	[6]

Δ_N	Δ_{NK}	Δ_K	δ_N	δ_K	Ref.
16.364(61)	−54.56(26)	87.49(27)	7.4799(77)	−2.864(25)	[4]
16.15(30)	−54.63(95)	84.85(163)	7.180(150)	−2.41(35)	[5]
16.42(19)	−52.70(55)	84.57(40)	7.57(14)	−3.14(20)	[6]

Φ_N	Φ_{NK}	Φ_{KN}	Φ_K	Ref.
0.0063(12)	−0.034(11)	0.031(26)	0.026(15)	[4]
0.0074(33)	−0.0472(161)	0.050(31)	−0.078(44)	[5]

φ_N	φ_{KN}	φ_K	Ref.
0.00305(31)	−0.0101(25)	0.0244(55)	[4]
−0.00132(167)	0.0046(52)	0.0067(80)	[5]

a) Very similar results given earlier [7], see also reproduction in [1], had been based on microwave absorption measurements of three rotational transitions [7], on both far-IR [8, 9] and mid-IR [10] LMR spectra, and on IMF spectroscopic work [11]. The results tabulated above [4] were derived from microwave data for five additional rotational transitions [4] and from the whole data set used earlier [7].

b) Constants given in [5] (in cm^{-1} units and with error limits equal to three standard deviations; see remark d)) were obtained from the $\tilde{A}\ ^2A_1$ (0, 4, 0)$\leftarrow\tilde{X}\ ^2B_1$ (0, 0, 0) absorption band [5] and from available LMR spectra [8 to 10]. The constant Δ^S_{NK} was fixed at zero, and an octic constant, L_K=0.00111(38) MHz, was given [5].

c) Constants given in [6] (in cm^{-1} units and with error limits equal to one standard deviation) were obtained from far-IR [8, 9] and mid-IR [10] LMR data.

d) The original data [5, 6] were converted from cm^{-1} into MHz units, and the error limits were given consistently as 2.5 standard deviations [7]. The negative signs of Φ_K and φ_N [5] were omitted in [7].

The next table gives spin-rotation interaction constants ε_{qq} and rotational constants A, B, and C of PH_2 and PD_2 (in the electronic and vibrational ground state), which were based on individual measurements and less extended analyses. They are listed in chronological order (for abbreviations, see above). All values are given in cm^{-1} and are thus to be compared with

$$\varepsilon_{aa}=-0.281150,\ \varepsilon_{bb}=-0.0819944,\ \text{and}\ \varepsilon_{cc}=-0.000260\ cm^{-1}$$
$$A=9.1324372,\ B=8.0838491,\ \text{and}\ C=4.2143785\ cm^{-1}$$

taken from [4] (see table above; converted from MHz into cm^{-1} using $c=2.9979246\times10^{10}$ cm/s).

species	ε_{aa} in cm^{-1}	ε_{bb} in cm^{-1}	ε_{cc} in cm^{-1}	method	Ref.	remark
PH_2	−0.292(10)	−0.067(20)	−0.023(20)	EL AB	[12, 13]	a)
	−0.270(12)	−0.077(14)	+0.005(13)	EL AB	[14]	b)
	−0.28160	−0.08055	−0.00063	FIR LMR	[9]	c)
	−0.28107(54)	−0.08171(29)	−0.00051(24)	IR LMR	[10]	d)
PD_2	−0.143(16)	−0.031(16)	0	EL AB	[15]	e)

species	A in cm^{-1}	B in cm^{-1}	C in cm^{-1}	method	Ref.	remark
PH_2	9.131(4)	8.091(4)	4.223(4)	EL AB	[12, 13]	a)
	9.1332(18)	8.0820(19)	4.2134(12)	EL AB	[14]	b)
	9.13274(25)	8.08348(21)	4.21388(14)	FIR LMR	[9]	c)
	9.13233(12)	8.08349(21)	4.21434(15)	IR LMR	[10]	d)
PD_2	4.8574(20)	4.0444(36)	2.1796(15)	EL AB	[15]	e)

a) From an analysis of the (0, 0, 0)←(0, 0, 0) absorption band [12, 13]. Only an average value $1/2(\varepsilon_{bb}+\varepsilon_{cc})$ was given in [13]. A value of ε_{aa} near zero was earlier reported [16]. For centrifugal distortion constants η (associated with ε_{qq}) and τ, see original papers [12, 13]. – b) From a reanalysis of the (0, 0, 0)←(0, 0, 0) absorption band measured in [13]. Similar values were obtained from the (0, 3, 0)←(0, 0, 0) band. For τ values, see original paper [14]. A weighted average of these two sets of constants [14] was reported in [10] (with conversion of τ's into Δ's and δ's). – c) The mid-IR LMR measurements due to [10] were also taken into account. For spin-rotation interaction constants in spherical coordinates and η, Δ, and δ values, see the original paper [9]. In a preliminary publication [8], FIR LMR data [8] were combined with optical data due to [14]. – d) For η, Δ, and δ values, see original paper [10]. – e) From absorption bands (v_2'=1 to 10)←(v_2''=0). ε_{cc} was fixed at zero. For τ values, see original paper [15].

Spin-rotation interaction and rotational constants were also measured together with different kinds of centrifugal distortion parameters for few **vibrationally excited** states. Data for $v_2=1$ were obtained from mid-IR LMR spectra (of the ν_2 band) [6, 10], for $v_2=2$ [17, 18] and 3 [18] from the respective optical emission bands (0, 0, 0)→(0, 2, 0) [17, 18] and (0, 0, 0)→(0, 3, 0) [18].

Rotation-vibration interaction constants $\alpha_2^X=[X(v_2=0)-X(v_2)]/v_2$ were derived from ground-state ($v_2=0$) rotational constants X = A, B, C due to [13] and excited-state rotational constants (for $v_2=2$ and 3) [18]:

α_2^A	α_2^B	α_2^C
−0.310 cm^{-1}	−0.170 cm^{-1}	+0.062$_6$ cm^{-1}

Two sets of α_2^X (with signs reversed) were extracted from two slightly different analytical bending potential energy functions for $\tilde{X}\ ^2B_1$ [19].

Excited State $\tilde{A}\ ^2A_1$

Spin-rotation interaction and rotational constants for the **vibrational ground state** (0, 0, 0) of $PH_2(\tilde{A}\ ^2A_1)$ were obtained by electronic absorption (EL AB) or electronic emission (EL EM) spectroscopy. For the types of centrifugal distortion constants included in the papers, see the remarks:

ε_{aa} in cm^{-1}	ε_{bb} in cm^{-1}	ε_{cc} in cm^{-1}	method	Ref.	remark
1.178(21)	0.041(13)	−0.059(12)	see [a)]	[11]	a)
1.187(16)	—	—	EL AB	[14]	b)
1.05(3)	−0.03(3)	−0.05(3)	EL EM	[18]	c)
1.154(13)	−0.006(20)	−0.008(20)	EL AB	[12]	d)

A in cm^{-1}	B in cm^{-1}	C in cm^{-1}	method	Ref.	remark
20.4012(66)	5.5963(15)	4.3025(14)	see [a)]	[11]	a)
20.408(7)	5.602(3)	4.295(3)	EL AB	[14]	b)
20.351(4)	5.609(2)	4.311(2)	EL EM	[18]	c)
20.376(4)	5.604(4)	4.316(4)	EL AB	[12]	d)

[a)] By refitting [11] the electronic absorption data of [14]. The original paper also gives Δ_K^S, $\Delta_{N,NK,K}$, $H_K(\Phi_K)$, and L_K [11]. – [b)] From a reanalysis of the (0, 0, 0)←(0, 0, 0) absorption band (measured in [13]). The original paper also gives η, τ's, H_K, and L_K [14]. – [c)] From (0, 0, 0)→(0, 2, 0) and →(0, 3, 0) emission bands. The original paper also gives τ's (and another set of quartic constants) and H(Φ) values [18]. Constants in [17] were based on (0, 0, 0)→(0, 2, 0). – [d)] Constants based on the (0, 0, 0)←(0, 0, 0) absorption band [13] were redetermined. The original paper also gives η, τ's, and H_K [12].

The spin-rotation interaction constant ε_{aa} and the rotational constants A, B, C (with τ's, H_K, and L_K) for the **vibrationally excited** levels $v_2'=1$ and 2 of $PH_2(\tilde{A}\ ^2A_1)$ were determined from the respective electronic absorption bands (0, v_2', 0)←(0, 0, 0) (see original paper) [14]. For levels $v_2' \geqq 3$ the rotational levels could no longer be fitted to the same Hamiltonian as for the lower levels. Spin splittings are now described by an effective constant ε, which depended appreciably on v_2' and on the quantum number K_a' (projection of N′ on the a axis). ε was approximately equal to $\varepsilon_{aa} \cdot K_a'$ for low v_2'. The rotational term values could be similarly fitted by a single effective rotational constant B_{eff} (replacing B and C). ε and B_{eff} were tabulated for $v_2'=2$ to 6 and K_a' up to 7 [14]. A theory of the Renner-Teller effect in a bent triatomic molecule undergoing large-amplitude bending vibrations yielded ε and B_{eff} up to $v_2'=9$ and $K_a'=7$ [19]. **Fig. 5** taken from [19] shows deviations of the calculated from the observed [14] ε values especially for $v_2'=5$ and 6 [19]; see also reproduction in [20] and further discussion in [21].

For vibrationally excited levels of $PD_2(\tilde{A}\ ^2A_1)$, ε and B_{eff} were determined from electronic absorption bands and calculated with the model described in [19]. Values were tabulated for $v_2'=1$ to 9 (K_a' up to 8) and $v_2'=10$, 11 ($K_a'=0$) [15]. Graphics of ε [15] were shown in [20].

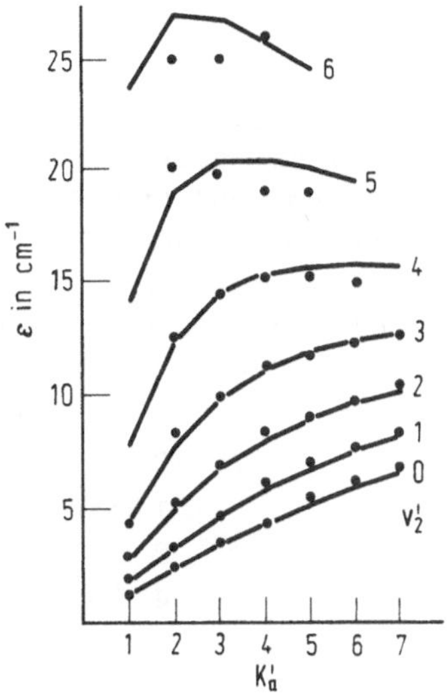

Fig. 5. Comparison of the observed (•) and calculated (—) spin-splitting constants ε for the $\tilde{A}\ ^2A_1$ state of PH_2.

References:

[1] Brown, J. M. (Landolt-Börnstein New Ser. Group II **14b** [1983] 171/291, 180, 282/4).
[2] Brown, J. M.; Sears, T. J. (J. Mol. Spectrosc. **75** [1979] 111/33, 121/5).
[3] Watson, J. K. G. (Vib. Spectra Struct. **6** [1977] 1/89, 33/5; J. Chem. Phys. **46** [1967] 1935/49, 1943).
[4] Kajita, M.; Endo, Y.; Hirota, E. (J. Mol. Spectrosc. **124** [1987] 66/71).
[5] Birss, F. W.; Lessard, G.; Thrush, B. A.; Ramsay, D. A. (J. Mol. Spectrosc. **92** [1982] 269/71).
[6] McKellar, A. R. W. (Faraday Discuss. Chem. Soc. No. 71 [1981] 63/76, 72/4).
[7] Endo, Y.; Saito, S.; Hirota, E. (J. Mol. Spectrosc. **97** [1983] 204/12).
[8] Davies, P. B.; Russell, D. K.; Thrush, B. A. (Chem. Phys. Lett. **37** [1976] 43/6).
[9] Davies, P. B.; Russell, D. K.; Thrush, B. A.; Radford, H. E. (Chem. Phys. **44** [1979] 421/6).
[10] Hills, G. W.; McKellar, A. R. W. (J. Chem. Phys. **71** [1979] 1141/9).

[11] Kakimoto, M.; Hirota, E. (J. Mol. Spectrosc. **94** [1982] 173/91, 177).
[12] Dixon, R. N.; Duxbury, G. (Chem. Phys. Lett. **1** [1967/68] 330/2).
[13] Dixon, R. N.; Duxbury, G.; Ramsay, D. A. (Proc. R. Soc. [London] A **296** [1967] 137/60, 160).
[14] Berthou, J.-M.; Pascat, B.; Guenebaut, H.; Ramsay, D. A. (Can. J. Phys. **50** [1972] 2265/76, 2266).
[15] Vervloet, M.; Berthou, J.-M. (Can. J. Phys. **54** [1976] 1375/82).
[16] Dixon, R. N. (Mol. Phys. **10** [1965/66] 1/6).
[17] Pascat, B.; Berthou, J.-M.; Guenebaut, H.; Ramsay, D. A. (C. R. Seances Acad. Sci. B **263** [1966] 1397/9).
[18] Pascat, B.; Berthou, J.-M.; Prudhomme, J.-C.; Guenebaut, H.; Ramsay, D. A. (J. Chim. Phys. Phys. Chim. Biol. **65** [1968] 2022/9).
[19] Barrow, T.; Dixon, R. N.; Duxbury, G. (Mol. Phys. **27** [1974] 1217/34, 1228/30).
[20] Duxbury, G. (Mol. Spectrosc. Chem. Soc. [London] **3** [1975] 497/573, 545/8).

[21] Jungen, C.; Hallin, K.-E. J.; Merer, A. J. (Mol. Phys. **40** [1980] 65/94, 69).

1.2.1.2.7 Asymmetry Parameter κ. Moments of Inertia I_X. Geometrical Structure

According to the rotational constants X=A, B, C given in the preceding chapter, the asymmetry parameter $\kappa=(2B-A-C)/(A-C)$ takes values of about +0.6 and −0.8 for PH_2 in its observed valence states $\tilde{X}\,^2B_1$ and $\tilde{A}\,^2A_1$. This shows the radical to be an **asymmetric top** in the ground state and a **near-prolate symmetric top** in the excited state.

The following values for κ, I_X, the inertial defect $\Delta=I_C-I_B-I_A$ (I_X and Δ in amu·Å²), the internuclear distance r (in Å), and the interbond angle α are based on rotational constants derived (from electronic spectra; see the preceding chapter) for the vibrational ground states of the indicated species:

species	$PH_2(\tilde{X})$		$PH_2(\tilde{A})$		$PD_2(\tilde{X})$
κ	+0.573	+0.57620	−0.838	−0.83848	+0.393
I_A	1.8457(4)	1.8466(8)	0.8260(3)	0.8275(2)	3.4704(14)
I_B	2.0858(5)	2.0840(10)	3.0089(16)	3.0088(22)	4.1681(36)
I_C	4.0009(11)	3.9928(35)	3.9249(27)	3.9067(35)	7.7345(53)
Δ	0.069(2)	0.0622	0.090(5)	0.0704	0.096(10)
r	1.418	1.417	1.389	1.389	1.417
α	91°42′	91°40′	123°12′	123°9′	91°45′
Ref.	[1]	[2]	[1]	[2]	[3]
remark	a)	b)	c)	b)	

a) Two electronic absorption bands yielded two slightly different sets of parameters. – b) Correcting earlier data given (with I_X and Δ in g·cm² units) in [4]. – c) A very similar set was previously given (with I_X and Δ in g·cm²) in [5].

Δ=0.06870 amu·Å² for the vibrational ground state of $PH_2(\tilde{X})$ and Δ=0.22776 for the excited state $v_2=1$ are based on an LMR spectrum of the ν_2 band [6].

Δ was also obtained from an expression given in [7] for Δ_{vib} (the vibrational state-dependent part of Δ; the corrections Δ_{cent} and Δ_{el} due to effects of centrifugal distortion and electronic interaction can be neglected) [5]. $\Delta_{vib}=0.117\times10^{-40}$ g·cm² (≙0.0704 amu·Å²) for $PH_2(\tilde{X})$ [5] and 0.080 amu·Å² for $PH_2(\tilde{A})$ [1], both vibrationally nonexcited, follows from measurements of the bending vibration ν_2 in $\tilde{X}$ [5] and $\tilde{A}$ [1] and from estimates for ν_3 and the Coriolis coupling constants ζ_{13} and ζ_{23} [1, 5]. For Δ_{vib} of PH_2 ($\tilde{X}$; $v_2=2$, 3), see [5].

Approximate interbond angles α (for both observed valence states $\tilde{X}\,^2B_1$ and $\tilde{A}\,^2A_1$), which were derived from earlier experimental work or from more qualitative theoretical concepts (e.g., the Walsh MO picture), are given in the chapter on the point group, p. 54. Results for the internuclear distance and the interbond angle, which were theoretically calculated by ab initio or semiempirical MO methods, are not reported explicitly. Ab initio calculations on r and α of a total of five valence states are in [8], of only the two observed valence states $\tilde{X}$ and $\tilde{A}$ in [9 to 11], of only the ground state in [12 to 20] and on α of both $\tilde{X}$ and $\tilde{A}$ in [21]. Semiempirical calculations were done on r and α of $\tilde{X}$ [20, 22] and on α of both $\tilde{X}$ and $\tilde{A}$ [23] as well as the ground state $\tilde{X}$ only [24].

References:

[1] Berthou, J. M.; Pascat, B.; Guenebaut, H.; Ramsay, D. A. (Can. J. Phys. **50** [1972] 2265/76, 2266).

[2] Dixon, R. N.; Duxbury, G. (Chem. Phys. Lett. **1** [1967/68] 330/2).

[3] Vervloet, M.; Berthou, J.-M. (Can. J. Phys. **54** [1976] 1375/82).

[4] Dixon, R. N.; Duxbury, G.; Ramsay, D. A. (Proc. R. Soc. [London] A **296** [1967] 137/60, 160).
[5] Pascat, B.; Berthou, J.-M.; Prudhomme, J.-C.; Guenebaut, H.; Ramsay, D. A. (J. Chim. Phys. Phys. Chim. Biol. **65** [1968] 2022/9).
[6] Hills, G. W.; McKellar, A. R. W. (J. Chem. Phys. **71** [1979] 1141/9).
[7] Oka, T.; Morino, Y. (J. Mol. Spectrosc. **8** [1962] 9/21, 9).
[8] Gu, J.-P.; Huang, M.-B.; Kong, F.; Liu, S.-H. (J. Mol. Struct. **201** [1989] 39/47 [THEOCHEM **60**]).
[9] Pope, S. A.; Hillier, I. H.; Guest, M. F. (Faraday Symp. Chem. Soc. No. 19 [1984] 109/23, 114).
[10] Müller, J.; Ågren, H.; Canuto, S. (J. Chem. Phys. **76** [1982] 5060/8).

[11] Perić, M.; Buenker, R. J.; Peyerimhoff, S. D. (Can. J. Chem. **57** [1979] 2491/7).
[12] Baird, N. C.; Kuhn, M.; Lauriston, T. M. (Can. J. Chem. **67** [1989] 1952/8).
[13] Alberts, I. L.; Handy, N. C. (J. Chem. Phys. **89** [1988] 2107/15).
[14] Nguyen, M. T. (J. Mol. Struct. **180** [1988] 23/9 [THEOCHEM **49**]).
[15] Cattani-Lorente, M.; Geoffroy, M.; Mishra, S. P.; Weber, J.; Bernardinelli, G. (J. Am. Chem. Soc. **108** [1986] 7148/53).
[16] Pople, J. A.; Luke, B. T.; Frisch, M. J.; Binkley, J. S. (J. Phys. Chem. **89** [1985] 2198/203).
[17] Hinchliffe, A.; Bounds, D. G. (J. Mol. Struct. **54** [1979] 231/8).
[18] Ball, J. R.; Thomson, C. (Int. J. Quantum Chem. **14** [1978] 39/53, 44).
[19] Hudson, A.; Treweek, R. F. (Chem. Phys. Lett. **39** [1976] 248/9).
[20] Hudson, A.; Wiffen, J. T. (Chem. Phys. Lett. **29** [1974] 113/5).

[21] So, S. P.; Richards, W. G. (Int. J. Quantum Chem. **11** [1977] 73/9).
[22] Kilcast, D.; Thomson, C. (J. Chem. Soc. Faraday Trans. II **68** [1972] 435/43).
[23] Jordan, P. C. (J. Chem. Phys. **41** [1964] 1442/9).
[24] Takahata, Y. (Chem. Phys. Lett. **59** [1978] 472/7).

1.2.1.2.8 Molecular Vibration. Coriolis Coupling. Force Constants

Fundamental Vibrations ν_i in cm^{-1}

Experimental data are available only for the symmetric stretching and bending vibrations, $\nu_1(A_1)$ and $\nu_2(A_1)$. The antisymmetric stretching vibration $\nu_3(B_1)$ has not been observed. Complete sets of ν_i (or of harmonic frequencies ω_i) were obtained from strictly theoretical calculations for PH_2 in both its electronic ground state $\tilde{X}\ ^2B_1$ and its observed excited state $\tilde{A}\ ^2A_1$ and also for the isotopic species PD_2.

Symmetric Stretching Vibration $\nu_1(A_1)$. Only two not very accurate values are available for $PH_2(\tilde{X}\ ^2B_1)$. $\nu_1 = 2270 \pm 80$ is based on the laser photoelectron spectrum of the PH_2^- ion (see p. 62). A small detachment peak at ~0.99 eV was assigned to the transition PH_2, $\tilde{X}\ ^2B_1(1, 0, 0) \leftarrow PH_2^-$, $\tilde{X}\ ^1A_1(0, 0, 0)$ (while the transition to (0, 0, 0) is at an "electron energy" of 1.27 ± 0.01 eV) [1]. This ν_1 value was also reported in a compilation of ground-state vibrational energy levels of polyatomic transient molecules [2]. $\nu_1 \approx 2310$ was inferred from the origin of a new band, which appeared in the Raman spectrum of PH_3 when PH_3 was heated from 25°C up to 632°C. The band was assigned to PH_2 [3] by comparison with $\nu_1 = 2295 \pm 20$. This value resulted from scaling a theoretical harmonic frequency of 2579 cm^{-1} taken from [4] by a factor of 0.89 (also given in [4]) [3]. ν_1 had earlier been expected to lie between 2308 (for HPO) and 2380 (for the PH molecule) [5].

Bending Vibration $\nu_2(A_1)$. Data for PH_2, PD_2, or PHD (point group C_s) in their electronic **ground state** have been obtained from electronic absorption (EL AB) and emission (EL EM) spectra, yielding $\Delta G(1/2)$ values, and also from matrix IR absorption and gas phase IR laser magnetic resonance (LMR) spectra:

species	$PH_2(\tilde{X})$				$PD_2(\tilde{X})$		$PHD(\tilde{X})$
phase	vapor			matrix	vapor	matrix	vapor
ν_2 in cm^{-1}	1101.9078(2)	1101.4	1102	1103	795.5	797	963.5
method	IR LMR	EL AB	EL EM	IR	EL EM	IR	EL EM
Ref.	[6]	[7]	[5]	[8, 9]	[5]	[8, 9]	[5]
remark	a)	b)	c)	d)	c)	d)	c)

a) Band origin obtained from an LMR spectrum after excitation by a CO_2 laser [6]. ν_2=1101.9086(2) from a fit [10] of these IR LMR data [6] and of far-IR LMR data [39, 40]. ν_2=1101.91 [2, 11]. – b) A hot band $(v_2'=1)\leftarrow(v_2''=1)$ was observed [7] and was combined with the 1←0 band reported by [12]. – c) Average $\Delta G(1/2)$ values taken from transitions $(v_2'=0$ to $3)\rightarrow(v_2''=0, 1)$ [5]. Earlier values were in [13], for PH_2 and PD_2 also in [14, 15]. – d) Vacuum UV photolysis of Ar–$PH_3(PD_3)$–N_2O deposits with mole ratios between 200:1:1 and 800:1:1 at 14 K [8, 9].

ν_2=1102.02 is derived from the two **vibrational constants** ω_2^0=1107.26 cm^{-1} and x_{22}=−5.24 cm^{-1}, which were obtained [16] by rotational analysis of two electronic emission bands, $(v_2'=0)\rightarrow(v_2''=2, 3)$ [16, 17], and of the 0←0 absorption band [7]. The vibrational constants ω_2=1112.5 cm^{-1} ($\triangleq\omega_2^0-2x_{22}$) and x_{22}=−5.2 cm^{-1} reported in [18] are apparently based on ω_2^0 and x_{22} taken from [16]; ω_2 and x_{22} were also derived from two analytical bending potential energy functions [18].

Data for PH_2, PD_2, and PHD (point group C_s) in their electronically **excited state** are given below:

species	$PH_2(\tilde{A})$				$PD_2(\tilde{A})$		$PHD(\tilde{A})$
phase	vapor		matrix		vapor	matrix	vapor
ν_2 in cm^{-1}	949.12	951	949(7)	962(25)	689.5	665(25)	830
method	EL AB	EL EM	EL AB	EL AB	EL EM	EL AB	EL EM
Ref.	[19]	[5, 13]	[20]	[11]	[5, 13]	[11]	[5, 13]
remark	a)	b)	c)	d)	b)	d)	b)

a) Difference of the $\tilde{A}\ ^2A_1$ term values $T_0(v_2'=1)$ and $T_0(v_2'=0)$ for the "rotational" quantum number $K_a'=0$ [19]. ν_2 also reported in [11, 21]. – b) Apparently average $\Delta G'(1/2)$ values from the transitions $(v_2'=0, 1)\rightarrow(v_2''=0$ to 3), $(v_2''=0$ to 6), and $(v_2''=0$ to 7) for PH_2, PD_2, and PHD, respectively [5, 13]. Similar data for PH_2 and PD_2 are 950.1 and 688 [14], 950.1 and 688.1 [15], and for PH_2 950 cm^{-1} [22]. – c) Photolysis of Ar–PH_3 (200:1) samples codeposited onto a 12 K precooled sapphire plate with O atoms [20]. ν_2 also reported in [21]. – d) Reported in a compilation [11] with reference to measurements in [8].

Antisymmetric Stretching Vibration $\nu_3(B_1)$. An experimental value is not available. A harmonic frequency $\omega_3=2294\pm30\ cm^{-1}$ was derived from the difference of the inertial defects Δ for $PH_2(\tilde{X}\ ^2B_1)$ in its vibrational ground state and excited state v_2=1 (see p. 72). An expression taken from [23] was used together with $\omega_2\approx\nu_2$=1101.9 cm^{-1} (see above) and an estimated Coriolis coupling constant ζ_{23} (see below) [6]. Conversely, ω_3=2310 cm^{-1} had been assumed (by comparison with analogous data for the PH_2 group in CH_3PH_2 and SiH_3PH_2 [24]) to derive Δ_{vib} for the vibrational states v_2=0, 2, and 3 [16].

Complete sets of the fundamental vibrations ν_i were obtained from ab initio MO-CI **calculations** for PH_2 and PD_2 in both their ground $\tilde{X}\ ^2B_1$ and excited states $\tilde{A}\ ^2A_1$. Symmetric and antisymmetric stretching and bending potential energy curves together with the respective vibrational levels were calculated within the Born-Oppenheimer approximation, i.e., neglecting the Renner-Teller effect. The values given below agree to within 25 cm^{-1} with the available experimental data [25]:

species	state	ν_1	ν_2	ν_3
PH_2	$\tilde{X}\ ^2B_1$	2330	1110	2495
	$\tilde{A}\ ^2A_1$	2430	973	2660
PD_2	$\tilde{X}\ ^2B_1$	1675	800	1785
	$\tilde{A}\ ^2A_1$	1756	710	1927

Harmonic frequencies were calculated for the PH_2 ground state with the Hartree-Fock [4, 26] and Møller-Plesset [27] methods.

Bending-Mode Vibrational Structure

Vibrational levels higher than $v_2''=1$ were observed for the electronic ground state $\tilde{X}\ ^2B_1$ of PH_2 up to $v_2''=7$, of PD_2 up to $v_2''=9$, and of PHD (point group C_s) up to $v_2''=8$ in electronic emission spectra of the transition $\tilde{A}\ ^2A_1 \rightarrow \tilde{X}\ ^2B_1$ [5, 13]. PH_2 and PD_2 levels for smaller ranges of v_2'' are given in [14 to 17]. All these levels are thus well below the ground state barrier to linearity of $H \approx 25100$ cm^{-1} given in [18]. Vibrational levels for the excited state $\tilde{A}\ ^2A_1$ ($T_0(v_2')$, referred to the ground state) were observed in the $^2A_1 \leftarrow {}^2B_1$ absorption spectrum of gaseous PH_2 up to $v_2'=10$ and of gaseous PD_2 up to $v_2'=12$ [12]. Smaller ranges of v_2' were covered in matrix absorption spectra [8, 20, 28] and in vapor phase emission spectra [5, 13 to 15]. Part of these excited-state vibrational levels thus go beyond the barrier to linearity of $\tilde{A}\ ^2A_1$, which is given by the H value above (since $^2B_1$ and $^2A_1$ are correlated with the same $^2\Pi_u$ state of linear PH_2, see p. 53) or by 6840 cm^{-1} [18] (if referred to the potential minimum of $\tilde{A}\ ^2A_1$).

The variation of the excited-state vibrational intervals $\Delta G(v_2'+1/2)$ of PH_2 and PD_2 [12] with increasing transition energy $1/2\{T_0(v_2')+T_0(v_2'+1)\}$, is shown in **Fig. 6** taken from [29, 30]. The minimum of $\Delta G(v_2'+1/2)$ has been called a "Dixon dip" [30] and occurs approximately at the barrier to linearity of the potential energy curve [29]. The Dixon dip thus is one consequence of the "quasi-linear" behavior of the radical. Another consequence is the need to introduce the "rotational" quantum number K (or K_a) into the vibrational analysis. K describes the rotational angular momentum about the inertial a axis, which becomes the molecular axis in the linear

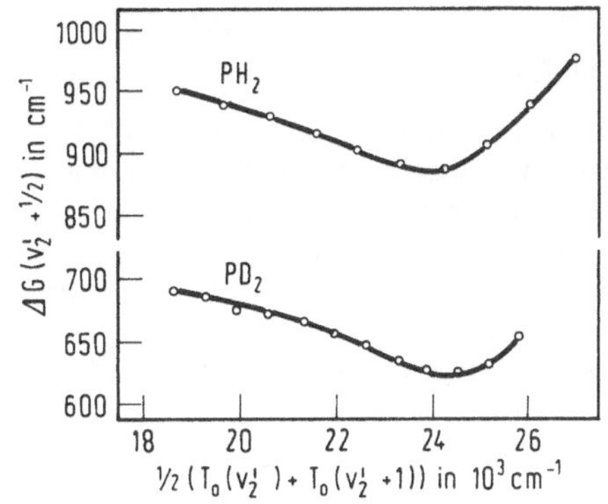

Fig. 6. The variation with transition energy of the observed vibrational intervals in the electronic absorption spectra of PH_2 and PD_2.

case [31]. Thus, rotational analyses have been performed to extract the "sublevel origins" $T_0(v_2', K')$ [18, 19] (or "vibronic" term values for v_2' levels [31]) from the observed absorption spectrum $\tilde{A}\ ^2A_1 \leftarrow \tilde{X}\ ^2B_1$ [12], the 3←0 absorption band of which had been reinvestigated in [19]. Observed vibronic levels of PH_2 are reported for $v_2'=0$ to 3 (with K′ up to 7), 4 (K′ up to 6), 5 (K′ up to 5), 6, 7 (K′ up to 4), 8, and 9 (K′=0) [18], for more limited ranges of v_2' and K′ also in [19, 31, 32]. Vibronic levels of PD_2 are given for $v_2'=1$ to 4 (K′ up to 8), 5 to 7 (K′ up to 7), 8 (K′ up to 6), 9 (K′ up to 4), 10, and 11 (K′=0) [33].

The vibrational structure was also studied by theoretical methods. Bending potential energy curves for $\tilde{X}\ ^2B_1$ and $\tilde{A}\ ^2A_1$ from ab initio MO-CI calculations were used within the Born-Oppenheimer approximation, i.e., without considering the Renner-Teller effect, to calculate vibrational levels for both electronic states of both PH_2 and PD_2 (see graphical representations of the potential curves in the original paper). The upper-state vibrational intervals $\Delta G(v_2'+1/2)$ were plotted against $\Delta v_2'$ for PH_2 (K′=0 and 1) and PD_2 (K′=0, 1, and 2); these five curves all showed the Dixon dip known from Fig. 6, p. 75 [25]. Three further calculations took the Renner-Teller effect into account. Analytical bending potential energy functions, fitted to known molecular parameters, were used to derive excited state vibronic levels $T_0(v_2', K')$ for PH_2 [18] and PD_2 [33]. Potential curves from ab initio MO-CI calculations were used to obtain the vibronic levels of the $^2\Pi_u$ state correlated with $\tilde{X}\ ^2B_1$ and $\tilde{A}\ ^2A_1$. For $K \geqq 1$, the levels in the neighborhood of the barrier to linearity cannot be assigned to a definite electronic state ($\tilde{X}$ or $\tilde{A}$) of the bent radical [34].

Coriolis Coupling Constants

The nonvanishing constants ζ_{13} and ζ_{23} obeying the relation $\zeta_{13}^2+\zeta_{23}^2=1$ [35], have not been measured or even calculated (because data for ω_1 and ω_3 are missing). $\zeta_{13}=0$ and $\zeta_{23}=1$ were assumed [16] in view of known data for the related H_2S molecule [36]. Similarly, $\zeta_{13}\approx 0$, $\zeta_{23}\approx 1$ [18] (referring to [35]) and $\zeta_{23}^2=1$ [6] (referring to [23]).

Force Constants

The stretching and bending force constants k_1 and k_2 have been estimated using data of the related PH_3 molecule or theoretically calculated. $k_1=3.10$ mdyn/Å for ground-state PH_2 was taken from PH_3 [7]. $k_1=7.260$ mdyn/Å and $k_2=0.01021$ mdyn/Å were calculated by an ab initio MO-SCF method [37]. Apparently more reliable data for k_2 are given below for the ground state $\tilde{X}\ ^2B_1$ and the excited state $\tilde{A}\ ^2A_1$ (for units, see also the remarks):

state	$\tilde{X}\ ^2B_1$		$\tilde{A}\ ^2A_1$		
k_2 in mdyn/Å	0.346	0.32	0.26	0.24	0.20
Ref.	[5]	[38]	[28]	[18]	[38]
remark	a)	b)	c)	d)	b)

a) Taken from PH_3. The original value 7.278×10^{-12} erg [5] was converted into 36590 cm^{-1} and used to construct an analytical bending potential energy function [18]. Division by $r^2=1.45^2\cdot Å^2$ led to the tabulated value. – b) From semiempirical calculations (using valence bond and "atoms in molecules" methods). Original values were 6.66×10^{-12} ($\tilde{X}$) and 4.15×10^{-12} erg ($\tilde{A}$). – c) Estimated from vibrational data for PH_3 [28]. The value was reported to be 5.45×10^{-12} erg in [38]. It corresponds to 27400 cm^{-1}. – d) By fitting an analytical potential energy function to certain observed vibronic levels. The original value was 25270 cm^{-1} [18].

References:

[1] Zittel, P. F.; Lineberger, W. C. (J. Chem. Phys. **65** [1976] 1236/43).
[2] Jacox, M. E. (J. Phys. Chem. Ref. Data **13** [1984] 945/1068, 953).
[3] Abraham, P.; Bekkaoui, A.; Soulière, V.; Bouix, J.; Monteil, Y. (J. Cryst. Growth **107** [1991] 26/31).
[4] Pople, J. A.; Luke, B. T.; Frisch, M. J.; Binkley, J. S. (J. Phys. Chem. **89** [1985] 2198/203).
[5] Guenebaut, H.; Pascat, B.; Berthou, J.-M. (J. Chim. Phys. Phys. Chim. Biol. **62** [1965] 867/77, 874/6).
[6] Hills, G. W.; McKellar, A. R. W. (J. Chem. Phys. **71** [1979] 1141/9).
[7] Dixon, R. N.; Duxbury, G.; Ramsay, D. A. (Proc. R. Soc. [London] A **296** [1967] 137/60, 138).
[8] Larzillière, M.; Jacox, M. E. (NBS Spec. Publ. [U.S.] 561-1 [1979] 529/43, 532, 537).
[9] Larzillière, M.; Jacox, M. E. (J. Mol. Spectrosc. **79** [1980] 132/50, 134/5).
[10] McKellar, A. R. W. (Faraday Discuss. Chem. Soc. No. 71 [1981] 63/76, 72/4).

[11] Jacox, M. E. (J. Phys. Chem. Ref. Data **17** [1988] 269/511, 283/4).
[12] Ramsay, D. A. (Nature **178** [1956] 374/5).
[13] Guenebaut, H.; Pascat, B. (C. R. Hebd. Seances Acad. Sci. **259** [1964] 2412/5).
[14] Guenebaut, H.; Pascat, B. (J. Chim. Phys. Phys. Chim. Biol. **61** [1964] 592/5).
[15] Pascat, B.; Guenebaut, H. (Bull. Univ. ARERS **2** [1963] 100/7).
[16] Pascat, B.; Berthou, J.-M.; Prudhomme, J.-C.; Guenebaut, H.; Ramsay, D. A. (J. Chim. Phys. Phys. Chim. Biol. **65** [1968] 2022/9).
[17] Pascat, B.; Berthou, J.-M.; Guenebaut, H.; Ramsay, D. A. (C.R. Seances Acad. Sci. B **263** [1966] 1397/9).
[18] Barrow, T.; Dixon, R. N.; Duxbury, G. (Mol. Phys. **27** [1974] 1217/34, 1225, 1228/9).
[19] Berthou, J.-M.; Pascat, B.; Guenebaut, H.; Ramsay, D. A. (Can. J. Phys. **50** [1972] 2265/76, 2269, 2276).
[20] Withnall, R.; McCluskey, M.; Andrews, L. (J. Phys. Chem. **93** [1989] 126/9).

[21] Jacox, M. E. (J. Phys. Chem. Ref. Data **19** [1990] 1387/546, 1397).
[22] Peyron, M. (Proc. Int. Symp. Mol. Struct. Spectrosc., Tokyo 1962, pp. B403-1/B403-4).
[23] Oka, T.; Morino, Y. (J. Mol. Spectrosc. **8** [1962] 9/21, 9).
[24] Linton, H. R.; Nixon, E. R. (Spectrochim. Acta **15** [1959] 146/55, 151, 155).
[25] Perić, M.; Buenker, R. J.; Peyerimhoff, S. D. (Can. J. Chem. **57** [1979] 2491/7).
[26] Gordon, M. S.; Gano, D. R.; Boatz, J. A. (J. Am. Chem. Soc. **105** [1983] 5771/5).
[27] Alberts, I. L.; Handy, N. C. (J. Chem. Phys. **89** [1988] 2107/15).
[28] McCarty, M., Jr.; Robinson, G. W. (J. Chim. Phys. Phys. Chim. Biol. **56** [1959] 723/30).
[29] Dixon, R. N. (Trans. Faraday Soc. **60** [1964] 1363/8).
[30] Duxbury, G. (Mol. Spectrosc. Chem. Soc. [London] **3** [1975] 497/573, 545).

[31] Dixon, R. N. (J. Phys. Colloq. [Paris] **32** [1971] C5a-147/C5a-154).
[32] Berthou, J.-M.; Pascat, B. (C.R. Seances Acad. Sci. C **271** [1970] 799/801).
[33] Vervloet, M.; Berthou, J.-M. (Can. J. Phys. **54** [1976] 1375/82).
[34] Perić, M. (Chem. Phys. Lett. **76** [1980] 573/8).
[35] Herschbach, D. R.; Laurie, V. W. (J. Chem. Phys. **40** [1964] 3142/53, 3146/7).
[36] Allen, H. C., Jr.; Plyler, E. K. (J. Chem. Phys. **25** [1956] 1132/6).
[37] Ball, J. R.; Thomson, C. (Int. J. Quantum Chem. **14** [1978] 39/53, 47).
[38] Jordan, P. C. (J. Chem. Phys. **41** [1964] 1442/9).
[39] Davies, P. B.; Russell, D. K.; Thrush, B. A. (Chem. Phys. Lett. **37** [1976] 43/6).
[40] Davies, P. B.; Russell, D. K.; Thrush, B. A.; Radford, H. E. (Chem. Phys. **44** [1979] 421/6).

1.2.1.2.9 Bond Dissociation Energy D(HP–H) in kcal/mol (if not otherwise stated)

$D_0 = 74.7 \pm 0.5$ is given in [1] referring to [2]. A similar but less precise value $D_0 = 74.2 \pm 2$ [2] was based upon an estimation of D(P–H) for the PH molecule (70.1±2 kcal/mol; see p. 20), a determination of $D_0(H_2P\text{–}H)$ for PH_3 (82.5±0.5 kcal/mol; see p. 174), and a PH_3 atomization energy of 227.1±0.4 kcal/mol (at 0 K) [2]. A remarkable agreement of E_i(PH), read from a photoelectron spectrum [3], with E_i(PH) determined from thermochemical data [2], was noticed after completing the data analysis [2]. This coincidence together with D(P–H) = 70.5 kcal/mol given in [1] apparently led to the specification of D_0 above [1].

$D = 67.6 \pm 3$ (≙2.93±0.15 eV) was obtained from the fluorescence excitation curve of the PH(A $^3\Pi_i$) photofragment from PH_3. The threshold for PH emission (10.24 eV) was combined with the excitation energy A $^3\Pi_i \leftarrow$ X $^3\Sigma^-$ (3.63 eV) and a $D(H_2P\text{–}H)$ value of 3.69 eV (≙85 kcal/mol) [4].

Estimates of 79 [5] (see also [7 to 9]) or ~67 kcal/mol [6] were based on thermochemical data of minor quality.

D_0 or D_e values from ab initio MO and semiempirical calculations are partially in remarkable agreement with the most recent experimental result (for abbreviations used for the theoretical methods, see the remarks):

D_0 in kcal/mol	75.3	75.4	75.69	—
D_e in kcal/mol	—	80.3	—	71.44
method	MP4+QCI	MP4	GVB	VB
Ref.	[10]	[11]	[12]	[13]
remark	a)	b)	c)	d)

a) Total energies (of PH_2 and PH) were calculated, and electron correlation was treated by Møller-Plesset perturbation theory up to fourth order and by quadratic configuration interaction (see [14]). The original paper gives the atomization energies of PH_2 and PH [10]. – b) Procedure according to remark a) but without the quadratic configuration interaction. Atomization energies of PH_2 and PH are given [11]. For D_0, see also [2]. – c) Generalized valence bond theory. The assumption $D(H_2P\text{–}H) = D(HP\text{–}H) + {}^1/_2 K_{pp} = D(P\text{–}H) + K_{pp}$ was made (with an exchange integral $K_{pp} = 10.84$ kcal/mol), and an experimental value of the PH_3 atomization energy was used (based, however, on a formation enthalpy of 7.0 kcal/mol at 0 K). – d) A valence bond method was combined with the method of "atoms in molecules" to calculate PH_2 and PH atomization energies.

An ab initio MO calculation without any electron correlation correction yielded $D_e = 1.92$ eV for the ground state $\tilde{X}$ $^2B_1$ and $D_e = 1.72$ eV for the excited state $\tilde{A}$ $^2A_1$ [9].

References:

[1] Berkowitz, J. (J. Chem. Phys. **89** [1988] 7065/76, 7072).
[2] Berkowitz, J.; Curtiss, L. A.; Gibson, S. T.; Greene, J. P.; Hillhouse, G. L.; Pople, J. A. (J. Chem. Phys. **84** [1986] 375/84, 382).
[3] Dyke, J. M.; Jonathan, N.; Morris, A. (Int. Rev. Phys. Chem. **2** [1982] 3/42, 22).
[4] Di Stefano, G.; Lenzi, M.; Margani, A.; Mele, A.; Nguyen Xuan, C. (J. Photochem. **7** [1977] 335/44, 343).
[5] Vedeneyev, V. I.; Gurvich, L. V.; Kondrat'yev, V. N.; Medvedev, V. A.; Frankevich, Ye. L. (Bond Energies, Ionization Potentials, and Electron Affinities, Arnold, London 1966, p. 83).
[6] Halmann, M.; Platzner, I. (J. Phys. Chem. **73** [1969] 4376/8).

[7] Friswell, N. J.; Gowenlock, B. G. (Adv. Free Radical Chem. **2** [1967] 1/45, 16).
[8] Okabe, H. (Photochemistry of Small Molecules, Wiley, New York 1978, p. 262).
[9] Gu, J.-P.; Huang, M.-B.; Kong, F.; Liu, S.-H. (J. Mol. Struct. **201** [1989] 39/47 [THEOCHEM **60**]).
[10] Curtiss, L. A.; Jones, C.; Trucks, G. W.; Raghavachari, K.; Pople, J. A. (J. Chem. Phys. **93** [1990] 2537/45).

[11] Pople, J. A.; Luke, B. T.; Frisch, M. J.; Binkley, J. S. (J. Phys. Chem. **89** [1985] 2198/203).
[12] Goddard, W. A., III; Harding, L. B. (Annu. Rev. Phys. Chem. **29** [1978] 363/96, 395).
[13] Jordan, P. C. (J. Chem. Phys. **41** [1964] 1442/9).
[14] Pople, J. A.; Head-Gordon, M.; Raghavachari, K. (J. Chem. Phys. **87** [1987] 5968/75).

1.2.1.3 Heat of Formation. Thermodynamic Functions

Formation enthalpy $\Delta_f H^\circ$ in kcal/mol (unless stated otherwise)

The formation enthalpies at 0 or 298 K which were derived from the bond dissociation energy D and the formation enthalpy of **phosphane** are shown in the table below together with the PH_3 data used (all values in kcal/mol; $\Delta_f H^\circ(H) = 51.6$ and 52.1 kcal/mol at 0 and 298 K). The source of the D value is also indicated (see also p. 174). For $\Delta_f H^\circ(PH_3)$, see the remarks below the table (see also p. 179).

T in K	$\Delta_f H^\circ$	$D(H_2P\text{–}H)$	$\Delta_f H^\circ(PH_3)$	source of D value	Ref.	remark
0	34.0±0.6	82.46±0.46	3.2	photoionization of PH_3 and PH_2	[1]	a)
298	33.1±2	83.9±3	1.3	electron impact ionization of PH_3 and PH_2	[2]	b)
0	40.5±1	85.1±1	7.0	fluorescence excitation of PH_3 photofragments	[3]	c)
0	29.5±1.5	77.9±1.5	3.2	literature	[4]	d)

a) Work was based on $\Delta_f H^\circ_{298}(PH_3) = 1.3$ kcal/mol from [5], yielding $\Delta_f H^\circ_0(PH_3) = 3.2$ kcal/mol (see, e.g. [4, 6]). – b) PH_3 formation enthalpy from [5]. The PH_2 value at 298 K [2] was converted into $\Delta_f H^\circ_0(PH_2) = 34.0 \pm 2$ in [1], apparently by applying a theoretically calculated difference of 0.9 kcal/mol between 298 and 0 K (see below). The original value [2] was erroneously ascribed to 0 K in [7]. It had earlier been changed to 37.3±2 in [3], apparently by using 7.4 instead of 3.2 kcal/mol as the PH_3 formation enthalpy (at 0 K), see next remark. – c) The PH_3 formation enthalpy was taken from [8], where the value due to [5] was now referred to the red instead of the white P modification as a standard. This value should, however, read 7.4 kcal/ mol, see, e.g. [9] (where nevertheless the white modification is again listed as a standard) and remark b). The original value for $\Delta_f H^\circ(PH_2)$ [3] was erroneously referenced to 298 K in [10] and was thought to be somewhat uncertain [11]. – d) $D(H_2P\text{–}H)$ is based on an upper limit of 78 kcal/mol given in [12]. The original paper gives $\Delta_f H^\circ_0 = 123.4 \pm 6.2$ kJ/mol [4].

The quite uncertain values $\Delta_f H^\circ_0 = 30.6 \pm 23$ and $\Delta_f H^\circ_{298} = 30.1 \pm 23$ in [8, 9] were based on an older estimate of 30 kcal/mol due to [13] and on another estimate due to [14]. (Different phosphorus reference states, however, were used in [8] and [9].) $\Delta_f H^\circ_{298} = 30 \pm 10$ [15].

Theoretically calculated formation enthalpies from fourth-order Møller-Plesset theory (MP4) are to be preferred to these latter estimates [16], namely $\Delta_fH^\circ_0 = 35.7$ and $\Delta_fH^\circ_{298} = 34.8 \pm 2$ [16], 37.3 [17], and 33.4 (139.8 kJ/mol) [18] (atomic data were taken from [19], where white P is used as the reference state).

A less reliable value, $\Delta_fH^\circ_{298} = 39.6 \pm 4.6$, was obtained from the appearance potential of the PH_2^+ ion from **diphosphane**, P_2H_4 (288.1±4.6 kcal/mol), which was combined with the formation enthalpies of PH_2^+ (253.6 kcal/mol; see p. 99) and P_2H_4 (5.0 kcal/mol). (An error of ±0.2 eV was estimated for the appearance potentials) [20]. The value, which was also reported in [21], however, must be seriously in error in view of the instability of P_2H_4 in conventional ion sources [2]. The appearance potential [20] had been earlier criticized and remeasured in [22]. A value of $\Delta_fH^\circ_0 = 40.5 \pm 4.6$ reported in [1], was derived apparently by adding 0.9 kcal/mol (see remark [b]) of the table above) to the value at 298 K in [20].

An estimate of $\Delta_fH \approx 24$ to 28 was based upon mass spectrometric work on triphosphane, P_3H_5, using an approximate P_3H_5 formation enthalpy of 14 to 18 kcal/mol due to [23] (and available data for PH_3 and P_2H_4) [24].

The following table lists the molar heat capacity C°_p, entropy S°, Gibbs free energy function $(G^\circ - H^\circ_{298})/T$ (all in $J \cdot mol^{-1} \cdot K^{-1}$) and enthalpy $H^\circ - H^\circ_{298}$ (in kJ/mol), taken from the JANAF tables [9], given in intervals of 100 K up to 6000 K for the ideal gas state at 0.1 MPa and based on estimated molecular parameters (partly taken from those of PH_3):

T in K	0	100	298	500	1000	3000	6000
C°_p	0	33.258	34.774	38.430	47.278	56.417	57.736
S°	0	175.957	212.652	231.466	261.000	319.071	358.730
$-(G^\circ - H^\circ_{298})/T$	∞	242.760	212.652	216.712	232.060	273.656	307.378
$H^\circ - H^\circ_{298}$	−10.006	−6.680	0	7.377	28.940	136.247	308.110

Similar values for C°_p, S°, and $H^\circ - H^\circ_0$ are compiled in another table up to 6000 K in intervals of 100 K [4].

References:

[1] Berkowitz, J.; Curtiss, L. A.; Gibson, S. T.; Greene, J. P.; Hillhouse, G. L.; Pople, J. A. (J. Chem. Phys. **84** [1986] 375/84, 382).
[2] McAllister, T.; Lossing, F. P. (J. Phys. Chem. **73** [1969] 2996/8).
[3] Di Stefano, G.; Lenzi, M.; Margani, A.; Mele, A.; Nguyen Xuan, C. (J. Photochem. **7** [1977] 335/44, 343).
[4] Glushko, V. P.; Gurvich, L. V.; Bergman, G. A.; Veits, I. V.; Medvedev, V. A.; Khachkuruzov, G. A.; Yungman, V. S. (Termodinamicheskie Svoistva Individual'nykh Veshchestv, Izd. Nauka, Moscow 1978, Vol. I, Pt. 1, p. 369, Pt. 2, p. 282).
[5] Gunn, S. R.; Green, L. G. (J. Phys. Chem. **65** [1961] 779/83).
[6] Wagman, D. D.; Evans, W. H.; Parker, V. B.; Schumm, R. H.; Halow, I.; Bailey, S. M.; Churney, K. L.; Nuttall, R. L. (J. Phys. Chem. Ref. Data **11** [1982] Suppl. No. 2, p. 2-73).
[7] Okabe, H. (Photochemistry of Small Molecules, Wiley, New York 1978, p. 377).
[8] Stull, D. R.; Prophet, H. (JANAF Thermochemical Tables, 2nd Ed., NSRDS-NBS-37 [1971]).
[9] Chase, M. W., Jr.; Davies, C. A.; Downey, J. R.; Frurip, D. J.; McDonald, R. A.; Syverud, A. N. (J. Phys. Chem. Ref. Data **14** [1985] Suppl. No. 1, p. 1289).
[10] Fraser, M. E.; Stedman, D. H. (J. Chem. Soc. Faraday Trans. I **79** [1983] 527/42, 535).
[11] Sam, C. L.; Yardley, J. T. (J. Chem. Phys. **69** [1978] 4621/7).
[12] Duewer, W. H.; Setser, D. W. (J. Chem. Phys. **58** [1973] 2310/20, 2316).

[13] Wiles, D. M.; Winkler, C. A. (J. Phys. Chem. **61** [1957] 902/3).
[14] Di Stefano, V. N.; Potter, R. L.; Fox, S. N. (The Thermodynamic Functions of Some Combustion Products Containing Phosphorus-I, American Cyanamide Company) cited according to [8].
[15] Vedeneyev, V. I.; Gurvich, L. V.; Kondrat'yev, V. N.; Medvedev, V. A.; Frankevich, Ye. L. (Bond Energies, Ionization Potentials, and Electron Affinities, Arnold, London 1966, p. 132).
[16] Pople, J. A.; Luke, B. T.; Frisch, M. J.; Binkley, J. S. (J. Phys. Chem. **89** [1985] 2198/203).
[17] Gordon, M. S.; Heitzinger, J. (J. Phys. Chem. **91** [1987] 2353/4).
[18] Wong, M. W.; Gill, P. M. W.; Nobes, R. H.; Radom, L. (J. Phys. Chem. **92** [1988] 4875/80).
[19] CODATA Task Group (J. Chem. Thermodyn. **10** [1978] 903/6).
[20] Saalfeld, F. E.; Svec, H. J. (Inorg. Chem. **3** [1964] 1442/3).

[21] Friswell, N. J.; Gowenlock, B. G. (Adv. Free Radical Chem. **2** [1967] 1/45, 16/7).
[22] Fehlner, T. P.; Callen, R. B. (Adv. Chem. Ser. **72** [1968] 181/90, 186).
[23] Fehlner, T. P. (J. Am. Chem. Soc. **90** [1968] 4817/22).
[24] Fehlner, T. P. (J. Am. Chem. Soc. **90** [1968] 6062/6).

1.2.1.4 Spectra

1.2.1.4.1 Electron Spin Resonance (ESR)

For g factors and nuclear coupling constants, see pp. 64 and 65.

An **isotropic** spectrum of PH_2 was first obtained, when a Kr matrix containing 2% PH_3 was irradiated at 4.2 K with γ rays from a ^{60}Co source. An 8 mT doublet arising from the ^{31}P splitting was found. The low-field component showed a triplet substructure with spacings of 1.8 mT, caused by the two protons. The high-field substructure was obscured by a line arising from P atoms (due to preferred dissociation via $PH_3 \rightarrow P + 3H$) [1]. The intensity ratio of the triplet was found to be 1:1:1 instead of an expected ratio of 1:2:1 and was explained analogous to an analysis [2] of the NH_2 radical. Free rotation of the PH_2 radicals about more than one axis is indicated by the isotropic nature of the spectrum and the symmetrical line shape [1]. Comparable strong PH_2 and P signals were observed in [3]. The PH_2 spectrum was found in a matrix containing PH_3 at a ratio $Kr:PH_3 \approx 5:1$ (in order to get the PH_4 radical). PH_3 was photolyzed at 4.2 K by the resonance radiation from an Ar lamp [4].

PD_2 signals in a Kr matrix were observed after substituting PD_3 for PH_3. The expected quintet pattern of each ^{31}P component (with relative intensities 1:2:3:2:1), however, was not resolved, since the spacings are six times smaller than those of the PH_2 triplet due to a smaller nuclear g factor [1].

A better PH_2 spectrum (with only very weak signals from atomic P) was obtained, when an Xe matrix with 2% PH_3 was γ-irradiated at 4.2 K. The 1:1:1 intensity distribution was found [3]. The PH_2 spectrum was observed at 4.2 and 77 K, when an Xe matrix with 1% PH_3 was γ-irradiated at 77 K. Spectra were studied on Xe matrices containing PH_3/HI mixtures (e.g., at a ratio 100:1:0.1) and irradiated by UV light with wavelengths of $\lambda > 250$ nm at various temperatures between 4.2 and 100 K. The triplet intensity ratio of 1:1:1 at 4.2 K was confirmed, while at higher temperatures the expected ratio 1:2:1 was approached [5].

A PH_2 spectrum with 1:2:1 intensity ratios of the two triplets was obtained, when PH_3 sorbed in a cancrinite matrix was photolyzed at 77 K in the far UV (Ar resonance lamp). This ratio pointed to a thermally equilibrated population of rotational states. PH_2 thus appeared to be trapped in pseudospherical cavities with minimum matrix perturbations [6].

An **anisotropic** spectrum showing a "parallel" ^{31}P hyperfine coupling and a near zero "perpendicular" coupling was observed, when PH_3 in a frozen concentrated aqueous solution of sulfuric acid (~90% H_2SO_4) was irradiated by ^{60}Co γ rays. The parallel (apparently with the axis directed normally to the radical plane, see [7]) features became better defined when solutions of D_2SO_4 in D_2O (giving PD_2 and a trace of PHD) were used. H bonding was assumed to prevent rotation and even libration of the radicals [8].

References:

[1] Morehouse, R. L.; Christiansen, J. J.; Gordy, W. (J. Chem. Phys. **45** [1966] 1747/51).
[2] McConnell, H. M. (J. Chem. Phys. **29** [1958] 1422).
[3] Jackel, G. S.; Gordy, W. (Phys. Rev. [2] **176** [1968] 443/52, 445).
[4] McDowell, C. A.; Mitchell, K. A. R.; Raghunathan, P. (J. Chem. Phys. **57** [1972] 1699/703).
[5] Shimokoshi, K.; Nakamura, K.; Sato, S. (Mol. Phys. **53** [1984] 1239/49).
[6] Raghunathan, P.; Sur, S. K. (Proc. Indian Acad. Sci. Chem. Sci. **92** [1983] 597/604).
[7] Wei, M. S.; Current, J. H.; Gendell, J. (J. Chem. Phys. **52** [1970] 1592/602); Nelson, W.; Jackel, G.; Gordy, W. (J. Chem. Phys. **52** [1970] 4572/8).
[8] Fullam, B. W.; Mishra, S. P.; Symons, M. C. R. (J. Chem. Soc. Dalton Trans. **1974** 2145/8).

1.2.1.4.2 Microwave Absorption

Absorption spectra of PH_2 were obtained when it was generated directly in the absorption cell by a glow discharge in a PH_3/O_2 mixture [1, 2]. Three rotational transitions, described by quantum numbers $N_{K_aK_c}$, within the electronic and vibrational ground state were observed in the mm-wavelength region (transition frequency $\nu < 300$ GHz). Altogether 34 hyperfine (hf) components were resolved [1]. Five more rotational transitions with 61 hf components were observed in the sub-mm-wavelength region ($\nu > 300$ GHz) [2]. Most microwave lines belonged to Q-branch transitions ($\Delta N = 0$) and obeyed the further selection rule $\Delta J = 0$ [1, 2]. One R-branch transition ($N = 1 \leftarrow 0$) consisting of two components with $J = 1/2 \leftarrow 1/2$ and $3/2 \leftarrow 1/2$ was found [2]. The tables in the original papers [1, 2] list the frequencies of all hf components, which are distinguished first by an intermediate angular momentum quantum number F_1 for the coupling between J and $I(^{31}P) = 1/2$ and then by another quantum number F for the coupling between F_1 and I(H) (where the total proton spin I(H) = 0 for (K_a, K_c) = (even, odd) or (odd, even), or I(H) = 1 for (K_a, K_c) = (even, even) or (odd, odd)). The following table shows the rotational transitions ($N_{K_aK_c}$), the fine structure components (J), the respective hf component numbers, and rough values of ν ($\nu < 300$ GHz [1], >300 GHz [2]):

$N_{K_aK_c}$	I(H)	J	hf	ν (full GHz)
$1_{10} \leftarrow 1_{01}$	0	$3/2 \leftarrow 1/2$*)	1	144
		$3/2 \leftarrow 3/2$	4	145
		$1/2 \leftarrow 1/2$	3	151 to 152
		$1/2 \leftarrow 3/2$*)	1	154
$2_{20} \leftarrow 2_{11}$	1	$5/2 \leftarrow 5/2$	10	181 to 182
		$3/2 \leftarrow 3/2$	11	188

$N_{K_aK_c}$	I(H)	J	hf	ν (full GHz)
$3_{30} \leftarrow 3_{21}$	0	7/2←7/2	2	244
		5/2←5/2	2	253
$4_{31} \leftarrow 4_{22}$	1	9/2←9/2	6	324
		7/2←7/2	6	327
$3_{21} \leftarrow 3_{12}$	0	7/2←7/2	2	330
		5/2←5/2	2	332
$4_{40} \leftarrow 4_{31}$	1	9/2←9/2	6	337
		7/2←7/2	6	348
$2_{11} \leftarrow 2_{02}$	1	5/2←5/2	10	351
		3/2←3/2	10	355
$1_{11} \leftarrow 0_{00}$	1	3/2←1/2	8	397
		1/2←1/2	5	404

*) Weak ($\Delta J = \pm 1$).

References:

[1] Endo, Y.; Saito, S.; Hirota, E. (J. Mol. Spectrosc. **97** [1983] 204/12).
[2] Kajita, M.; Endo, Y.; Hirota, E. (J. Mol. Spectrosc. **124** [1987] 66/71).

1.2.1.4.3 IR Laser Magnetic Resonance (LMR)

Far-IR LMR was observed for eight rotational transitions ($N_{K_aK_c}$) of PH_2 in its electronic and vibrational ground state with laser lines of gaseous H_2O [1], CH_3OH, N_2H_4, $C_2H_2F_2$, and HCOOH [2]. The PH_2 radical was obtained by the reaction $PH_3 + F$ [1, 2]. Most of the observed Zeeman components were doublets arising from the $^{31}P(I=1/2)$ hyperfine (hf) splitting [1, 2]. Narrow triplets due to the total spin I(H) = 1 of the two protons were found for the Zeeman components of three rotational transitions (with K_a, K_c = even, even or odd, odd) [2]. The following table lists the rotational transitions, the laser lines used (with their wavelengths λ_L and frequencies ν_L taken from [3]), the number of Zeeman components observed, and the magnetic field strengths H applied [1, 2]:

$N_{K_aK_c}$	I(H)	laser	λ_L in μm	ν_L in GHz	Zeeman	H in G
$2_{12} \leftarrow 1_{01}$	0	HCOOH	458.5	653.8214	2	519, 563
$5_{42} \leftarrow 5_{33}$	1	HCOOH	432.6	692.9505	1	5425
$6_{61} \leftarrow 6_{52}$	0	HCOOH	418.6	716.1564	9	6910 to 14340
$6_{52} \leftarrow 6_{43}$	0	$C_2H_2F_2$	407.3	736.0596	14	4720 to 11390
$4_{13} \leftarrow 4_{04}$	1	N_2H_4	331.7	903.8894	15	1087 to 4722
$5_{33} \leftarrow 5_{24}$	1	N_2H_4	331.7	903.8894	1	3277
$6_{43} \leftarrow 5_{32}$	0	H_2O	118.6	2527.9520	9	7860 to 12580
$7_{53} \leftarrow 6_{42}$	0	CH_3OH	96.5	3105.9368	19	1107 to 8837

Mid-IR LMR was observed for the bending vibration ν_2 (at 1102 cm^{-1}; see p. 74) of PH_2 in its electronic ground state with CO_2 laser radiation around 9.4 μm (from the "II-band"). The PH_2 radical was generated in a flow system by the reaction PH_3+H [4]. Spectra attributable to PH_2 were found for 30 out of 67 tested laser lines of $^{12}C^{16}O_2$ and $^{12}C^{18}O_2$. A table in the original paper lists for twenty-three CO_2 laser lines ranging from the $^{12}C^{16}O_2$ P(34) line at 1033.48800 cm^{-1} to the $^{12}C^{18}O_2$ R(38) line at 1106.95490 cm^{-1} altogether 113 Zeeman components with their resonant fields, hf splittings, and changes of quantum numbers $N_{K_aK_c}$, J, and M. The ^{31}P doublet splitting was given for 36 out of the 113 resonances. Proton hf splitting was observed for only one resonance, which was detected with the $^{12}C^{18}O_2$ R(6) line at 1088.43783 cm^{-1} and assigned to $N_{K_aK_c}=0_{00}\leftarrow 1_{11}$, $J=1/2\leftarrow 1/2$, $M=1/2\leftarrow -1/2$ [4]. The wave numbers used [4] for the $^{12}C^{16}O_2$ and $^{12}C^{18}O_2$ laser lines were taken from [5] and [6], respectively; they should, however, be replaced by more recent data in [7]; see also comments in [8].

References:

[1] Davies, P. B.; Russell, D. K.; Thrush, B. A. (Chem. Phys. Lett. **37** [1976] 43/6).
[2] Davies, P. B.; Russell, D. K.; Thrush, B. A.; Radford, H. E. (Chem. Phys. **44** [1979] 421/6).
[3] Radford, H. E.; Petersen, F. R.; Jennings, D. A.; Mucha, J. A. (IEEE J. Quantum Electron. **13** [1977] 92/4); Petersen, F. R.; Evenson, K. M.; Jennings, D. A.; Wells, J. S.; Goto, K.; Jiménez, J. J. (IEEE J. Quantum Electron. **11** [1975] 838/43).
[4] Hills, G. W.; McKellar, A. R. W. (J. Chem. Phys. **71** [1979] 1141/9).
[5] Petersen, F. R.; McDonald, D. G.; Cupp, J. D.; Danielson, B. L. (Phys. Rev. Lett. **31** [1973] 573/6).
[6] Freed, C.; Ross, A. H. M.; O'Donnell, R. G. (J. Mol. Spectrosc. **49** [1974] 439/53, 445/6).
[7] Freed, C.; Bradley, L. C.; O'Donnell, R. G. (IEEE J. Quantum Electron. **16** [1980] 1195/206, 1196/7).
[8] McKellar, A. R. W. (Faraday Discuss. Chem. Soc. No. 71 [1981] 63/76, 72/4).

1.2.1.4.4 IR Absorption

Matrix-isolated $PH_2(PD_2)$ absorbs at 1103 (797) cm^{-1}; see "Bending Vibration", p. 74.

1.2.1.4.5 Raman Scattering

A Raman shift of ~2310 cm^{-1} was supposed for gaseous PH_2; see "Symmetric Stretching Vibration", p. 73.

1.2.1.4.6 Electronic Absorption and Emission. Lifetimes. Quenching

Absorption spectra of gaseous PH_2 [1 to 6] or PD_2 [1, 2, 7] have been obtained during the flash photolysis (FP) of PH_3 or PD_3. Absorption spectra of matrix-isolated PH_2 [8 to 10] or PD_2 [9] are also available. Absorption features of the gaseous species can also be seen in laser-induced fluorescence (LIF) spectra, when measuring the fluorescence signal as a function of the exciting laser wavelength. Thus, the hyperfine (hf) structure of rovibronic absorption lines was observed using a continuous wave (cw) dye laser [11, 12]. Pulsed dye lasers were also used [13 to 15], allowing the determination of the radiative lifetimes τ_R and quenching rate constants k_Q [13, 15]. Emission spectra of gaseous PH_2 were first obtained, when PH_3 was reacted with atomic H or N (from high-voltage discharges through H_2 or N_2) or with discharged He [16 to 23] and later, when PH_3 was photolyzed in the vacuum UV [24 to 26]. The spectra of PD_2 [16, 19 to 21] and PHD [20, 21] were obtained in an "atomic flame" PD_3+H. All spectra referred to above arise from the electronic transition $\tilde{A}\ ^2A_1 \leftrightarrow \tilde{X}\ ^2B_1$ between two states, which are correlated with a common $^2\Pi_u$ state in the linear radical configuration (see p. 53). The vibrational structure is due to excitation of only the bending mode ν_2 in each state.

Absorption

Flash photolysis of PH_3 and PD_3 through a 6m-long absorption path yielded complex spectra between 3600 and 5500 Å [1, 2]. Groups of strong lines, probably Q branches ($\Delta N=0$), were clustered at regular intervals along the spectra. The most intense part of the PH_2 spectrum extends from ~4500 to ~4550 Å [1]. An extinction coefficient was later determined in that region [27]. The long-wavelength edges of the strong groups were used for a vibrational analysis. Thus, eleven bands of PH_2 (and thirteen bands of PD_2) were identified and assigned as the transitions $\tilde{A}\ ^2A_1$, $v_2'=0$ to 10(12)$\leftarrow\tilde{X}\ ^2B_1$, $v_2''=0$ [1, 2]. The band heads are given in a table on p. 86. The PH_2 bands with $v_2'=1$ to 6 (5202 to 4201 Å, erroneously 4501 Å in the original paper) were also recorded in [3], those with $v_2'=3$ to 7 (4740 to 4050 Å) also in [4]. The longer-wavelength part of the spectrum was again recorded through a 24m-long path. Two hot bands $v_2=0\leftarrow1$ and $1\leftarrow1$ were found. The $v_2=0\leftarrow0$ band was subjected to a rotational analysis. 464 lines were assigned (for wave numbers $\bar{\nu}$, quantum numbers $N_{K_aK_c}$, and calculated relative intensities, see a table in the original paper). 373 of these lines belonged to branches with $\Delta K_a=\pm1$ (branches ^PP, ^PQ, ^PR, ^RQ, and ^RR, extending from 18140 to 19030 cm^{-1}), 89 lines belonged to branches with $\Delta K_a=\pm3$ (branches ^NP, extending from 18040 to 18170 cm^{-1}, ^TQ, and ^TR, from 18410 to 18920 cm^{-1}), and 2 lines belonged to a branch with $\Delta K_a=+5$ ($^VR_{2,N-2}$ at ~18960 cm^{-1}; the full designation $^{\Delta K_a}\Delta N_{K_aK_c}$ shows that $K_a=2$ and $K_c=N-2$). Strong lines near the band center belong to the branches $^RQ_{0,N}$ (with a head consisting of $5_{15}\leftarrow5_{05}$ and $6_{16}\leftarrow6_{06}$ at 18282 cm^{-1}) and $^PQ_{1,N}$ (with a head consisting of $2_{02}\leftarrow2_{12}$ and $3_{03}\leftarrow3_{13}$ at ~18272 cm^{-1}). A doublet structure due to the electron spin and an asymmetry doubling were obvious especially for ^RR branches [5]. The PH_2 band $v_2=3\leftarrow0$ was reinvestigated through a 16 m-long pathlength. Rotational analyses were performed for all bands, $v_2'=1$ to $8\leftarrow v_2''=0$ (using also the earlier data [1]), and ~1000 lines were assigned. The analysis of the $0\leftarrow0$ band [5] was extended [6]; for a preliminary report, see [28]. The PD_2 bands $v_2'=1$ to $10\leftarrow v_2''=0$ were reinvestigated and subjected to a rotational analysis (lines are available from the source mentioned in the footnote[1]) [7].

A cw dye laser excitation spectrum of the $0_{00}\leftarrow1_{10}$ line from the subbranch $^PP_{1,N-1}$ of the $0\leftarrow0$ band revealed the fine structure (electron spin doublet) and also the ^{31}P hf structure [11]. That study [11] was improved by using an intermodulated fluorescence (IMF) method due to [29]. The hf structure was resolved for 26 lines of the $0\leftarrow0$ band (12 with a total ^1H spin I(H) = 0 and 14 with I(H)=1) [12].

Pulsed dye laser excitation spectra were obtained for bands of the $v_2''=0$ progression, i.e., $v_2'=0$ [14, 15], 1 to 5 [13, 15], 4 [14], and 6 [15], covering shorter or longer spectral ranges ($\Delta\bar{\nu}=40$ to 60 cm^{-1} [13] or 230 to 280 cm^{-1} [14], $\Delta\lambda=10$ to 20 Å [15]). The heads of the branches $^PQ_{1,N}$ (with $K_a'=0$) and $^RQ_{0,N}$ (with $K_a'=1$) were seen as strong features in almost all cases and were termed shortly $^PQ_1$ and $^RQ_0$. The $v_2'=0$ spectrum in [15] showed only $^RQ_0$, while $^PQ_1$ was hardly perceptible. Another $v_2'=0$ spectrum was not explicitly assigned, although all

[1] A complete list is available from the Depositary of Unpublished Data, Natl. Science Library, Natl. Res. Council of Canada in Ottawa [6].

464 lines due to [5] could be easily identified [14]. For $v_2'=1$ to 4, both $^PQ_1$ and $^RQ_0$ were observed [15], while $^RQ_0$ was not clearly seen in another $v_2'=4$ spectrum [13]. The $v_2'=4$ spectrum in [14] was not analyzed. The $v_2'=5$ spectrum in [15] showed $^PQ_1$, the branches $^PP_{1,N-1}$ and $^PP_{2,N-2}$, and additional features, which were assumed to result from a transition $v_2'=6 \leftarrow v_2''=1$. The spectrum agreed with that in [13], where no assignments had been made [15]. The $v_2'=6$ spectrum showed only $^PQ_1$ [15]. The heads $^RQ_0$ and $^PQ_1$ from the laser excitation spectrum of PH_2 [15] are given in the table below together with the heads from flash photolysis absorption spectra of PH_2 and PD_2 (apparently $^PQ_1$ values) [1, 2] (all data in Å):

v_2'	0	1	2	3	4	5
PH_2 $^RQ_0$ [15]	5468.0	5197.7	4954.4	4735.3	4534.8	—
$^PQ_1$ [15]	—	5200.8	4958.2	4740.3	4542.9	—
$^PQ_1$ [1, 2]	5471.4	5200.6	4958.8	4740.4	4543.2	4364.3
PD_2 $^PQ_1$ [1, 2]	5471.5	5272.3	5088.6	4919.3	4761.5	4615.1

v_2'	6	7	8	9	10	11*)
PH_2 $^PQ_1$ [1, 2]	4201.3	4050.3	3907.0	3768.7	3634.9	—
PD_2 $^PQ_1$ [1, 2]	4479.3	4353.0	4235.5	4125.9	4021.9	3921.9*)

*1) $v_2'=12$: 3823.6 Å.

Matrix-isolated PH_2 and PD_2 absorption spectra showed several bands of the $v_2''=0$ progression, shifting by $\Delta\bar{\nu}$ to the red. The PH_2 bands $v_2'=0$ to 7 ($\Delta\bar{\nu}=62$ cm^{-1} for the $0 \leftarrow 0$ band) were observed when Ar–PH_3 (200:1) samples were codeposited onto a 12 K sapphire plate with O atoms (in order to obtain PO_2 and PO_3 radicals), photolyzed, and annealed [10]. PH_2 bands $v_2'=0$ to 6 ($\Delta\bar{\nu}\approx 100$ cm^{-1}) and PD_2 bands $v_2'=2$ to 9 were obtained when Ar–PH_3(PD_3) (200 to 500:1) samples were photolyzed in the vacuum UV at 14 K [9]. The bands $v_2'=1$ to 5 had been observed earlier after condensing an electrically discharged Ar/PH_3 mixture at 4.2 K [8].

Emission

A PH_2 spectrum covering the range 440 to 770 nm was excited by irradiating PH_3 with an ArF excimer laser (193.3 nm). The smaller range 450 to 600 nm was plotted, and bands belonging to six sequences, i.e., $v_2'-v_2''=4$ through -1, were indicated (each sequence contained 5, 6, or 7 members) [26]. Photolysis of PH_3 in the vacuum UV yielded spectra in the range 390 to 630 nm with bands $v_2' \rightarrow v_2''=7 \rightarrow 0$ to $1 \rightarrow 3$ [24] or 420 to 700 nm with bands $6 \rightarrow 0$ to $0 \rightarrow 3$ [25]. Excitation of PH_3(PD_3) by discharged He led to PH_2, PD_2, and PHD bands belonging to altogether fourteen sequences, $v_2'-v_2''=5$ through -8. From a complete list of wavelengths, wave numbers, relative intensities, and assignments, the following wavelengths (in Å) of bands with $v_2'=0$ or $v_2''=0$ (usually the most intense in each sequence) are given [21]:

$v_2' \rightarrow v_2''$	$5 \rightarrow 0$	$4 \rightarrow 0$	$3 \rightarrow 0$	$2 \rightarrow 0$	$1 \rightarrow 0$	$0 \rightarrow 0$	$0 \rightarrow 1$
PH_2	—	4537.65*)	4735.45	4954.4	5197.3	5467.35	5818.5
PD_2	4614.65	4761.3	4918.9	5088.6	5269.95	5468.3	5719.7
PHD	—	—	4817.8	5015.44	5231.4	5467.8*)	5772.95

$v_2' \rightarrow v_2''$	$0 \rightarrow 2$	$0 \rightarrow 3$	$0 \rightarrow 4$	$0 \rightarrow 5$	$0 \rightarrow 6$	$0 \rightarrow 7$	$0 \rightarrow 8$
PH_2	6212.55	6660	7173.1	7763.1	8448.1	—	—
PD_2	5991.6	6287.6	6612.4	6971.2	7366.2	7805.3	8296.1
PHD	6110.2	6485.15	6905.95	7381.75	7920.8	8541.65	—

*) Uncertain.

Rotational analyses were performed for two intense emission bands of PH_2, i.e., for 0→2 [22, 23] and 0→3 [23].

Lifetimes. Quenching

The radiative lifetimes τ_R and quenching rate constants k_Q have been measured for a few vibrational levels v_2' of the upper state $\tilde{A}\,^2A_1$ and two quenchers Q, i.e., for $v_2'=0$, 1, 3, and 4 and $Q=PH_3$ [26], for $v_2'=0$ through 4 (for $v_2'=1$, 2, 3 also for $K_a'=0$ or 1) and $Q=H_2$ [13], and for $v_2'=0$ through 4, $K_a'=0$ or 1, and $Q=PH_3$ [15]. The results do not agree very well as shown in the following tables for τ_R and k_Q; (maximum) uncertainties in units of the last digit are added in parentheses:

τ_R in μs

K_a'	v_2' 0	1	2	3	4	Ref.	remark
0	—	3.00(218)	2.44(54)	2.57(108)	1.08(22)	[15]	a)
0	—	3.6	4.3	5.0	—	[13]	b)
1	5.69(346)	4.81(711)	3.28(236)	1.76(22)	0.89(10)	[15]	a)
1	—	4.0	2.6	~10.0	—	[13]	b)
not spe-	3.8	3.4	3.8	4.0	0.31	[13]	c)
cified	2.2(1)	2.4(1)	—	2.2(1)	2.1(1)	[26]	—

a) Lifetimes in $v_2'=1$ and 2 are longer for $K_a'=1$ than for $K_a'=0$. Renner-Teller coupling between $\tilde{A}\,^2A_1$ and nonemitting upper vibrational levels of $\tilde{X}\,^2B_1$ may be an explanation [15]. – b) Error limits between 0.5 and 0.9 μs; higher for $v_2'=3$. – c) $v_2'=0$ excited from $v_2''=1$.

k_Q in $\mu s^{-1} \cdot Torr^{-1}$ for $Q=PH_3$

K_a'	v_2' 0	1	2	3	4	Ref.
0	—	10.8(9)	12.5(5)	16.4(9)	23.2(13)	[15]
1	7.8(16)*)	10.4(8)	12.0(8)	16.8(6)	23.3(10)	[15]
—	5.5(3)	5.3(2)	—	6.4(3)	6.8(4)	[26]

*) $15(3)\times10^{10}$ $L \cdot mol^{-1} \cdot s^{-1}$.

k_Q in 10^{10} $L \cdot mol^{-1} \cdot s^{-1}$ for $Q=H_2$ [13]

K_a'	v_2' 0	1	2	3	4
0	—	2.90(36)	5.13(18)	13.96(109)	—
1	—	2.73(22)	3.70(55)	13.80(164)	—
—	1.24(22)	3.13(27)	6.09(64)	13.4(11)	16.2(18)

References:

[1] Ramsay, D. A. (Nature **178** [1956] 374/5).
[2] Ramsay, D. A. (Ann. N.Y. Acad. Sci. **67** [1956/57] 485/98, 492/5).
[3] Norrish, R. G. W.; Oldershaw, G. A. (Proc. R. Soc. [London] A **262** [1961] 1/9).

[4] Kley, D.; Welge, K. H. (Z. Naturforsch. **20a** [1965] 124/31).
[5] Dixon, R. N.; Duxbury, G.; Ramsay, D. A. (Proc. R. Soc. [London] A **296** [1967] 137/60, 148/60).
[6] Berthou, J.-M.; Pascat, B.; Guenebaut, H.; Ramsay, D. A. (Can. J. Phys. **50** [1972] 2265/76).
[7] Vervloet, M.; Berthou, J.-M. (Can. J. Phys. **54** [1976] 1375/82).
[8] McCarty, M., Jr.; Robinson, G. W. (J. Chim. Phys. Phys. Chim. Biol. **56** [1959] 723/30).
[9] Larzillière, M.; Jacox, M. E. (NBS Spec. Publ. [U.S.] 561-1 [1979] 529/43, 532/4).
[10] Withnall, R.; McCluskey, M.; Andrews, L. (J. Phys. Chem. **93** [1989] 126/9).

[11] Curl, R. F.; Endo, Y.; Kakimoto, M.; Saito, S.; Hirota, E. (Chem. Phys. Lett. **53** [1978] 536/8).
[12] Kakimoto, M.; Hirota, E. (J. Mol. Spectrosc. **94** [1982] 173/91).
[13] Huie, R. E.; Long, N. J. T.; Thrush, B. A. (J. Chem. Soc. Faraday Trans. II **74** [1978] 1253/62).
[14] Baugh, D.; Koplitz, B.; Xu, Z.; Wittig, C. (J. Chem. Phys. **88** [1988] 879/87).
[15] Nguyen Xuan, C.; Margani, A. (J. Chem. Phys. **93** [1990] 136/46).
[16] Pascat, B.; Guenebaut, H. (Bull. Univ. ARERS **2** [1963] 100/7).
[17] Guenebaut, H.; Pascat, B. (C.R. Hebd. Seances Acad. Sci. **256** [1963] 677/80).
[18] Guenebaut, H.; Pascat, B. (C.R. Hebd. Seances Acad. Sci. **256** [1963] 2850/3).
[19] Guenebaut, H.; Pascat, B. (J. Chim. Phys. Phys. Chim. Biol. **61** [1964] 592/5).
[20] Guenebaut, H.; Pascat, B. (C.R. Hebd. Seances Acad. Sci. **259** [1964] 2412/5).

[21] Guenebaut, H.; Pascat, B.; Berthou, J.-M. (J. Chim. Phys. Phys. Chim. Biol. **62** [1965] 867/77).
[22] Pascat, B.; Berthou, J.-M.; Guenebaut, H.; Ramsay, D. A. (C.R. Seances Acad. Sci. B **263** [1966] 1397/9).
[23] Pascat, B.; Berthou, J.-M.; Prudhomme, J. C.; Guenebaut, H.; Ramsay, D. A. (J. Chim. Phys. Phys. Chim. Biol. **65** [1968] 2022/9).
[24] Di Stefano, G.; Lenzi, M.; Margani, A.; Mele, A.; Nguyen Xuan, C. (J. Photochem. **7** [1977] 335/44).
[25] Di Stefano, G.; Lenzi, M.; Margani, A.; Nguyen Xuan, C. (J. Chem. Phys. **68** [1978] 959/63).
[26] Sam, C. L.; Yardley, J. T. (J. Chem. Phys. **69** [1978] 4621/7).
[27] Zakhar'in, V. I.; Nadtochenko, V. A.; Sarkisov, O. M.; Teitel'boim, M. A. (Dokl. Akad. Nauk SSSR **263** [1982] 127/30; Dokl. Phys. Chem. [Engl. Transl.] **262/267** [1982] 168/70).
[28] Berthou, J.-M.; Pascat, B. (C.R. Seances Acad. Sci. C **271** [1970] 799/801).
[29] Sorem, M. S.; Schawlow, A. L. (Opt. Commun. **5** [1972] 148/51).

1.2.1.5 Chemical Reactions

Two gas-phase reactions of PH_2 have been studied in some detail: (1) The reaction with molecular O_2, which shows a variety of product channels and which is of interest for the oxidation of phosphane (a branched chain process; see p. 236). (2) The reaction with another PH_2 radical, which occurs mainly by recombination to form P_2H_4 rather than by disproportionation to form PH_3 and a PH radical and which is important for the photolysis of PH_3 (see p. 206). Second-order rate constants (k) are valid for 298 K (if not otherwise stated) and given in $cm^3 \cdot molecule^{-1} \cdot s^{-1}$, reaction enthalpies ΔH in kcal/mol or kJ/mol. See also under "Chemical Reactions of PH_3" (p. 199) for further reactions.

Reactions with Elements

PH_2 + H → Products. The exothermic bimolecular H-abstraction reaction $PH_2 + H \rightarrow PH + H_2$, for which $\Delta H = -97$ kJ/mol [7] or −22.1 kcal/mol was given [12], possibly is the main channel for the formation of PH [13], observed during the flash photolysis of PH_3 (and explained by PH_2 disproportionation) [14]. The reaction was also proposed for the PH_3 photolysis at different pressure ranges [15, 16] and for the PH_3 + H [17] and $PH_3 + N_2O$ [18] reactions. A rate constant of $k = 6.2 \times 10^{-11}$ exp(−318/T) was estimated with a modified bond-energy bond-order (BEBO) method (see [19]) using empirical corrections (to obtain agreement with the experimental data in the case of the PH_3 + H reaction) [20].

The recombination reaction $PH_2 + H + M \rightarrow PH_3 + M$ was also proposed for the PH_3 photolysis [15, 16] and PH_3 + H reaction [17]. A high-pressure limit of the recombination rate constant, $k^{\infty}_{rec} = 3.7 \times 10^{-10}$ exp(−340/T) $cm^3 \cdot molecule^{-1} \cdot s^{-1}$, was derived [20] from the Rice-Ramsperger-Kassel-Marcus (RRKM) theory of the activated complex, modified in [21].

$PH_2 + H_2 \rightarrow PH_3$ + H. Energy barriers ΔE_f and ΔE_b for the forward and backward reaction and the energy change $\Delta E = \Delta E_b - \Delta E_f$ were calculated by several ab initio MO methods, including Møller-Plesset, multi-configuration(MC)-SCF, and polarization configuration interaction (POL-CI, see [1]) treatments. The POL-CI method yielded $\Delta E_f = 26.2$, $\Delta E_b = 5.6$, and $\Delta E = -19.8$ kcal/mol (endothermic reaction) [2].

$PH_2 + Cl_2 \rightarrow PH_2Cl$ + Cl. This reaction was proposed to be a chain-propagating step of the branched-chain reaction of PH_3 with Cl_2. Branching may occur by formation of excited PH_2Cl^* [8].

PH_2 + O → Products. An upper limit of the rate constant for homogeneous removal of PH_2 by O of 1×10^{-10} $cm^3 \cdot molecule^{-1} \cdot s^{-1}$ was used in a study of the reaction of PH_3 with O atoms (at 298 K) [22]. Two [23] or four [7] product channels have been proposed:

channel	→ PH + OH¹⁾	→ HPO + H	→ $PO + H_2$	→ $P + H_2O$²⁾
ΔH in kcal/mol [23]	−21	−29	—	—

¹⁾ Considered as a route of PO formation via PH + O → PO + H [24]. − ²⁾ Not a "basic" channel [7].

$PH_2 + O_2$ → Products. The rate constants k were measured at 300 K by flash photolysis (FP) of $PH_3/O_2/N_2$ mixtures (0.35:3.5 to 17:250 to 270 Torr) and PH_3/O_2 mixtures (0.35:1.5 Torr), above the upper and below the lower explosion limit, followed by absorption of radiation near 455 nm (from a pulsed dye laser) by PH_2. $k = (1.2 \pm 0.3) \times 10^{-13}$ at a total pressure of ~270 Torr (i.e., above the upper explosion limit) was derived from the dependence of the effective rate constant $k_{eff} = d\,\ln([PH_2]_0/[PH_2])/dt$ on the O_2 partial pressure. $k = (0.8 \pm 0.2) \times 10^{-13}$ at 1.85 Torr (i.e., below the lower explosion limit) was obtained from a computer calculation of k_{eff} with known rate constants for $PH_3 + H \rightarrow PH_2 + H_2$ (see p. 233) and $2PH_2 + M \rightarrow P_2H_4 + M$ (see below). The agreement of both k values points to a pressure-independent reaction $PH_2 + O_2 \rightarrow$ products. The step $PH_2 + O_2 \rightarrow PO_2 + H_2$ is probably responsible for the observed rapid decay of PH_2 [3, 4]. A later measurement of the removal of rotationally thermalized PH_2 ($\tilde{X}\,^2B_1$, v = 0) in the presence of O_2 by laser-induced fluorescence (LIF) under pseudo-first-order conditions gave $k = 2.7 \times 10^{-13}$. The removal of vibrationally excited PH_2 by O_2 was also investigated [5].

Four exothermic product channels of the **bimolecular** reaction $PH_2 + O_2$ are listed below together with available data for k and ΔH:

channel	→ $PO_2 + H_2$	→ HPO + OH	→ $PO + H_2O$	→ H_2PO + O
$10^{13} \cdot k$ [3, 4]	1.2±0.3	<4	—	—
ΔH in kJ/mol [3]	−418	−137	−361	<0
remark	a), b)	b), c)	d)	e)

[a] Probably the main reaction channel; thus, k was taken from the overall reaction (see text on p. 89) [3, 4]. – [b] Reaction channel involves an atomic rearrangement; thus, it can be assumed that an intermediate complex H_2POO^* isomerizes to $H_2P(O)O^*$ and then decays into the products [3, 4]. – [c] Channel not responsible for rapid PH_2 decay in the presence of PH_3, since $PH_3 + OH$ reforms PH_2. k was estimated on the assumption that a quasi-stationary OH concentration in the "chain reaction" $PH_2 + O_2 \rightarrow HPO + OH$, $PH_3 + OH \rightarrow PH_2 + H_2O$ is achieved [3, 4]. The channel is part of a mechanism for PH_3/O_2 flash photolysis [6]; see also discussion in [7]. – [d] Channel not responsible for rapid PH_2 decay in the presence of PH_3 since $PO + O_2 \rightarrow PO_2 + O$, $PH_3 + O \rightarrow PH_2 + OH$ reform PH_2 [3, 4]. The channel was introduced into the mechanism [6] (mentioned in [c]) as a "more realistic" one [7]. – [e] Channel cannot explain the rapid PH_2 decay in the presence of PH_3, since $PH_3 + O \rightarrow PH_2 + OH$ reforms PH_2 [3, 4].

The endothermic reaction $PH_2 + O_2 \rightarrow PH + HO_2$ with ΔH given to be 142 [7] or 128 kJ/mol (31 kcal/mol) [3, 4] is probably not involved in the flash photolysis of PH_3/O_2 mixtures [3, 4, 6, 7].

The **termolecular** reaction $PH_2 + O_2 + M \rightarrow PH_2O_2 + M$ was introduced to serve as a termination reaction into the mechanism of the flash photolysis of PH_3/O_2 mixtures [6, 7], however, the formation of PH_2O_2 by collisional deactivation of the intermediate complex H_2POO^* is ruled out, because $k(PH_2 + O_2)$ was found to be pressure-independent (see above) [3, 4].

$PH_2 + O_3 \rightarrow$ Products. The reaction $PH_2 + O_3 \rightarrow HPOH + O_2$ was expected to govern the fate of PH_2 in the troposphere (by comparison with the chemically similar NH_2 radical) [9]. $\Delta H = -176.6$ kJ/mol was given for the reaction $PH_2 + O_3 \rightarrow P + H_2O + O_2$ [10].

$PH_2 + N \rightarrow PN + H_2$. The reaction (with participation of a third body M) was proposed to be a PN-forming step in the reaction of PH_3 with active nitrogen; $\Delta H = -162.7$ kcal/mol [25].

Reactions with Compounds

$PH_2 + OH \rightarrow PH + H_2O$. The reaction may be important for the $PH_3 + N_2O$ reaction [18].

$PH_2 + HO_2 \rightarrow$ Products. A rate constant of $k < 5 \times 10^{-11}$ was proposed. The reaction cannot explain the PH_2 decay kinetics in the presence of O_2 in flash-photolyzed PH_3/O_2 mixtures above the upper explosion limit (where HO_2 may be formed from $O_2 + H + M \rightarrow HO_2 + M$) [3, 4].

$PH_2 + NH_2 + M \rightarrow NH_2PH_2 + M$. Upper limits of a pseudo-second-order rate constant k at 175 K were estimated from the absence of NH_2PH_2 as a photoproduct in the photolysis of PH_3/NH_3 (1:1) mixtures at 11 Torr. The limits depended on the wavelength λ of the NH_2PH_2 UV absorption maximum, i.e., $k < 5 \times 10^{-10}$ for $\lambda \leqq 235$ nm and $k < 2 \times 10^{-10}$ for $\lambda > 240$ nm [16]. A high-pressure limit of the recombination rate constant of $k_{rec}^{\infty} = 1.0 \times 10^{-10} \exp(-18/T)$ cm$^3 \cdot$molecule$^{-1} \cdot s^{-1}$ was derived [20] from the modified RRKM theory of the activated complex [21]. That expression for the temperature dependence yields $k_{rec}^{\infty} = 0.9 \times 10^{-10}$ at 175 K in good agreement with the two upper limits from above [16]. The rate is expected to be comparable to that of the PH_2 recombination (see below) [26].

$PH_2 + NO \rightarrow$ Products. The removal of rotationally thermalized PH_2 in its electronic and vibrational ground state was studied at 300 K by LIF in the presence of NO under pseudo-first-order conditions. NO was essentially unreactive ($k < 10^{-14}$) [5].

$2PH_2 \rightarrow$ Products. The pseudo-second-order overall rate constant k is defined by $-d[PH_2]/dt = 2k[PH_2]^2$. It is made up by the pressure-independent constant k_{dis} for the disproportionation $2PH_2 \rightarrow PH_3 + PH$ and by the pressure-dependent constant k_{rec} for the recombination $2PH_2 + M \rightarrow P_2H_4 + M$ [13, 27]. The disproportionation reaction is exothermic, since the bond dissociation energy of PH_3 is greater than that of PH_2 (see pp. 78, 174, and [28]). This reaction channel was proposed (and favored over recombination) to account for the PH formation during the flash photolysis of PH_3 [14]. It is part of several mechanisms describing the radiolysis [28, 29], photolysis [11, 30, 31], and reaction with N_2O of PH_3 [18]. The recombination reaction was proposed to account for the detection of P_2H_4 during the photolysis of PH_3 at 206 nm [11, 16, 26, 31] and at other wavelengths [11], and for the detection of $P_2H_4(P_2D_4)$ in the photolysis of $PH_3(PD_3)$ at 147 nm [15]. For the reaction $2PH_2 \rightarrow P_2 + 2H_2$ [14, 25] a ΔH value of −26.2 kcal/mol was reported [25].

The **overall** rate constant k was measured at 298 K by following the PH_2 absorption from pulsed dye laser radiation at 455 nm during the flash photolysis of PH_3 (0.3 to 0.5 Torr) in the presence of N_2 (3 to 610 Torr). k increased by a factor of ~8 when the N_2 pressure was increased from 3 to 610 Torr [13]. A few explicit data for k were later given [27]:

PH_3 in Torr	0.30	0.31	0.31	0.29	0.30	0.48	0.30	0.65
N_2 in Torr	3.3	8.2	25	50	91	175	410	650
$10^{11} \cdot k$	0.84	1.67	1.77	2.55	3.34	4.31	5.57	6.0

Extrapolation of k to an infinite pressure yielded $k^{\infty} = (7 \pm 3) \times 10^{-11}$ [13].

The **disproportionation** rate constant, which is pressure-independent, was estimated to be $k_{dis} < 0.15\ k_{rec}$ from the eightfold increase found for the overall rate constant $k_{dis} + k_{rec}$ [13]. $k_{dis}/k_{rec} \approx 0$ was obtained by extrapolating the mole ratio $4n(P_4)/n(P_2H_4)$ measured for PH_3 photolysis to zero reaction time, where it becomes k_{dis}/k_{rec} [11]. For critical remarks concerning the P_4, see, however [26].

The **recombination** rate constant was derived from the k values under the two assumptions $k_{dis} = 0$ (i.e., $k_{rec} = k$) and $k_{dis}/k^{\infty} = 0.05$, and was plotted against the pressure p; see original papers [13, 27]. Recombination was confirmed to be the main reaction channel since equal yields of H_2 (manometrically) and of P_2H_4 (by UV absorption at 230 nm) were obtained [13]. Two numerical values apparently based on [13] were reported as $k_{rec} = (1.0 \pm 0.4) \times 10^{-11}$ at ~2 Torr and $(5 \pm 2) \times 10^{-11}$ at ~270 Torr [27]. $k_{rec} = (5.4 \pm 2.4) \times 10^9\ L \cdot mol^{-1} \cdot s^{-1}$ ($\triangleq 0.90 \times 10^{-11}\ cm^3 \cdot molecule^{-1} \cdot s^{-1}$) was directly measured by following the P_2H_4 absorption during flash photolysis of a PH_3/H_2 mixture (1 Torr : 200 Torr) at room temperature [11, 26].

The third-order rate constant k°_{rec} (i.e., the low-pressure limit of $k_{rec}/[M]$; in $cm^6 \cdot molecule^{-2} \cdot s^{-1}$) was obtained from the plots of k_{rec} against p. Thus, $k^{\circ}_{rec} > 7 \times 10^{-29}$ for $k_{dis} = 0$ [13, 27] and $k^{\circ}_{rec} > 4 \times 10^{-29}$ for $k_{dis} = 0.05 \cdot k^{\infty}$ [27]. A numerical value for $k^2_w/k^{\circ}_{rec} \cdot [M]$, where k_w = rate constant for dissociation of PH_2 at the wall, was obtained from the photolysis of PH_3 at 147 nm [15].

The RRKM theory of the activated complex was used to calculate the p dependence of k_{rec} and its low- and high-pressure limits. k°_{rec}(298 K) = 18×10^{-29} and $21 \times 10^{-29}\ cm^6 \cdot molecule^{-2} \cdot s^{-1}$ followed from two models for the activated complex using a modification [32] of RRKM. The modification [33] yielded k^{∞}_{rec}(298 K) = $6.7 \times 10^{-11}\ cm^3 \cdot molecule^{-1} \cdot s^{-1}$ [27]; for similar values, see [13]. $k^{\infty}_{rec} = 2.8 \times 10^{-11} \exp(-30/T)$ [20] (yielding k^{∞}_{rec}(298 K) = 2.5×10^{-11}) was based on another method [21].

$PH_2 + P_2H_3 \rightarrow PH_3 + P_2H_2$. The reaction was assumed to take part in the conversion of P_2H_4 to red phosphorus [11, 31].

$PH_2 + P_2H_4 \rightarrow PH_3 + P_2H_3$. This reaction was proposed for the conversion of P_2H_4, formed in the flash photolysis of PH_3/O_2 mixtures, into red phosphorus [11].

$PH_2 + CH_3 + M \rightarrow CH_3PH_2 + M$. The formation of CH_3PH_2 was mass spectrometrically observed [34]. A high-pressure recombination rate constant $k^{\infty}_{rec} = 1.2 \times 10^{-10} \exp(-37/T)$ was calculated with the RRKM theory [20] using a method taken from [21].

$PH_2 + SiH_3 + M \rightarrow SiH_3PH_2 + M$. $k^{\circ}_{rec} \cdot [M] < 10^{-10}$ $cm^3 \cdot molecule^{-1} \cdot s^{-1}$ was estimated on the basis of collision theory and of comparisons with similar systems. A numerical value for $k^{\circ}_{rec} \cdot [M]/k_w$ was obtained (for k_w, see p. 91) from quantum yields measured for the 147-nm photolysis of PH_3/SiH_4 mixtures [35].

References:

[1] Hay, P. J.; Dunning, T. J., Jr. (J. Chem. Phys. **64** [1976] 5077/87).
[2] Gordon, M. S.; Gano, D. R.; Boatz, J. A. (J. Am. Chem. Soc. **105** [1983] 5771/5).
[3] Zakhar'in, V. I.; Nadtochenko, V. A.; Sarkisov, O. M.; Teitel'boim, M. A. (Khim. Fiz. **1982** No. 8, pp. 1068/74; C.A. **101** [1984] No. 101094).
[4] Zakhar'in, V. I.; Nadtochenko, V. A.; Sarkisov, O. M.; Teitel'boim, M. A. (Oxid. Commun. **4** [1983] 443/9).
[5] Baugh, D.; Koplitz, B.; Xu, Z.; Wittig, C. (J. Chem. Phys. **88** [1988] 879/87).
[6] Norrish, R. G. W.; Oldershaw, G. A. (Proc. R. Soc. [London] A **262** [1961] 10/8).
[7] Aleksandrov, E. N.; Arutyunov, V. S.; Dubrovina, I. V.; Kozlov, S. N. (Fiz. Goreniya Vzryva **18** No. 4 [1982] 73/8; Combust. Explos. Shock Waves [Engl. Transl.] **18** [1982] 451/5).
[8] Azatyan, V. V.; Gagarin, S. G.; Zakhar'in, V. I.; Kalkanov, V. A.; Kolbanovskii, Yu. A. (Kinet. Katal. **26** [1985] 222/6; Kinet. Catal. [Engl. Transl.] **26** [1985] 191/4).
[9] Fritz, B.; Lorenz, K.; Steinert, W.; Zellner, R. (EUR-7624 [1982] 192/202, 197/9; C.A. **96** [1982] No. 223993).
[10] Fraser, M. E.; Stedman, D. H. (J. Chem. Soc. Faraday Trans. I **79** [1983] 527/42, 540).

[11] Ferris, J. P.; Benson, R. (J. Am. Chem. Soc. **103** [1981] 1922/7).
[12] Aleksandrov, E. N.; Arutyunov, V. S.; Dubrovina, I. V.; Kozlov, S. N. (Izv. Akad. Nauk SSSR Ser. Khim. **31** [1982] 22/5; Bull. Acad. Sci. USSR Div. Chem. Sci. [Engl. Transl.] **31** [1982] 15/8).
[13] Zakhar'in, V. I.; Nadtochenko, V. A.; Sarkisov, O. M.; Teitel'boim, M. A. (Dokl. Akad. Nauk SSSR **263** [1982] 127/30; Dokl. Phys. Chem. [Engl. Transl.] **262/267** [1982] 168/70).
[14] Norrish, R. G. W.; Oldershaw, G. A. (Proc. R. Soc. [London] A **262** [1961] 1/9).
[15] Blazejowski, J.; Lampe, F. W. (J. Phys. Chem. **85** [1981] 1856/64).
[16] Ferris, J. P.; Khwaja, H. (Icarus **62** [1985] 415/24).
[17] Lee, J. H.; Michael, J. V.; Payne, W. A.; Whytock, D. A.; Stief, L. J. (J. Chem. Phys. **65** [1976] 3280/3).
[18] Harris, D. G.; Chou, M. S.; Cool, T. A. (J. Chem. Phys. **82** [1985] 3502/15, 3512).
[19] Gilliom, R. D. (J. Am. Chem. Soc. **99** [1977] 8399/402).
[20] Kaye, J. A.; Strobel, D. F. (Geophys. Res. Lett. **10** [1983] 957/60).

[21] Troe, J. (J. Chem. Phys. **75** [1981] 226/37).
[22] Hamilton, P. A.; Murrells, T. P. (J. Chem. Soc. Faraday Trans. II **81** [1985] 1531/41, 1537/8).
[23] Davies, P. B.; Thrush, B. A. (Proc. R. Soc. [London] A **302** [1968] 243/52).
[24] Clyne, M. A. A.; Heaven, M. C. (Chem. Phys. **58** [1981] 145/50).
[25] Wiles, D. M.; Winkler, C. A. (J. Phys. Chem. **61** [1957] 902/3).
[26] Ferris, J. P.; Bossard, A.; Khwaja, H. (J. Am. Chem. Soc. **106** [1984] 318/24).
[27] Zakhar'in, V. I.; Nadtochenko, V. A.; Teitel'boim, M. A. (Khim. Fiz. **1983** No. 1, pp. 61/5; C.A. **101** [1984] No. 98481).
[28] Buchanan, J. W.; Hanrahan, R. J. (Radiat. Res. **42** [1970] 244/54, 250/2).
[29] Buchanan, J. W.; Hanrahan, R. J. (Radiat. Res. **44** [1970] 296/304).
[30] Prinn, R. G.; Lewis, J. S. (Science **190** [1975] 274/6).

[31] Ferris, J. P.; Benson, R. (Nature **285** [1980] 156/7).
[32] Troe, J. (J. Chem. Phys. **66** [1977] 4758/75).
[33] Quack, M.; Troe, J. (Ber. Bunsenges. Phys. Chem. **81** [1977] 329/37).
[34] McAllister, T.; Lossing, F. P. (J. Phys. Chem. **73** [1969] 2996/8).
[35] Blazejowski, J.; Lampe, F. W. (J. Photochem. **16** [1981] 105/20).

1.2.2 PH_2^+

CAS Registry Numbers: PH_2^+ *[12339-26-3]* Phosphorus(1+), dihydro-; PD_2^+ *[65756-21-0]* Phosphoniumyl-d_2

1.2.2.1 Formation and Detection

From PH_2

The PH_2^+ ion was mass spectrometrically detected when PH_2 was photoionized in the wavelength range λ=1270 to 1100 Å (9.7 to 11.3 eV; PH_2 from $C_6H_5CH_2PH_2$ pyrolysis) [1] and down to λ=940 Å (13.2 eV; PH_2 from PH_3+H) [2]. A threshold at ~1261.5 Å (9.824 eV) in the plots of relative photoion yields against λ corresponds to the formation of PH_2^+ in its $\tilde{X}\ ^1A_1$ ground state [1, 2]. A step-like structure at ~1178 Å (10.525 eV) was assigned to direct ionization forming the excited state $\tilde{a}\ ^3B_1$ of PH_2^+. Broad autoionizing features were observed between 1259 and 1113 Å, which probably are Rydberg states of PH_2 (see p. 57) forming the excited PH_2^+ states $\tilde{a}\ ^3B_1$ and $\tilde{A}\ ^1B_1$ [2]. Photoelectron (PE) spectroscopic work using HeI radiation (21.21 eV) on PH_2 (from the PH_3+F reaction) showed a sharp band at 9.84 eV and a broad band at ~11.0 to 11.5 eV, which were assigned to formation of the PH_2^+ ion in its $\tilde{X}\ ^1A_1$ and $\tilde{a}\ ^3B_1$ states [3]. The PH_2^+ ion was also obtained by electron bombardment of PH_2 (from $C_6H_5CH_2PH_2$ pyrolysis) and mass spectrometrically detected. The ionization efficiency curve between 9.4 and ~11 eV showed a threshold at 9.83 eV [4].

From PH_3

Photon, electron, and ion impact on PH_3 and electron impact on PD_3 were used.

Photoionization mass spectrometry (PIMS) led to a PH_2^+ appearance potential of 13.40±0.02 eV at 0 K (a measured value of 13.36±0.02 eV was corrected to account for the internal thermal energy of PH_3) [1].

Dipole (e, e+ion) coincidence spectroscopy was used to simulate PIMS. PH_2^+ ions from **high-energy electron** impact (8 keV) were mass spectrometrically measured in coincidence with the forward scattered electrons as a function of energy loss (equivalent photon energy range up to 50 eV). The appearance potential of 13.0±1 eV and the smoothness of the oscillator strength curve for PH_2^+ pointed to formation from PH_3 ionized in the 2e MO [5]. An earlier PE spectrum of PH_3, taken with NeI radiation (16.8 eV), had been similarly interpreted: A vibrational structure in the second PE band, corresponding to 2e ionization, disappeared at an ionization energy well correlated with the PH_2^+ appearance potential obtained by conventional mass spectrometric work; this points to a decay PH_3^+ ($2e^{-1}$)→PH_2^++H [6]. Dipole (e, e+ion) spectroscopy also revealed the formation of PH_2^+ at an equivalent photon energy in the region of excitation and ionization of P2p and P2s electrons (129.5 to 218.9 eV) [7].

The formation of PH_2^+ by **low-energy electron** impact with mass spectrometric detection is dealt with on p. 211 ("Mass Spectrum of PH_3"). Appearance potentials were measured (a way of obtaining the PH_2^+ formation enthalpy) and varied between 13.2 [8, 9] and 14.4 eV [10].

(Another "high" value of 14.4 eV [11] was corrected into 13.4 eV [12].) See also the tables of appearance potentials in [5, 13, 14]. A reliable value seems to be 13.47±0.05 eV [4]. An ionization efficiency curve for PH_2^+ over a larger energy range (13 to 37 eV, with an appearance potential of 13.4 eV and a flat maximum at ~18 eV) is given in [15]. Absolute partial ionization cross sections for PH_2^+ and PD_2^+ (from a PH_3/PD_3 mixture) were measured up to 180 eV giving appearance potentials of 14.2±0.2 and 14.3±0.2 eV [13].

Ion cyclotron resonance (ICR) has also been used as a detection method. Single-resonance spectra of PH_3 bombarded with 23 eV [16] or 70 eV [17] electrons showed the PH_2^+ ion. It appeared at 18 eV [17].

Thermal energy charge transfer reactions with several **cations** (Ar^+, CO^+, CO_2^+, Kr^+) yielding PH_2^+ were observed with the double resonance technique. Product distribution and rate constants (not for Kr^+) were measured (see p. 217) [18]. PH_2^+ from the ion-molecule reaction $PH_3 + PH^+ \rightarrow PH_2^+ + PH_2$ had been earlier observed by ICR [16].

From Other Sources

Electron impact on P_2H_4 or SiH_3PH_2 yielded the PH_2^+ ion. Its appearance potential from P_2H_4 is 15.3±0.5 eV and points to the formation route $P_2H_4 \rightarrow PH_2^+ + PH + H + e^-$ (with no) or $\rightarrow PH_2^+ + P + H_2 + e^-$ (with some excess energy) [9]. An older value, 12.5 eV [12], is probably due to P_2H_4 decomposition [9]. The appearance potential from SiH_3PH_2 is 13.1 eV [12].

Another ion-molecule reaction leading to PH_2^+ is $H_2 + P^+ + M \rightarrow PH_2^+ + M$. A ternary rate constant of 7.5×10^{-30} $cm^6 \cdot molecule^{-2} \cdot s^{-1}$ was measured at 300 K for M = He (0.48 Torr) by the selected ion flow tube (SIFT; see [19]) method [20]. A rate constant of 4.1×10^{-29} at 80 K points to a temperature dependence of $T^{-1.3}$ [21].

The decomposition of doubly charged cations leading to the PH_2^+ ion was theoretically treated. Activation barriers and reaction enthalpies were calculated by ab initio MO methods at the MP3 level for $CH_2PH_3^{2+}$ [22], at the MP4 level for $HCPH_2^{2+}$ [23], and at the multireference CI level for PH_3^{2+} [24].

References:

[1] Berkowitz, J.; Curtiss, L. A.; Gibson, S. T.; Greene, J. P.; Hillhouse, G. L.; Pople, J. A. (J. Chem. Phys. **84** [1986] 375/84).
[2] Berkowitz, J.; Cho, H. (J. Chem. Phys. **90** [1989] 1/6).
[3] Dyke, J. M.; Jonathan, N.; Morris, A. (Int. Rev. Phys. Chem. **2** [1982] 3/42, 24).
[4] McAllister, T.; Lossing, F. P. (J. Phys. Chem. **73** [1969] 2996/8).
[5] Zarate, E. B.; Cooper, G.; Brion, C. E. (Chem. Phys. **148** [1990] 277/88, 285).
[6] Potts, A. W.; Price, W. C. (Proc. R. Soc. [London] A **326** [1971/72] 181/97, 193).
[7] Zarate, E. B.; Cooper, G.; Brion, C. E. (Chem. Phys. **148** [1990] 289/97).
[8] Wada, Y.; Kiser, R. W. (Inorg. Chem. **3** [1964] 174/7).
[9] Fehlner, T. P.; Callen, R. B. (Adv. Chem. Ser. **72** [1968] 181/90, 184/6).
[10] Fischler, J.; Halmann, M. (J. Chem. Soc. **1964** 31/6).

[11] Saalfeld, F. E.; Svec, H. J. (Inorg. Chem. **2** [1963] 46/50).
[12] Saalfeld, F. E.; Svec, H. J. (Inorg. Chem. **3** [1964] 1442/3).
[13] Märk, T. D.; Egger, F. (J. Chem. Phys. **67** [1977] 2629/35).
[14] Franklin, J. L.; Dillard, J. G.; Rosenstock, H. M.; Herron, J. T.; Draxl, K.; Field, F. H. (NSRDS-NBS-26 [1969] 167).
[15] Morrison, J. D.; Traeger, J. C. (Int. J. Mass Spectrom. **11** [1973] 277/88, 283).
[16] Eyler, J. R. (Inorg. Chem. **9** [1970] 981/2).

[17] Holtz, D.; Beauchamp, J. L.; Eyler, J. R. (J. Am. Chem. Soc. **92** [1970] 7045/55, 7047).
[18] Chau, M.; Bowers, M. T. (Chem. Phys. Lett. **44** [1976] 490/4).
[19] Smith, D.; Adams, N. G. (in: Bowers, M. T.; Gas Phase Ion Chemistry, Vol. 1, Academic, New York 1979, pp. 1/44; Adv. At. Mol. Phys. **24** [1988] 1/49, 3/4).
[20] Smith, D.; McIntosh, B. J.; Adams, N. G. (J. Chem. Phys. **90** [1989] 6213/9).

[21] Adams, N. G.; McIntosh, B. J.; Smith, D. (Astron. Astrophys. **232** [1990] 443/6).
[22] Yates, B. F.; Bouma, W. J.; Radom, L. (J. Am. Chem. Soc. **108** [1986] 6545/54).
[23] Wong, M. W.; Yates, B. F.; Nobes, R. H.; Radom, L. (J. Am. Chem. Soc. **109** [1987] 3181/7).
[24] Pope, S. A.; Hillier, I. H.; Guest, M. F.; Kendric, J. (Chem. Phys. Lett. **95** [1983] 247/9).

1.2.2.2 Point Group. Electronic States

The **ground state** ... $4a_1^2\ 2b_2^2\ 5a_1^2$, $\tilde{X}\ ^1A_1$ with an approximate bond angle of 90° (point group C_{2v}) was inferred from the photoion yield curve of PH_2^+ from PH_2. The threshold region showed a relatively sharp onset and is similar to the photoelectron (PE) spectrum of NH_2 in the region of its first excited state (a $^1A_1$ state). The electron configuration results from the removal of the unpaired $2b_1$ electron of the PH_2 radical [1 to 3]. A sharp band in the PE spectrum of PH_2 (from PH_3+F) was earlier ascribed to the $\tilde{X}\ ^1A_1$ state of PH_2^+ [4]. This state was also assumed to be involved in an electronic spectrum (of PH_2^+), which was obtained by detection of fragment ions arising from predissociation of the laser-excited upper state. States were assigned analogous to those of SiH_2 [5].

The **first excited state** ... $4a_1^2\ 2b_2^2\ 5a_1\ 2b_1$, $\tilde{a}\ ^3B_1$ with an approximate bond angle of 120° was inferred from the photoion yield curve mentioned above [1, 3]. The term value T_0=0.70 eV (referred to the ground state $\tilde{X}\ ^1A_1$ of the ion) follows from the adiabatic ionization potentials of PH_2 to the ionic states $\tilde{a}\ ^3B_1$ and $\tilde{X}\ ^1A_1$ [3]. An earlier estimate is $T_0 \geqq 0.71$ eV [1]. Both values were later reported in two compilations of energy levels of polyatomic transient molecules to be $T_0 \approx 5650\ cm^{-1}$ [6] and $\geqq 5730\ cm^{-1}$ [7]. The assignment of the second, broad band in the PE spectrum of PH_2 at an ionization energy between 11.0 and 11.5 eV to $\tilde{a}\ ^3B_1$ [4] seems to be plausible, since the vibrational structure extracted from the spectrum agrees with a scaled theoretical bending frequency (see below) [3]. The lowest observed [4] vibrational member is at an ionization energy of 11.08 eV and would thus imply T_0=1.24 eV. It is possible, however, that still lower vibrational members exist, which are masked by PE bands due to P, P_2, and PH_3 [3]. T_e=0.87 eV is based [3] on ionization potentials of PH_2 which were calculated by an ab initio MO method using MP4 perturbation theory and including a modification to give singlet-triplet separations [8]. T_e=0.59 eV from multi-reference double-excitation (MRD) CI calculations [9] is not in agreement with the respective bending potential energy curves given there too [9]; see Fig. 7, p. 96, from which T_e=0.75 eV may be read [3].

The related **excited singlet state** ... $4a_1^2\ 2b_2^2\ 5a_1\ 2b_1$, $\tilde{A}\ ^1B_1$ with an approximate bond angle of 120° may be found at T_0=1.92 eV. This term value was derived from the convergence limit (11.74 eV, referred to the PH_2 ground state) of a Rydberg series of PH_2. Such a series, comprising also vibrationally excited levels, was inferred from the autoionization features of the photoion yield curve of PH_2^+ from PH_2 (for a detailed discussion, see the original paper) [3]. The state is probably involved in the first electronic spectrum of PH_2^+, observed between 420 and 670 nm monitoring the P^+ ions from predissociation of the laser-excited upper state. The spectrum is analogous to that of SiH_2 and is assigned to $\tilde{A}\ ^1B_1 \leftarrow \tilde{X}\ ^1A_1$. Another predissociation spectrum, which was observed via PH^+ ions between 590 and 640 nm, is probably associated with Ã $^1B_1$ as lower state [5]. Term values from theoretical calculations (in parentheses) were T_e=2.05 (MRD-CI) [9], 2.1 (CI), and 2.6 eV (GVB) [10].

Another excited singlet ...$4a_1^2\ 2b_2^2\ 2b_1^2$, $\tilde{B}$ $^1A_1$ (for the electron configuration, see correlation scheme below) was presumed in a predissociation spectrum between 590 and 640 nm, which was assigned to $\tilde{B}$ $^1A_1 \leftarrow$ Ã $^1B_1$ [5].

A **correlation scheme** for the low-lying electronic states of the linear and bent conformations of the ion with approximate bond angles α is given below [9]; for a few states, see also [11]. The energetic order applies to the linear case:

linear PH_2^+		bent PH_2^+					α
$3\sigma_g^2\ \ 2\pi_u^4$	$^1\Sigma_g^+$	$4a_1^2$		$5a_1^2$	$2b_1^2$	$^1A_1$	30°
$3\sigma_g^2\ 2\sigma_u\ 2\pi_u^3$	$^{1,3}\Pi_g$	$4a_1^2$	$2b_2$	$5a_1$	$2b_1^2$	$^{1,3}B_2$	90°
		$4a_1^2$	$2b_2$	$5a_1^2$	$2b_1$	$^{1,3}A_2$	30°
$3\sigma_g^2\ 2\sigma_u^2\ 2\pi_u^2$	$^1\Sigma_g^+$	$4a_1^2$	$2b_2^2$		$2b_1^2$	$\tilde{B}$ $^1A_1$	180°
	$^1\Delta_g$	$4a_1^2$	$2b_2^2$	$5a_1$	$2b_1$	Ã $^1B_1$	120°
		$4a_1^2$	$2b_2^2$	$5a_1^2$		$\tilde{X}$ $^1A_1$	90°
	$\tilde{X}$ $^3\Sigma_g^-$	$4a_1^2$	$2b_2^2$	$5a_1$	$2b_1$	ã $^3B_1$	120°

Bending **potential energy curves** for eight of these nine states of bent PH_2^+ were calculated with an MRD-CI method assuming a P–H distance of 1.40 Å (an optimized distance was also used for $^{1,3}A_2$) and are shown in **Fig. 7** due to [9], see also [12]. The respective barriers to linearity of the states $\tilde{X}$ $^1A_1$ and ã $^3B_1$ are 2.83 and 0.84 eV [9].

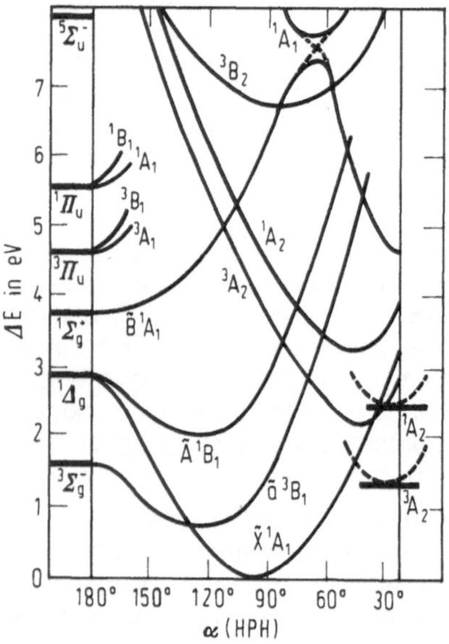

Fig. 7. Calculated MRD-CI bending potential energy curves for PH_2^+.

Total energies were calculated with several ab initio methods (characterized in parentheses) for $\tilde{X}$ $^1A_1$, ã $^3B_1$, and Ã $^1B_1$ (CI and GVB) [10], for $\tilde{X}$ $^1A_1$ and ã $^3B_1$ (MP3 [13] and MP4 [1, 14, 15]), and for $\tilde{X}$ $^1A_1$ (MP4) [16].

Charge distributions were derived from ab initio GVB wave functions for the lowest two states $\tilde{X}\ ^1A_1$ and $\tilde{a}\ ^3B_1$. The positive charge is essentially located at the heavy atom (contrary to the NH_2^+ ion) [10]. Atomic charges, the dipole moment, and the magnetic hyperfine coupling constants were also calculated with an MO-SCF method (for a $^3B_1$ ground state, however) [17].

An **ionization potential** of 19.2 eV was obtained from MR-CI calculations [11, 18]. For orbital energies (pointing to a $^3B_1$ ground state), see [17].

References:

[1] Berkowitz, J.; Curtiss, L. A.; Gibson, S. T.; Greene, J. P.; Hillhouse, G. L.; Pople, J. A. (J. Chem. Phys. **84** [1986] 375/84).
[2] Berkowitz, J. (J. Chem. Phys. **89** [1988] 7065/76, 7066).
[3] Berkowitz, J.; Cho, H. (J. Chem. Phys. **90** [1989] 1/6).
[4] Dyke, J. M.; Jonathan, N.; Morris, A. (Int. Rev. Phys. Chem. **2** [1982] 3/42, 22, 24).
[5] Edwards, C. P.; Jackson, P. A.; MacLean, C. S.; Sarre, P. J. (Bull. Soc. Chim. Belg. **92** [1983] 605/6).
[6] Jacox, M. E. (J. Phys. Chem. Ref. Data **19** [1990] 1387/546, 1396).
[7] Jacox, M. E. (J. Phys. Chem. Ref. Data **17** [1988] 269/511, 282).
[8] Pople, J. A.; Curtiss, L. A. (J. Phys. Chem. **91** [1987] 3637/9).
[9] Bruna, P. J.; Hirsch, G.; Buenker, R. J.; Peyerimhoff, S. D. (NATO ASI Ser. B **90** [1983] 309/54, 315, 319, 321).
[10] Harrison, J. F. (J. Am. Chem. Soc. **103** [1981] 7406/13).

[11] Pope, S. A.; Hillier, I. H.; Guest, M. F. (Faraday Symp. Chem. Soc. No. 19 [1984] 109/23, 112, 119).
[12] Bruna, P. J.; Peyerimhoff, S. D. (Bull. Soc. Chim. Belg. **92** [1983] 525/46, 536/7, 539).
[13] Yates, B. F.; Bouma, W. J.; Radom, L. (J. Am. Chem. Soc. **108** [1986] 6545/54).
[14] Pople, J. A.; Curtiss, L. A. (J. Phys. Chem. **91** [1987] 155/62).
[15] MacLagan, R. G. A. R. (Chem. Phys. Lett. **163** [1989] 349/53).
[16] Nguyen, M. T.; Hegarty, A. F.; McGinn, M. A.; Ruelle, P. (J. Chem. Soc. Perkin Trans. II **1985** 1991/7).
[17] Ball, J. R.; Thomson, C. (Int. J. Quantum Chem. **14** [1978] 39/53, 42, 47/8).
[18] Pope, S. A.; Hillier, I. H.; Guest, M. F.; Kendric, J. (Chem. Phys. Lett. **95** [1983] 247/9).

1.2.2.3 Geometrical Structure. Vibrations. Dissociation

The ground-state **geometry** is not very different from that of PH_2 in its ground state (see p. 72), as may be inferred from the threshold region of the photoion yield curve of PH_2^+ from PH_2 [1]. Rough bond angles α of several states are given in the correlation scheme on p. 96. For calculated geometrical parameters, see below.

The ground state **bending frequency** $\nu_2(A_1)$ may be ~1200 cm^{-1}, a characteristic separation between different bands in the predissociation spectrum of the $\tilde{A}\ ^1B_1 \leftarrow \tilde{X}\ ^1A_1$ transition (see above) [2]. The excited triplet state $\tilde{a}\ ^3B_1$ may show a bending frequency of ~900 cm^{-1} (derived from features in the photoion yield curve just mentioned) [1] or of ~925 cm^{-1} (inferred [1] from the PH_2 photoelectron spectrum given in [3]). These two values are in reasonable agreement with a calculated frequency of 1071 cm^{-1} [4, 5], if a scaling factor of 0.89 is applied [1] (giving $\nu_2 = 953\ cm^{-1}$).

Theoretical geometrical and vibrational data from HF calculations [1, 5] are given below (internuclear distance r in Å; fundamental vibrations ν_i in cm^{-1}; almost identical values are in [6]):

state	r	α	$\nu_1(A_1)$	$\nu_2(A_1)$	$\nu_3(B_2)$
ã $^3B_1$	1.387	121.4°	2680	1071	2755
$\tilde{X}$ $^1A_1$	1.400	93.4°*)	2668	1274	2682

*) 94.4° in [5 to 7].

Geometrical parameters for both states were also calculated (HF and complete active space (CAS) SCF) in [7], for both states and $\tilde{A}$ $^1B_1$ (GVB) in [8]. A bond angle for $\tilde{X}$ $^1A_1$ was calculated (CNDO/2) in [9]. Geometrical parameters and force constants from earlier MO-SCF calculations [10] are for a $^3B_1$ state (supposedly the ground state).

Dissociation. Electronic states of the dissociation products $P^+ + H_2$, $PH^+ + H$, and $PH + H^+$ have been correlated with states of PH_2^+ in [11] (in a figure of bending potential energy curves).

The dissociation path $PH_2^+ \rightarrow P^+ + H_2$ was observed at the injection of PH_2^+ into He [12].

A bond dissociation energy of D=106 kcal/mol (≙4.6 eV) for $PH_2^+ \rightarrow PH^+ + H$ was derived from the difference between the appearance potentials of PH^+ and PH_2^+ from PH_3 (0.1 eV) and the H_2 bond dissociation energy (4.5 eV) [13]. A value of D=2.0 eV is based on deviating appearance potentials [14].

D=160.2 kcal/mol (≙6.95 eV) for $PH_2^+ \rightarrow PH + H^+$ was derived from the formation enthalpies of PH_2^+ and PH (taken from [4]) and the ionization potential of the H atom [12]. D=243 kcal/mol (≙10.55 eV) for $PH_2^+ \rightarrow PH(A\ ^3\Pi) + H^+$ is tabulated in [15].

References:

[1] Berkowitz, J.; Cho, H. (J. Chem. Phys. **90** [1989] 1/6).
[2] Edwards, C. P.; Jackson, P. A.; MacLean, C. S.; Sarre, P. J. (Bull. Soc. Chim. Belg. **92** [1983] 605/6).
[3] Dyke, J. M.; Jonathan, N.; Morris, A. (Int. Rev. Phys. Chem. **2** [1982] 3/42, 22, 24).
[4] Berkowitz, J.; Curtiss, L. A.; Gibson, S. T.; Greene, J. P.; Hillhouse, G. L.; Pople, J. A. (J. Chem. Phys. **84** [1986] 375/84).
[5] Pople, J. A.; Curtiss, L. A. (J. Phys. Chem. **91** [1987] 155/62).
[6] MacLagan, R. G. A. R. (Chem. Phys. Lett. **163** [1989] 349/53).
[7] Pope, S. A.; Hillier, I. H.; Guest, M. F. (Faraday Symp. Chem. Soc. No. 19 [1984] 109/23, 114).
[8] Harrison, J. F. (J. Am. Chem. Soc. **103** [1981] 7406/13).
[9] Takahata, Y. (Chem. Phys. Lett. **59** [1978] 472/7).
[10] Ball, J. R.; Thomson, C. (Int. J. Quantum Chem. **14** [1978] 39/53, 44, 47).

[11] Bruna, P. J.; Hirsch, G.; Buenker, R. J.; Peyerimhoff, S. D. (NATO ASI Ser. B **90** [1983] 309/54, 321).
[12] Smith, D.; McIntosh, B. J.; Adams, N. G. (J. Chem. Phys. **90** [1989] 6213/9).
[13] Wada, Y.; Kiser, R. W. (Inorg. Chem. **3** [1964] 174/7).
[14] Saalfeld, F. E.; Svec, H. J. (Inorg. Chem. **2** [1963] 46/50).
[15] Nguyen Xuan, C.; Tamanini, M.; Di Stefano, G.; Margani, A. (Gazz. Chim. Ital. **116** [1986] 243/53, 252).

1.2.2.4 Heat of Formation

Formation enthalpy Δ_fH in kcal/mol

$\Delta_fH_0^\circ = 260.6 \pm 0.6$ was derived from the appearance potential $AP(PH_2^+/PH_3) = 13.40 \pm 0.02$ eV (see above) and from a PH_3 formation enthalpy of 3.2 kcal/mol at 0 K [1]. An "experimental" value of $\Delta_fH_{298}^\circ = 261.5$ was apparently obtained by adding a difference of 0.9 kcal/mol between 298 and 0 K (taken from theoretical calculations) to $\Delta_fH_0^\circ$ [2]. $\Delta_fH_{298}^\circ = 259.8 \pm 2$ is based on $AP = 13.47 \pm 0.05$ eV and a PH_3 formation enthalpy of 1.3 kcal/mol at 298 K [3]. Two lower values, 253.6 [4] and 254 kcal/mol [5, 6], were derived from AP = 13.2 eV. Compare also a table of AP's and related $\Delta_fH_{298}^\circ$ values in [14].

Theoretically calculated formation enthalpies from ab initio MO methods, including the Møller-Plesset perturbation theory up to the fourth order, are in good agreement with experimental data [1, 2]. A modified procedure (to give singlet-triplet separations) yielded $\Delta_fH_0^\circ = 260.9$ and $\Delta_fH_{298}^\circ = 261.8$ [7].

References:

[1] Berkowitz, J.; Curtiss, L. A.; Gibson, S. T.; Greene, J. P.; Hillhouse, G. L.; Pople, J. A. (J. Chem. Phys. **84** [1986] 375/84).
[2] Pople, J. A.; Curtiss, L. A. (J. Phys. Chem. **91** [1987] 155/62).
[3] McAllister, T.; Lossing, F. P. (J. Phys. Chem. **73** [1969] 2996/8).
[4] Saalfeld, F. E.; Svec, H. J. (Inorg. Chem. **3** [1964] 1442/3).
[5] Wada, Y.; Kiser, R. W. (Inorg. Chem. **3** [1964] 174/7).
[6] Fehlner, T. P.; Callen, R. B. (Adv. Chem. Ser. **72** [1968] 181/90, 184/6).
[7] Pople, J. A.; Curtiss, L. A. (J. Phys. Chem. **91** [1987] 3637/9).

1.2.2.5 Chemical Reactions

Reactions of PH_2^+ with altogether sixteen molecular species have been investigated with the selected ion flow tube (SIFT; see [1]) method. The results are given first (in a table) [2]. Reactions with four of these molecules were also studied by ion cyclotron resonance (ICR). The latter results are reported in a separate section.

The table below lists the ion product distributions and reaction rate constants k at 300 K obtained from SIFT measurements. The reactants are in the order of decreasing proton affinity A_p (the proton detachment energy via $PH_2^+ \rightarrow PH + H^+$ amounts to 160.2 kcal/mol; see above). The ionization potentials E_i of the reactants are also given (the recombination energy via $PH_2^+ + e^- \rightarrow PH_2$ amounts to 9.82 eV; see p. 61). k is given in $cm^3 \cdot molecule^{-1} \cdot s^{-1}$; for ternary association reactions, it represents an effective binary rate constant at an He pressure of 0.48 Torr [2]:

reactant	A_p in kcal/mol	E_i in eV	ionic products (%)	$10^9 \cdot k$
CH_3NH_2 [a)]	214.1	8.97	$CH_2NH_2^+$ (62), $CH_3NH_3^+$ (38)	1.7
NH_3 [a), b)]	204.0	10.166	PNH_3^+ (81), NH_4^+ (19)	2.0
PH_3 [a), b)]	188.6	9.98	$P_2H_3^+$ (78), P_2H^+ (17), PH_4^+ (5)	1.1
CH_3CCH	182	10.36	PCH_2^+ (44), $C_2H_5^+$ (43), PCH_4^+ (7), $PC_3H_4^+$ (6)	1.6

reactant	A_p in kcal/mol	E_i in eV	ionic products (%)	$10^9 \cdot k$
CH_3OH	181.9	10.85	CH_2OH^+ (65), PCH_4^+ (8), H_2PO^+ (27)	2.0
HCN [a]	171.4	13.59	H_2CN^+ (72), $HCN \cdot PH_2^+$ (28)	1.4 [c]
H_2S	170.2	10.47	H_2PS^+ (100)	1.5
H_2O [a]	165.1	12.62	H_2PO^+ (67), H_3O^+ (33)	0.49
C_2H_4	162.6	10.51	PCH_2^+ (12), $PC_2H_4^+$ (88)	1.2
C_2H_2	153.3	11.41	$PC_2H_2^+$ (100) [d]	1.4
COS	151	11.184	H_2PS^+ (100)	0.99
CO	141.6	14.013	$CO \cdot PH_2^+$ (100)	0.0029 [e]
CH_4 [b]	132.0	12.62	PCH_4^+ (100)	1.1
CO_2	130.9	13.769	$CO_2 \cdot PH_2^+$ (100)	0.0075 [f]
H_2 [b]	101.3	15.43	PH_4^+ (100)	0.0011 [g]
O_2	100.9	12.059	PO^+ (100)	0.078

[a] Proton transfer from PH_2^+ to the reactant was observed. – [b] For ICR measurements, see below. – [c] Ternary association rate constant $k^* > 3.4 \times 10^{-27}$ $cm^6 \cdot molecule^{-2} \cdot s^{-1}$. – [d] The easy production of $PC_2H_2^+$ suggests a low formation enthalpy and a cyclic (aromatic) structure of $PC_2H_2^+$ as predicted in [3]. – [e] $k^* = 1.9 \times 10^{-28}$. – [f] $k^* = 5.1 \times 10^{-28}$. – [g] $k^* = 6.7 \times 10^{-29}$.

Proton transfer from singlet PH_2^+ ($\tilde{X}\ ^1A_1$) to a singlet state molecule requires spin inversion, if a singlet state ion and $PH(^3\Sigma^-)$ are formed. The upper part of the table shows that proton transfer was observed in only five of nine possible cases [2].

A reaction with N_2 was not detected by the SIFT measurement [2].

The association reactions with CO and H_2 were also measured with the SIFT method at 80 K. The rate constants $k^*(80\ K) = 2.9 \times 10^{-27}$ and 3.6×10^{-28} $cm^6 \cdot molecule^{-2} \cdot s^{-1}$ together with $k^*(300\ K)$ (see footnotes [e] and [g] of the table above) exhibit proportionality of k^* with $T^{-2.1}$ and $T^{-1.3}$ [4].

ICR measurements were performed at room temperature with NH_3, PH_3, CH_4, and H_2. Rate constants k are given in $cm^3 \cdot molecule^{-1} \cdot s^{-1}$, reaction enthalpies ΔH in kcal/mol.

Reaction with NH_3. Ion product distributions were 72% for PNH_3^+ (+H_2) and 28% for NH_4^+ (+PH). $10^9 \cdot k = 1.7 \pm 10\%$. $\Delta H = -49$ for the proton transfer channel (formation of NH_4^+) [5].

Reaction with PH_3. The proton transfer channel (PH_4^+ formation) was not observed [5 to 7]; PH_3^+ was found to be a product in [7]. Yields of $P_2H_3^+$ and P_2H^+ were 30 and 70% [5] or 26 and 74% [6]. $10^9 \cdot k = 0.90 \pm 30\%$ [5] or 0.84 [6]. Reaction enthalpies are not well known: $\Delta H = -30$ [5] or -43 [6] for the product channel $P_2H_3^+ + H_2$, $\Delta H = -4$ [5] or 0 [6] for $P_2H^+ + 2H_2$.

Reaction with CH_4. $10^9 \cdot k = 0.58 \pm 10\%$ for the formation of $PCH_4^+ + H_2$ [5]. The reaction was previously observed in [6].

A reaction with H_2 was not detected; an upper limit of $10^9 \cdot k < 0.01$ was given. For isotope exchange with D_2: $10^9 \cdot k < 0.04$ [5].

The reaction with C_2H_2, forming the cyclic $PC_2H_2^+$ ion (see footnote [d] of the table), was predicted by several theoretical methods to be exothermic [8].

Bombardment of a Cu–Be alloy dynode (an electron multiplier material) by PH_2^+ and PD_2^+ ions caused secondary electron emission. Deuterated ions gave a smaller emission coefficient [9].

References:

[1] Smith, D.; Adams, N. G. (in: Bowers, M. T.; Gas Phase Ion Chemistry, Vol. 1, Academic, New York 1979, pp. 1/44; Adv. At. Mol. Phys. **24** [1988] 1/49, 3).
[2] Smith, D.; McIntosh, B. J.; Adams, N. G. (J. Chem. Phys. **90** [1989] 6213/9).
[3] Bews, J. R.; Glidewell, C. (J. Mol. Struct. **104** [1983] 105/14, 107 [THEOCHEM **13**]).
[4] Adams, N. G.; McIntosh, B. J.; Smith, D. (Astron. Astrophys. **232** [1990] 443/6).
[5] Thorne, L. R.; Anicich, V. G.; Huntress, W. T. (Chem. Phys. Lett. **98** [1983] 162/6).
[6] Holtz, D.; Beauchamp, J. L.; Eyler, J. R. (J. Am. Chem. Soc. **92** [1970] 7045/55).
[7] Eyler, J. R. (Inorg. Chem. **9** [1970] 981/2).
[8] MacLagan, R. G. A. R. (Chem. Phys. Lett. **163** [1989] 349/53).
[9] Märk, T. D. (Proc. 7th Int. Vac. Congr., Vienna 1977, Vol. 2, pp. 1341/4 from C.A. **88** [1978] No. 98031).

1.2.3 PH_2^{2+}

CAS Registry Numbers: PH_2^{2+} *[65756-16-3]* Phosphorus(2+), dihydro-; PD_2^{2+} *[65756-20-9]*

The PH_2^{2+} ion formed by electron impact on PH_3 was mass spectrometrically detected. An ionization efficiency curve between 30 and 50 eV led to appearance potentials of 34.0 eV (square root plot) or 32.7 eV (linear extrapolation) [1]; both values are tabulated in [2]; for the first value, see also [3]. Partial PH_3 ionization cross sections for PH_2^{2+} and PD_3 ionization cross section ratios, PD_2^{2+}/PD_3^+, were obtained for electron energies up to 180 eV and showed appearance potentials of 35.0±0.5 eV for PH_2^{2+} and 34.9±0.5 eV for PD_2^{2+} (square root plots) [4]. The PH_2^{2+} ion was also mass spectrometrically detected after ionization of PH_3 by high-energy electrons (8 keV). Oscillator strengths were obtained by dipole (e,e+ion) spectroscopy simulating photoionization [5].

PH_2^{2+} formation enthalpies of 733 or 703 kcal/mol at 298 K are based on the appearance potentials of 34.0 or 32.7 eV from above [2].

Low-lying electronic states are expected to be ... $4a_1^2\ 2b_2^2\ 5a_1$, $\tilde{X}\ ^2A_1$ with an approximate bond angle α of 120° and ... $4a_1^2\ 2b_2^2\ 2b_1$, $\tilde{A}\ ^2B_1$ with $\alpha \approx 180°$, both correlating with ... $3\sigma_g^2\ 2\sigma_u^2\ 2\pi_u$, $^2\Pi_u$ in the linear configuration [6]. Geometrical parameters from a complete active space (CAS) SCF calculation are given below [6]:

state	r(P–H) in Å	α
$\tilde{A}\ ^2\Pi_u$	1.496	180.0°
$\tilde{X}\ ^2A_1$	1.483	124.8°

A "deprotonation" energy of 48.1 kcal/mol and an activation barrier of 42.8 kcal/mol were calculated with a CI method for the strongly exothermic dissociation reaction $PH_2^{2+} \rightarrow PH^+ + H^+$ [6, 7].

PH_2^{2+} and PD_2^{2+} ions were used to excite secondary electron emission from a Cu–Be target. The emission coefficients are smaller for PD_2^{2+} [8].

References:

[1] Fischler, J.; Halmann, M. (J. Chem. Soc. **1964** 31/6).
[2] Franklin, J. L.; Dillard, J. G.; Rosenstock, H. M.; Herron, J. T.; Draxl, K.; Field, F. H. (NSRDS-NBS-26 [1969] 168).
[3] Halmann, M. (Top. Phosphorus Chem. **4** [1967] 49/84, 71).
[4] Märk, T. D.; Egger, F. (J. Chem. Phys. **67** [1977] 2629/35).
[5] Zarate, E. B.; Cooper, G.; Brion, C. E. (Chem. Phys. **148** [1990] 277/88).
[6] Pope, S. A.; Hillier, I. H.; Guest, M. F. (Faraday Symp. Chem. Soc. No. 19 [1984] 109/23, 112, 114).
[7] Pope, S. A.; Hillier, I. H.; Guest, M. F.; Kendric, J. (Chem. Phys. Lett. **95** [1983] 247/9).
[8] Märk, T. D. (Proc. 7th Int. Vac. Congr., Vienna 1977, Vol. 2, pp. 1341/4 from C.A. **88** [1978] No. 98031).

1.2.4 PH_2^-

CAS Registry Number *[13937-34-3]* Phosphide (H_2P^{1-})

1.2.4.1 Formation and Detection

In the Vapor Phase

The PH_2^- ion was mass spectrometrically detected after resonance electron capture by PH_3 in the electron energy range up to 11 eV. Appearance potentials of 2.3±0.1 and 5.2±0.2 eV were measured [1]. Measurements up to 75 eV confirmed these appearance potentials (at 2.2 and 5.2 eV, both ±0.1 eV) and revealed another resonance capture process at 8 eV. The lowest value was assigned to $PH_3 + e^- \rightarrow PH_2^- + H$, the higher values to unknown excited ionic states. Ion pair formation via $PH_3 + e^- \rightarrow PH_2^- + H^+ + e^-$ was observed at 15.8±0.2 eV [2]. PH_2^- ions, formed at the first resonance electron capture maximum, were also detected by ion cyclotron resonance (ICR) spectroscopy [3, 4]. PD_2^- and PHD^- ions were prepared by resonance capture in PD_3, which was generated by deuteration of PH_3 with acidified D_2O, and detected by ICR (in a 5:1 ratio after reacting for 43 h). The formation of PD_2^- was by a factor of ~10 smaller than that of PH_2^- (under identical conditions) [4]. ICR detection was also applied to PH_2^- ions formed at 6.3 [5], 7.2 (or 7.3), and 9.6 eV and to PD_2^- ions formed at 7 eV [6].

Beams of PH_2^- ions (and others) were produced by burning PH_3 with N_2O in a low-pressure glow discharge chamber. The ions were mass spectrometrically detected [7].

PH_2^- was formed via ion-molecule reactions of the type $PH_3 + M^- \rightarrow PH_2^- + MH$. ICR detection was used for M = OH, NH_2 [6, 8], CH_3O, C_2H_5O, *n*-C_3H_7O [6], and SiH_3 [8]. The flowing afterglow technique was used for M = OH and NH_2 [9]. Intrinsic barriers (see [10, 11]) for the ion-molecule reactions $CH_3PH_2 + M^- \rightarrow PH_2^- + CH_3M$ were calculated for M = H, F, and CCH with the second-order Møller-Plesset perturbation theory [12].

In Liquid Solvents

Ammonia. NH_4Br was added to a solution containing both Na_4P_2 and metallic Na. Formation of PH_2^- according to $P_2^{4-} + 4NH_4^+ + 2e^- \rightarrow 2PH_2^- + 4NH_3$ was detected by its conversion to $CH_3PH_2 + I^-$, when the solution was treated with CH_3I (and also through its reaction with NH_4^+ to $PH_3 + NH_3$) [13]. A stoichiometric amount of PH_3 was distilled into a suspension of $NaNH_2$ in NH_3. Formation of PH_2^- according to $PH_3 + NH_2^- \rightarrow PH_2^- + NH_3$ was indicated by the yellow color of the solution. ^1H NMR spectra were not conclusive (see below). PH_2^- was also formed when

NaOH was used instead of $NaNH_2$ [14]. The reaction of PH_3 with excess $NaNH_2$ was used in later work on ^1H and ^{31}P NMR spectra. The presence of PH_2^- was inferred from its ^{31}P NMR triplet. The reaction of excess PH_3 with K (or Na) was also studied [15]. Dissolving KPH_2 in an NH_3/ND_3 (1:2) mixture yielded PH_2^-, PHD^-, and PD_2^- ions, which exhibited characteristic ^{31}P NMR multiplets, i.e., triplets, sextets, and quintets [16].

Additional details concerning the synthesis of solutions of PH_2^- and the isolation of solid phosphides under various experimental conditions are given in the Sections 1.3.1.5.5, p. 233, and 1.3.1.5.7, p. 255.

Dimethylformamide (DMF). A solution of KPH_2 (from the reaction of PH_3 with K in NH_3) has a yellow color and a conductivity, comparable to solutions of $KClO_3$ or KNO_3. Thus, KPH_2 is expected to form the ions PH_2^- and K^+. IR spectroscopy revealed a P–H stretching vibration (see below). ^1H or ^{31}P NMR spectra showed a doublet or a triplet [17]. Solutions of MPH_2 with M = Li through Cs were also prepared. H exchange between PH_2^- and MPH_2 was inferred from the NMR spectra [18].

Dimethyl Sulfoxide (DMSO). When PH_3 was introduced into a suspension of KOH in DMSO, PH_2^- ions formed according to $PH_3 + 2\,KOH \rightarrow PH_2^- + K^+ + KOH \cdot H_2O$, giving rise to a yellow color. They were detected by treatment with CH_3I (see the section on ammonia above) [19]. The concentration of PH_2^- in a $DMSO/H_2O$ mixture decreases with increasing H_2O content. The ^{31}P NMR spectrum points to rapid proton transfer (equilibrium $PH_3 + OH^- \rightleftharpoons PH_2^- + H_2O$) [20].

Ethylene Glycol Dimethyl Ether (monoglyme). PH_3 reacts with $LiP(C_2H_5)_2$ in mono- or diglyme (diethylene glycol dimethyl ether) to $LiPH_2 + HP(C_2H_5)_2$. $LiPH_2$ in these solvents was also obtained in the disproportionation equilibria $2(CH_3)_{3-k}H_kSiPHLi \rightleftharpoons LiPH_2 + [(CH_3)_{3-k}H_kSi]_2PLi$ (k = 0 to 3) and other reactions [21]. The formation of $LiPH_2$ by the reaction of PH_3 with n-C_4H_9Li in monoglyme was described in [22]; see also [23].

In Crystalline MPH_2

PH_2^- ions were detected by IR absorption spectra of MPH_2. Pure and solvent-free crystalline MPH_2 could be isolated at room temperature after reacting excess PH_3 gas with M or MNH_2 in liquid NH_3 (M = K, Rb, Cs) at −35°C. X-ray diffraction methods did not allow to locate the H atoms within the PH_2^- ion; librations and quasi-free rotations of the ions were assumed [24]. X-ray diffraction on single crystals of KPH_2 and $RbPH_2$ (obtained from solutions in DMF) was done by [25].

References:

[1] Ebinghaus, H.; Kraus, K.; Müller-Duysing, W.; Neuert, H. (Z. Naturforsch. **19a** [1964] 732/6).
[2] Halmann, M.; Platzner, I. (J. Phys. Chem. **73** [1969] 4376/8).
[3] Smyth, K. C.; McIver, R. T., Jr.; Brauman, J. I.; Wallace, R. W. (J. Chem. Phys. **54** [1971] 2758/9).
[4] Smyth, K. C.; Brauman, J. I. (J. Chem. Phys. **56** [1972] 1132/42).
[5] Wyatt, R. H.; Holtz, D.; McMahon, T. B.; Beauchamp, J. L. (Inorg. Chem. **13** [1974] 1511/7).
[6] Holtz, D.; Beauchamp, J. L.; Eyler, J. R. (J. Am. Chem. Soc. **92** [1970] 7045/55).
[7] Zittel, P. F.; Lineberger, W. C. (J. Chem. Phys. **65** [1976] 1236/43).
[8] Brauman, J. I.; Eyler, J. R.; Blair, L. K.; White, M. J.; Comisarow, M. B.; Smyth, K. C. (J. Am. Chem. Soc. **93** [1971] 6360/2).
[9] Anderson, D. R.; Bierbaum, V. M.; DePuy, C. H. (J. Am. Chem. Soc. **105** [1983] 4244/8).
[10] Marcus, R. A. (J. Phys. Chem. **72** [1968] 891/9; J. Am. Chem. Soc. **91** [1969] 7224/5).

[11] Cohen, A. O.; Marcus, R. A. (J. Phys. Chem. **72** [1968] 4249/56).
[12] Shi, Z.; Boyd, R. J. (J. Am. Chem. Soc. **113** [1991] 2434/9).
[13] Evers, C.; Street, E. H., Jr.; Jung, S. L. (J. Am. Chem. Soc. **73** [1951] 5088/91).
[14] Birchall, T.; Jolly, W. L. (Inorg. Chem. **5** [1966] 2177/80).
[15] Sheldrick, G. M. (Trans. Faraday Soc. **63** [1967] 1065/70).
[16] Wasylishen, R. E.; Burford, N. (Can. J. Chem. **65** [1987] 2707/12).
[17] Knoll, F.; Bergerhoff, G. (Monatsh. Chem. **97** [1966] 808/19, 809/11, 818).
[18] Knoll, F.; Wazer, van, J. R. (J. Inorg. Nucl. Chem. **31** [1969] 2620/3).
[19] Jolly, W. L. (Inorg. Synth. **11** [1968] 124/6).
[20] Langhans, K. P.; Stelzer, O.; Svara, J.; Weferling, N. (Z. Naturforsch. **45b** [1990] 203/11).

[21] Fritz, G.; Schäfer, H.; Hölderich, W. (Z. Anorg. Allg. Chem. **407** [1974] 266/86, 270, 280).
[22] Schäfer, H.; Fritz, G.; Hölderich, W. (Z. Anorg. Allg. Chem. **428** [1977] 222/4).
[23] Baudler, M.; Glinka, K. (Inorg. Synth. **27** [1990] 227/35).
[24] Jacobs, H.; Hassiepen, K. M. (Z. Anorg. Allg. Chem. **531** [1985] 108/18, 109, 115).
[25] Bergerhoff, G.; Schultze-Rhonhof, E. (Acta Crystallogr. **15** [1962] 420).

1.2.4.2 Electronic Structure

The electronic **ground state** was given to be ... $4a_1^2\ 2b_2^2\ 5a_1^2\ 2b_1^2$, $\tilde{X}\ ^1A_1$ [1] by comparison with the isoelectronic, bent H_2S molecule (point group C_{2v}; see "Sulfur" Suppl. Vol. 4a/b, 1983, p. 53). This point group may also be inferred from a laser photoelectron spectrum of PH_2^- which indicates similar geometries of the anion and the PH_2 radical (see also p. 107) [1]. Calculations of orbital energies ε_i by an MO-SCF method (the original paper gives values for all nine orbitals $1a_1$ through $2b_1$) [2] and by an extended Hückel (EH) method [3] gave the same order of the four valence orbitals.

Total energies E_t of PH_2^- have been calculated with the Møller-Plesset perturbation theory up to the fourth [4 to 7] or second order [8], with a CI method involving single and double excitations (CISD) [8], with the perturbed Hartree-Fock method (coupled HF scheme) [9], with an MO-SCF method [2], and also with a united-atom (Ar) approximation [10].

Net atomic **charges** of about −0.2 at each H were calculated with an ab initio MO-SCF method [2], with the semiempirical CNDO/2 method [11], and with another semiempirical method using localized bond orbitals for CI [12]. A lower value came from an EH calculation [3]. A radial electron density distribution was calculated within the united-atom approximation [10]. Two different dipole moments were obtained with an MO-SCF calculation (yielding also quadrupole and octupole moments) [2] and with the electron propagator theory (EPT) [13].

For the **ionization potential** E_i, see data on the electron affinity of PH_2 (lying at ~1.25 eV), p. 62. An ionization potential in aqueous solution, $E_i = 4.18 \pm 0.15$ eV, was derived from the gas-phase value (1.25 eV) and the free energy of hydration of the PH_2^- ion (see p. 111) [14].

For the **proton affinity**, see the section on the acidity of PH_3, p. 199 (a gas-phase value of 370.9 kcal/mol or 16.1 eV was reported in [4]).

An Li^+ affinity has been calculated in [15].

The **electric field gradients** at P and H, the anisotropic (dipolar) **magnetic hyperfine coupling** constants of ^{31}P and ^1H, and the **diamagnetic susceptibility** were calculated by an ab initio MO-SCF method [2].

References:

[1] Zittel, P. F.; Lineberger, W. C. (J. Chem. Phys. **65** [1976] 1236/43).
[2] Ball, J. R.; Thomson, C. (Int. J. Quantum Chem. **14** [1978] 39/53, 42, 44, 47, 50).
[3] Issleib, K.; Gründler, W. (Theor. Chim. Acta **6** [1966] 64/72).
[4] Gronert, S. (J. Am. Chem. Soc. **113** [1991] 6041/8).
[5] Pople, J. A.; Schleyer, P. von R.; Kaneti, J.; Spitznagel, G. W. (Chem. Phys. Lett. **145** [1988] 359/64).
[6] Spitznagel, G. W.; Clark, T.; Schleyer, P. von R.; Hehre, W. J. (J. Comput. Chem. **8** [1987] 1109/16).
[7] Gordon, M. S.; Davis, L. P.; Burggraf, L. W.; Damrauer, R. (J. Am. Chem. Soc. **108** [1986] 7889/93).
[8] Lohr, L. L.; Ponas, S. H. (J. Phys. Chem. **88** [1984] 2992/7).
[9] Lazzeretti, P.; Rossi, E.; Taddei, F.; Zanasi, R. (J. Chem. Phys. **77** [1982] 408/14).
[10] Banyard, K. E.; Hake, R. B. (J. Chem. Phys. **43** [1965] 2684/9).

[11] Takahata, Y. (Chem. Phys. Lett. **59** [1978] 472/7).
[12] Friedemann, R.; Gründler, W.; Issleib, K. (Tetrahedron **26** [1970] 2861/7).
[13] Ortiz, J. V. (Int. J. Quantum Chem. Symp. No. 22 [1988] 431/6).
[14] Pearson, R. G. (J. Am. Chem. Soc. **108** [1986] 6109/14).
[15] Del Bene, J. E.; Shavitt, I. (Int. J. Quantum Chem. Symp. No. 24 [1990] 365/73).

1.2.4.3 **Nuclear Magnetic Resonance** (NMR)

The NMR data tabulated below were all measured on solutions of alkali metal salts MPH_2 in various liquid solvents, specifically NH_3 [1 to 4] and dimethylformamide (DMF) [5 to 7]. However, it is not known, whether MPH_2 in these solutions is completely ionized into $M^+ + PH_2^-$; rather one has to assume an equilibrium between PH_2^- (doubly connected P) and MPH_2 (triply connected P), see a compilation of ^{31}P NMR data [8]. Solutions were obtained not only by direct dissolution of MPH_2, but also by reacting PH_3 with $NaNH_2$ suspended in liquid NH_3. This latter method may suffer from the additional difficulty of proton exchange between PH_2^- and PH_3, see [1, 2].

The shifts $\delta(^1H)$ and $\delta(^{31}P)$ and spin-spin coupling constants $J(^{31}P,^1H)$ were derived from the ^1H NMR doublets (splitting by ^{31}P with $I=1/2$) and the ^{31}P NMR triplets (splitting by two protons), which were observed in the experimental systems described above, and are listed in the table below regardless of the difficulties in assigning the spectra. $\delta(^1H)$ is referred to TMS and is positive for low-field shifts; see a compilation of chemical shifts of protons directly bonded to P [10]. $\delta(^{31}P)$ is referred to 85% H_3PO_4 (also positive for low-field shifts); see a review on ^{31}P NMR spectra [11]. For details and supplemental information, see the remarks below the table:

$-\delta(^1H)$ in ppm	$-\delta(^{31}P)$ in ppm	$J(^{31}P,^1H)$ in Hz	experimental system	Ref.	remarks
–	279.4	138.24(1)	KPH_2 dissolved in NH_3–ND_3 (1:2)	[4]	a)
–	280(3)	139(2)	KPH_2 dissolved in NH_3	[2]	b)
–	279(2)	139	KPH_2 dissolved in NH_3	[3]	

$-\delta(^1H)$ in ppm	$-\delta(^{31}P)$ in ppm	$J(^{31}P,^1H)$ in Hz	experimental system	Ref.	remarks
—	272(2)	—	KPH_2 dissolved in $N(CH_3)_3$	[3]	
1.12	255.3	136.8, 139.0	KPH_2 dissolved in DMF	[5, 6]	c)
1.36	—	140(1)	KPH_2 dissolved in DMF	[10, 12]	d)
1.487(4)	—	138.71(7)	PH_3 reacted with $NaNH_2$ in NH_3	[2]	e)
1.64	—	139.3	PH_3 reacted with $NaNH_2$ in NH_3	[1]	f)
1.23	283	151(1)	pure $LiPH_2 \cdot 1$ monoglyme	[9]	g)

[a] D isotope effects on ^{31}P NMR at 20°C were examined: $\delta(^{31}P) = -2.7613(2)$ and $-5.5197(2)$ ppm for PHD^- (exhibiting a sextet) and PD_2^- (a quintet), both refer to PH_2^-. $J(^{31}P, ^1H) = 138.93(1)$ Hz for PHD^-, $J(^{31}P, ^2D) = 21.74(1)$ and 21.85(1) Hz for PHD^- and PD_2^- [4]. – [b] 5% KPH_2 at 22°C. $\delta(^{31}P) = -393 \pm 3$ ppm from external P_4O_6 [2] was combined with $\delta(^{31}P) = 113.0$ ppm for P_4O_6 with respect to 85% H_3PO_4 [11]. J was also given for 2% $NaPH_2$ in NH_3 at +21 to −60°C [2]. – [c] The first J value from the ^1H and the second from the ^{31}P spectrum [5, 6]. For the ^{31}P data, see also the compilation in [8] (which also lists unpublished data of V. Mark). $\delta(^1H)$ and $\delta(^{31}P)$ depend on the concentration of KPH_2; a pronounced dependence of $\delta(^1H)$ on the temperature of MPH_2 solutions (M = Li through Cs) was observed. These effects are attributed to aggregation with the solvent in the case of $\delta(^1H)$ and to dissociative ionization of KPH_2 in the case of $\delta(^{31}P)$ [7]. – [d] Taken from [7] (see foregoing remark). – [e] "4% $NaPH_2$ in NH_3" at 22°C; the original paper gives $\tau = 11.487 \pm 0.004$, τ and J also at +2 and −23°C, and J for "2% $NaPH_2$ in NH_3" at +22 to −58°C [2]. – [f] The value for $\delta(^1H)$ was not reported in a more recent compilation [12]. $\delta(^1H)$ of PH_3 changed by 3.28 ppm to higher magnetic field upon removal of a proton from PH_3 to give PH_2^- [1]. – [g] $\tau = 11.23$.

$J(^{31}P, ^1H) = 130.56$ Hz was theoretically calculated by a parameterized MO-SCF method assuming a P–H distance of 1.42 Å and an interbond angle of 125° [13].

The anisotropies of $J(^{31}P,^1H)$ and $J(^1H,^1H)$ have been theoretically investigated assuming several geometries of the PH_2^- ion [14].

A ^{31}P nuclear shielding $\sigma = 663.5$ ppm and an anisotropy $\Delta\sigma = 306.2$ ppm were obtained by an ab initio calculation in the gauge including atomic orbital (GIAO) method for an optimized geometry [15]. An experimental value of $\sigma = 607.8$ ppm was based [15] on the shift measured in [4] and on the absolute shielding scale ($\sigma = 328.4$ ppm for PO_4^{3-}) given in [16].

The diamagnetic shielding at P and H was calculated by an ab initio MO-SCF method for the calculated equilibrium geometry [17].

References:

[1] Birchall, T.; Jolly, W. L. (Inorg. Chem. **5** [1966] 2177/80).

[2] Sheldrick, G. M. (Trans. Faraday Soc. **63** [1967] 1065/70).

[3] Fluck, E.; Novobilsky, V., unpublished results from Fluck, E. (Fortschr. Chem. Forsch. **35** [1973] 1/64, 27).

[4] Wasylishen, R. E.; Burford, N. (Can. J. Chem. **65** [1987] 2707/12).

[5] Bergerhoff, G.; Knoll, F. (Angew. Chem. **77** [1965] 1016; Angew. Chem. Int. Ed. Engl. **4** [1965] 968/9).

[6] Knoll, F.; Bergerhoff, G. (Monatsh. Chem. **97** [1966] 808/19, 809/10).
[7] Knoll, F.; Wazer, van, J. R. (J. Inorg. Nucl. Chem. **31** [1969] 2620/3).
[8] Mark, V.; Dungan, C. H.; Crutchfield, M. M.; Wazer, van, J. R. (Top. Phosphorus Chem. **5** [1967] 227/457, 233, 236, 239).
[9] Schäfer, H.; Fritz, G.; Hölderich, W. (Z. Anorg. Allg. Chem. **428** [1977] 222/4).
[10] Houalla, D.; Marty, R.; Wolf, R. (Z. Naturforsch. **25b** [1970] 451/61, 459).

[11] Crutchfield, M. M.; Dungan, C. H.; Wazer, van, J. R. (Top. Phosphorus Chem. **5** [1967] 1/74, 7, 59).
[12] Brazier, J. F.; Houalla, D.; Loenig, M.; Wolf, R. (Top. Phosphorus Chem. **8** [1976] 99/192, 110).
[13] Cowley, A. H.; White, W. D. (J. Am. Chem. Soc. **91** [1969] 1917/21).
[14] Lazzeretti, P.; Rossi, E.; Taddei, F.; Zanasi, R. (J. Chem. Phys. **77** [1982] 408/14).
[15] Chesnut, D. B.; Rusiloski, B. E. (Chem. Phys. **157** [1991] 105/10).
[16] Jameson, C. J.; De Dios, A.; Jameson, A. K. (Chem. Phys. Lett. **167** [1990] 575/82).
[17] Ball, J. R.; Thomson, C. (Int. J. Quantum Chem. **14** [1978] 39/53, 50).

1.2.4.4 Geometry. Vibrations. Dissociation

X-ray diffraction of KPH_2 or $RbPH_2$ single crystals gave an **ionic radius** of 2.12 or 2.13 Å [1].

The **geometrical structure** of gaseous PH_2^- in its $\tilde{X}\ ^1A_1$ ground state appears to be similar to that of ground-state PH_2 (with an internuclear distance of r=1.42 Å and an interbond angle of α=92°; see p. 72). This was inferred from a sharp increase of the photodetachment cross section at threshold, measured by ion cyclotron resonance [2, 3] and from the predominance of the (0, 0, 0)←(0, 0, 0) transition in the PH_2, $\tilde{X}\ ^2B_1 \leftarrow PH_2^-$, $\tilde{X}\ ^1A_1$ photoelectron spectrum [4]. r=1.34±0.05 Å and α=92°±5° were taken from the isoelectronic H_2S molecule (and used to calculate the thermodynamic functions of PH_2^-, see p. 109) [5]. r and α have also been theoretically calculated by several ab initio MO methods, i.e., at an MP2 [6, 7], a CEPA (coupled electron pair approximation) [8], and an HF level [9 to 15]. r was also obtained from a united-atom approximation [16]; α was also calculated by a semiempirical (CNDO/2) method [17] and estimated by extended Hückel calculations [18].

α = 92.7° for PH_2^- dissolved in dimethylformamide (DMF) was derived from the chemical shift of the ^{31}P NMR signal (see above) [19] using an approximate method taken from [20]. A different value, α=125°, was obtained [21] by comparing a measured coupling constant $J(^{31}P, ^1H)$ in liquid ammonia [22] with a calculated one [21].

X-ray structure determinations of crystalline $LiPH_2 \cdot DME$ at −120±3°C gave two P–H internuclear distances of 1.31 and 1.38 Å and an HPH interbond angle of 85° [23]. Earlier work on $LiPH_2 \cdot DME$ at 23±2°C [24] and on crystalline KPH_2 at room temperature [25] could not determine the position of the H atoms.

Complete sets of the **fundamental vibrations** ν_1 (symmetric stretching), ν_2 (bending), and ν_3 (antisymmetric stretching), all in cm^{-1}, were obtained from IR absorption bands of MPH_2 (M = K, Rb, Cs) in KBr pellets (and assigned to PH_2^- by comparison with known vibrational frequencies of PH_3, NH_3, NH_2^-, H_2O, and H_2S) [25] (see table on the next page).

ν_1 = 2210 cm^{-1} was measured by IR absorption of KPH_2 dissolved in DMF [19].

Ab initio MO calculations at the CEPA level yielded fundamental frequencies for PH_2^- and PD_2^- (in parentheses) of 2187 (1591), 1069 (773), and 2182 (1590) cm^{-1} (for harmonic frequencies ω_i and combinations, see the original paper) [8]. Larger values were obtained at the HF level [8, 10, 11] (in [10, 11] only ω_i for PH_2^-).

M	ν_1 in cm^{-1}	ν_2 in cm^{-1}	ν_3 in cm^{-1}
K	2230 vs	1085 vs	2170 s, sh
Rb	2230 vs	1080 vs	2160 s, sh
Cs	2230 vs	1080 vs	2140 s, sh

Stretching and bending force constants were calculated by an ab initio MO-SCF method [15].

Average or bond **dissociation energies** were obtained for the processes $PH_2^- \rightarrow P^- + 2H$, $\rightarrow PH^- + H$, and $PH^{2-} + H^+$ from extended Hückel calculations [18].

References:

[1] Bergerhoff, G.; Schultze-Rhonhof, E. (Acta Crystallogr. **15** [1962] 420).
[2] Smyth, K. C.; McIver, R. T., Jr.; Brauman, J. I.; Wallace, R. W. (J. Chem. Phys. **54** [1971] 2758/9).
[3] Smyth, K. C.; Brauman, J. I. (J. Chem. Phys. **56** [1972] 1132/42, 1141).
[4] Zittel, P. F.; Lineberger, W. C. (J. Chem. Phys. **65** [1976] 1236/43).
[5] Glushko, V. P.; Gurvich, L. V.; Bergman, G. A.; Veits, I. V.; Medvedev, V. A.; Khachkuruzov, G. A.; Yungman, V. S. (Termodinamicheskie Svoistva Individual'nykh Veshchestv, Izd. Nauka, Moscow 1978, Vol. I, Book 1, p. 369, Book 2, p. 283).
[6] Del Bene, J. E.; Shavitt, I. (J. Phys. Chem. **94** [1990] 5514/8).
[7] Nguyen, M. T. (J. Mol. Struct. **180** [1988] 23/9 [THEOCHEM **49**]).
[8] Botschwina, P. (J. Chem. Soc. Faraday Trans. II **84** [1988] 1263/76, 1272/3; NATO-ASI Ser. C **193** [1987] 261/70, 264/5).
[9] Gronert, S. (J. Am. Chem. Soc. **113** [1991] 6041/8).
[10] Pople, J. A.; Schleyer, P. von R.; Kaneti, J.; Spitznagel, G. W. (Chem. Phys. Lett. **145** [1988] 359/64).

[11] Spitznagel, G. W.; Clark, T.; Schleyer, P. von R.; Hehre, W. J. (J. Comput. Chem. **8** [1987] 1109/16).
[12] Nordholm, S. (Austral. J. Chem. **40** [1987] 1465/70).
[13] Gordon, M. S.; Davis, L. P.; Burggraf, L. W.; Damrauer, R. (J. Am. Chem. Soc. **108** [1986] 7889/93).
[14] Lazzeretti, P.; Rossi, E.; Taddei, F.; Zanasi, R. (J. Chem. Phys. **77** [1982] 408/14).
[15] Ball, J. R.; Thomson, C. (Int. J. Quantum Chem. **14** [1978] 39/53, 44, 47).
[16] Banyard, K. E.; Hake, R. B. (J. Chem. Phys. **43** [1965] 2684/9).
[17] Takahata, Y. (Chem. Phys. Lett. **59** [1978] 472/7).
[18] Issleib, K.; Gründler, W. (Theor. Chim. Acta **6** [1966] 64/72).
[19] Knoll, F.; Bergerhoff, G. (Monatsh. Chem. **97** [1966] 808/19, 809/11).
[20] Muller, N.; Lauterbur, P. C.; Goldenson, J. (J. Am. Chem. Soc. **78** [1956] 3557/61).

[21] Cowley, A. H.; White, W. D. (J. Am. Chem. Soc. **91** [1969] 1917/21).
[22] Sheldrick, G. M. (Trans. Faraday Soc. **63** [1967] 1065/70).
[23] Becker, G.; Hartmann, H.-M.; Schwarz, W. (Z. Anorg. Allg. Chem. **577** [1989] 9/22, 14, 18).
[24] Jones, R. A.; Koschmieder, S. U.; Nunn, C. M. (Inorg. Chem. **26** [1987] 3610/1).
[25] Jacobs, H.; Hassiepen, K. M. (Z. Anorg. Allg. Chem. **531** [1985] 108/18, 111, 115).

1.2.4.5 Heat of Formation. Thermodynamic Functions

The formation enthalpy $\Delta_f H^\circ_{298} = 4.5 \pm 3$ kcal/mol is based on a PH_2^- proton affinity of 370.4 kcal/mol (see p. 199) and on the PH_3 (1.3 kcal/mol, see p. 179) and H^+ formation enthalpies [1, 2]. $\Delta_f H^\circ_0 = 7$ kJ/mol (≙1.7 kcal/mol) was given in an earlier compilation [3]. $\Delta_f H^\circ_0 = 0.8 \pm 6.4$ kJ/mol was based on a PH_2 formation enthalpy of 123.4 ± 6.2 kJ/mol (see p. 79) and a PH_2 electron affinity of 122.6 ± 1.0 kJ/mol or 1.271 ± 0.010 eV (see p. 62) [4].

Thermodynamic functions have been calculated assuming a PH_2^- structure (1.34 Å, 92°; see above) in the ideal gas state between 298.15 and 6000 K at intervals of 100 K. Selected values for the molar heat capacity C°_p, entropy S° (both in $J \cdot mol^{-1} \cdot K^{-1}$), and enthalpy $H^\circ - H^\circ_0$ (in kJ/mol) are given below [4]:

T in K	298.15	400	800	1500	2500	4000	6000
C°_p	34.123	35.500	42.339	50.769	55.046	56.891	57.605
S°	205.170	215.380	242.041	271.388	298.537	324.901	348.131
$H^\circ - H^\circ_0$	9.960	13.502	29.046	62.037	115.364	199.609	314.243

References:

[1] Bartmess, J. E.; Scott, J. A.; McIver, R. T., Jr. (J. Am. Chem. Soc. **101** [1979] 6046/56).
[2] Bartmess, J. E.; McIver, R. T., Jr. (in: Bowers, M. T.; Gas Phase Ion Chemistry, Vol. 2, Academic, New York 1979, pp. 87/121, 98).
[3] Rosenstock, H. M.; Draxl, K.; Steiner, B. W.; Herron, J. T. (J. Phys. Chem. Ref. Data **6** Suppl. No. 1 [1977] 736/73, 755).
[4] Glushko, V. P.; Gurvich, L. V.; Bergman, G. A.; Veits, I. V.; Medvedev, V. A.; Khachkuruzov, G. A.; Yungman, V. S. (Termodinamicheskie Svoistva Individual'nykh Veshchestv, Izd. Nauka, Moscow 1978, Vol. I, Book 1, p. 369, Book 2, p. 283).

1.2.4.6 Chemical Reactions

In the Gas Phase

Relative **photodetachment** cross sections have been measured in the wavelength range 725 to 1020 nm (1.71 to 1.22 eV) by ion cyclotron resonance (ICR) spectroscopy [1].

For the reaction with the H^+ **ion**, see the section on the proton affinity of the PH_2^- ion (1.3.1.5.1.1, p. 199).

Reactions with **neutral molecules** have been investigated by the flowing afterglow (FA) technique (see [2, 3]) and earlier by ICR spectroscopy. The reaction usually starts with an initial nucleophilic attack of PH_2^- on the neutral, followed by intramolecular proton transfer and/or expulsion of a neutral fragment [4]. The table on p. 110 lists the rate constants k at 298 K (with an estimated error of ±25% for the FA measurements [4]), efficiencies k/k_{ADO} (with k_{ADO} calculated by the average-dipole-orientation theory of [5]), products (neutral products were not detected), and branching ratios. k and the branching ratio depend on the total pressure, when adducts are formed. Molecules, for which no reaction could be observed, are listed below the table.

reactant	k in $cm^3 \cdot molecule^{-1} \cdot s^{-1}$	k/k_{ADO}	products (branching ratio)	method	Ref.	remarks
O_2	1.5×10^{-12}	0.002	$PO_2^- + H_2$	FA	[4]	—
N_2O	2.3×10^{-12}	0.002	$H_2PO^- + N_2$ (0.92) $PN_2^- + H_2O$ (0.08)	FA	[4]	a)
NO_2	1.3×10^{-9}	~1	$NO_2^- + PH_2$	FA	[4]	—
H_2S	—	—	$SH^- + PH_3$	ICR	[6]	b)
SO_2	2.9×10^{-11}	0.018	$H_2PSO_2^-$ (0.40) $SO_2^- + PH_2$ (0.32) $PO_2^- + H_2S$ (or $PS^- + H_2O_2$) (0.22) $PSO^- + H_2O$ (0.05)	FA	[4]	c)
CO_2	$<1\times10^{-12}$	<0.001	$H_2PCO_2^-$	FA	[4]	—
HCN	—	—	$CN^- + PH_3$	ICR	[7]	—
CH_3Cl	1.1×10^{-10}	0.056	$Cl^- + H_2PCH_3$	FA	[4]	—
CH_3Br	8.2×10^{-10}	0.47	$Br^- + H_2PCH_3$	FA	[4]	—
CH_3I	1.3×10^{-9}	0.77	$I^- + H_2PCH_3$	FA	[4]	—
OCS	3.2×10^{-11}	0.023	$H_2PS^- + CO$ (0.89) $PCO^- + H_2S$ (0.09) H_2PCOS^- (0.02)	FA	[4]	d)
CS_2	1.4×10^{-10}	0.096	$PCS^- + H_2S$ (0.60) $PS^- + H_2CS$ (0.29) $H_2PCS_2^-$ (0.11)	FA	[4]	e)
$(CH_3)_3B$	—	—	$(CH_3)_2B{=}CH_2^- + PH_3$	ICR	[8]	—
$(CH_3)_3SiCl$	1.8×10^{-9}	~1	$Cl^- + H_2PSi(CH_3)_3$	FA	[4]	—
PH_3	—	—	$P_2H^- + 2H_2$ (~0.5) $P_2H_3^- + H_2$ (<0.5)	ICR	[7]	f)
$(CH_3O)_3P$	$\leqq1\times10^{-12}$	$\leqq$0.001	$(CH_3O)_2PO^- + H_2PCH_3$	FA	[9]	—
AsH_3	—	—	$AsH_2^- + PH_3$	ICR	[6]	g)

a) The main product H_2PO^- may result from direct attack of PH_2^- on the O of N_2O with displacement of N_2 or from initial nucleophilic attack on the N end and formation of a four-membered cyclic intermediate [4]. – b) The reaction $PD_2^- + H_2S \rightarrow SH^- + PHD_2$ was earlier observed by ICR [7]. – c) Kinetic plots of the loss of PH_2^- indicate the formation of a species with a mass-to-charge ratio of 33, which could be SH^- (or HO_2^-) from $PO_2^- + H_2S \rightleftharpoons SH^- + HPO_2$ (or $PS^- + H_2O_2 \rightleftharpoons HO_2^- + HPS$, possibly followed by $HO_2^- + PH_3 \rightarrow PH_2^- + H_2O_2$) [4]. The charge transfer channel was earlier observed mass spectrometrically [10]. – d) It was not possible to decide whether the conjugate acid-base pair $SH^- + HPCO$ was formed in addition to $PCO^- + H_2S$, because PH_2^- and SH^- have the same mass-to-charge ratio. The formation of $\leqq$5% SH^- was implied by the kinetic plots [4]. – e) The formation of the conjugate acid-base pair $SH^- + HPCS$ is also possible. PS_2^- forms as a secondary product from the apparently highly reactive primary species PS^- and from CS_2 [4]. Charge transfer was earlier observed with a mass spectrometer [10]. – f) An upper limit of 1×10^{-11} $cm^3 \cdot molecule^{-1} \cdot s^{-1}$ was placed on the rate constants for both channels [7]. – g) See also later ICR work in [11].

No reaction was detected by FA [4] or ICR [6] methods with H_2O [6], D_2O [4], NH_3 [6], CO, oxirane, CH_3OCH_3 [4], SiH_4 [6], and $Si(CH_3)_4$ [4].

Purely theoretical work has been done on the reaction of PH_2^- with CH_3CH_2Cl [12], on bimolecular complexes of PH_2^- with H_2O [13] and NH_3 [14] (all ab initio MO), and on the anion-radical association reaction $PH_2^- + CH_3$ [15] (MINDO/3).

In Liquid Solutions

A hypothetical free energy of hydration of $\Delta G^\circ = -65 \pm 2$ kcal/mol has been derived for the PH_2^- ion from its gas-phase proton affinity and the aqueous pK_a value of PH_3 [16].

The characteristic reactions of the PH_2^- ion in various solvents comprise nucleophilic substitution of halide ligands and proton exchange reactions. In some cases both types of reactions are observed either as parallel or consecutive reactions. Thus KPH_2 and SiH_3Br react in dimethyl ether ultimately giving $P(SiH_3)_3$, presumably via $SiH_3Br + PH_2^- \rightarrow SiH_3PH_2 + Br^-$ and $SiH_3PH_2 + PH_2^- \rightarrow SiH_3PH^- + PH_3$ as the initial two of a whole sequence of reactions of these types [17 to 19]. For further examples described in this volume, see Section 1.3.1.3.4, p. 121, and Section 1.3.1.3.7, p. 126. Reactions of PH_2^- with white phosphorus (P_4) depend strongly upon the solvent, the reactant ratios, and the reaction conditions. With $LiPH_2$ in ether solvents, PH_3 is evolved and Li_3P_7 or Li_2P_{16} may be obtained; see Section 1.3.1.3.2, p. 118, and [20]. Solutions of KPH_2 in dimethylformamide are decomposed by O_2 giving mixtures of different oxoacids of phosphorus [20].

References:

[1] Smyth, K. C.; Brauman, J. I. (J. Chem. Phys. **56** [1972] 1132/42), Smyth, K. C.; McIver, R. T., Jr.; Brauman, J. I.; Wallace, R. W. (J. Chem. Phys. **54** [1971] 2758/9).
[2] Ferguson, E. E.; Fehsenfeld, F. C.; Schmeltekopf, A. L. (Adv. At. Mol. Phys. **5** [1969] 1/56).
[3] Bierbaum, V. M.; De Puy, C. H.; Shapiro, R. H. ; Stewart, J. H. (J. Am. Chem. Soc. **98** [1976] 4229/35), De Puy, C. H.; Bierbaum, V. M. (Acc. Chem. Res. **14** [1981] 146/53).
[4] Anderson, D. R.; Bierbaum, V. M.; De Puy, C. H. (J. Am. Chem. Soc. **105** [1983] 4244/8).
[5] Su, T.; Bowers, M. T. (Int. J. Mass Spectrom. Ion Phys. **12** [1973] 347/56).
[6] Brauman, J. I.; Eyler, J. R.; Blair, L. K.; White, M. J.; Comisarow, M. B.; Smyth, K. C. (J. Am. Chem. Soc. **93** [1971] 6360/2).
[7] Holtz, D.; Beauchamp, J. L.; Eyler, J. R. (J. Am. Chem. Soc. **92** [1970] 7045/55, 7052).
[8] Murphy, M. K.; Beauchamp, J. L. (J. Am. Chem. Soc. **98** [1976] 1433/40).
[9] Anderson, D. R.; De Puy, C. H.; Filley, J.; Bierbaum, V. M. (J. Am. Chem. Soc. **106** [1984] 6513/7).
[10] Ebinghaus, H.; Kraus, K.; Müller-Duysing, W.; Neuert, H. (Z. Naturforsch. **19a** [1964] 732/6).

[11] Wyatt, R. H.; Holtz, D.; McMahon, T. B.; Beauchamp, J. L. (Inorg. Chem. **13** [1974] 1511/7).
[12] Gronert, S. (J. Am. Chem. Soc. **113** [1991] 6041/8).
[13] Del Bene, J. E. (J. Phys. Chem. **92** [1988] 2874/80).
[14] Del Bene, J. E. (J. Comput. Chem. **10** [1989] 603/15).
[15] Sweeney, C. W. (Int. J. Mass Spectrom. Ion Processes **71** [1986] 37/58, 46).
[16] Pearson, R. G. (J. Am. Chem. Soc. **108** [1986] 6109/14).
[17] Amberger, E.; Boeters, H. D. (Chem. Ber. **97** [1964] 1999/2004).
[18] Glidewell, C.; Sheldrick, G. M. (J. Chem. Soc. A **1969** 350/1).
[19] Fluck, E. (Fortschr. Chem. Forsch. **35** [1973] 1/64, 36/7).
[20] Knoll, F.; Bergerhoff, G. (Monatsh. Chem. **97** [1966] 808/19).

1.3 PH_3 and Ions

The dimer $(PH_3)_2$, its monocation, and the protonated species $P_2H_7^+$ will be covered in the forthcoming "Phosphorus" Suppl. Vol. C 2.

1.3.1 PH_3 Phosphane, Phosphine

CAS Registry Numbers: PH_3 *[7803-51-2]* Phosphine; PH_2D *[13587-50-3]*; PHD_2 *[13780-29-5]*; PD_3 *[13537-03-6]*; PH_2T *[22011-09-2]*; PHT_2 *[22011-07-0]*; PT_3 *[16890-43-0]*; PD_2T *[22011-10-5]*; PDT_2 *[22011-08-1]*; PHDT *[99001-47-5]*; $^{32}PH_3$ *[14921-84-7]*

1.3.1.1 General Remarks and General References

PH_3 is that phosphorus-hydrogen compound for which by far the most information is available regarding its chemistry and physical properties. The scientific interest in this compound is mostly due to the fact that it is the phosphorus analog of ammonia exhibiting, however, a quite different chemical behavior and also substantial differences in its major physical properties compared to those of NH_3. Phosphane is also of some industrial importance either itself or as an intermediate in chemical, electronic, and food industries (cf. 1.3.1.6, p. 293).

Results puplished prior to 1960 are covered in "Phosphor" C, 1965, pp. 16/50. Major aspects of the chemistry of PH_3, its most important physical data, and its industrial significance have been reviewed or are included in reviews dealing with phosphorus compounds.

Bettermann, G.; Krause, W.; Riess, G.; Hofmann, T.; Inorganic Phosphorus Compounds, Ullmann's Encycl. Ind. Chem. 5th Ed. A **19** [1991] 527/43.

Harnisch, H.; Heymer, G.; Klose, W.; Schrödter, K.; Phosphor und Phosphorverbindungen (Phosphide, Phosphorwasserstoff und Hypophosphite); in: Winnacker-Küchler, Chemische Technologie, Anorganische Technologie I, 4th Ed., Vol. 2, Hanser, München 1982, pp. 204/67.

Boeing, I. A.; Crutchfield, M. M.; Heitsch, C. W.; Phosphorus Compounds (Phosphines and Phosphine Derivatives), Kirk-Othmer Encycl. Chem. Technol. 3rd Ed. **17** [1982] 490/539.

Fluck, E.; The Chemistry of Phosphine, Fortschr. Chem. Forsch. **35** [1973] 1/64.

Toy, A. D. F.; Phosphorus. Phosphorus Hydrides and Phosphonium Compounds, in: Trotman-Dickenson, A. F.; Comprehensive Inorganic Chemistry, Vol. 2, Pergamon, Oxford 1973, pp. 389/545.

Fluck, E.; Novobilsky, V.; Die Chemie des Phosphins, Fortschr. Chem. Forsch. **13** [1969/70] 125/66.

Osadchenko, I. M.; Tomilov, A. P.; Phosphorus Hydrides, Usp. Khim. **38** [1969] 1089/107; Russ. Chem. Rev. [Engl. Transl.] **38** [1969] 495/504.

Jolly, W. L.; Norman, A. D.; Hydrides of Groups IV and V, Prep. Inorg. React. **4** [1968] 1/58.

1.3.1.2 Occurrence

Phosphane has been detected spectroscopically to be a minor component of the atmospheres of several planets by using different techniques. It is now considered one of the constituents of most of the model atmospheres of these planets. In the following, a series of publications dealing with the occurrence of PH_3 in planetary atmospheres are listed for each planet and by the major topic of the paper.

Jupiter. Abundance ratio of PH_3 [1 to 9]; optical studies of its atmosphere in the UV [10, 11] and IR regions (2300 to 1800 cm^{-1} with so-called 5 μm window [2, 4, 8, 12 to 16], 1200 to 900 cm^{-1} [5, 6, 9, 16 to 19], and 980 to 744 cm^{-1} [20]); FIR and sub-mm brightness temperature [21, 22]; photochemistry [23 to 34]; model atmosphere [7, 18, 35 to 42].

Mars. Abundance ratio of PH_3 [43].

Neptune. FIR and sub-mm brightness temperature [21, 22]; model atmosphere [35].

Saturn. Abundance ratio of PH_3 [1, 44, 45]; optical studies of its atmosphere in the UV [46], IR (4000 to 1700 cm^{-1} [47], 2229 to 1994 cm^{-1} [48], and 1300 to 900 cm^{-1} [44, 45, 49 to 51]), FIR [52 to 54], and mm region [55]; FIR and sub-mm brightness temperature [21, 22]; photochemistry [26, 56, 57]; model atmosphere [35, 44, 58].

Uranus. 5 μm spectrum [59]; FIR and sub-mm brightness temperature [21, 22]; model atmosphere [35, 60, 61].

A search for PH_3 in **interstellar** and **circumstellar** sources is described and the results are discussed with respect to the depletion of P in dense and diffuse interstellar clouds [62]. Laboratory experiments on the gas-phase ion-molecule chemistry of phosphorus using the ion cyclotron resonance technique predicted that the PH_3 molecule is formed only with difficulty in dense interstellar clouds [63].

For a discussion of the role of PH_3 in the depletion of P in **volcanic gases,** see [64]. For the occurrence of PH_3, see also Section 1.3.1.3.7, p. 126.

References:

[1] Fegley, B., Jr.; Prinn, R. G. (Astrophys. J. **326** [1988] 490/508 from C.A. **108** [1988] No. 154045).

[2] Drossart, P.; Encrenaz, T.; Tokunaga, A. T. (Icarus **60** [1984] 613/20).

[3] Lellouch, E.; Combes, F.; Encrenaz, T. (Astron. Astrophys. **140** [1984] 216/9).

[4] Drossart, P.; Encrenaz, T.; Kunde, V.; Hanel, R.; Combes, M. (Icarus **49** [1982] 416/26).

[5] Knacke, R. F.; Kim, S. J.; Ridgway, S. T.; Tokunaga, A. T. (Astrophys. J. **262** [1982] 388/95).

[6] Kunde, V.; Hanel, R.; Maguire, W.; Gautier, D.; Baluteau, J. P.; Marten, A.; Chedin, A.; Husson, N.; Scott, N. (NASA-TM-83891 [1982] 1/70; C.A. **97** [1982] No. 82298; Astrophys. J. **263** [1982] 443/67).

[7] Encrenaz, T.; Combes, M.; Zeau, Y. (Astron. Astrophys. **84** [1980] 148/53).

[8] Larson, H. P.; Treffers, R. R.; Fink, U. (Astrophys. J. **211** [1977] 972/9).

[9] Combes, M.; Encrenaz, T.; Berezne, J.; Vapillon, L.; Zeau, Y. (Astron. Astrophys. **50** [1976] 287/92).

[10] Wagener, R.; Caldwell, J.; Owen, T. (Icarus **66** [1986] 188/91).

[11] Owen, T.; Caldwell, J.; Rivolo, A. R.; Moore, V.; Lane, A. L.; Sagan, C.; Hunt, G.; Ponnamperuma, C. (Astrophys. J. **236** [1980] L39/L42).

[12] Drossart, P.; Lellouch, E.; Bezard, B.; Maillard, J. P.; Tarrago, G. (Icarus **83** [1990] 248/53 from C.A. **113** [1990] No. 215594).

[13] Lellouch, E.; Drossart, P.; Encrenaz, T. (Icarus **77** [1989] 457/65 from C.A. **110** [1989] No. 182271).
[14] Bjoraker, G. L.; Larson, H. P.; Kunde, V. G. (Astrophys. J. **311** [1986] 1058/72; Icarus **66** [1986] 579/609).
[15] Beer, R.; Taylor, F. W. (Icarus **40** [1979] 189/92).
[16] Hanel, R.; Conrath, B.; Flasar, M.; Kunde, V.; Lowman, P.; Maguire, W.; Pearl, J.; Pirraglia, J.; Samuelson, R.; Gautier, D.; Gierasch, P.; Kumar, S.; Ponnamperuma, C. (Science **204** [1979] 972/6).
[17] Hanel, R.; Conrath, B.; Flasar, M.; Herath, L.; Kunde, V.; Lowman, P.; Maguire, W.; Pearl, J.; Pirraglia, J.; Samuelson, R.; Gautier, D.; Gierasch, P.; Horn, L.; Kumar, S.; Ponnamperuma, C. (Science **206** [1979] 952/6).
[18] Encrenaz, T.; Combes, M.; Zeau, Y. (Astron. Astrophys. **70** [1978] 29/36).
[19] Encrenaz, T.; Combes, M.; Zeau, Y. (Infrared Phys. **17** [1977] 551/5).
[20] Tokunaga, A. T.; Knacke, R. F.; Ridgway, S. T.; Wallace, L. (Astrophys. J. **232** [1979] 603/15).

[21] Hildebrand, R. H.; Loewenstein, R. F.; Harper, D. A.; Orton, G. S.; Keene, J.; Whitcomb, S. E. (Icarus **64** [1985] 64/87).
[22] Macy, W. W., Jr.; Sinton, W. M.; Beichman, C. A. (Icarus **42** [1980] 68/70).
[23] Ferris, J. P.; Khwaja, H. (Icarus **62** [1985] 415/24).
[24] Ferris, J. P.; Bossard, A.; Khwaja, H. (J. Am. Chem. Soc. **106** [1984] 318/24).
[25] Kaye, J. A.; Strobel, D. F. (Icarus **55** [1983] 399/419).
[26] Strobel, D. F. (Int. Rev. Phys. Chem. **3** [1983] 145/76).
[27] Ferris, J. P.; Morimoto, J. Y.; Benson, R.; Bossard, A. (Origins Life **12** [1982] 261/5).
[28] Ferris, J. P.; Morimoto, J. Y.; Benson, R. (Origin Life Proc. 3rd ISSOL Meet., Jerusalem 1980 [1981], pp. 101/5; C.A. **96** [1982] No. 126761).
[29] Raulin, F.; Ponnamperuma, C. (Origin Life Proc. 3rd ISSOL Meet., Jerusalem 1980 [1981], pp. 107/14; C.A. **96** [1982] No. 176588).
[30] Ferris, J. P.; Morimoto, J. Y. (Icarus **48** [1981] 118/26).

[31] Visconti, G. (Icarus **45** [1981] 638/52).
[32] Strobel, D. F. (Astrophys. J. **214** [1977] L97/L99).
[33] Ridgway, S. T.; Wallace, L.; Smith, G. R. (Astrophys. J. **207** [1976] 1002/6).
[34] Ridgway, S. T. (Bull. Am. Astron. Soc. **6** [1974] 376).
[35] Bezard, B.; Gautier, D.; Marten, A. (Astron. Astrophys. **161** [1986] 387/402).
[36] Lewis, J. S.; Fegley, B. (Space Sci. Rev. **39** [1984] 163/92).
[37] Atreya, S. K.; Donahue, T. M. (Rev. Geophys. Space Phys. **17** [1979] 388/96).
[38] Howland, G. R.; Harteck, P.; Reeves, R. R. (Icarus **37** [1979] 301/6).
[39] Barshay, S. S.; Lewis, J. S. (Icarus **33** [1978] 593/611).
[40] Larson, H. P.; Fink, U.; Treffers, R. R. (Astrophys. J. **219** [1978] 1084/92).

[41] Prinn, R. G.; Lewis, J. S. (Science **190** [1975] 274/6).
[42] Lewis, J. S. (Icarus **10** [1969] 393/409).
[43] Maguire, W. C. (Icarus **32** [1977] 85/97).
[44] Courtin, R.; Gautier, D.; Marten, A.; Bezard, B.; Hanel, R. (Astrophys. J. **287** [1984] 899/916).
[45] Bregman, J. D.; Lester, D. F.; Rank, D. M. (Astrophys. J. **202** [1975] L55/L56).
[46] Moore, V.; Hunt, G. E.; Caldwell, J.; Owen, T.; Encrenaz, T.; Combes, M. (Geophys. Res. Lett. **10** [1983] 1196/9).
[47] Larson, H. P.; Fink, U.; Smith, H. A.; Davis, D. S. (Astrophys. J. **240** [1980] 327/37).
[48] Noll, K. S.; Larson, H. P. (Icarus **89** [1991] 168/89 from C.A. **114** [1991] No. 189468).

[49] Hanel, R.; Conrath, B.; Flasar, F. M.; Kunde, V.; Maguire, W.; Pearl, J.; Pirraglia, J.; Samuelson, R.; Herath, L.; Allison, M.; Cruikshank, D.; Gautier, D.; Gierasch, P.; Horn, L.; Koppany, R.; Ponnamperuma, C. (Science **212** [1981] 192/200).
[50] Tokunaga, A. T.; Dinerstein, H. L.; Lester, D. F.; Rank, D. M. (Icarus **42** [1980] 79/85, **48** [1981] 540).

[51] Gillett, F. C.; Forrest, W. J. (Astrophys. J. **187** [1974] L37/L39).
[52] Haas, M. R.; Erickson, E. F.; Goorvitch, D.; McKibbin, D. D.; Rank, D. M. (Icarus **64** [1985] 549/56).
[53] Haas, M. R.; Erickson, E. F.; Goorvitch, D.; McKibbin, D. D.; Rank, D. M. (NASA-CP-2353 [1984] 76/80; C.A. **103** [1985] No. 95714).
[54] Encrenaz, T.; Combes, M. (Astron. Astrophys. **61** [1977] 387/90).
[55] Lellouch, E.; Encrenaz, T.; Combes, F. (Astron. Astrophys. **135** [1984] 371/6).
[56] Kaye, J. A.; Strobel, D. F. (Icarus **59** [1984] 314/35).
[57] Kaye, J. A.; Strobel, D. F. (Geophys. Res. Lett. **10** [1983] 957/60).
[58] Fegley, B.; Prinn, R. G. (Astrophys. J. **299** [1985] 1067/77).
[59] Orton, G. S.; Kaminski, C. D. (Icarus **77** [1989] 109/17 from C.A. **110** [1989] No. 144198).
[60] Fegley, B.; Prinn, R. G. (Astrophys. J. **307** [1986] 852/65).

[61] Fegley, B.; Prinn, R. G. (Nature **318** [1985] 48/50).
[62] Turner, B. E.; Tsuji, T.; Bally, J.; Guelin, M.; Cernicharo, J. (Astrophys. J. **365** [1990] 569/85).
[63] Thorne, L. R.; Anicich, V. G.; Prasad, S. S.; Huntress, W. T., Jr. (Astrophys. J. **280** [1984] 139/43).
[64] Mambo, V. S.; Yoshida, M.; Matsuo, S. J. (Volcanol. Geotherm. Res. **46** [1991] 37/47 from C.A. **115** [1991] No. 117924).

1.3.1.3 Formation and Preparation

Many formation reactions and the principal methods for the preparation of PH_3 both on laboratory and industrial scales are known for a long time and are already described or mentioned in "Phosphor" C, 1965, pp. 7/16.

The following sections cover the literature after 1960 on reactions leading to the formation of PH_3, including a large number of patents. Laboratory procedures are not described in detail. Important criteria for choosing a particular laboratory method are, besides others, the purity and the hazards of the resulting PH_3 (traces of P_2H_4 impurities cause PH_3 to become highly inflammable) or whether a constant stream of the compound or only a limited amount are to be obtained. Therefore, some of the convenient laboratory procedures mentioned in the following sections, such as the hydrolysis of certain metal phosphides (cf. 1.3.1.3.5, p. 123), the hydrolysis of PH_4I (cf. 1.3.1.3.6, p. 125), and the pyrolysis of H_3PO_3 (cf. 1.3.1.3.4, p. 122), are classified according to these aspects. Disproportionation reactions of P in alkaline or acidic solutions are mostly used for the manufacture of PH_3 in the chemical industry (cf. 1.3.1.3.2, p. 118). In the fumigation of foodstuffs and animal feed, e.g., in grain silos for pest control, the PH_3 is slowly released under controlled conditions by hydrolyzing selected metal phosphide formulations (cf. 1.3.1.3.10, p. 131).

1.3.1.3.1 Electrochemical Formation

PH_3 is formed by cathodic reduction of **white phosphorus** under nitrogen. Polarographic investigations (mercury drop electrode) show that in protic media (methanol or ethanol solution containing benzene) the irreversible cathode reaction

$$P_4 + 12H^+ + 12e^- \rightarrow 4PH_3$$

takes place where reduction proceeds through intermediate formation of a mercury-phosphorus alloy, P_4Hg_x [1], on the cathode surface [2]. In a 0.1M LiCl solution of ethanol containing 2 to 3% benzene, the reduction proceeds at −1.88 to −1.90 V (22 to 23°C) relative to the mercury pool at the bottom of the cell [3].

Investigations of the electrochemical reduction in acidic aqueous solution (e.g., 10 wt% HCl, H_2SO_4, H_3PO_4, CH_3COOH, or a KH_2PO_4-Na_2HPO_4 buffer of pH 4.58) at graphite, zinc, cadmium, mercury, titanium, tin, iron, or copper cathodes (current density < 24 A/dm², t = 50 to 90°C) show that the highest current efficiency of 65 to 67% for PH_3 is obtained with a zinc cathode (18 to 20 A/dm²) in an H_3PO_4 catholyte (75 to 85°C). The phosphane yield increases with increasing H_3PO_4 concentration from 12.5% (10 wt%) to 57% (40 wt%). The addition of phosphorus to the electrolyte causes the potential of the zinc cathode to shift to a higher negative value, probably due to the formation of a zinc phosphide film which is assumed to be an intermediate in the reduction of P_4 to PH_3. The zinc cathode has serious disadvantages, as it is attacked rapidly during electrolysis [4]. Part of the phosphorus oxidized to phosphorus oxoacids (mainly phosphonic and phosphinic acids) was found in the catholyte [5]. If an acid-resistant graphite cathode is used, the addition of metal ions increases the phosphane yield (% in parentheses) in the order Sb^V (3), Ba^{II} (10.3), Tl^{III} (16.5), Hg^{II} (23), Pb^{II} (26), Cd^{II} (43.2), Zn^{II} (56) [4]. Investigation of the electrochemical reduction of phosphorus to phosphane on a lead cathode showed that the reaction is not a single-stage process, rather it is accompanied by intermediate formation of higher phosphanes which are subsequently reduced to PH_3. Thus, a lead cathode is not recommended for the preparation of PH_3 [6].

Many patents describe the electrochemical formation of PH_3 where molten white phosphorus is reduced at 50 to 100°C in aqueous solutions of alkali salts and acids (HCl, H_2SO_4, H_3PO_4, CH_3COOH) on metal cathodes with a high hydrogen overvoltage, e.g., Bi, Hg, Zn, Cd, Al, Sn, Pb, Ni, Fe, Ag, Te, Au, Co, and alloys such as Woods' metal [7] or black phosphorus (current 0.2 to 5 A, current density 60 to 300 A/m², and voltage 1 to 46 V) [7 to 10]; further references are [11 to 19]. Phosphorus-laden conductive particles can also serve as the cathode [20]. The mixture is usually stirred and reducing gases like hydrogen, hydrogen sulfide, or the cathode gas are admitted into the cathode compartment [21, 22]. A vertical electrolysis cell involving a turbulent flow of the catholyte was used for adding an emulsion of white phosphorus to improve the electrolysis efficiency [33]. To prevent formation of other phosphorus hydrides and impurities, elemental phosphorus can be dispersed in aqueous solutions of methanol, ethanol, butanol, allyl alcohol, benzyl alcohol, glycol, or in carbon disulfide [16]. Addition of soluble salts of Bi, Ca, Ba, Zn, Cd, Hg, or Pb [9, 23] is recommended in order to stabilize the activity of the cathode. Riffled, ribbed, or grooved cathode surfaces [24] or a cathode fitted flush with the diaphragm above the anode [19, 25] are proposed in order to eliminate the harmful effect of spongy, loose deposits of metals and phosphorus, which may form during electrolysis. If a graphite anode [7, 14, 22] is not used, a diaphragm is necessary to avoid reaction of PH_3 with the gas (Cl_2 or O_2) forming at the anode [7, 9, 10]. The diaphragm may be made of porous sintered glass [7], porous alundum, porous porcelain, resin-impregnated wool felt, or various other separators of the types normally employed in lead storage batteries [21, 22, 26]. Blockage of the pores by adhesion of phosphorus is prevented by coating the diaphragm with glass fabric or other nonwettable materials predominantly organic polymers [26, 28]. Recommended materials for the cell vessel are glass [9, 10, 26, 28], glazed ceramic, tantalum, titanium, hard rubber, polyethylene, polyurethane, or rigid materials coated with phenol formaldehyde resin [26, 28]. For investigations in alkaline solutions, see [34, 35]. A review of PH_3 formation by electrolysis is given in [29].

A **black phosphorus** electrode [7, 30] has a low hydrogen overvoltage, and electrolysis in aqueous solution is accompanied by reduction of the cathode material with formation of phosphane. The largest PH_3 yield of 27.5% (calculated from the loss of the electrode weight) at 12.5% current efficiency was obtained in 1M K_2HPO_4 (pH 7.5) at a current density of 0.07 to 0.09 A/cm² and at 20°C. Altering the pH to either the acidic or basic side reduces the PH_3 yield. An increase in catholyte temperature from 20 to 55°C lowers the PH_3 yield by a factor of 5 to 6. Decreasing the current density also lowers the current efficiency [30].

Cathodic polarization of *n*-type **InP** single crystals in 1N H_2SO_4 leads to phosphane and metallic indium [31] according to $InP+3H^++3e^- \rightarrow PH_3+In$. In 1N KOH or 1N KF, formation of PH_3 and indium is only observed at current densities $\geqq 10^{-2}$ A/cm². Anodic dissolution of *p*-InP in KOH solution yields PH_3, probably by disproportionation of the intermediately formed H_3PO_3. Phosphane also forms in acidic (1N H_2SO_4, 1N $NH_4F \cdot HF$, 1N HF) or neutral electrolytes (0.8N KF, 1N NH_4F) [31].

Phosphinic acid, $\mathbf{H_3PO_2}$, or phosphonic acid, $\mathbf{H_3PO_3}$, can be reduced at zinc, mercury, or lead cathodes (3 to 6 V, 1 to 11 mA/cm²) to give PH_3 [32].

References:

[1] Brago, I. N.; Tomilov, A. P. (Elektrokhimiya **4** [1968] 697/700; Sov. Electrochem. [Engl. Transl.] **4** [1968] 623/4).
[2] Tomilov, A. P.; Brago, I. N.; Osadchenko, I. M. (Elektrokhimiya **4** [1968] 1153/6; Sov. Electrochem. [Engl. Transl.] **4** [1968] 1039/41).
[3] Tomilov, A. P.; Osadchenko, I. M. (Zh. Anal. Khim. **21** [1966] 1498/9; J. Anal. Chem. USSR [Engl. Transl.] **21** [1966] 1331/2).
[4] Osadchenko, I. M.; Tomilov, A. P. (Zh. Prikl. Khim. [Leningrad] **43** [1970] 1255/61; J. Appl. Chem. USSR [Engl. Transl.] **43** [1970] 1262/8).
[5] Osadchenko, I. M.; Tomilov, A. P. (Zh. Prikl. Khim. [Leningrad] **42** [1969] 1404/5; J. Appl. Chem. USSR [Engl. Transl.] **42** [1969] 1323/4).
[6] Shandrinov, N. Ya.; Tomilov, A. P. (Elektrokhimiya **4** [1968] 237/9; Sov. Electrochem. [Engl. Transl.] **4** [1968] 209/11).
[7] Gordon, I. (U.S. 3109785 [1960/63] 1/4; C.A. **60** [1964] 2552).
[8] Osadchenko, I. M.; Tomilov, A. P. (Usp. Khim. **38** [1969] 1098/107; Russ. Chem. Rev. [Engl. Transl.] **38** [1969] 495/504, 499/502).
[9] Price, D. T.; Gordon, I. (Ger. 1112722 [1960/62] 1/5; C.A. **56** [1962] 11366).
[10] Price, D. T.; Gordon, I. (Brit. 889639 [1959/60] 1/11).

[11] Gordon, I. (U.S. 3109791 [1960/63] 1/4; C.A. **60** [1964] 2552).
[12] Gordon, I. (U.S. 3109792 [1960/63] 1/5; C.A. **60** [1964] 2552).
[13] Gordon, I. (U.S. 3109793 [1960/63] 1/4; C.A. **60** [1964] 2552).
[14] Gordon, I. (U.S. 3109794 [1960/63] 1/4; C.A. **60** [1964] 2552).
[15] Gordon, I. (U.S. 3109795 [1960/63] 1/4; C.A. **60** [1964] 2552).
[16] Miller, G. T. (U.S. 3109790 [1960/63] 1/4; C.A. **60** [1964] 2552).
[17] Price, D. T.; Gordon, I. (U.S. 3109787 [1960/63] 1/7; C.A. **60** [1964] 2552).
[18] Miller, G. T.; Steingart, J. (U.S. 3109788 [1960/63] 1/4; C.A. **60** [1964] 2552).
[19] Miller, G. T.; Steingart, J. (U.S. 3109789 [1960/63] 1/5; C.A. **60** [1964] 2552).
[20] Haycock, E. W.; Rhodes, P. R. (U.S. 3404076 [1965/68] 1/5; C.A. **69** [1968] No. 112921).

[21] Miller, G. T. (Ger. 1210425 [1964/66] 1/5; C.A. **64** [1966] 13767).
[22] Miller, G. T. (U.S. 3312610 [1963/67] 1/4; C.A. **67** [1967] No. 39625).
[23] Miller, G. T.; Steingart, J. (U.S. 3109786 [1960/63] 1/3; C.A. **60** [1964] 2552).
[24] Miller, G. T. (Ger. 1210426 [1964/66] 1/8; C.A. **64** [1966] 12198).
[25] Miller, G. T.; Steingart, J. (U.S. 3262871 [1963/66] 1/5; C.A. **65** [1966] 11776).
[26] Miller, G. T. (U.S. 3337433 [1963/67] 1/4; C.A. **68** [1968] No. 35369).
[27] Miller, G. T. (Ger. 1210424 [1964/66] 1/4; C.A. **64** [1966] 13767).
[28] Miller, G. T. (Brit. 1042391 [1964/66] 1/5; C.A. **65** [1966] 18168).
[29] Presnov, A. E.; Tomilov, A. P.; Promonenkov, B. K.; Fioshin, M. Ya. (Itogi Nauki Elektrokhim. **6** [1971] 293/326, 294/9; C.A. **77** [1972] No. 28046).
[30] Chernykh, I. N.; Zubova, E. V.; Savranskii, V. V.; Tomilov, A. P. (Elektrokhimiya **16** [1980] 1797/800; Sov. Electrochem. [Engl. Transl.] **16** [1980] 1476/9).

[31] Erusalimchik, I. G.; Levin, D. M. (Elektrokhimiya **16** [1980] 1854/6; Sov. Electrochem. [Engl. Transl.] **16** [1980] 1522/4).
[32] Baudler, M.; Schellenberg, D. (Z. Anorg. Allg. Chem. **340** [1965] 113/25).
[33] Tsuchiya, H.; Otsujii, A.; Sakagami, Y. (Jpn. Kokai Tokkyo Koho 01-139781 [1987/89] 1/5 from C.A. **112** [1990] No. 65485).
[34] Barry, M. L.; Tobias, C. W. (Electrochem. Technol. **4** [1966] 502/6).
[35] Barry, M. L. (UCRL-11573 [1964]; C.A. **63** [1965] 3894).

1.3.1.3.2 Formation from Phosphorus and Hydrogen or Hydrogen Compounds

Phosphane ist obtained by reacting phosphorus vapor with **hydrogen** (3 to 4% in N_2; 10 L/h) in an electric discharge (20 to 32 kV) at 330 to 400°C [1]. It also forms by subjecting gaseous or finely divided phosphorus to a plasma stream of hydrogen (mole ratio $P_4:H_2=1:6$ to 1:30; 1.34 to 25.5 mmol H_2/min) obtained by irradiation preferably in the microwave range 1000 to 5000 MHz. The reaction zone is cooled to −78°C, and the pressure ranges from 0.01 to 0.05 atm. The rate of formation is 2.78 mmol PH_3/h for solid red phosphorus and 1.103 mmol/h for white phosphorus vapor [2] the best yields being 68.8% for red phosphorus and 25.6% for white phosphorus; see also [3 to 6]. Small amounts of PH_3 (2 to 3%) were detected when white phosphorus in cyclohexane solution was exposed to γ-rays (^{60}Co, 100 to 3000 Curie) at 25°C. Phosphorus acts as a scavenger for the free hydrogen atoms and C_6H_{11} radicals from the radiolysis of cyclohexane [7]. Vaporization of a red-phosphorus target (purity of 99.99%) in an atmosphere of hydrogen (10^4 to 10^5 Pa) by emission from a transversely excited atmospheric-pressure (TEA) CO_2 laser with a pulse energy of up to 2 J yields PH_3. The magnitude of the mass removed from the target amounts to 10^{-5} g/J and depending on reaction conditions (e.g., energy flux and H_2 pressure), the maximum conversion was 4.7% based on the vaporized phosphorus. Thermodynamic calculations suggest that PH_3 formation is obviously due to non-equilibrium concentrations of, for example, P, P_2, H, and H_2^* particles [8]. For ab initio MO calculations related to the reactions $P_2+3H_2 \rightarrow 2PH_3$, see [9].

White phosphorus disproportionates under certain conditions yielding PH_3 and phosphorus compounds in higher oxidation states [10]. Reaction of white phosphorus with **water** in 57% hydroiodic acid containing glacial acetic acid at 100 to 130°C gives PH_3 free of P_2H_4 with 89% yield according to $2P+3H_2O \rightarrow PH_3+H_3PO_3$ [11]. In NaOH solution (1 to 10 M), PH_3 is formed at a white phosphorus electrode (silver amalgam or graphite matrix). The rate of formation at the silver amalgam matrix was investigated at 40 and 60°C depending upon the concentration of NaOH and the current density. When current is applied, less PH_3 is formed than without it [10, 12].

Red phosphorus in a highly reactive form (containing 0.1 to 4% white phosphorus [13] or about 15% phosphorus extractable by CS_2 [14]) when reacted with water and phosphoric acid at about 280°C in a silica or graphite vessel gave PH_3 via $8P + 12H_2O \rightarrow 5PH_3 + 3H_3PO_4$. The gaseous reaction product contained ≧90 vol% PH_3 [13, 14]. The reactive red phosphorus was obtained either by heating white phosphorus vapor in an electric arc and quenching the vapor in water [14] or by first heating white phosphorus in phosphoric acid to 280°C [13]. A single-step reaction is possible if white phosphorus and steam (P_4 : H_2O = 1 : 0.1 to 1 : 10) in the presence of ammonia (P_4 : NH_3 = 1 : 0.05 to 1 : 5) are heated by passing the vapors through a stainless-steel reactor over an active carbon catalyst (BET surface >10 m^2/g) at 280 to 350°C (residence time 0.5 to 50 s). Unreacted NH_3 is subsequently removed by washing the gas stream with aqueous H_3PO_4. The diluted acid also removes unreacted phosphorus (about 70%), which can be recycled. The PH_3 from the washing vessel is 90 to 99% pure. The yield of condensed PH_3 is 94% if the ratio of the starting materials is 2 mol P_4/h : 3 mol H_2O/h : 3 mol NH_3/h [15].

A dispersion of white phosphorus in water containing TiO_2, calcium phosphate or phosphite, activated carbon, talc, $BaSO_4$, or pulp reacts with sodium hydroxide solution at 75 to 90°C under nitrogen to form PH_3 and $NaHPO_2$ with 94.9 and 84.8% yield, respectively [16]. PH_3 yields up to 100% at a 25% conversion of phosphorus from the reaction $4P + 3H_2O + 3NaOH \rightarrow PH_3 + 3NaH_2PO_2$ are obtained at 50°C in aqueous methanol or ethanol solution with at least 70 vol% alcohol. By-products such as H_2, P_2H_4, and Na_2HPO_3 decrease, especially in absolute alcohol. The PH_3 yield increases with increasing NaOH : P mole ratio; a working range of 0.6 to 0.75 is preferred [17]. PH_3 yields can be improved by heating the residue of the reaction with 35% phosphoric acid at 400°C (cf. 1.3.1.3.4, p. 122) [18]. The reaction of white phosphorus dissolved in an aliphatic alcohol (e.g., pentanol or hexanol) [19, 20] or in alcohol mixtures (e.g., 35% n-pentanol, 15% 2-methylbutanol, 25% n-octanol, and 25% n-decanol [21]) with 50% [19, 20], 65% [20], or 70% [21] aqueous NaOH under nitrogen preferably at 60°C [19] or 70°C [20] yields a phosphorus conversion of about 30%, if carried out on a large scale. The PH_3 contains hydrogen as a by-product [19]. A two-stage, continuous production in two connected reactors is possible in the presence of surface-active agents [21]. Commercial red phosphorus in the presence of water liberates PH_3 in the ppm range [22].

PH_3 forms among other phosphorus compounds upon reacting white phosphorus with **hydrogen iodide** in anhydrous carbon disulfide in the range from −45°C to ambient temperature [23].

Reaction of white phosphorus with **$LiPH_2$** yields PH_3 and Li_3P_7 [24 to 26] or Li_2P_{16} [24, 27] depending on the reactant ratios and reaction conditions.

References:

[1] Ershov, V. A.; Smirnova, N. A.; Nikolaeva, I. A. (Zh. Prikl. Khim. [Leningrad] **47** [1974] 2385/9; J. Appl. Chem. USSR [Engl. Transl.] **47** [1974] 2455/8).

[2] Mezey, E. J.; Laughlin, R. G. (U.S. 3652437 [1969/72] 1/6; C.A. **77** [1972] No. 37217).

[3] Naitoh, M.; Katoh, M.; Soga, T.; Jimbo, T.; Umeno, M. (Proc. Jpn. Symp. Plasma Chem. **2** [1989] 203/7 from C.A. **112** [1990] No. 89421).

[4] Naitoh, M.; Umeno, M. (Jpn. J. Appl. Phys. Pt. 2 **26** [1987] L1538/L1539).

[5] Umeno, M.; Sakai, S.; Naito, M.; Uchiyama, H. (Jpn. Kokai Tokkyo Koho 01-68475 [89-68475] [1987/89] 1/4 from C.A. **112** [1990] No. 67219).

[6] Marinace, J. C. (Eur. Appl. 245600 [1987] 1/8; C.A. **108** [1988] No. 29875).

[7] Asmus, K.-D.; Henglein, A.; Meissner, G.; Perner, D. (Z. Naturforsch. **19b** [1964] 549/57).

[8] Akinfiev, N. N.; Kuznetsova, T. V.; Skachkov, A. N.; Sosnine, G. F. (Khim. Vys. Energ. **18** [1984] 63/7; High Energy Chem. [Engl. Transl.] **18** [1984] 49/52).

[9] Pietro, W. J.; Francl, M. M.; Hehre, W. J.; De Frees, D. J.; Pople, J. A.; Binkley, J. S. (J. Am. Chem. Soc. **104** [1982] 5039/48).
[10] Barry, M. L.; Tobias, C. W. (Electrochem. Technol. **4** [1966] 502/6).

[11] Kestner, M. O.; Teliszczak, P. J. (U.S. 4207300 [1978/80] 1/3; C.A. **93** [1980] No. 116696).
[12] Barry, M. L. (UCRL-11573 [1964]; C.A. **63** [1965] 3894).
[13] Lowe, E. J.; Ridgway, F. A. (Brit. 990918 [1962/65] 1/5; Fr. 1352605 [1963/64] 1/5; C.A. **61** [1964] 9198; Ger. 1219911 [1963/67] 1/4; U.S. 3371994 [1963/68] 1/3).
[14] Lowe, E. J.; Ridgway, F. A. (Brit. 1175522 [1965/69] 1/4; Brit. 1175523 [1965/69] 1/4; C.A. **72** [1970] No. 57284; Fr. 1504213 [1966/67] 1/5; C.A. **69** [1968] No. 88476; U.S. 4013756 [1974/77] 1/4; C.A. **86** [1977] No. 157750).
[15] Hestermann, K.; Heymer, G. (Ger. Offen. 2721374 [1977/78] 1/7; C.A. **90** [1979] No. 139765).
[16] Ouchi, Y.; Tsutsumi, Y. (Jpn. 71-43621 [1967/71] 1/4; C.A. **76** [1972] No. 115623).
[17] Cummins, R. W. (U.S. 2977192 [1958/61] 1/3; C.A. **1961** 18039).
[18] Cummins, R. W. (U.S. 3116109 [1961/63] 1/3; C.A. **60** [1964] 10263).
[19] Hestermann, K.; Stenzel, J.; Heymer, G.; May, C. (Ger. Offen. 2549084 [1975/77] 1/8; C.A. **87** [1977] No. 55236).
[20] Stenzel, J.; Heymer, G.; May, C. (Ger. Offen. 2632316 [1976/78] 1/15; C.A. **88** [1978] No. 173006).

[21] Elsner, G.; Klose, W.; May, C.; Heymer, G. (Ger. Offen. 2840147 [1978/80] 1/11; C.A. **93** [1980] No. 28738).
[22] McCallum, J. J.; Alder, J. F. (J. Chem. Technol. Biotechnol. **44** [1989] 121/33).
[23] Schmidt, M.; Schröder, H. H. J. (Z. Anorg. Allg. Chem. **378** [1970] 185/91).
[24] Baudler, M.; Glinka, K. (Inorg. Synth. **27** [1990] 227/35).
[25] Baudler, M. (Angew. Chem. **94** [1982] 520/39, 526/8).
[26] Baudler, M.; Faber, W. (Chem. Ber. **113** [1980] 3394/5).
[27] Baudler, M.; Exner, O. (Chem. Ber. **116** [1983] 1268/70).

1.3.1.3.3 Formation from Phosphorus Compounds and Hydrogen or Hydrides

The energetics of the mostly hypothetical formation reactions of PH_3 from H_2 and the molecules P_2H_4, PN, PHO, PF_5, POF_3, PSF_3, HCP, H_2CPH, H_3CPH_2, and $(CH_3)_3PO$ were predicted by ab initio MO calculations at the STO 3-21G and STO 3-21G* level [1].

Phosphane is obtained in nearly quantitative yield by reduction of **PF_3** or **POF_3** with $LiAlH_4$ in diethyl ether [2].

PH_3 is also obtained by the reaction of **PCl_3** with LiH [3, 4], $LiAlH_4$ [5], $(i\text{-}C_4H_9)_2AlH$ [6], $NaAlH_2(OCH_2CH_2OCH_3)_2$ [5], or a mixture of NaH and $NaBH_4$ [7]. If solid LiH is used, the hydride is mixed with an inert powder like quartz and whirled by an inert-gas stream in a fluidized-bed reactor at about 200°C where the PCl_3 vapor is supplied from below [3]. In another process, LiH is first formed by the reaction of H_2 with Li obtained by electrolyzing an LiCl/KCl mixture, the chlorine evolved at the anode being used to prepare PCl_3. The lithium hydride passes into a separate chamber where it reacts with this PCl_3 [4]. The formation reactions of PH_3 in solution with $LiAlH_4$, $NaAlH_2(OCH_2CH_2OCH_3)_2$ [5], or an $NaH/NaBH_4$ mixture [7] are carried out in ether (ambient temperature), benzene (ambient temperature) [5], or in a diglyme/mineral oil mixture (reactor cooled with liquid nitrogen to keep the temperature below 25°C) [7] while PCl_3 is added dropwise. $\mathbf{POCl_3}$ is not reduced to PH_3 by LiH at 35°C [8].

PH_3 is evolved on warming a 1:2 mixture of $\mathbf{(NPCl_2)_3}$ and $LiAlH_4$ in diethyl ether from 0 to 20°C. There is no reduction of $(NPCl_2)_3$ by LiH, even in boiling toluene [8].

Only traces of PH_3 are formed by reacting $\mathbf{PSF_3}$ or $\mathbf{PSCl_3}$ with hydrazine hydrate added at liquid nitrogen temperature and warmed to ambient temperature [9]. However, PH_3 forms almost quantitatively by the reaction of PSF_3 with $LiAlH_4$ in tetrahydrofuran or $NaBH_4$ in diglyme [10]. Small amounts of PH_3 are evolved when $\mathbf{P_4O_{10}}$ reacts with $LiAlH_4$ in the absence of a solvent at 165°C [11] or when a hydroborate solution and either **phosphinates** or **phosphonates** are added to an acid [12].

Organophosphorus compounds such as $\mathbf{(C_6H_5O)_3PO}$, $\mathbf{(\textit{n}\text{-}C_4H_9O)_3PO}$, $\mathbf{(C_6H_5O)_2P(O)Cl}$, $\mathbf{\textit{p}\text{-}[(C_6H_5O)_2P(O)O]_2C_6H_4}$, and $\mathbf{(C_6H_5CH_2O)_2P(O)ONa}$ are reduced by $LiAlH_4$ in tetrahydrofuran at ambient or higher temperatures to give PH_3. Triphenylphosphane does not react under these conditions [13].

References:

[1] Pietro, W. J.; Francl, M. M.; Hehre, W. J.; DeFrees, D. J.; Pople, J. A.; Binkley, J. S. (J. Am. Chem. Soc. **104** [1982] 5039/48).
[2] Kalbandkeri, R. G.; Padma, D. K.; Murthy, A. R. V. (Indian J. Chem. A **23** [1984] 875).
[3] Siemens-Schuckertwerke A.-G. (Neth. Appl. 6504634 [1965] 1/5; C.A. **64** [1966] 13803).
[4] Litz, L. M.; Ring, S. A. (U.S. 3163590 [1961/64] 1/11; C.A. **62** [1965] 8710).
[5] Peet, J. H. J. (School Sci. Rev. **54** [1973] 534/5).
[6] Bochkarev, E. P.; Dudina, N. B.; Lifanova, T. A.; Khripunov, V. M.; Yushkov, Yu. V. (Nauchn. Tr. Nauchn. Issled. Proekt. Inst. Redkometal. Prom. **33** [1971] 64/6; C.A. **76** [1972] No. 74336).
[7] Gordon, R. G. (Belg. 890356 [1981/82] 1/14; C.A. **96** [1982] No. 199887).
[8] Levin, V. V.; Rumyantseva, Z. G.; Mironova, V. V. (Izv. Akad. Nauk SSSR Neorg. Mater. **3** [1967] 586; Inorg. Mater. [Engl. Transl.] **3** [1967] 523/4).
[9] Padma, D. K.; Vijayalakshmi, S. K.; Murthy, A. R. V. (Curr. Sci. **47** [1978] 153/5).
[10] Padma, D. K.; Vijayalakshmi, S. K.; Murthy, A. R. V. (J. Indian Inst. Sci. **59** [1977] 108/10).
[11] Bellama, J. M.; MacDiarmid, A. G. (Inorg. Chem. **7** [1968] 2070/2).
[12] Macomber, J. D. (unpublished results from Jolly, W. L.; J. Am. Chem. Soc. **83** [1961] 335/7, 337).
[13] Chumachenko, M. N.; Serebryakova, I. F. (Izv. Akad. Nauk SSSR Ser. Khim. **1971** 2314/6; Bull. Acad. Sci. USSR Div. Chem. Sci. [Engl. Transl.] **1971** 2191/3).

1.3.1.3.4 Formation by Thermal Decomposition of PH-Containing Compounds

Most of the higher phosphanes of the general formula P_xH_y starting with P_2H_4 are thermally and photolytically unstable at room temperature or above, releasing PH_3 upon decomposition. The pyrolysis of, e.g., P_2H_4 or P_3H_5 at high temperatures is accompanied by the formation of large amounts of PH_3 and can be used to obtain pure PH_3 (cf. 1.3.1.3.9, p. 129) and higher phosphanes. In solution most of the P_xH_y compounds can be metallated at low temperatures, e.g., with LiBu or $LiPH_2$, in the latter case with direct formation of PH_3. The resulting, partially metallated species quite often rearrange in solution upon warming to ambient temperature to give PH_3 and fully metallated phosphorus compounds. Details of the usually very complicated parallel and consecutive reactions occurring during pyrolysis, photolysis, and metallation of the polyphosphanes will be described in the forthcoming "Phosphorus" Suppl. Vol. C 2 [1 to 14]; for some earlier results, see "Phosphor" C, 1965, pp. 52/3. When the pressure above $LiPH_2 \cdot$ monoglyme is reduced below about 0.5 Torr by pumping, PH_3 and monoglyme are slowly split off leaving Li_2PH [15].

PH_2-containing silyl- or germylphosphanes quite often release PH_3 when they are decomposed thermally or photolytically. Redistribution reactions of this type initiated by Lewis acids are covered in Section 1.3.1.3.7, p. 126. Phosphane is formed, e.g., during the thermal decomposition of $SiH(PH_2)_3$ (70°C; 3 h; 11% dec.), $SiH_2(PH_2)_2$ (95°C; 6 h; 5% dec.) [16], $(CH_3)_2Si(PH_2)_2$ (190 to 200°C; 70 h; 43% dec.) [17], and SiH_3PH_2 (300°C) [18]. GeH_3PH_2 slowly decomposes already above −40°C evolving PH_3 [16, 19], and the decomposition of $R_2Ge(PH_2)_2$ to give PH_3 can be catalyzed by Hg ($R=CH_3$; 100°C; 24 h; 100% dec.) [20] or initiated by small amounts of O_2 ($R=CH_3$, C_2H_5; 120 to 140°C; 72 to 96 h; 90 to 95% dec.) [21].

One of the most versatile laboratory methods for preparing PH_3, namely the pyrolysis of H_3PO_3 (phosphonic acid) already mentioned in "Phosphor" C, 1965, p. 13, is described in detail in [22]. In this method the acid is heated in vacuum first to 200 and finally to 350°C to obtain about 0.1 mol PH_3 (97% yield) containing small amounts of P_2H_4 [22]. However, in the original paper H_3PO_3 is heated under 400 to 500 Torr N_2. If the total pressure above the molten mixture of acids is held at this range all the time during the PH_3 evolvement by controlled pump-off of the PH_3/N_2 mixture, PH_3 comes off readily and smoothly at much lower temperatures. About 40 mL of liquid phosphane essentially free of P_2H_4 can thus be obtained from about 400 g of H_3PO_3 [23]. Phosphonic acid wastes from chlorination of fatty acids can be treated with phosphoric acid at 350°C to give phosphane (23.7% yield) and polyphosphoric acids [24].

PH_3 is formed along with H_2, $Na_4P_2O_7$, and $NaPO_3$ during the thermal decomposition of NaH_2PO_2 at 300°C under N_2 [35]. The formation of pure phosphane with 95.5% yield by introducing NaH_2PO_2 into anhydrous H_3PO_4 (molar ratio 1:2.5 to 1:10) at 135 to 150°C and subsequent heating to 220°C was claimed in a patent [25]. When PH_3 is prepared by disproportionation of P_4 in aqueous NaOH solution (cf. 1.3.1.3.2, p. 119), its yield can be substantially improved by heating the residual sodium salts in 35% H_3PO_4 to 400°C. The reactions probably proceed via $2NaH_2PO_2 \cdot H_2O + H_3PO_4 \rightarrow PH_3 + 2NaH_2PO_4 + H_2O$ and $4Na_2HPO_3 \cdot 5H_2O + 5H_3PO_4 \rightarrow PH_3 + 8NaH_2PO_4 + 20H_2O$, whereby the dihydrogenphosphate is converted to glassy polyphosphates [26]. PH_3 is formed during the pyrolysis of molten $SnHPO_3$ at 325°C [27]. Similarly, pyrolysis of $Sc_2(HPO_3)_3 \cdot xH_2O$ or the dehydrated phosphonate at 600 to 1000°C under He yields PH_3 along with $Sc_4(P_2O_7)_3$ and $ScPO_4$ [28]. PH_3 is also obtained together with Na_3PS_4 and S by the decomposition of Na_2HPS_3 between 150 and 500°C [29].

PH_3 was found to be one of the volatile thermolysis products of the arylphosphonous acids $RC_6H_4P(O)(OH)H$ with $R=H$ or CH_3 in chlorobenzene in sealed tubes at 200 to 220°C after 20 h [30]. Pyrolysis studies of i-$C_4H_9PH_2$ and t-$C_4H_9PH_2$, intended to investigate the possible use of these compounds as precursors in the OMVPE process, revealed that PH_3 is one of the first decomposition products [33, 34].

The energetics of the decomposition of PH_5 [31] and $CH_2{=}PH_3$ [32] to give PH_3 were evaluated by ab initio MO studies at different computational levels.

References:

[1] Baudler, M. (Angew. Chem. **99** [1987] 429/51, 435/7).
[2] Baudler, M.; Heumüller, R. (Z. Anorg. Allg. Chem. **559** [1988] 49/56).
[3] Baudler, M.; Heumüller, R.; Hahn, J. (Z. Anorg. Allg. Chem. **529** [1985] 7/14).

[4] Baudler, M.; Heumüller, R.; Germeshausen, J.; Hahn, J. (Z. Anorg. Allg. Chem. **526** [1985] 7/14).
[5] Baudler, M.; Heumüller, R.; Langerbeins, K. (Z. Anorg. Allg. Chem. **514** [1984] 7/17).
[6] Baudler, M.; Heumüller, R.; Düster, D.; Germeshausen, J.; Hahn, J. (Z. Anorg. Allg. Chem. **518** [1984] 7/13).
[7] Baudler, M.; Heumüller, R. (Z. Naturforsch. **39b** [1984] 1306/14).
[8] Baudler, M. (Angew. Chem. **94** [1982] 520/39).
[9] Baudler, M.; Ternberger, H.; Faber, W.; Hahn, J. (Z. Naturforsch. **34b** [1979] 1690/7).
[10] Callen, R. B.; Fehlner, T. P. (J. Am. Chem. Soc. **91** [1969] 4122/8).

[11] Fehlner, T. P. (J. Am. Chem. Soc. **90** [1968] 6062/6).
[12] Fehlner, T. P. (J. Am. Chem. Soc. **90** [1968] 4817/22).
[13] Fehlner, T. P. (J. Am. Chem. Soc. **89** [1967] 6477/82).
[14] Fehlner, T. P. (J. Am. Chem. Soc. **88** [1966] 1819/21).
[15] Schäfer, H.; Fritz, G.; Hölderich, W. (Z. Anorg. Allg. Chem. **428** [1977] 222/4).
[16] Norman, A. D.; Wingeleth, D. C. (Inorg. Chem. **9** [1970] 98/103).
[17] Fritz, G.; Uhlmann, R.; Hölderich, W. (Z. Anorg. Allg. Chem. **442** [1978] 86/90).
[18] Elliot, L. E.; Estacio, P.; Ring, M. A. (Inorg. Chem. **12** [1973] 2193/4).
[19] Norman, A. D.; Wingeleth, D. C.; Heil, C. A. (Inorg. Synth. **15** [1974] 177/81).
[20] Dahl, A. R.; Norman, A. D. (J. Am. Chem. Soc. **92** [1970] 5525/7).

[21] Dahl, A. R.; Norman, A. D.; Shenav, H.; Schaeffer, R. (J. Am. Chem. Soc. **97** [1975] 6364/70).
[22] Gokhale, S. D.; Jolly, W. L. (Inorg. Synth. **9** [1967] 56/8).
[23] Martin, D. R.; Dial, R. E. (J. Am. Chem. Soc. **72** [1950] 852/6).
[24] Heymer, G.; Scheibitz, W.; Spott, H. (Ger. Offen. 2314232 [1973/74] 1/2; C.A. **82** [1975] No. 61348).
[25] Issleib, K.; Gans, W.; Krech, F. (Ger. [East] 214593 [1983/84] 1/5; C.A. **103** [1985] No. 24441).
[26] Cummins, R. W. (U.S. 3116109 [1961/63] 1/3; C.A. **60** [1964] 10263).
[27] Donaldson, J. D.; Moser, W.; Simpson, W. B. (J. Chem. Soc. **1964** 323/6).
[28] Petrů, F.; Muck, A. (Collect. Czech. Chem. Commun. **36** [1971] 3774/9).
[29] Seel, F.; Zindler, G. (Z. Anorg. Allg. Chem. **470** [1980] 167/70).
[30] Chauzov, V. A.; Kostina, L. P. (Zh. Obshch. Khim. **61** [1991] 1266/7; J. Gen. Chem. USSR [Engl. Transl.] **61** [1991] 1150).

[31] Kutzelnigg, W.; Wasilewski, J. (J. Am. Chem. Soc. **104** [1982] 953/60).
[32] Mitchell, D. J.; Wolfe, S.; Schlegel, H. B. (Can. J. Chem. **59** [1981] 3280/92).
[33] Larsen, C. A.; Chen, C. H.; Kitamura, M.; Stringfellow, G. B.; Brown, D. W.; Robertson, A. J. (Appl. Phys. Lett. **48** [1986] 1531/3).
[34] Chen, C. H.; Larsen, C. A.; Stringfellow, G. B.; Brown, D. W.; Robertson, A. J. (J. Cryst. Growth **77** [1986] 11/8).
[35] Kobayashi, M. (Nippon Kagaku Zasshi **81** [1960] 1838/40 from C.A. **56** [1962] 11198).

1.3.1.3.5 Formation by Hydrolysis of Metal Phosphides

The interrelation between structure and hydrolytic behavior of metal phosphides is reviewed in [1]. Of the numerous phosphides that release PH_3 when hydrolyzed, Ca_3P_2 and AlP have been predominantly used to prepare the compound on a laboratory scale.

When Ca_3P_2 is hydrolyzed with water, the resulting PH_3 is contaminated with relatively large amounts of P_2H_4, with higher phosphanes, and with H_2. A convenient laboratory procedure for the preparation of pure phosphane was developed in which the generation of the crude material is followed by the thermal decomposition of P_2H_4 and higher phosphane by-products at 450°C to give additional PH_3. Thus, starting with about 400 g of phosphide, about 40 g of doubly distilled, essentially P_2H_4-free PH_3 can be synthesized. Schematic drawings of the apparatus are displayed in the original papers [2, 3]. The evolution of PH_3 from slaked carbide lime containing 1 to 2% Ca_3P_2 was found to be a heterogeneous first-order reaction [4]. Hydrolysis of annealed mixtures of Mg_3P_2 with Mg_3As_2 or Mg_3N_2 also yield large amounts of PH_3 [5]. The production of a continuous PH_3 stream from Mg_3P_2 and H_2O in an inert gas such as CO_2 (about 5 vol% PH_3) is described in [23].

A versatile laboratory method for the synthesis of PH_3 by hydrolyzing **AlP** first with water and then with 6 M aqueous H_2SO_4 solution is described in [6]. The authors point out that this method is useful for the production of phosphane in quantities up to 5 g and that it can be used to introduce the compound in the gas phase directly into a subsequent reaction. However, the nonpurified PH_3 then contains small amounts of P_2H_4. A schematic drawing of the apparatus used for preparing and purifying the PH_3 is depicted in the original paper [6]. A very convenient procedure for synthesizing even larger quantities of PH_3 in the laboratory is to hydrolyze 20-g portions of AlP with neat water; the author emphasizes that the formation of P_2H_4 impurities can be avoided almost completely by using stoichiometrically pure AlP, i.e., containing no excess phosphorus and thus no P–P bonds. The apparatus and the synthetic procedure are described in great detail. Pressure changes in the apparatus, which can cause spontaneous decomposition of PH_3 into the elements, are to be avoided. Pressure changes can, for example, be caused by the rapid uptake of PH_3 in the condensation tube or it may result from a blockage in the drying tube [7]. Formation of PH_3 by addition of water to a suspension of AlP in methanol, ethanol, *i*-propanol, methyl ethyl ketone, or acetone is described in [8]. For the effect of water and temperature on the release of phosphane from AlP-containing formulations used for fumigation, see [9]; see also [10]. Dissolution of **InP** in 1.5 to 9 M aqueous HCl solution also yields PH_3 [11].

Zn_3P_2 is not hydrolyzed by neat water, but only in acidic or basic solution or at pH 7 in a phosphate buffer solution. The rate of PH_3 formation was found to depend mainly on the pH and temperature [12]. Finely dispersed Zn_3P_2, obtained by reduction of $Zn_3(PO_4)_2$ with carbon powder or petroleum coke at 1000 to 1300°C, is hydrolyzed by 20 to 85 wt% H_3PO_4 at 20 to 200°C to give PH_3 (>95 vol%) and H_2. The regenerated zinc phosphate can be dehydrated at around 250°C and recycled [13]. Zn_3P_2 mixed with clay oxide soils of volcanic origin (pH 5.4, 5.7, or 6.9) at various moisture levels decomposes with liberation of variable amounts of PH_3 [12].

Dilute acids hydrolyze LaP_2 to give PH_3, P_2H_4, and solid yellow phosphanes. The slower hydrolysis in dilute bases begins with polycrystalline fragments of LaP_2 yielding only PH_3 [14].

Phosphides obtained by melting ferrophosphorus with aluminium at 1350°C under argon [15] or with calcium silicide at 1250 to 1450°C [16] were claimed to react with water to give phosphane [15, 16] which is free of P_2H_4 [16]; see also [17, 24]. Hydrolysis of an amalgam containing 0.3% sodium and 0.14% phosphorus at 60°C with water or alcohol yields 49% PH_3 [18]. Ferrosilicon alloys (45 to 65 wt% Si) containing traces of phosphide decompose into a powder which liberates PH_3 in moist air [19]. Traces of PH_3 also form by machining spheroidal graphite iron in moist air [20] and by etching phosphorus-containing carbon steels [21]; see also [22].

References:

[1] von Schnering, H. G. (in: Rheingold, A. L.; Homoatomic Rings, Chains, and Macromolecules of Main-Group Elements, Elsevier, Amsterdam 1977, pp. 317/48, 322/4).
[2] Baudler, M.; Ständecke, H.; Dobbers, J. (Z. Anorg. Allg. Chem. **353** [1967] 122/6).
[3] Klement, R. (in: Brauer, G.; Handbuch der Präparativen Anorganischen Chemie, Vol. 1, Enke, Stuttgart 1975, pp. 510/3).
[4] Vladimirov, V. A.; Vorob'eva, I. V.; Stepanova, L. V. (Fosfornaya Prom-st. **1978** No. 4, pp. 28/32; C. A. **90** [1979] No. 173763).
[5] Royen, P.; Rocktäschel, C. (Z. Anorg. Allg. Chem. **346** [1966] 290/4).
[6] Mariott, R. C.; Odom, J. D.; Sears, C. T., Jr. (Inorg. Synth. **14** [1973] 1/4).
[7] Fluck, E. (Fortschr. Chem. Forsch. **35** [1973] 1/64, 17/21).
[8] Kuznetsov, E. V.; Valetdinov, R. K.; Zavlin, P. M. (U.S.S.R. 125551 [1960] 1/2; C. A. **1960** 15874).
[9] Banks, H. J. (J. Stored Prod. Res. **27** [1991] 41/56).
[10] Samejima, S. (Jpn. Kokai Tokkyo Koho 63-215503 [88-215503] [1987/88] 1/3 from C. A. **110** [1989] No. 78762).

[11] Notten, P. H. L. (J. Electrochem. Soc. **131** [1984] 2641/4).
[12] Hilton, H. W.; Robison, W. H. (J. Agric. Food Chem. **20** [1972] 1209/13).
[13] Stenzel, J.; Heymer, G. (Ger. Offen 2639941 [1976/78] 1/9; C. A. **88** [1978] No. 193925).
[14] von Schnering, H. G.; Wichelhaus, W.; Schulze Nahrup, M. (Z. Anorg. Allg. Chem. **412** [1975] 193/201).
[15] Scott, M. J.; Stenzel, J. A. (U.S. 3650732 [1969/72] 1/3; C. A. **76** [1972] No. 144001).
[16] Stenzel, J. (Ger. Offen. 2524259 [1975/76] 1/9; C. A. **86** [1977] No. 75423).
[17] Tom, G. M.; McManus, J. V. (PCT Int. Appl. 90-11246 [1990] 1/54 from C. A. **114** [1991] No. 84843).
[18] Minklei, A. O. (U.S. 3387934 [1964/68] 1/2; C. A. **69** [1968] No. 44976).
[19] Wefers, K. (Metall [Berlin] **17** [1963] 446/51, 451).
[20] Mathew, G. G. (Ann. Occup. Hyg. **4** [1961] 19/35; C. A. **61** [1964] 6256).

[21] Vdovenko, I. D.; Vakulenko, L. I.; Yakovleva, L. A.; Shedenko, L. I., Bereshlavskaya, A. N.; Yatsenko, V. S. (Ukr. Khim. Zh. [Russ. Ed.] **50** [1984] 418/21; Sov. Prog. Chem. [Engl. Transl.] **50** No. 4 [1984] 88/91).
[22] Mosher, E.; Skoeld, R. (Lubr. Eng. **45** [1989] 445/50; C. A. **112** [1990] No. 11331).
[23] Dörnemann, M.; Reif, H. (Ger. Offen. 3618297 [1987] 1/4; C. A. **108** [1988] No. 200241).
[24] Banks, H. J.; Waterford, C. J. (PCT Int. Appl. 91-19671 [1990/91] 1/60 from C. A. **116** [1992] No. 154887).

1.3.1.3.6 Formation by Hydrolysis of PH_4I

Two versatile laboratory methods for obtaining a continuous stream of P_2H_4-free and thus not spontaneously inflammable PH_3 by the well-known hydrolysis of PH_4I (cf. "Phosphor" C, 1965, pp. 13/4) were described again in more detail. Hydrolyzing 7.3 g PH_4I with dilute aqueous KOH produces about 1 L PH_3. When moist ether is used to hydrolyze 10 g PH_4I, the gas evolution lasts for about 8 h yielding a continuous stream of ether-containing phosphane. This reaction can even be carried out in a small Kipp apparatus.

Klement, R. (in: Brauer, G.; Handbuch der Präparativen Anorganischen Chemie, Vol. 1, Enke, Stuttgart 1975, p. 513).

1.3.1.3.7 Formation by Miscellaneous Reactions

On warming a P_2H_4/HF mixture from liquid N_2 temperature to 0°C, PH_3 vigorously forms along with other products [1].

The cleavage of the Si–P bond in SiH_3PH_2 with water, acids, bases, liquid NH_3, or alcohols to give PH_3 was already mentioned in "Phosphor" C, 1965, p. 10, and reviewed in [2]. Analogous Si–P cleavage reactions with H-acidic compounds were used to convert phosphorus in $SiH_2(PH_2)_2$, $SiH(PH_2)_3$, $Si_2H_5PH_2$, $(SiH_3)_2PH$, $(SiH_3)_3P$, or its Si-methylated derivatives into PH_3 [3 to 7, 27]. The Ge–P bond in GeH_3PH_2 can be cleaved with H_2S to give PH_3 [8]. A complete cleavage of all Ge–P bonds, e.g., in $(CH_3)_2Ge(PH_2)H$, $R_2Ge(PH_2)_2$ with R = CH_3, C_2H_5, $[(CH_3)_2GePH_2]_2PH$, $[(CH_3)_2GePH_2]_3P$ or $(R_2Ge)_6P_4$ with R = CH_3, C_2H_5, converting all P atoms into PH_3 was achieved with anhydrous HCl at room temperature [9, 10]. Lewis acid-catalyzed redistribution reactions of SiH_3PH_2, $Si_2H_5PH_2$, and GeH_3PH_2 with BF_3, BCl_3, B_2H_6, or B_5H_9 [11] or of $(CH_3)_3SiPH_2$ with $AlCl_3$ [12] yield PH_3 and the corresponding trisilyl- or trigermylphosphanes. Rearrangement reactions of this type, initiated by Lewis acids or thermally (cf. 1.3.1.3.4, p. 122), along with E–P cleavage reactions (E = Si, Ge; see above) are assumed to be responsible for the formation of PH_3 as a by-product in reactions of the phosphanylating reagents $M[Al(PH_2)_4]$ (M = Li, Na) with a variety of halogenated silicon or germanium compounds; see, e.g. [3, 6, 7, 13 to 15]. The formation of relatively large amounts of PH_3, when SiH_3Br is reacted with KPH_2 [16] or $(CH_3)_3SiCl$ with $NaPH_2$ [17], is attributed to transmetalation reactions between the metal phosphides and the initially formed PH-containing silylphosphanes [18].

Treating $(CH_3O)_2P(O)H$ with $(4,3,5\text{-}(HO)Br_2C_6H_2)_2C(CH_3)_2$ at 160°C gave some PH_3 along with $(CH_3)_2O$, $CH_3OP(O)(H)OH$, and $[CH_3OP(O)(H)O]_2PO$ [19]. The thermal oxidative degradation of hydraulic fluids comprising hydrocarbons containing triphenyl phosphate at 400 to 420°C yielded small amounts of PH_3 [20]. The compound was found to be one of the volatile products during the thermal decomposition of pentaerythritol diphosphate between 280 and 900°C [21]. Appreciable amounts of PH_3 were released together with H_2, when a mixture of diethyl phosphite and finely divided Na in toluene was warmed from liquid N_2 temperature to 0°C [22]. Hydrolysis of tris(ethoxycarbonyl)phosphane with 1M aqueous NaOH or HCl solutions proceeded readily at 100°C giving PH_3 with 81.5 and 23% yield, respectively [23].

For ab initio MO calculations related to the hypothetical reaction of $H_2C{=}PH_3$ with H_2CO to give PH_3 and oxirane, see [24].

Traces of PH_3 were detected during the pyrolysis up to 700°C of certain clays in vacuum [25]. Bacteriological reduction of small amounts of inorganic phosphates to phosphane is assumed to occur in open-air sewage treatment plants or in sediments of shallow waters [26]. PH_3 was also detected in ocean sediments and in the mud of Hamburg's harbor [28]. The occurrence of PH_3 in the biosphere, in particular in human beings, cattle, and certain fishes, was investigated in [29].

References:

[1] Seel, F.; Velleman, K. (Z. Anorg. Allg. Chem. **385** [1971] 123/30).
[2] Fritz, G. (Angew. Chem. **78** [1966] 80/5; Angew. Chem. Int. Ed. Engl. **5** [1966] 53/7).
[3] Norman, A. D. (J. Am. Chem. Soc. **90** [1968] 6556/7).
[4] Gokhale, S. D.; Jolly, W. L. (Inorg. Chem. **4** [1965] 596/7).
[5] Ebsworth, E. A. V.; Glidewell, C.; Sheldrick, G. M. (J. Chem. Soc. A **1969** 352/3).
[6] Norman, A. D.; Wingeleth, D. C. (Inorg. Chem. **9** [1970] 98/103).
[7] Norman, A. D. (Inorg. Chem. **9** [1970] 870/4).
[8] Drake, J. E.; Riddle, C. (J. Chem. Soc. A **1968** 2709/15).
[9] Dahl, A. R.; Norman, A. D. (J. Am. Chem. Soc. **92** [1970] 5525/7).

[10] Dahl, A. R.; Norman, A. D.; Shenav, H.; Schaeffer, R. (J. Am. Chem. Soc. **97** [1975] 6364/70).

[11] Wingeleth, D. E.; Norman, A. D. (Phosphorus Sulfur Relat. Elem. **39** [1988] 123/9).
[12] Fritz, G.; Emül, R. (Z. Anorg. Allg. Chem. **416** [1975] 19/31).
[13] Norman, A. D. (Chem. Commun. **1968** 812/3).
[14] Fritz, G.; Schäfer, H. (Z. Anorg. Allg. Chem. **385** [1971] 243/55).
[15] Baudler, M.; Scholz, G.; Tebbe, K.-F.; Fehér, M. (Angew. Chem. **101** [1989] 352/4).
[16] Amberger, E.; Boeters, H. (Angew. Chem. **74** [1962] 32/3; Angew. Chem. Int. Ed. Engl. **1** [1962] 52).
[17] Leffler, A. J.; Teach, E. G. (J. Am. Chem. Soc. **82** [1960] 2710/2).
[18] Glidewell, C.; Sheldrick, G. M. (J. Chem. Soc. A **1969** 350/1).
[19] Troev, K.; Simeonov, M. (Phosphorus Sulfur Relat. Elem. **19** [1984] 363/7).
[20] Paciorek, K. L.; Kratzer, R. H.; Kaufman, J.; Nakahara, J. H. (PB-267658 [1976] 1/108, 55, 78, 88, 92, 96, 104, 105, 107; C.A. **88** [1978] No. 11329).

[21] Camino, G.; Martinasso, G.; Costa, L. (Polym. Degrad. Stab. **27** [1990] 285/96; C.A. **113** [1990] No. 7415).
[22] Kosolapoff, G. M.; Brown, A. D., Jr. (J. Chem. Soc. D **1969** 1266).
[23] Frank, A. W.; Drake, G. L., Jr. (J. Org. Chem. **36** [1971] 3461/4).
[24] Volatron, F.; Eisenstein, O. (J. Am. Chem. Soc. **109** [1987] 1/14).
[25] Heller-Kallai, L.; Miloslavski, I.; Aizenshtat, Z.; Halicz, L. (Am. Mineral. **73** [1988] 376/82).
[26] Dévai, I.; Felföldy, L.; Wittner, I.; Plósz, S. (Nature [London] **333** [1988] 343/5).
[27] Bürger, H.; Goetze, U. (J. Organomet. Chem. **12** [1968] 451/7).
[28] Gassmann, G.; Schorn, F. (Naturwissenschaften **80** [1993] 78/80).
[29] Gassmann, G.; Glindemann, D. (Angew. Chem. **105** [1993] 749/51; Angew. Chem. Int. Ed. Engl. **105** [1993] 761).

1.3.1.3.8 Formation of Isotopomers

Earlier results are covered in "Phosphor" C, 1965, p. 31, and "Phosphor" B, 1964, pp. 133, 153. See also the Section 1.3.1.5.3 on "Isotope Exchange" on p. 222 in this volume.

PD_3 is prepared analogously to PH_3 by hydrolysis of Ca_3P_2 with D_2O [1, 2] or of AlP with a D_2SO_4/D_2O mixture [3], or by thermal decomposition of D_3PO_3 [4, 5]. The perdeuterated phosphane is also obtained by cleaving all Si–P bonds in $(R_3Si)_3P$ (R = H, CH_3) with D_2O [6, 7] or all Ge–P bonds in $(R_2Ge)_6P_4$ (R = CH_3, C_2H_5) with DCl [8, 9].

PH_2D can be prepared by hydrolyzing AlP with the appropriate aqueous H_2SO_4/D_2SO_4 mixture [10]. For a suggestion on how to form PH_2D and **PHD_2** via suitable sequences of Si–P cleavage reactions with H_2O and D_2O, see [6]; see also [7].

PH_2T is formed upon neutron irradiation of $PH_3/^3He$ mixtures [11] (cf. 1.3.1.5.1.5, p. 215).

Radioactive **$^{32}PH_3$** molecules can be obtained by irradiation of PH_3 (cf. 1.3.1.5.1.5, p. 215) or $(CH_3)_3P$ with thermal neutrons [12 to 14] or by hydrolysis of $Al^{32}P$ [15] or $K_3{}^{32}P$ [16 to 18]. Hydrogen abstraction by recoil ^{32}P atoms from H-containing molecules also leads to the formation of $^{32}PH_3$ [19, 20]; see also Section 1.3.1.5.1.5, p. 215. The reactivity of the H-containing molecules towards ^{32}P atoms decreases in the order $PH_3 \sim SiH_4 \gg C_2H_6 > (CH_3)_4C \sim H_2 > CH_4 > C_2H_4$. Acetylene gives almost no $^{32}PH_3$ when reacted with ^{32}P [20]. The separation of $^{32}PH_3$ from the other H-containing molecules could be achieved by standard gas chromatographic techniques [21, 22].

References:

[1] Guenebaut, H.; Pascat, B. (C. R. Hebd. Seances Acad. Sci. **256** [1963] 677/80).
[2] Arlinghaus, R. T.; Andrews, L. (J. Chem. Phys. **81** [1984] 4341/51, 4341).
[3] Mariott, R. C.; Odom, J. D.; Sears, C. T., Jr. (Inorg. Synth. **14** [1973] 1/4).
[4] Wickramaaratchi, M. A.; Setser, D. W. (J. Phys. Chem. **87** [1983] 64/72).
[5] Ashfold, M. N. R.; Dixon, R. N.; Stickland, R. J. (Chem. Phys. Lett. **111** [1984] 226/33).
[6] Jolly, W. L.; Norman, A. D. (Prep. Inorg. React. **4** [1968] 1/58, 34/5).
[7] Bürger, H.; Goetze, U. (J. Organomet. Chem. **12** [1968] 451/7).
[8] Dahl, A. R.; Norman, A. D. (J. Am. Chem. Soc. **92** [1970] 5525/7).
[9] Dahl, A. R.; Norman, A. D.; Shenav, H.; Schaeffer, R. (J. Am. Chem. Soc. **97** [1975] 6364/70).
[10] Legon, A. C.; Willoughby, L. C. (Chem. Phys. **85** [1984] 443/50).

[11] Castiglioni, M.; Volpe, P. (Radiochim. Acta **34** [1983] 165/8).
[12] Halmann, M. (Proc. Chem. Soc. **1960** 289/90).
[13] Kugel, L.; Halmann, M. (Bull. Res. Counc. Isr. A **11** [1962] 205).
[14] Halmann, M.; Kugel, L. (J. Inorg. Nucl. Chem. **25** [1963] 1343/50).
[15] Robinson, J. R.; Bond, E. J. (J. Stored Prod. Res. **6** [1970] 133/46).
[16] Bogdanov, R. V.; Murin, A. N. (Zh. Obshch. Khim. **35** [1965] 916/23; J. Gen. Chem. USSR [Engl. Transl.] **35** [1965] 921/7).
[17] Murin, A. N.; Banasevich, S. N.; Bogdanov, R. V. (Izv. Akad. Nauk SSSR Ser. Khim. **1961** 1433/7; Bull. Acad. Sci. USSR Div. Chem. Sci. [Engl. Transl.] **1961** 1334/7).
[18] Bogdanov, R. V.; Murin, A. N. (Zh. Obshch. Khim. **37** [1967] 2425/31; J. Gen. Chem. USSR [Engl. Transl.] **37** [1967] 2308/12).
[19] Zeck, O. F.; Gennaro, G. P.; Tang, Y.-N. (J. Chem. Soc. Chem. Commun. **1974** 52/3).
[20] Zeck, O. F.; Gennaro, G. P.; Tang, Y.-N. (J. Am. Chem. Soc. **97** [1975] 4498/503).

[21] Gennaro, G. P.; Tang, Y.-N. (J. Inorg. Nucl. Chem. **36** [1974] 259/62).
[22] Zeck, O. F.; Ferrieri, R. A.; Copp, C. A.; Gennaro, G. P.; Tang, Y.-N. (J. Inorg. Nucl. Chem. **41** [1979] 785/9).

1.3.1.3.9 Purification of PH_3

Phosphane obtained by one of the laboratory methods described in the previous sections is generally purified by low-temperature fractional condensation in vacuum [1 to 5]. The compound passes through a trap held at −120°C [1, 4], −126°C [5], −131°C [3] and condenses completely at −196°C [1 to 5]. Purification of larger amounts is achieved by low-temperature rectification [6 to 9], distilling in a packed column filled with activated zeolite [10] (see also [11, 12]), or large-scale chromatographic separation [13]. Molecular sieves, in particular zeolites, were used to remove H_2O [7, 14 to 17, 37, 38]. The removal of H_2O by ZnO-based moldings is described in [18, 19].

Separation of PH_3 from P_2H_4, P_3H_3, and P_3H_5 is possible by gas chromatography on glass beads (60 to 80 mesh) with silicon oil [20]. Quantitative removal of P_2H_4 and higher phosphanes from crude PH_3 is possible by thermal decomposition at 450°C in an empty pyrolysis tube [1, 4] or at 150 to 250°C in a column filled with activated carbon, zeolite, or silica gel, preferably for a residence time of 5 to 20 s to give phosphane, phosphorus, and hydrogen [21]. The removal of P_2H_4 and higher phosphanes from PH_3 by using certain mixtures of organic solvents is described in [22].

Purification of PH_3 containing small amounts of AsH_3 (100 ppm) is possible by passing the gas at a pressure of 20 at through a stainless steel tube over activated carbon. The purified PH_3 contains 0.3 to 0.7 ppm AsH_3. The column can be used again after heating it at 100 to 130°C for 3 h in vacuum [23]; see also [15]. Highly pure phosphane can also be obtained by heating crude PH_3 gas above 240°C but below its thermolysis temperature in the presence of certain arsenide-forming metals or their phosphides [24].

The removal of other impurities by means of zeolites is described in [7, 16, 38]. For a gas-chromatographic study of impurities in PH_3, see [25]. Passage of PH_3 through activated carbon before rectification, then drying out the moisture by freezing at −30 to −50°C, and finally passing it through silica gel and NaA zeolite is recommended in [7]. An absorption-adsorption process for the purification of PH_3 uses an aqueous suspension of activated C [26]. The removal of oxygen by reaction with various solids was patented [27 to 31]. The use of a hydrogenated getter metal for the purification of PH_3 is claimed in [32]. Impurities like oxidants, protic acids, or compounds that can be metalated may be removed by lithiated styrene-divinylbenzene copolymers or lithiated aromatic hydrocarbons supported on cross-linked styrene-based polymers [33, 34]. For the separation of PH_3 from GeH_4, which has nearly the same volatility as PH_3, by the addition of excess HCl, see [35]. The separation of a PH_3/GeH_4 mixture by permeation through a membrane is described in [36]. See also [39].

References:

[1] Klement, R. (in: Brauer, G.; Handbuch der Präparativen Anorganischen Chemie, Vol. 1, Enke, Stuttgart 1975, pp. 505/66, 510/3).

[2] Fluck, E. (Fortschr. Chem. Forsch. **35** [1973] 1/64, 18/21).

[3] Mariott, R. C.; Odom, J. D.; Sears, C. T. (Inorg. Synth. **14** [1973] 1/4).

[4] Baudler, M.; Ständecke, H.; Dobbers, J. (Z. Anorg. Allg. Chem. **353** [1967] 122/6).

[5] Gokhale, S. D.; Jolly, W. L. (Inorg. Synth. **9** [1967] 56/8).

[6] Devyatykh, G. G.; Zorin, A. D.; Frolov, I. A.; Kedyarkin, V. M.; Balabanov, V. V.; Petrik, A. G.; Malorossiyanov, V. S.; Galkin, P. N.; Levinson, D. I.; Ymapol'skii, A. L.; Gur'yanov, A. N.; Rumovskaya, I. V.; Kuznetsova, T. S. (Metody Poluch. Anal. Veshchestv Osoboi Chist. Tr. Vses. Konf., Gorkiy 1968 [1970], pp. 42/50; C.A. **75** [1971] No. 111184).

[7] Pospelov, V. P.; Efremov, A. A.; Potepalov, V. P.; Velichko, A. P.; Fedorov, V. A.; Grinberg, E. E.; Egurnov, V. Ya.; Potolokov, N. A. (U.S.S.R. 497224 [1974/75] 1/2; C.A. **84** [1976] No. 107915).

[8] Frolov, I. A.; Zaburdyaev, V. S.; Kulakov, S. I.; Sokolov, E. B. (Sb. Nauchn. Tr. Probl. Mikroelektron. No. 21 [1975] 77/83; C.A. **84** [1976] No. 166780).

[9] Frolov, I. A. (Zh. Prikl. Khim. [Leningrad] **51** [1978] 721/31; J. Appl. Chem. USSR [Engl. Transl.] **51** [1978] 709/11).

[10] Ota, S.; Demura, Y.; Shimada, S. (Jpn. Kokai Tokkyo Koho 62-138313 [87-138313] [1985/87] 1/7 from C.A. **107** [1987] No. 137031).

[11] Vorotyntsev, V. M.; Mochalov, G. M.; Balabanov, V. V. (Vysokochist. Veshchestva **1991** No. 6, pp. 126/30; High Purity Subst. [Engl. Transl.] **5** [1991] 1050/4).

[12] Balabanov, V. V.; Vorotyntsev, V. M.; Zhdanov, E. A.; Mikhailova, N. N.; Stepanov, V. M.; Fidel'man, A. R.; Yan'kov, S. V. (Vysokochist. Veshchestva **1990** No. 6, pp. 60/6; High Purity Subst. [Engl. Transl.] **4** [1990] 1025/30).
[13] Cygelman, S. (Solid State Technol. **14** [1971] 45/8; C.A. **74** [1971] No. 77853).
[14] Kitahara, K.; Shimada, T.; Iwata, K. (Jpn. Kokai Tokkyo Koho 03-040902 [91-040902] [1989/91] 1/4 from C.A. **115** [1991] No. 11796).
[15] Tsuchiya, H.; Otsuji, A. (Jpn. Kokai Tokkyo Koho 01-164711 [89-164711] [1987/89] 1/4 from C.A. **112** [1990] No. 39256).
[16] Tolmachev, A. M.; Kuznetsova, T. A.; Egorov, E. N.; Orlov, V. Yu. (Vysokochist. Veshchestva **1988** No. 2, pp. 65/70 from C.A. **109** [1988] No. 28029).
[17] Morozov, V. I.; Efremov, A. A.; Zel'venskii, Ya. D. (Tr. Inst. Mosk. Khim. Tekhnol. Inst. im. D.I. Mendeleeva No. 85 [1975] 82/3; C.A. **86** [1977] No. 31373).
[18] Kitahara, K.; Shimada, T.; Iwata, K. (Jpn. Kokai Tokkyo Koho 03-040901 [91-040901] [1989/91] 1/5 from C.A. **115** [1991] No. 11795).
[19] Kitahara, K.; Shimada, T.; Iwata, K. (Jpn. Kokai Tokkyo Koho 03-075202 [91-075202] [1989/91] 1/4 from C.A. **115** [1991] No. 74671).
[20] Baudler, M.; Ständecke, H.; Kemper, M. (Z. Anorg. Allg. Chem. **388** [1972] 125/36).

[21] Kurze, R.; Ober, D.; Piske, B.; Zobel, D.; Gisbier, D.; Ebersbach, K.-H. (Ger. [East] 200422 [1981/83] 1/12; C.A. **99** [1983] No. 124975).
[22] Sonobe, K.; Suganuma, S.; Shimura, S. (Jpn. 73-41438 [1968/73] 1/4 from C.A. **81** [1974] No. 5148).
[23] Nippon Synthetic Chemical Industry Co., Ltd. (Jpn. Kokai Tokkyo Koho 84-45913 [1982/84] 1/5 from C.A. **100** [1984] No. 212476).
[24] Iso, A. (Jpn. Kokai Tokkyo Koho 85-260412 [1984/85] 1/4 from C.A. **105** [1986] No. 45806).
[25] Ivanova, N. T.; Vislykh, N. A.; Voevodina, V. V. (Vysokochist. Veshchestva **1990** No. 5, pp. 198/203; High Purity Subst. [Engl. Transl.] **4** [1990] 953/7).
[26] Efremov, A. A.; Morozov, V. I.; Fedorov, V. A. (Vysokochist. Veshchestva **1991** No. 1, pp. 115/23; High Purity Subst. [Engl. Transl.] **5** [1991] 94/100).
[27] Kitahara, K.; Shimada, T.; Iwata, K. (Jpn. Kokai Tokkyo Koho 03-012303 [91-012303] [1989/91] 1/5 from C.A. **114** [1991] No. 219599).
[28] Kitahara, K.; Shimada, T.; Iwata, K. (Jpn. Kokai Tokkyo Koho 03-012304 [91-012304] [1989/91] 1/5 from C.A. **114** [1991] No. 219600).
[29] Kitahara, K.; Shimada, T.; Iwata, K. (Jpn. Kokai Tokkyo Koho 02-144114 [90-144114] [1988/90] 1/4 from C.A. **113** [1990] No. 203183).
[30] Kitahara, K.; Shimada, T.; Iwada, K.; Akita, N. (Eur. Appl. 361386 [1990] 1/11 from C.A. **112** [1990] No. 219898).

[31] Kitahara, K.; Shimada, T.; Iwada, K. (Jpn. Kokai Tokkyo Koho 02-204306 [90-204306] [1989/90] 1/5 from C.A. **114** [1991] No. 46026).
[32] Succi, M.; Boffito, C.; Solcia, C. (Eur. Appl. 470936 [1990/92] 1/9 from C.A. **116** [1992] No. 197120).
[33] Tom, G. M.; Brown, D. W. (U.S. 4761395 [1987/88] 1/8 from C.A. **109** [1988] No. 152468).
[34] Tom, G. M. (U.S. 4659552 [1986/87] 1/3 from C.A. **107** [1987] No. 42565).
[35] Jolly, W. L. (The Synthesis and Characterization of Inorganic Compounds, Prentice Hall, Englewood Cliffs, N.J., 1970, p. 174).
[36] Vorotyntsev, V. M.; Drozdov, P. N.; Nosyrev, S. A.; Pripisnov, A. E. (Vysokochist. Veshchestva **1987** No. 4, pp. 137/41 from C.A. **107** [1987] No. 219636).

[37] Efremov, A. A.; Morozov, V. I.; Fedorov, V. A. (Vysokochist. Veshchestva **1991** No. 3, pp. 115/21; High Purity Subst. [Engl. Transl.] **5** [1991] 468/74).
[38] Efremov, A. A.; Fedorov, V. A. (Vysokochist. Veshchestva **1990** No. 2, pp. 72/83; High Purity Subst. [Engl. Transl.] **4** [1990] 255/67).
[39] Devyatykh, G. G.; Vlasov, S. M.; Tsinovoi, Yu. N. (Fiz. Khim. **42** [1968] 2745/50; Russ. J. Phys. Chem. [Engl. Transl.] **42** [1968] 1460/3).

1.3.1.3.10 Handling, Storage, and Safety Aspects

Phosphane must be prepared and handled in an O_2-free atmosphere because of its flammability. That means that air has to be carefully excluded during its synthesis. Crude PH_3 obtained according to one of the synthetic procedures described in the preceeding sections, with only a few exceptions already mentioned, generally contains more or less P_2H_4 which causes the gas mixture to ignite spontaneously at ambient temperature in the presence of air. The purification of PH_3 is described in detail in Section 1.3.1.3.9, p. 128. Excess phosphane left after any reaction ought to be burnt in a flame or decomposed by means of some oxidizing agent; see, e.g., Section 1.3.1.5.4, p. 222.

PH_3 can be liquified and stored in steel cylinders. The P_2H_4 content of purified PH_3 was found to increase during storage. However, the gas did not ignite spontaneously even after months upon exposure to air, but it was generally found to be spontaneously inflammable after one year of storage [1, 2]. Storage of PH_3 is also possible by adsorbing the gas into the microcavities of synthetic zeolites. The adsorption capacity of a (Zn, K)A zeolite was found to be 15% at 21 Torr and 23°C. The gas is desorbed at higher temperatures [3]. Sorption was suggested to account for most of the loss in contact with materials like glass, stainless steel, Teflon, nylon, and PVC. Permeation of PH_3 through silicon rubber tubing was found to be excessive and to render this material unsuitable for use with atmospheres containing this gas [4]. The permeation through other materials was also investigated in detail [26].

PH_3 gas used for fumigation is predominantly produced from formulations of metallic phosphides (usually phosphides of Al, Mg, or Zn as tablets, pellets, or small sachets of powder) which contain additional materials for regulating the release of the gas and suppressing its flammability. Such additives are ammonium carbamate [5, 7], ammonium bicarbonate [5], ammonium stearate [7], paraffin [5, 6], or zeolites (pore size 1 nm) [8] or a synergistic mixture of benzoic acid and ammonium carbamate [9]. After the phosphane has evolved from the formulation, the residue that remains consists mainly of aluminium or magnesium hydroxide still containing small amounts of undecomposed phosphide. The handling of the phosphide formulations is described in detail in [5]. A reactor to produce a continuous PH_3 stream from Mg_3P_2 and H_2O in an inert gas such as CO_2 containing only about 5 vol% PH_3 has been described. This method enables fumigation with a constant PH_3 concentration over a longer period [10].

Avoiding phosphane may not be a simple matter, because it may form quite unexpectedly (cf. 1.3.1.3.11 "Sources of Poisoning", p. 133). Personnel working with phosphorus chemicals should be cautioned about the possibility of its formation from initial products or by-products. Control measures include ventilation, which should be in downward direction because the gas is heavier than air [11]. Integrated safety systems for laboratories using MOCVD are described in [12, 13]; see also [22 to 24]. The vapor hazard index advises use of the gas only under monitored conditions if in prolonged regular use [14]. Personal respirator protection is recommended, including canister-type gas masks for lower levels of contamination and, otherwise, self-contained breathing apparatus [5, 11, 15 to 18]. For general handling procedures, see also [25].

PH_3 or materials containing reactive phosphides must be stored in a cool, dry, isolated area, away from acute fire hazards and powerful oxidizing materials. Containers should be kept closed and plainly labeled [11, 19]. For packing standards of PH_3, see also [20, 25]. An apparatus for handling gas leaks during delivery is described in [21]. Explosion limits are discussed in Section 1.3.1.5.5, p. 237.

References:

[1] Baudler, M.; Ständeke, H.; Dobbers, J. (Z. Anorg. Allg. Chem. **353** [1967] 122/6).
[2] Klement, R. (in: Brauer, G.; Handbuch der Präparativen Anorganischen Chemie, Vol. 1, Enke, Stuttgart 1975, pp. 505/66, 510/3).
[3] Sillmon, R. S.; Freitas, J. A., Jr. (Appl. Phys. Lett. **56** [1990] 174/6).
[4] Waterford, C. J.; Winks, R. G. (J. Stored Prod. Res. **22** [1986] 25/7).
[5] Bond, E. J. (FAO Plant Prot. Bull. **54** [1984] 96/108).
[6] Kapp, W. (U.S. 3917823 [1974/75] 1/6; C.A. **84** [1976] No. 26907).
[7] Friemel, W. (Eur. Appl. 82-5652 [1982] 1/17; C.A. **100** [1984] No. 212459).
[8] Moog, A.; Kapp, W. (Eur. 342471 [1988/89] 1/17 from C.A. **112** [1990] No. 72342).
[9] Herrmann, R.; Toepe, W.; Barysch, G. (Ger. [East] 288491 [1989/91] 1/3 from C.A. **115** [1991] No. 226186).
[10] Dörnemann, M.; Reif, H. (Ger. Offen. 3618297 [1986/87] 1/14; C.A. **108** [1988] No. 200241).

[11] Parmeggiani, L. (Encyclopaedia of Occupational Health and Safety, 3rd Ed., Vol. II, International Labour Office, Geneva 1983, p. 1681).
[12] Hess, K. L.; Riccio, R. J. (J. Cryst. Growth **77** [1986] 95/100).
[13] Kaufmann, L. M.; Heuken, M.; Tilders, R.; Heime, K.; Jürgensen, H.; Heyen, M. (J. Cryst. Growth **93** [1988] 279/84).
[14] Pitt, M. J. (Chem. Ind. [London] **1982** 804/6).
[15] Schäfer, H. K. (in: Winnacker-Küchler, Chemische Technologie, Allgemeines, 4th Ed., Vol. 1, Hanser, München 1984, pp. 656/724, 691, 705).
[16] Mackison, F. W.; Stricoff, R. S.; Partridge, L. J., Jr. (DHHS [NIOSH] Publ. [U.S.] No. 81-123 [1978] 1/4).
[17] Kloos, E. J.; Spinetti, L.; Raymond, L. D. (Bur. Mines Inf. Circ. No. 8291 [1966] 1/7).
[18] Oettel, H. (Ullmanns Encykl. Tech. Chem. 3rd Ed. **13** [1962] 571/3).
[19] Austenat, L.; Zink, C. (Handbuch Stoffdaten zur Störfall-Verordnung, Vol. II, Schmidt, Berlin 1986, Merkblatt No. 120 Umweltbundesamt).
[20] United States Dept. of Transportation (Fed. Regist. **55** [1990] No. 52402729 from C.A. **116** [1992] No. 135528).

[21] Shinagawa, T. (Jpn. Kokai Tokkyo Koho 63-291619 [88-291619] [1987/88] 1/4 from C.A. **111** [1989] No. 63239).
[22] Hoffman, K.; Brubaker, S.; Nath, P.; Di Dio, G.; Izu, M.; Doehler, J. (AIP Conf. Proc. No. 166 [1988] 138/44 from C.A. **109** [1988] No. 193014).
[23] Fthenakis, V. M.; Moskowitz, P. D. (Sol. Cells **22** [1987] 303/17).
[24] Fthenakis, V. M.; Moskowitz, P. D. (BNL-51989 [1986] 37 pp. from C.A. **108** [1988] No. 26391).
[25] Anonymous (Dangerous Prop. Ind. Mater. Rep. **6** [1986] 103/7; C.A. **104** [1986] No. 192013).
[26] Kashi, K. P.; Muthu, M.; Majumder, S. K. (Pestic. Sci. **8** [1977] 492/6).

1.3.1.3.11 Toxicity

A review of older publications is given in "Phosphor" B, 1964, pp. 91/3. Reviews of more recent publications can be found in [1 to 8]. For short reviews of the physiology of PH_3, see e.g. [9, 10].

Limiting Values. Phosphane is very toxic to all forms of animal life, hence exposure of humans even in small amounts should be avoided as much as possible [5]. Poisoning is caused directly by inhalation or indirectly by ingestion of, e.g., AlP-, Mg_3P_2-, or Zn_3P_2-containing formulations which evolve PH_3 on contact with water or acids [11 to 13]. PH_3 gas is not absorbed through the skin [4]; see also [17]. Slightly different values for the threshold limit value (time weighted average) (TLV-TWA) for an eight-hour daily exposure in a five-day week are (vol-ppm = mL/m³): 0.1 vol-ppm (0.15 mg/m³, MAK value) [14], 0.3 vol-ppm (0.4 mg/m³) [15 to 20] or 0.1 mg/m³ [21]. A short-time exposure to 2 vol-ppm should not exceed 5 min [14]. The maximum concentration for a 30-min exposure, which is said to cause no serious harm to humans, is 200 vol-ppm. A concentration of about 2000 vol-ppm in air is assumed to be lethal within a very short time [3, 22]; see also [17]. Results of setting emergency exposure indexes by several companies are discussed in [23].

Detection. The smell of phosphane is described as garlic-like [24] or oniony [25]. The published odor threshold values vary from 0.01 to 200 vol-ppm [17, 24, 26] and depend on the method of preparation and purification [24]. Thus, the odor of PH_3 is generally not a reliable indicator to avoid toxic concentrations [27, 28] except lethal concentrations [28].

The odor seems to be due to the presence of by-products formed along with phosphane (P_2H_4 as the source of the odor was mentioned without proof [11, p. 209]). These by-products may be preferentially adsorbed during fumigation, and under some conditions the odor may disappear even when insecticidally effective concentrations of PH_3 are still present in the gas phase [4]. For a quick detection and determination of PH_3 concentrations, see Section 1.3.1.3.12, p. 135.

Sources of Poisoning. The liberation of phosphane has to be taken into account in many cases, a few of which are given here. Poisoning by PH_3 can occur during the preparation or manufacture of the gas or when using leaky steel cylinders. Phosphides and alloys containing phosphorus evolve PH_3 in humid atmosphere; see Section 1.3.1.3.5, p. 123. An important source of poisoning is technical, impure acetylene which is formed by the action of water on calcium carbide containing calcium phosphide as an impurity [1]. When machining gray cast iron with an aqueous coolant, the air-borne concentration of PH_3 in the breathing zone of the operator can reach 0.7 vol-ppm, thus exceeding the limits given above. Attempts to prevent PH_3 release by adding, e.g., $KMnO_4$ or Cu citrate to the coolant failed [29].

Symptoms of Poisoning and First Aid. Depending on the amount of PH_3 inhaled, symptoms may occur immediately or several hours (up to 24 h) after exposure. Feeling of fatigue, ringing in the ears, nausea, pressure in the chest, and uneasiness were mentioned to be the major symptoms after slight or mild poisoning. All of these symptoms should normally disappear in fresh air. However, consulting a physician is always advised. More severe symptoms caused by poisoning with higher PH_3 concentrations which may even lead to death (immediately or after several days) are described in [4]; see also [1, 7, 8, 11, 17, 18, 30 to 32]. Therapy can only be symptomatically, since no specific antidote is known for PH_3 [4]; see also [5, 20, 30].

References:

[1] Bettermann, G.; Krause, W.; Riess, G.; Hofmann, T. (Ullmann's Encycl. Ind. Chem. 5th Ed. A **19** [1991] 527/43, 542).

[2] Zaebst, D. D.; Blade, L. M.; Burroughs, G. E.; Morrelli-Schroth, P.; Woodfin, W. J. (Appl. Ind. Hyg. **3** [1988] 146/54; C. A. **108** [1988] No. 226178).

[3] Henschler, D. (Gesundheitsschädliche Arbeitsstoffe, Verlag Chemie, Weinheim 1987, pp. 1/3).
[4] Bond, E. J. (FAO Plant Prot. Bull. No. 54 [1984] 96/108).
[5] Braker, W.; Mossman, A. L.; Siegel, D. (Effects of Exposure to Toxic Gases - First Aid and Medical Treatment, 2nd Ed., Matheson, Lyndhurst, N.J., 1977, pp. 70/2).
[6] Fluck, E. (Fortschr. Chem. Forsch. **35** [1973] 1/64, 17).
[7] Klimmer, O. R. (Arch. Toxikol. **24** [1969] 164/87).
[8] Oettel, H. (Ullmanns Encykl. Tech. Chem. 3rd Ed. **13** [1962] 571/3).
[9] Garry, V. F.; Griffith, J.; Danzl, T. J.; Nelson, R. L.; Whorton, E. B.; Krueger, L. A.; Cervenka, J. (Science **246** [1989] 251/5).
[10] Berners-Price, S. J.; Sadler, P. J. (Structure Bonding [Berlin] **70** [1988] 27/102, 79/80).

[11] Zipf, K. E.; Arndt, T.; Heintz, R. (Arch. Toxikol. **22** [1967] 209/22).
[12] Singh, S.; Dilawari, J. B.; Vashist, R.; Malhotra, H. S.; Sharma, B. K. (Br. Med. J. **290** [1985] 1110/1).
[13] Krüger, P. (Notfallmed. **8** [1982] 1520/32 from DIMDI).
[14] Maximale Arbeitsplatzkonzentrationen und Biologische Arbeitsstofftoleranzwerte (Mitt. Senatskomm. Prüf. Gesundheitsschädlicher Arbeitsstoffe Deut. Forschungsgem. No. 28 [1992] 64).
[15] United States Occupational Safety and Health Administration (Fed. Regist. **54** [1989] No. 2332/983, p. 2949; C.A. **110** [1989] No. 218230).
[16] Mackison, F. W.; Stricoff, R. S.; Partridge, L. J., Jr. (DHHS [NIOSH] Publ. [U.S.] No. 81-123 [1978] 1/4).
[17] Anonymous (Dangerous Prop. Ind. Mater. Rep. **6** [1986] 103/7; C.A. **104** [1986] No. 192013).
[18] Sax, N. I. (Dangerous Properties of Industrial Materials, 7th Ed., Reinhold, New York 1989, Vol. III, pp. 2768/9).
[19] Anonymous (Cah. Notes Doc. No. 121 [1985] 473/508, 491).
[20] Permeggiani, L. (Encyclopaedia of Occupational Health and Safety, 3rd Ed., Vol. 2, International Labour Office, Geneva 1983, p. 1681).

[21] Pazynich, V. M.; Mazur, I. A.; Podloznyi, A. V.; Chinchevich, V. I.; Mandrichenko, B. E.; Rudenko, L. M.; Gazina, V. M.; Gagara, V. F.; Torgun, V. P.; et al. (Gig. Sanit. **1984** No. 1, pp. 13/5 from C.A. **100** [1984] No. 186789).
[22] Austenat, L.; Zink, C. (Handbuch Stoffdaten zur Störfall-Verordnung, Vol. II, Schmidt, Berlin 1986, Merkblatt No. 120 Umweltbundesamt).
[23] European Chemical Industry Ecology and Toxicology Centre (Tech. Rep.-ECETOC No. 43 [1991] 1/74 from C.A. **115** [1991] No. 286144).
[24] Fluck, E. (J. Air Pollut. Control Assoc. **26** [1976] 795; C.A. **86** [1977] No. 46889).
[25] Ruth, J. H. (Am. Ind. Hyg. Assoc. J. **47** [1986] A142/A152, A149).
[26] Dumas, T.; Bond, E. J. (J. Stored Prod. Res. **10** [1974] 67/8, **3** [1967] 389/92).
[27] Amoore, J. E.; Hautala, E. (J. Appl. Toxicol. **3** [1983] 272/90, 280/1).
[28] Rousselin, X.; Falcy, M. (Cah. Notes Doc. No. 124 [1986] 331/44, 341; C.A. **105** [1986] No. 177745).
[29] König, R. (Zentralbl. Arbeitsmed. Arbeitsschutz Prophyl. Ergon. **36** [1986] 365/7 from C.A. **106** [1987] No. 107267).
[30] Moeschlin, S. (Klinik und Therapie der Vergiftungen, Thieme, Stuttgart 1986, pp. 232/3).

[31] Bouzoubaa, A.; Benkiran, A.; Chraibi, N.; El Makhlauf, A. (J. Toxicol. Clin. Exp. **8** [1988] 193/9; C.A. **110** [1989] No. 35060).
[32] Misra, U. K.; Bhargava, S. K.; Nag, D.; Kidwai, M. M.; Lal, M. M. (Toxicol. Lett. **42** [1988] 257/63 from C.A. **109** [1988] No. 175586).

1.3.1.3.12 Analysis

Several older methods for the detection of PH_3 are given in "Phosphor" B, 1964, pp. 339/41. More recent reviews covering the sampling and analysis of PH_3 can be found in [1 to 5, 71].

A great variety of analytical methods has been applied for the detection and quantitative analysis of PH_3, including GC, GC-MS, column or paper chromatography, electrochemical, colorimetric, spectrophotometric, and titration methods and labeling techniques and neutron activation; see [1, 3] and below.

The development of methods for the determination of PH_3 in air is described in [6]; see also [7 to 19]. The monitoring of PH_3, in particular in the working areas of the microelectronic processing facilities, is discussed in [1, 20 to 24]; see also [25].

Further references related to the detection and determination of PH_3, most of them published in the 80's or more recently, deal with GC [12, 19, 26 to 35] (see also [15, 36 to 40]), electronic sensors [8 to 11, 41], detector tubes [14, 42] (see also [43, 44]), indicator strips [45], electrochemical methods [46, 47], mass spectrometry [28, 48], AAS [49], flame spectroscopy [50, 51], ICP-AES [21, 23], aerosol ionization gas analyzer [52], spectrophotometry [53 to 56], colorimetry [7, 16, 18, 23, 57, 58], laser-induced photofragment emission [59], laser-induced breakdown spectroscopy [60], photochemical reaction/light-scattering [61], FT-IR [62], submillimeter spectroscopy [63], and IR laser Stark spectroscopy [13]. For the determination of impurities in PH_3 by GC or GC-MS, see [64 to 68] or by AAS [49, 69].

The reaction of PH_3 with $HgCl_2$ in aqueous solution to give a compound of the composition Hg_3PCl_3 and HCl is the basis for several chemical detection and determination methods [70] and is described in detail in Section 1.3.1.5.7, p. 256.

References:

[1] Verstuyft, A. W. (Am. Ind. Hyg. Assoc. J. **39** [1978] 431/7; C.A. **89** [1978] No. 79350).
[2] Greifer, B.; Taylor, J. K. (PB-251220 [1976] 1/32; C.A. **87** [1977] No. 122138).
[3] Dhaliwal, G. S. (Bull. Grain Technol. **11** [1973] 214/9; C.A. **82** [1975] No. 63680).
[4] Fluck, E. (Fortschr. Chem. Forsch. **35** [1973] 1/64, 17).
[5] Fetchin, J. A.; Grayson, M. (Chem. Anal. [N.Y.] **37** [1972] 371/83, 373/4).
[6] Barrett, W. J.; Dillon, H. K. (DHEW [NIOSH] Publ. [U.S.] No. 78-177 [1978] 1/102, 50/83, 94/102; C.A. **90** [1979] No. 11651).
[7] Ono-Ogasawara, M.; Furuse, M.; Matsumura, Y. (Ind. Health **28** [1990] 175/84; C.A. **114** [1991] No. 213129).
[8] Toda, K. (Bunseki Kagaku **39** [1990] 611/5 from C.A. **114** [1991] No. 149145).
[9] Eguchi, K.; Kayser, P.; Menil, F.; Lucat, C.; Aucouturier, J. L.; Portier, J. (Sens. Actuators B **2** [1990] 193/7 from C.A. **114** [1991] No. 68135).
[10] Nakanishi, K.; Sunada, Y. (Hyomen Gijutsu **40** [1989] 335/6 from C.A. **112** [1990] No. 68809).

[11] Eguchi, K.; Cauhape, J. S.; Menil, F.; Lucat, C.; Videau, J. J. (Sens. Actuators **17** [1989] 319/25 from C.A. **112** [1990] No. 15648).
[12] Ivanova, N. T.; Vislykh, N. A.; Voevodina, V. V. (Vysokochist. Veshchestva **1989** No. 6, pp. 102/7; High Purity Subst. [Engl. Transl.] **3** [1989] 1011/6).
[13] Zamaraeva, L. A.; Chesnokov, E. N. (Zh. Prikl. Spektrosk. **48** [1988] 929/35; J. Appl. Spectrosc. [Engl. Transl.] **48** [1988] 598/603).
[14] Hery, M. (Staub-Reinhalt. Luft **47** [1987] 140/2; C.A. **107** [1987] No. 182415).
[15] Qi, X. (Sepu **5** [1987] 243/5 from C.A. **107** [1987] No. 222265).
[16] Krasuska, E. (Pr. Cent. Inst. Ochr. Pr. **35** No. 124 [1985] 13/22 from C.A. **104** [1986] No. 94386).

[17] Kokk, H.; Bukovskii, M. I.; Must, M.; Kaart, K.; Kolesnik, M. I. (U.S.S.R. 963949 [1981/82] from C.A. **98** [1983] No. 77529).
[18] Bukovskii, M. I.; Vostrikov, V. I.; Kokk, H.; Must, M.; Kaart, K. (Metody Anal. Kontrolya Kach. Prod. Khim. Prom. **1978** No. 3, pp. 64/6 from C.A. **90** [1979] No. 60224).
[19] Dumas, T.; Bond, E. J. (J. Chromatogr. **206** [1981] 384/6).
[20] Ponsold, B.; Kath, H. (Z. Gesamte Hyg. Ihre Grenzgeb. **37** No. 2 [1991] 58/63, 63/5; C.A. **115** [1991] No. 238747).

[21] Hetland, S.; Martinsen, I.; Radziuk, B.; Thomassen, Y. (Anal. Sci. **7** [1991] 1029/32 from C.A. **116** [1992] No. 200033).
[22] Sartori, P.; Pahlmann, W. (Schriftenr. Bundesanst. Arbeitsschutz Gefährliche Arbeitsst. GA **36** [1990] 1/130; C.A. **114** [1991] No. 170311).
[23] Mortensen, G.; Søndergaard, K.; Pedersen, B.; Rietz, B. (Anal. Lett. **22** [1989] 1791/806).
[24] McMahon, R.; Denenberg, B. A. (Sol. Cells **19** [1987] 367/74).
[25] Leichnitz, K. (Gefahrstoffanalytik: Meßtechnische Überwachung von MAK- und TRK-Werten – Emissionskontrolle – Prozeßgasanalyse, ecomed Verlagsgesellschaft, Landsberg 1991).
[26] Chasteen, T. G.; Fall, R.; Birks, J. W.; Martin, H. R.; Glinski, R. J. (Chromatographia **31** [1991] 342/6; C.A. **115** [1991] No. 21320).
[27] Vorotyntsev, V. M.; Drozdov, P. N.; Nosyrev, S. A.; Krylov, V. A.; Ezheleva, A. E. (Vysokochist. Veshchestva **1989** No. 4, pp. 236/9 from C.A. **112** [1990] No. 15715).
[28] Etienne, C. de Saint; Mettes, J. (Analyst [London] **114** [1989] 1649/53 from C.A. **112** [1990] No. 191120).
[29] Ezheleva, A. E.; Malygina, L. S.; Krylov, V. A. (Vysokochist. Veshchestva **1987** No. 3, pp. 214/8 from C.A. **108** [1988] No. 30915).
[30] Ezheleva, A. E.; Baturina, N. M.; Kazakov, V. P.; Krylov, V. A. (Zavod. Lab. **53** [1987] 18/20 from C.A. **107** [1987] No. 50978).

[31] Scudamore, K. A.; Goodship, G. (Pestic. Sci. **17** [1986] 385/95 from C.A. **105** [1986] No. 132262).
[32] Hill, D. (Anal. Methods Pestic. Plant Growth Regul. **15** [1986] 135/60, 145).
[33] Nozoye, H.; Someno, K. (Bunseki Kagaku **34** [1985] 508/9 from C.A. **103** [1985] No. 152878).
[34] Devyatykh, G. G.; Ezheleva, A. E.; Krylov, V. A.; Baturina, N. M.; Malygina, L. S.; Lazarev, V. A. (Zh. Anal. Khim. **40** [1985] 2161/4; J. Anal. Chem. USSR [Engl. Transl.] **40** [1985] 1703/6).
[35] Bossard, A.; Kamga, R.; Raulin, F. (J. Chromatogr. **330** [1985] 400/2).
[36] Saitoh, H. (Koatsu Gasu **29** No. 1 [1992] 36/42 from C.A. **116** [1992] No. 112445).
[37] Saitoh, H.; Sakai, Y. (Nippon Sanso Giho No. 9 [1990] 61/7 from C.A. **115** [1991] No. 14362).
[38] Echnig, W.; Gloeckl, D.; Schwarz, G. W. (Ger. [East] 282764 [1989/90] 1/5 from C.A. **115** [1991] No. 41047).
[39] Otsuka, T. (Jpn. Kokai Tokkyo Koho 01-013454 [89-013454] [1987/89] 1/9 from C.A. **111** [1989] No. 126245).
[40] Abe, T.; Miyagawa, H.; Ito, M.; Ikeda, K.; Murakami, M.; Yanagawa, N. (Jpn. Kokai Tokkyo Koho 62-204157 [87-204157] [1986/87] 1/12 from C.A. **108** [1988] No. 231141).

[41] Henrion, L.; Derost, G.; Barraud, A.; Ruaudel-Teixier, A. (Sens. Actuators **17** [1989] 493/8 from C.A. **112** [1990] No. 68827).

[42] Leichnitz, K. (Prüfröhrchen Taschenbuch, Drägerwerk AG, Lübeck 1982, pp. 132/3, 209).
[43] Kitahara, K.; Akita, N.; Shimada, T.; Sasaki, K.; Hiramoto, T. (Jpn. Kokai Tokkyo Koho 02-032254 [90-032254] [1988/90] 1/6 from C.A. **113** [1990] No. 102630).
[44] Yoneda, N.; Iwamoto, N.; Nakamura, M. (Jpn. Kokai Tokkyo Koho 02-080941 [90-080941] [1988/90] 1/3 from C.A. **113** [1990] No. 8530).
[45] Kashi, K. P.; Muthu, M. (Pestic. Sci. **6** [1975] 511/4 from C.A. **84** [1976] No. 131275).
[46] Soedingresko, A. (Ger. Offen. 3917334 [1989/90] 1/2 from C.A. **114** [1991] No. 258884).
[47] Whitson, P. E. (Eur. Appl. 239190 [1987] 1/10 from C.A. **108** [1988] No. 105593).
[48] Hille, J. (J. Chromatogr. **502** [1990] 265/74, 269/72).
[49] Scharf, H.; Hahn, E.; Emrich, G. (Z. Chem. **30** [1990] 107/8).
[50] Steindorf, W.; Geserick, M.; Ardelt, W. (Chem. Tech. [Leipzig] **25** [1973] 611/3; C.A. **80** [1974] No. 43766).

[51] Mavrodineanu, R.; Boiteux, H. (Flame Spectroscopy, Wiley, New York 1965, pp. 62/3).
[52] Grosse, H.-J.; Krabbe, R.; Döring, H.-R.; Adler, J.; Leschke, W. (Chem. Tech. [Leipzig] **38** [1986] 394/6; C.A. **105** [1986] No. 196353).
[53] Rangaswamy, J. R.; Muthu, M. (J. Assoc. Off. Anal. Chem. **68** [1985] 205/8 from C.A. **102** [1985] No. 165346).
[54] Rangaswamy, J. R. (J. Assoc. Off. Anal. Chem. **67** [1984] 117/22; C.A. **100** [1984] No. 155299).
[55] Rezchikov, V. G.; Skachkova, I. N.; Kuznetsova, T. S.; Khrushcheva, V. V. (Zh. Anal. Khim. **40** [1985] 263/6; J. Anal. Chem. USSR [Engl. Transl.] **40** [1985] 217/20).
[56] Krivtsun, V. M.; Kuritsyn, Yu. A.; Snegirev, E. P.; Zasavitskii, I. I.; Shotov, A. P. (Zh. Prikl. Spektrosk. **43** [1985] 571/6; J. Appl. Spectrosc. [Engl. Transl.] **43** [1985] 1096/100).
[57] Evers, W. (Ger. Offen. 4020753 [1989/91] 1/4 from C.A. **115** [1991] No. 222307).
[58] Dévai, I.; Felföldy, L.; Wittner, I.; Plósz, S. (Nature **333** [1988] 343/5).
[59] Donnelly, V. M.; Karlicek, R. F. (Proc. SPIE Intern. Soc. Opt. Eng. No. 476 [1984] 102/9; C.A. **101** [1984] No. 238003).
[60] Cheng, E. A. P.; Fraser, R. D.; Eden, J. G. (Appl. Spectrosc. **45** [1991] 949/52).

[61] Tao, H.; Miyazaki, A.; Bansho, K. (Anal. Chem. **58** [1986] 202/7).
[62] Herget, W. F.; Levine, S. P. (Appl. Ind. Hyg. **1** [1986] 110/2 from C.A. **105** [1986] No. 84357).
[63] Andreev, B. A.; Ezheleva, A. E.; Kazakov, V. P.; Krupnov, A. F.; Krylov, V. A. (Zh. Anal. Khim. **41** [1986] 1622/5; C.A. **105** [1986] No. 217985).
[64] Yoshida, H.; Nishina, A.; Hirano, M. (Koatsu Gasu **29** No. 2 [1992] 140/7 from C.A. **116** [1992] No. 206837).
[65] Maiorov, V. I.; Morozova, L. N.; Molodyk, A. D. (Vysokochist. Veshchestva **1990** No. 4, pp. 145/50; High Purity Subst. [Engl. Transl.] **4** [1990] 710/5).
[66] Reents, W. D., Jr. (Anal. Chim. Acta **237** [1990] 83/90).
[67] Ivanova, N. T.; Vislykh, N. A.; Voevodina, V. V. (Vysokochist. Veshchestva **1990** No. 5, pp. 198/203; High Purity Subst. [Engl. Transl.] **4** [1990] 953/7).
[68] Ecknig, W.; Glöckl, D.; Schulz, B.; Schön, H. (Chem. Tech. [Leipzig] **37** [1985] 214/8; C.A. **103** [1985] No. 81016).
[69] Cui, X.; Xu, X.; Yan, X.; Lang, W. (Bandaoti Xuebao **10** [1989] 945/51 from C.A. **113** [1990] No. 164556).
[70] Banks, H. J. (J. Stored Prod. Res. **23** [1987] 213/21; C.A. **108** [1988] No. 70521).

[71] Anonymous (Dangerous Prop. Ind. Mater. Rep. **6** [1986] 103/7; C.A. **104** [1986] No. 192013).

1.3.1.3.13 Removal

It may be necessary to remove PH_3 from gases either because of its toxicity or its interference in reactions of the main component of the mixture. Especially, the concentration of PH_3 remaining in gaseous or condensed wastes has to be reduced to very low levels because of its toxicity and odor. The results of earlier investigations are described in "Phosphor" C, 1965, pp. 15/6. The present section summarizes the removal of PH_3 by physical and chemical methods which do not specifically involve oxidations. Quantitative investigations on the sorption of PH_3 by solids are described in Section 1.3.1.5.11, pp. 287/93. Removal by oxidation is treated in Section 1.3.1.5.4, pp. 222/33.

From H_2. The reversibility of the adsorption of PH_3 in H_2 on activated carbon was described in [1]. The adsorption of PH_3 from a mixture with H_2 on zeolite CaA was studied in [2].

From N_2 and Other Inert Gases. Adsorption isotherms for PH_3 in mixtures with N_2 at 293 K were determined on various types of silica gel, activated carbon, and zeolites [3]. The removal from inert gases can be achieved by contact with silica gels [4], zeolites [4, 5], or activated carbon impregnated with CH_3CN, picric acid, or maleic acid [4]. The adsorption on activated carbon pretreated with iodine compounds and sulfates or nitrates of ammonium and metals was used in [6]. Cu- or Ag-modified molecular sieves [7] or a soda-lime bed, which may contain nitrite salts, were also used as adsorbents [8]. The removal of PH_3 by an aqueous solution of LiCl and $HClO_4$ was mentioned [9]. Its removal by oxidative methods is described in Section 1.3.1.5.4, p. 222.

From C_2H_2. The concentration of PH_3 in C_2H_2 produced by hydrolysis of impure CaC_2 has to be lowered for most applications; industrially important methods are reviewed in [10, 11]. The well-known scrubbing with diluted and concentrated H_2SO_4, followed by scrubbing with an NaOH solution, is described in [12, 13]. The rate of the absorption of PH_3 by about 80 to 90% aqueous H_2SO_4 at 28°C as a function of PH_3 partial pressure is described in [55]. The effectiveness of concentrated H_2SO_4 increases upon addition of $\leqq$0.5% of Cu^{II} salts [14, 15] or 0.007% of $HgSO_4$ [14, 16]. Scrubbing with 70% H_3PO_4 is described in [17]. The adsorption of PH_3 to 11 brands of activated carbon is promoted by pretreatment with acids [18]. The pretreatment of the activated carbon with inorganic acids, which increases the capacity for adsorption, is also claimed in a patent [19]. The removal of PH_3 from C_2H_2 by a large number of oxidative methods is described in Section 1.3.1.5.4, p. 222.

From CO. Fast thermolysis decomposes PH_3 at 1000 to 1100 K yielding H_2 and red phosphorus [20]. The concentration of PH_3 in CO can also be reduced by scrubbing with CH_3OH at ~240 K [21].

From Silanes. A summary of the purification of SiH_4, including the removal of PH_3, is given in "Silicon" Suppl. Vol. B 1, 1982, pp. 76/7. The removal of small amounts of PH_3 in SiH_4 succeeds with zeolites [22, 23] and in some cases with zeolites which are cation-exchanged with the bivalent cations of Mg [24, 25], Pb, Mn, Co, and Cd [25]. Contact with Cu on an alumina carrier [26] or with a metalated macroreticular polymer [27] was also used. A solution of $NaAlH_4$ can also remove PH_3 selectively from SiH_4 [28].

Adsorption of PH_3 impurities in silanes on zeolites at low temperatures [29] is the method usually proposed. Cation-exchanged zeolites containing Ag [30] or the divalent ions of Ca [31], Cd [32], or Mn [33, 34] were employed. PH_3 impurities in halosilane/H_2 mixtures can also be removed by adsorption on zeolite [35].

From GeH_4. The adsorption of PH_3 on the zeolite CaA from a GeH_4/PH_3 mixture was studied in [23].

From AsH_3. The selective adsorption of PH_3 in AsH_3 on Polysorb 1 was investigated in [36].

From H_2Se and H_2Te. A thermodynamic analysis of the removal of PH_3 impurities from H_2Se and H_2Te was carried out (cf. 1.3.1.5.5, p. 239) [56].

From Air. PH_3 in air can be adsorbed on activated carbon containing 2% KI [37]. The catalytic oxidation is described in Section 1.3.1.5.4, p. 222.

From Waste Gases. The removal of PH_3 from furnace gases is reviewed in [38]. The various methods for the removal of PH_3 (and other noxious compounds) from the effluent gases of the electronics industry are surveyed in [39]. Decomposition is achieved by electrical discharges [40, 41] and by exposure to a radiofrequency plasma [42]. Gases free of O_2 can be pyrolyzed in the presence of red phosphorus [43]. Catalytic decomposition, said to be cheaper than adsorption methods, is achieved by Pt on Al_2O_3, by CrO (sic) at 490 K [44], and by Pd on Al_2O_3 [45]. If the gas mixture is contacted with a zeolite at or below ambient temperature, the recovery of adsorbed PH_3 is possible [46]. Other suitable media for the adsorption of phosphane such as Al_2O_3, silica gel, activated carbon, kieselguhr [47], or halides of Si, Ti, and Sn, were suggested [48]. A PCl_3 solution was mentioned in [49]. Noxious compounds including PH_3 can be removed upon contact with $Ca(OH)_2$ containing KOH [50], with beds of Si followed by an alkaline solid, and with CuO in a heated column [51]. PH_3 is scrubbed with diluted HNO_3 [52] or with H_2SO_4 in a multistage process [53]. The removal of PH_3 from waste gases by oxidative methods is described in Section 1.3.1.5.4, p. 222.

From Waste Water. The removal of PH_3 from an aqueous solution by various zeolites is described in [54]. Several methods for the removal of PH_3 from waste water by oxidation are described in Section 1.3.1.5.4, p. 222.

References:

[1] Colabella, J. M.; Stall, R. A.; Sorenson, C. T. (J. Cryst. Growth **92** [1988] 189/95).
[2] Orlov, V. Yu.; Matveev, I. V.; Egorov, E. N.; Tolmachev, A. M. (Vysokochist. Veshchestva **1988** No. 2, pp. 71/4; High Purity Subst. [Engl. Transl.] **2** [1988] 245/8).
[3] Tkach, O. D. (Izv. Vyssh. Uchebn. Zaved. Khim. Khim. Tekhnol. **17** [1974] 1653/4 from C.A. **82** [1975] No. 103609).
[4] Bochkarev, E. P.; Khripunov, V. M. (Poluch. Anal. Veshchestv Osoboi Chist. Mater. Vses. Konf., Gorkiy, USSR, 1963 [1966], pp. 287/9).
[5] Tkach, O. D. (U.S.S.R. 413968 [1972/74] from C.A. **81** [1974] No. 107950).
[6] Takeda Chemical Industries, Ltd. (Jpn. Kokai Tokkyo Koho 85-71040 [1983/85] 5 pp. from C.A. **103** [1985] No. 165491).
[7] Schön, H.; Röthe, K. P.; Thiele, H. U.; Schedone, D.; Winkel, A. (Ger. [East] 218562 [1983/85] 9 pp.; C.A. **103** [1985] No. 73242).
[8] Foster, D. I.; Scott, P. T.; Pierick, M. W.; Shaddock, C. L., Jr. (U.S. 4442077 [1982/84] 4 pp.; C.A. **100** [1984] No. 212371).
[9] Dorfman, Ya. A.; Emelyanova, V. S.; Kisanova, T. N. (U.S.S.R. 889066 [1978/81] from C.A. **96** [1982] No. 145326).
[10] Miller, S. A. (Acetylene, Vol. 1, Benn, London 1965, pp. 336/41).

[11] Beller, H.; Wilkinson, J. M., Jr. (Kirk-Othmer Encycl. Chem. Technol. 2nd Ed. **1** [1963] pp. 171/211, 177/8).
[12] Cervinka, M.; Malkova, Z.; Kosar, P.; Kus, M.; Capek, J. (Czech. 253088 [1985/88] 5 pp. from C.A. **110** [1989] No. 59924).
[13] Lassmann, E. (Ger. 2549399 [1975/77] 4 pp. from C.A. **87** [1977] No. 133854).
[14] Slizovskaya, L. V.; Leites, I. L.; Strizhevskii, I. I. (Khim. Prom-st. [Moscow] **45** [1969] 524/6).
[15] Sokol'skii, D. V.; Dorfman, Ya. A.; Rakitskaya, T. L. (U.S.S.R. 618123 [1972/78] from C.A. **90** [1979] No. 5877).

[16] Strizhevskii, I. I.; Stolyarova, L. V. (U.S.S.R. 171861 [1964/65] from C.A. **63** [1965] 16118).
[17] Zimmermann, J.; Meurer, P.; Liebig, W. (Ger. Offen. 3439368 [1984/86] 11 pp.; C.A. **105** [1986] No. 99868).
[18] Bogdanov, V. M.; Moiseichuk, O. V.; Shumyatskii, Yu. I. (Zh. Prikl. Khim. [Leningrad] **60** [1987] 1119/23; J. Appl. Chem. USSR [Engl. Transl.] **60** [1987] 1055/8).
[19] Bogdanov, V. M.; Shumyatskii, Yu. I.; Moiseichuk, O. V.; Suchkova, Z. A.; Neshumova, S. P.; Fedorovskaya, V. V.; Chugunova, G. I. (U.S.S.R. 1181692 [1983/85] from C.A. **104** [1986] No. 36169).
[20] Munday, T. F.; Walden, J. (Ger. Offen. 2854086 [1977/79] 9 pp. from C.A. **91** [1979] No. 93790).

[21] Dorfman, Ya. A.; Yukht, I. M.; Levina, L. V.; Polimbetova, G. S.; Petrova, T. V.; Emel'yanova, V. S. (Usp. Khim. **60** [1991] 1190/228; Russ. Chem. Rev. [Engl. Transl.] **60** [1991] 605/27).
[22] Ito, M.; Miyagawa, H.; Abe, T.; Inoue, K.; Ikeda, K.; Murakami, M. (Jpn. Kokai Tokkyo Koho 62-191413 [87-191413] [1986/87] 4 pp. from C.A. **108** [1988] No. 58883).
[23] Orlov, V. Yu.; Matveev, I. V.; Egorov, E. N.; Tolmachev, A. M. (Vysokochist. Veshchestva **1988** No. 2, pp. 71/4; High Purity Subst. [Engl. Transl.] **2** [1988] 245/8).
[24] Schöllner, R. (Ger. [East] 150599 [1980/81] 10 pp. from C.A. **96** [1982] No. 145334).
[25] Ohgushi, T.; Yusa, A.; Kinoshita, K.; Yatsurugi, Y. (Bull. Chem. Soc. Jpn. **51** [1978] 419/21).
[26] Pacaud, B.; Popa, J. M.; Cartier, C. B. (Eur. Appl. 314542 [1988/89] 6 pp. from C.A. **111** [1989] No. 99877).
[27] Tom, G. M. (Eur. Appl. 299488 [1988/89] 5 pp. from C.A. **110** [1989] No. 157158).
[28] Smith, I. L.; Nelson, G. E. (U.S. 4532120 [1983/85] 6 pp. from C.A. **103** [1985] No. 125925).
[29] Ito, M.; Miyagawa, H.; Abe, T.; Ikeda, K.; Murakami, M.; Yanagawa, N. (Jpn. Kokai Tokkyo Koho 62-212217 [87-212217] [1986/87] 5 pp. from C.A. **108** [1988] No. 115181).
[30] Kawasaki, K.; Tazawa, S. (Jpn. Kokai Tokkyo Koho 86-53106 [1984/86] 4 pp. from C.A. **105** [1986] No. 8735).

[31] Ito, M.; Miyagawa, H.; Abe, T.; Ikeda, K.; Murakami, M.; Yanagawa, N. (Jpn. Kokai Tokkyo Koho 62-212215 [87-212215] [1986/87] 6 pp. from C.A. **108** [1988] No. 115134).
[32] Parent, J. C.; Renaudin, M. H. (Eur. Appl. 419358 [1990/91] 3 pp. from C.A. **114** [1991] No. 210001).
[33] Ito, M.; Miyagawa, H.; Abe, T.; Ikeda, K.; Murakami, M.; Yanagawa, N. (Jpn. Kokai Tokkyo Koho 63-147815 [88-147815] [1986/88] 5 pp. from C.A. **109** [1988] No. 131702).
[34] Ito, M.; Miyagawa, H.; Abe, T.; Inoue, K.; Ikeda, K.; Murakami, M. (Jpn. Kokai Tokkyo Koho 63-25210 [88-25210] [1986/88] from C.A. **109** [1988] No. 15753).
[35] Prigge, H.; Rurlaender, R.; Hoffmann, H.; Bortner, H. P. (Ger. Offen. 3843313 [1988/90] 5 pp. from C.A. **113** [1990] No. 222841).
[36] Molodyk, A. D.; Popkova, I. A.; Morozova, L. N.; Maiorov, V. I.; Sakodynskii, K. I. (Vysokochist. Veshchestva **1991** No. 5, pp. 182/6; High Purity Subst. [Engl. Transl.] **5** [1991] 918/21).
[37] Friemel, W.; Ehret, R.; Barth, V.; Munzel, M. (S. African 87-1742 [1986/87] 15 pp. from C.A. **109** [1988] No. 165736).
[38] Koverya, V. M.; Monin, V. Ya.; Belyakov, V. P. (Khim. Prom-st. Ser. Okhr. Okruz. Sredy Ratsion. Ispol'z. Prir. Resur. **1986** No. 5/66, 41 pp. from Ref. Zh. Khim. **1987** No. 11 I 703).

[39] Khanna, A. K.; Gupta, R. (AIP Conf. Proc. No. 166 [1988] 89/98 from C.A. **110** [1989] No. 12682).
[40] Minoshima, H.; Oe, T.; Miura, A.; Itaya, R. (Jpn. Kokai Tokkyo Koho 02-253823 [1989/90] 9 pp. from C.A. **114** [1991] No. 128235).

[41] Ogawa, S.; Oe, T.; Sato, A.; Itaya, R. (Jpn. Kokai Tokkyo Koho 01-148329 [1987/89] 7 pp. from C.A. **112** [1990] No. 222607).
[42] Baumann, J. A.; Schachter, R.; Viscogliosi, M. (U.S. 4867952 [1988/89] 4 pp. from C.A. **112** [1990] No. 104183).
[43] Rybakov, V. V.; Belyakov, B. P.; Dyatlov, A. I.; Teplov, L. I. (U.S.S.R. 1551402 [1988/90] from C.A. **113** [1990] No. 8836).
[44] Kikuchi, Y. (Jpn. Kokai Tokkyo Koho 63-200820 [1987/88] 4 pp. from C.A. **110** [1989] No. 28546).
[45] Kitahara, K.; Akita, N.; Hiramoto, T.; Shimada, T. (Jpn. Kokai Tokkyo Koho 63-283752 [1987/88] 5 pp. from C.A. **110** [1989] No. 198443).
[46] Jödden, K.; Heymer, G.; Klose, W. (Ger. Offen. 2746910 [1977/79] 8 pp.; C.A. **91** [1979] No. 59609).
[47] Sadakata, T.; Fuje, N. (Jpn. Kokai Tokkyo Koho 63-218236 [1987/88] 3 pp. from C.A. **110** [1989] No. 81776).
[48] Prigge, H.; Rurlaender, R.; Schwab, M.; Bortner, H. P.; Englmuller, A. (Eur. Appl. 329149 [1989] 9 pp. from C.A. **112** [1990] No. 14638).
[49] Kaneko, S. (Jpn. Kokai Tokkyo Koho 79-45682 [1977/79] 2 pp. from C.A. **91** [1979] No. 78456).
[50] Morikawa, Y.; Nishizono, M. (Eur. Appl. 261950 [1987/88] 8 pp. from C.A. **108** [1988] No. 22623).

[51] Smith, J. R. (PCT Int. Appl. 89-11905 [1989] 22 pp. from C.A. **112** [1990] No. 124436).
[52] Sawada, A.; Ebe, K. (Jpn. Kokai Tokkyo Koho 03-109922 [1989/91] 5 pp. from C.A. **115** [1991] No. 141558).
[53] Sontag, H. J.; Kunde, F.; Schraufstetter, M. (Eur. Appl. 399085 [1989/90] 7 pp. from C.A. **114** [1991] No. 85195).
[54] Chu, P.; Dwyer, F. G. (U.S. 4401572 [1979/83] 7 pp.; C.A. **99** [1983] No. 178200).
[55] Chandrasekaran, K.; Sharma, M. M. (Chem. Eng. Sci. **32** [1977] 275/80).
[56] Gladyshev, V. P.; Kovaleva, S. V.; Dubinina, L. K. (Izv. Vyssh. Uchebn. Zaved. Khim. Khim. Tekhnol. **24** [1981] 1462/4; C.A. **96** [1982] No. 111184).

1.3.1.4 Physical Properties

1.3.1.4.1 Molecular Properties

1.3.1.4.1.1 Electronic Structure

The electron configuration of the pyramidal molecule (point group C_{3v}; see "Phosphor" C, 1965, p. 17) in its electronic ground state $\tilde{X}\ ^1A_1$ is given (in a one-electron picture) in two more recent papers dealing with the absorption of synchrotron radiation [1] and with dipole (e, e) spectroscopy simulating photoabsorption [2, 3] as follows:

$$\underbrace{\underset{\text{P1s}}{(1a_1)^2}\ \ \underset{\text{P2s}}{(2a_1)^2}\ \ \underset{\text{P2p}}{\underbrace{(1e)^4\ (3a_1)^2}}}_{\text{core}}\qquad \underbrace{\underset{\text{P–H }\sigma\text{ bonding}}{\underbrace{(4a_1)^2\ (2e)^4}}\ \ \underset{\text{P lone pair}}{(5a_1)^2}}_{\text{valence}}$$

This configuration is basic to all discussions of excitation and ionization processes of valence and core electrons (see pp. 144, 149) and was also confirmed by theoretical calculations of orbital energies (see the literature gathered on p. 150). The valence shell configuration was earlier given in [4, 5].

The inner valence orbital $4a_1$ is a bonding combination of P3s and H1s. The P3s character was experimentally confirmed by the behavior of the angular distribution (asymmetry) parameter β for photoionization [6 to 8]. Significant P3s-H1s overlap was found in several theoretical calculations: ab initio SCF-MO [9], SCF-Xα-SW (scattered wave) [10], SCM (self-consistent multi-polar)-Xα-DV (discrete variational) [11]. A participation of P3p in $4a_1$ was, however, assumed in the early MO picture of Walsh [4] and was supported in an interpretation [12] of earlier SCF-MO results [13] by a localized orbital picture.

The outer-valence orbital 2e is a bonding combination of P3p and H1s; see the qualitative MO picture [4] and the later ab initio SCF-MO [12, 13], SCF-Xα-SW [10], and SCM-Xα-DV [11] calculations. A predominance of P3p was also inferred from a Cooper minimum in the asymmetry parameter β [7]. A lower bonding character (compared to 1e of NH_3) was derived from the Jahn-Teller splitting of the 2e photoelectron band of PH_3 [14] and also from the behavior of β [6]. A certain π-bonding character of 2e was considered in [15].

The $5a_1$ lone-pair orbital has mainly P3p character. Ab initio SCF-MO calculations yielded 88% localization of $5a_1$ on P [9, 16] and a P3p character of 73% [9] or 72% [16]; see also a review on photoelectron spectra and bonding in phosphorus compounds [17]. Similar P3p percentages were 71% (SCF-MO) [13, 18, 19] and 67% (SCM-Xα-DV) [11]. The P3p character was also derived from a Cooper minimum in β [7, 8] and from an SCF-Xα-SW calculation [10]. A pure P3s character had been assumed in the simple Walsh [4] picture. This model was later modified by allowing for a mixing of $3s(a_1)$ and $3p(a_1)$ orbitals [20]. An $sp^{0.8}$ hybridization of the lone-pair orbital was given in a localized-orbital description [12]. An increased s character (compared to NH_3) was inferred from a consideration of lone-pair and core ionization potentials (for the pyramidal as well as the planar PH_3 molecule) [21]. The s or p character of $5a_1$ is also related to the derivation of a P–H bond moment from the total dipole moment (see p. 153).

Electron density maps based on theoretical calculations (methods in parentheses) are given in [22] (SCF-MO [23]; also for the highest occupied MO $5a_1$), in [24] (SCF and CI; also for the three valence orbitals), in [10] (SCF-Xα-SW; for the valence orbitals and the total valence shell), in [25] (SCF and SCGF [self-consistent group function]), and in [26] (united atom). The Laplacian $\nabla^2\rho$ of the charge density ρ showed four local concentrations of electronic charge in the valence shell of the central P atom in accordance with the VSEPR (valence shell electron pair repulsion) model [27]; for this latter model and its application to PH_3, see [28 to 31].

A negative (effective) atomic charge $q=-0.24$ on P (compared to $q=0$ for elemental phosphorus) was determined from the ionization potentials of the P1s and P2p core electrons and the KLL Auger transition energy [32, 33]. An appreciable negative charge on P was also inferred from the displacement of the $PK\alpha_1$ X-ray line (at 2013.6 eV) by -0.99 ± 0.02 eV relative to red phosphorus [34]. Atomic charges taken from Mulliken population analyses (MPA) or modified procedures give no consistent picture of the charge distribution, since even their signs are contradictory. This ambiguity is, however, not too astonishing in view of both a strong basis-set dependence of the theoretical results and a small absolute value of the measured dipole moment (see p. 153). A negative charge on P (given in parentheses) was obtained by the following calculations: SCF-MO (−0.56 [23], −0.32 [18, 19], −0.17 [15], −0.0744 [35]), SCF-MO with MCB (modified Christoffersen and Baker [36]) procedure replacing MPA (−0.181) [37], spherical density basis functions with the Yáñez et al. [38] population analysis replacing MPA (−0.03) [39]. A positive charge on P was apparently calculated more often: SCF-MO

(+0.2437 [40], +0.159 [37], +0.13 [41], +0.102 [42], +0.073 to +0.045 [9], +0.0631 [43], +0.06 [44]), HF (+0.162) with corrections by MP2, MP4, and quadratic CI [45], SCF-MO including the correlation (+0.52 and +0.34) [46], valence-only SCF + CI (0.21) [47], MP3 (+0.038) [48]. Different signs were obtained by a least-squares fit of either the electric field or the electron density and were ascribed to the bad quality of the wave functions used [49].

The lowest empty molecular orbitals are $6a_1$ (a Rydberg-state MO; for a contour map from an SCF-Xα-SW calculation, see the original paper) [10] or 3e, followed by $6a_1$, 4e, and $7a_1$ (composition of 3e is 36% P3p, 23% P3d, 30% H1s, and 11% H2p from an SCM-Xα-DV calculation) [11].

The influence of P3d orbitals or functions has been investigated in almost all theoretical papers cited in the foregoing text. Another paper [50] was especially devoted to this topic.

References:

[1] Ishiguro, E.; Iwata, S.; Mikuni, A.; Suzuki, Y.; Kanamori, H.; Sasaki, T. (J. Phys. B **20** [1987] 4725/39, 4728).
[2] Zarate, E. B.; Cooper, G.; Brion, C. E. (Chem. Phys. **148** [1990] 277/88, 278).
[3] Zarate, E. B.; Cooper, G.; Brion, C. E. (Chem. Phys. **148** [1990] 289/97).
[4] Walsh, A. D. (J. Chem. Soc. **1953** 2296/301).
[5] Walsh, A. D.; Warsop, P. A. (in: Mangini, A.; Advances in Molecular Spectroscopy, Vol. 2, Pergamon, Oxford 1962, pp. 582/91).
[6] Cauletti, C.; Piancastelli, M. N.; Adam, M. Y. (J. Mol. Struct. **174** [1988] 135/40).
[7] Cauletti, C.; Adam, M. Y.; Piancastelli, M. N. (J. Electron Spectrosc. Relat. Phenom. **48** [1989] 379/87).
[8] Adam, M. Y.; Cauletti, C.; Piancastelli, M. N.; Svensson, A. (J. Electron Spectrosc. Relat. Phenom. **50** [1990] 219/27).
[9] Lehn, J. M.; Munsch, B. (Mol. Phys. **23** [1972] 91/107, 97/8).
[10] Norman, J. G., Jr. (J. Chem. Phys. **61** [1974] 4630/5).

[11] Xiao, S.-X., Trogler, W. C.; Ellis, D. E.; Berkovitch-Yellin, Z. (J. Am. Chem. Soc. **105** [1983] 7033/7).
[12] Guest, M. F.; Hillier, I. H.; Saunders, V. R. (J. Chem. Soc. Faraday Trans. II **68** [1972] 867/73).
[13] Hillier, I. H.; Saunders, V. R. (Trans. Faraday Soc. **66** [1970] 2401/7).
[14] Potts, A. W.; Price, W. C. (Proc. R. Soc. [London] A **326** [1971/72] 181/97, 193).
[15] Robert, J.-B.; Marsmann, H.; Schaad, L. J.; Van Wazer, J. R. (Phosphorus Relat. Group V Elem. **2** [1972] 11/8).
[16] Lehn, J. M.; Munsch, B. (J. Chem. Soc. D **1969** 1327/9).
[17] Bock, H. (Pure Appl. Chem. **44** [1975] 343/72, 349/51).
[18] Hillier, I. H.; Saunders, V. R. (J. Chem. Soc. D **1970** 316/8).
[19] Hillier, I. H.; Saunders, V. R. (J. Chem. Soc. A **1970** 2475/7).
[20] Elbel, S.; Bergmann, H.; Enßlin, W. (J. Chem. Soc. Faraday Trans. II **70** [1974] 555/9).

[21] Eyermann, C. J.; Jolly, W. L. (J. Phys. Chem. **87** [1983] 3080/2).
[22] Boyd, D. B. (J. Chem. Phys. **52** [1970] 4846/57, 4853/6).
[23] Boyd, D. B.; Lipscomb, W. N. (J. Chem. Phys. **46** [1967] 910/9).
[24] Petke, J. D.; Whitten, J. L. (J. Chem. Phys. **59** [1973] 4855/66, 4857/9).
[25] Cook, D. B.; Palmieri, P. (Chem. Phys. Lett. **3** [1969] 219/22).
[26] Banyard, K. E.; Hake, R. B. (J. Chem. Phys. **43** [1965] 2684/9).
[27] Bader, R. F. W.; MacDougall, P. J.; Lau, C. D. H. (J. Am. Chem. Soc. **106** [1984] 1594/605, 1597/9).

[28] Gillespie, R. J. (Molecular Geometry, Van Nostrand-Reinhold, London 1972, pp. 56/8).
[29] Hargittai, I.; Baranyi, A. (Acta Chim. Acad. Sci. Hung. **93** [1977] 279/88).
[30] Schmiedekamp, A.; Cruickshank, D. W. J.; Skaarup, S.; Pulay, P.; Hargittai, I.; Boggs, J. E. (J. Am. Chem. Soc. **101** [1979] 2002/10).

[31] Ahlrichs, R. (Chem. Uns. Zeit **14** [1980] 18/24).
[32] Nefedov, V. I.; Yarzhemskii, V. G. (Zh. Strukt. Khim. **29** No. 6 [1988] 104/11; J. Struct. Chem. [Engl. Transl.] **29** [1988] 911/7).
[33] Nefedov, V. I.; Yarzhemsky, V. G.; Chuvaev, A. V.; Trishkina, E. M. (J. Electron Spectrosc. Relat. Phenom. **46** [1988] 381/404, 389).
[34] Dolenko, G. N.; Krupoder, S. A.; Mazalov, L. N. (Zh. Strukt. Khim. **20** [1979] 334/6; J. Struct. Chem. [Engl. Transl.] **20** [1979] 279/81).
[35] Fantucci, P.; Polezzo, S. (Mol. Phys. **37** [1979] 831/41, 836).
[36] Christoffersen, R. E.; Baker, K. A. (Chem. Phys. Lett. **8** [1971] 4/9).
[37] Edgecombe, K. E.; Boyd, R. J. (J. Chem. Soc. Faraday Trans. II **83** [1987] 1307/15).
[38] Yáñez, M.; Stewart, R. F.; Pople, J. A. (Acta Crystallogr. A **34** [1978] 641/8).
[39] Escudero, F.; Yáñez, M. (Mol. Phys. **45** [1982] 617/28, 624).
[40] Rothenberg, S.; Young, R. H.; Schaefer, H. F., III (J. Am. Chem. Soc. **92** [1970] 3243/50).

[41] Demuynck, J.; Veillard, A. (J. Chem. Soc. D **1970** 873/4).
[42] Barthelat, J. C.; Durand, P.; Serafini, A. (Mol. Phys. **33** [1977] 159/80, 170).
[43] Desmeules, P. J.; Allen, L. C. (J. Chem. Phys. **72** [1980] 4731/48, 4734).
[44] Kollman, P.; McKelvey, J.; Johansson, A.; Rothenberg, S. (J. Am. Chem. Soc. **97** [1975] 955/65, 956).
[45] Kraka, E.; Gauss, J.; Cremer, D. (J. Mol. Struct. **234** [1991] 95/126, 115 [THEOCHEM **80**]).
[46] Wallmeier, H.; Kutzelnigg, W. (J. Am. Chem. Soc. **101** [1979] 2804/14, 2808).
[47] Trinquier, G.; Daudey, J. P.; Caruana, G.; Madaule, Y. (J. Am. Chem. Soc. **106** [1984] 4794/9).
[48] Yabushita, S.; Gordon, M. S. (Chem. Phys. Lett. **117** [1985] 321/5).
[49] Hall, G. G.; Smith, C. M. (Int. J. Quantum Chem. **25** [1984] 881/90, 889).
[50] Ball, E. E.; Ratner, M. A.; Sabin, J. R. (Chem. Scr. **12** [1977] 128/41).

1.3.1.4.1.2 Electronically Excited States

Electronic excitation energies E are given below in two tables. The first covers excitation of valence electrons, $5a_1$ and 2e, with E lying between 6 and 10 eV (vacuum UV region). The second covers excitation of core electrons, P2p and P2s, with E lying between 132 and 136 eV, at 142 and 162 eV, and at 190 eV (soft X-ray region). Assignments of electronic transitions, especially for the soft X-ray region, are limited by a poor understanding of the nature (valence or Rydberg) and energetic order of the respective unoccupied orbitals. Assignments have been also based on the transferability of the term values $T=E_i-E$, with E_i = ionization potential for the excited electron, thus, $T>0$ for the pre-edge ("discrete") and <0 for the post-edge ("continuous") regions. E was measured by optical or X-ray photoabsorption (PAS or XAS), by electron energy loss spectroscopy (EELS), or by multiphoton ionization (MPI) spectroscopy.

Excitation of Valence Electrons

The adiabatic (ad) or vertical (vert) excitation energies E (in eV and cm^{-1}) given in the table below are based on recent EELS [1], PAS [2], and MPI [3] measurements. Thus, the term numbering (Ã, B̃, ...) does not agree with that in an older table [4]. The electronic transitions and term designations are partly given for the point group D_{3h} (of the planar molecule) or for C_s; the respective ground-state valence-electron configurations would be $(a_1')^2 (e')^4 (a_2'')^2$ [5, 6] or $(5a')^2 (6a')^2 (2a'')^2 (7a')^2$ [7], compared to $(4a_1)^2 (2e)^4 (5a_1)^2$ for C_{3v}. The term values were derived from $E_i(ad) = 9.87$ and 12.50 eV for $5a_1$ and 2e (see p. 149) and $E_i\,(vert) = 10.58$ eV for $5a_1$ [1]. Details, additional work (also on PD_3), and different assignments are dealt with in the subsections on individual states or transitions adjoining the table:

E in eV	6.00	6.89	≦7.57	≦7.86
E in cm^{-1}	48500	55500	≦61099	≦63392
type	vert	vert	ad	ad
transition	$a_1'(8a') \leftarrow 5a_1$	$4s \leftarrow 5a_1$	$4p(e) \leftarrow 5a_1$	$4p(a_1) \leftarrow 5a_1$
state	ã $^3A_2''(^3A')$	Ã $^1A_1(^1A_2'')$	B̃ ^1E	C̃ $^1A_1$
T in eV	4.58	3.69	≧2.30	≧2.01
T in cm^{-1}	37000	29700	≧18550	≧16300
method	EELS	EELS	MPI	MPI
Ref.	[1]	[1]	[3]	[3]

E in eV	≦8.46	≦8.81	9.95
E in cm^{-1}	≦68188	≦71000	80370
type	ad	ad	ad
transition	$3d \leftarrow 5a_1$(?)	$5p \leftarrow 5a_1$(?)	$4p \leftarrow 2e$
state	—	—	—
T in eV	≧1.41	≧1.06	2.55
T in cm^{-1}	≧11400	≧8550	20600
method	MPI	MPI	PAS
Ref.	[3]	[3]	[2]

Triplet ã $^3A_2''$ (D_{3h}). The loss spectrum for an incident electron energy of 20 eV and a scattering angle of 5° showed a broad tail on the low-energy side of the main peak at 6.89 eV. This band with an estimated maximum around 6 eV was interpreted to be the conjugated triplet state of the singlet state at 6.89 eV. D_{3h} symmetry was assumed in view of theoretical results (see below). The term value of 4.58 eV (based on E = 6.00 eV) resembles that of a core-excited Rydberg state (4.49 eV [8], see p. 147) [1]. Ab initio calculations using C_s symmetry yielded for the state $(5a')^2 (6a')^2 (2a'')^2 7a' 8a'$, $^3A'$ the total energy and a planar structure. The latter points to a non-Rydberg character of the state in view of a nonplanar structure of the ground-state PH_3^+ ion (see p. 309). The non-Rydberg nature follows also from a population analysis yielding small P4s contributions. The state is dissociative with a vanishing barrier [7].

Singlet Ã $^1A_1$. The state is given in the original paper [1] as $^1A_2''$ (D_{3h}), apparently following the assignment for the conjugated triplet $^3A_2''$ (see above). It was observed in two loss spectra at incident electron energies of 60 eV (scattering angle Θ = 0°) and 20 eV (Θ = 5°). The absence of vibrational structure was in accord with the known predissociative character (see below) [1]. The state was long known from a continuous absorption band with a maximum at 1800 Å [6, 9]. Slightly different vertical excitation energies are based on that wavelength: E = 55600 cm^{-1}

(for both PH_3 and PD_3) [6], 55550 (PH_3) and 55000 cm^{-1} (PD_3), with term values T=30000 and 31000 cm^{-1} (in a table), E=55700 cm^{-1} with T=29800 cm^{-1} (in the text) [10], E=6.82 eV with T=31000 cm^{-1} (PH_3) [11] (misprinted as 8.82 eV in [7]), and E=6.90 eV or 55700 cm^{-1} [12]. These energies were confirmed by a measurement with synchrotron radiation yielding maximum absorption at 1798 Å [2]. Maximum absorption at 1910 Å [13], corresponding to E= 52400 cm^{-1}, was not confirmed in [10]. An "absorption feature" at ~7 eV was observed by dipole (e, e) spectroscopy and ascribed to the $\tilde{A}\ {}^1A_1$ state [14]. An adiabatic excitation energy of ≦47847 cm^{-1} was obtained from the photoabsorption measurement using synchrotron radiation [2]. An uncertain value of 43000 cm^{-1} had been based on the older literature [4].

The terminating orbital was given as 4s (Rydberg) [10], apparently in view of the term value being similar to that of atomic P (32200 cm^{-1} for 4s←3p). Ab initio HF and CI calculations (for C_s symmetry) showed that the predissociation process of $\tilde{A}\ {}^1A_1$ involved a Rydberg-valence transformation. The HOMO 8a′ (C_s) had P4s character at the equilibrium geometry, but changed into H1s when the P–H bond length was increased [7]. Rydberg-valence mixing of the terminating orbital was supposed to be responsible for the lowering of the term value T by 0.8 eV against T=4.49 eV of a core-excited Rydberg state (see above and p. 147) [1]. An a_1(4s) Rydberg orbital (instead of an antibonding a_1^* valence orbital) had been proposed in [5, 6] to account for earlier observations of a discrete (instead of a continuous) band structure. The discrete structure was, however, not confirmed [15] (see also "Phosphor" C, 1965, p. 23), but a more recent MPI study using a dye laser revealed almost twenty bands at 351 to 398 nm, which could be assigned as (2+1) excitation via the $\tilde{A}$ state [16].

Singlet $\tilde{B}\ {}^1E$. Resonance-enhanced MPI spectra (2-photon excitation below 330 nm) showed a vibrational progression (in v_2') for both PH_3 and PD_3. The energies of the first bands, 61099 cm^{-1} for PH_3 and 61249 cm^{-1} for PD_3, were taken to be the (adiabatic) excitation energies, but the true electronic origins possibly were not identified. The quantum defect δ=1.565 pointed to a p character of the terminating orbital [3]. Photoabsorption measurements using synchrotron radiation gave 61825 cm^{-1} for the "first" PH_3 band (of a 4p←$5a_1$ transition) [2]. Energies from conventional photoabsorption work [6, 9] were 62801 cm^{-1} for PH_3 (see also [4]) and 62865 cm^{-1} for PD_3. The transition had, however, been assumed to be 4p(a_1)←$5a_1$, i.e., leading to a ${}^1A_1$ state [6]. Term values of T=17530 cm^{-1} for PH_3 and ~17900 cm^{-1} for PD_3 (for a transition "4p←$5a_1$") had been based [10] upon these latter energies [6, 9]; T=17900 cm^{-1} [11]. A long, single progression observed by EELS between 8 and 9 eV was more recently attributed to a transition 4p←$5a_1$ [1].

Singlet $\tilde{C}\ {}^1A_1$. Resonance-enhanced MPI spectra (2-photon excitation below 330 nm) showed another vibrational progression (in v_2') for both PH_3 and PD_3. The energies of the first bands, 63392 cm^{-1} for PH_3 and 63100 cm^{-1} for PD_3, were taken to be the (adiabatic) excitation energies, but the true electronic origins possibly were not reached. A quantum defect δ=1.42 pointed to a p character of the Rydberg orbital [3]. A similar vibrational progression beginning at 64034 cm^{-1} was observed by photoabsorption using synchrotron radiation. This energy had, however, been related to the transition at 61825 cm^{-1} (now attributed to $\tilde{B}\ {}^1E$; see above) by assuming an additional excitation of the symmetric stretching frequency v_1' [2].

Transition 3d←$5a_1$(?). Resonance-enhanced MPI spectra (3-photon excitation) revealed vibrational progressions beginning at 68188 cm^{-1} (PH_3) and 68040 cm^{-1} (PD_3). Contrary to the interpretation in the table, these bands might also belong to vibronic levels ($\Delta v_n > 0$ with n=1, 3, or 4) of the $\tilde{B}$ and $\tilde{C}$ states or of a triplet state [3]. Two-photon excitation pointed to a shorter PD_3 progression (in the same region), which did not agree very well with the 3-photon progression [3], but which resembled an earlier progression beginning at 67532 cm^{-1} [9] (69716 cm^{-1} had been given in [6]). A term value of ~13300 cm^{-1} (for a transition 3p←$5a_1$) had been based upon E=67532 cm^{-1} [10, 11].

Transition 5p←5a$_1$(?). Resonance-enhanced MPI spectra (3-photon excitation) revealed a progression in ν_2' for both PH_3 (first band at 71000 cm^{-1}) and PD_3 (71622 cm^{-1}). The assignment to 5p←5a$_1$ was based on quantum defects; it contradicted, however, to a more ready 3-photon (than 2-photon) excitation [3]. MPI measurements (3-photon excitation) of four PH_3 bands in this region pointed to an adiabatic energy of ≦75567 cm^{-1} [16]. The PD_3 progression had been earlier observed with its first band at 74946 cm^{-1} [6, 9]; see also E=75000 cm^{-1} (for a "$\tilde{D}$" state) in [4]. A term value of ~5800 cm^{-1} for a transition 5p←5a$_1$ was based on this earlier measurement [10].

Transition 4p←2e. Photoabsorption using synchrotron radiation revealed a vibrational progression (in ν_2') beginning at 80370 cm^{-1} and another one (with additional excitation of the stretching vibration ν_1') beginning at 82713 cm^{-1}. The tentative assignment was based on quantum defects [2].

Vertical excitation energies of the singlet and triplet states $^1A_1$ and $^3A_1$, resulting from transitions of the 5a$_1$ electron to the orbitals 4s through 7s, 4p to 6p, 3d, and 4d, were calculated by a one-center expansion approximation [17].

Excitation of Core Electrons

Excitation energies E (in eV) of four broad and four sharp features in the 2p pre-edge region, of two continuum features in the 2p post-edge region, and of a feature in the 2s pre-edge region are listed in the first table below. They are mainly based on two high-resolution photoabsorption measurements using synchrotron radiation [18, 19]. Similar values had been measured earlier [20]. A few data from an EELS investigation [8] have been added. Doublets observed in the 2p pre-edge region, arising from $2p_{1/2}$-$2p_{3/2}$ spin-orbit splitting, are listed below each other. Assignments to virtual valence or Rydberg orbitals are contradictory; only the most recent ones [18] are reported in the table (see also the second table below). The term values T (in eV) are based on the ionization potentials of 137.05 ($2p_{3/2}$), 137.95 ($2p_{1/2}$), and 194.88 eV (2s) [8] (see p. 150):

E in eV [18]	131.86 132.77	132.56 133.34	134.48 135.31	134.92 135.77	142	162	–
E in eV [19][1)]	131.95 132.9	132.67 133.48	134.58 135.40	135.04 135.87	141.0	156.7 [8]	189.8[2)]
transition [18] ...	s←2p	e*←2p	p(a$_1$)←2p	p(e)←2p	ka$_1$←2p	ke←2p	σ*(e)←2s
T in eV [18]	5.19 5.18	4.49 4.61[3)]	2.57 2.64	2.13 2.18	−4.5[4)]	−24.4[4)]	5.21 [8]

[1)] The original table [19] contains even more absorption features. – [2)] E=189.67 [8], leading to the reported T value. – [3)] T=4.49 eV [8]. – [4)] T values given for excitation of $2p_{3/2}$ [18].

The assignment of the two broad and the two sharp spin-orbit doublets in the 2p pre-edge region to certain terminating orbitals has been controversely discussed since the first investigations [20]. The various proposals are chronologically arranged in the following table:

year	1972	1975	1975	1979	1985	1987	1990
broad	4s 3d	σ* 4s	σ*(a$_1$) σ*(e)	σ*(4s) σ*(3p)	σ*(e) σ*(a$_1$)	σ*(e) σ*(e)	s e*
sharp	5s 4d	4p 3d	4s (or 5s) 3d	5s 3d	4s 3d	4s 4p	p(a$_1$) p(e)
Ref.	[20]	[11]	[21]	[22]	[8]	[19]	[18]
remark	a)	b)	c)	d)	e)	f)	g)

a) Based on a united-atom treatment (yielding only Rydberg orbitals) [20]. – b) Based on an assumed transferability of term values from valence to core regions [11]. – c) From the "Z+1 analogy", i.e., based on HFR calculations on the series ArH^+, ClH_2^+, and PH_4^+ (representing core-excited HCl, H_2S, and SiH_4) [21]. – d) Supporting the assignments in [21] by photoabsorption of solid PH_3 and calculations on SiH_4 [22]. – e) The ordering of the two σ^* orbitals was inverted compared with that in [21, 22] in view of SCM(self-consistent multipolar)-Xα-DV(discrete variational) calculations [23] for PH_3 [8]. – f) Ligand-field splitting of σ^*(e) was assumed [19]. – g) Based on MS-Xα calculations. The energetic ordering of the Rydberg s orbital and the antibonding e* orbital depends strongly on slight changes of the atomic radii [18].

References:

[1] Ben Arfa, M.; Tronc, M. (Chem. Phys. **155** [1991] 143/8).
[2] Xia, T. J.; Wu, C. Y. R.; Judge, D. L. (Phys. Scr. **41** [1990] 870/3).
[3] Ashfold, M. N. R.; Stickland, R. J.; Tutcher, B. (Mol. Phys. **65** [1988] 1455/71, 1457/65).
[4] Herzberg, G. (Molecular Spectra and Molecular Structure, Vol. 3, Electronic Spectra and Electronic Structure of Polyatomic Molecules, Van Nostrand-Reinhold, New York – Cincinnati – Toronto – London – Melbourne 1966, p. 610).
[5] Walsh, A. D. (J. Chem. Soc. **1953** 2296/301).
[6] Walsh, A. D.; Warsop, P. A. (in: Mangini, A.; Advances in Molecular Spectroscopy, Vol. 2, Pergamon, Oxford 1962, pp. 582/91, 583/4).
[7] Müller, J.; Ågren, H.; Canuto, S. (J. Chem. Phys. **76** [1982] 5060/8).
[8] Sodhi, R. N. S.; Brion, C. E. (J. Electron Spectrosc. Relat. Phenom. **37** [1985] 97/123, 103, 114).
[9] Humphries, C. M.; Walsh, A. D.; Warsop, P. A. (Discuss. Faraday Soc. No. 35 [1963] 148/57, 151).
[10] Robin, M. B. (Higher Excited States of Polyatomic Molecules, Vol. I, Academic, New York 1974, pp. 231/6).

[11] Robin, M. B. (Chem. Phys. Lett. **31** [1975] 140/4).
[12] Robin, M. B. (Higher Excited States of Polyatomic Molecules, Vol. III, Academic, New York 1985, pp. 159/63).
[13] Halmann, M. (J. Chem. Soc. **1963** 2853/6).
[14] Zarate, E. B.; Cooper, G.; Brion, C. E. (Chem. Phys. **148** [1990] 277/88, 281).
[15] Mayor, L.; Walsh, A. D.; Warsop, P. (J. Mol. Spectrosc. **10** [1963] 320).
[16] Lu, Q.; He, S.; Yu, S.; Kong, F. (Wuli Huaxue Xuebao **3** [1987] 345/50 from C.A **107** [1987] No. 186486).
[17] Hatano, Y. (Chem. Phys. Lett. **56** [1978] 314/7).
[18] Liu, Z. F.; Cutler, J. N.; Bancroft, G. M.; Tan, K. H.; Cavell, R. G.; Tse, J. S. (Chem. Phys. Lett. **172** [1990] 421/9).
[19] Ishiguro, E.; Iwata, S.; Mikuni, A.; Suzuki, Y.; Kanamori, H.; Sasaki, T. (J. Phys. B **20** [1987] 4725/39, 4732).
[20] Hayes, W.; Brown, F. C. (Phys. Rev. [3] A **6** [1972] 21/30, 27).

[21] Schwarz, W. H. E. (Chem. Phys. **11** [1975] 217/28, 221).
[22] Friedrich, H.; Sonntag, B.; Rabe, P.; Butscher, W.; Schwarz, W. H. E. (Chem. Phys. Lett. **64** [1979] 360/6).
[23] Xiao, S.-X.; Trogler, W. C.; Ellis, D. E.; Berkovitch-Yellin, Z. (J. Am. Chem. Soc. **105** [1983] 7033/7).

1.3.1.4.1.3 Ionization Potentials E_i in eV

Valence Orbitals

Selected values of the **adiabatic** (ad) ionization potential of the **outer-valence** orbital $5a_1$ from different methods of measurement are given in order of increasing magnitude in the table below (PES = photoelectron spectroscopy, PIMS = photoionization mass spectrometry, also simulated by dipole (e, e+ion) spectroscopy, PA = photoabsorption). Other adiabatic and the **vertical** (vert) values for $5a_1$ and values for the other outer-valence orbital 2e are reported in separate sections on PES (also simulated by binary (e, 2e) spectroscopy) and MS work added to the table:

E_i in eV	9.868(5)	9.870(2)	9.97(2)	9.982	10.0(10)
method	PES	PIMS	MS	PA	dipole (e, e+ion)
Ref.	[1]	[2]	[3]	[4]	[5]
remark	a)	b)	c)	d)	e)

a) HeI radiation. – b) He and Ar continua and hydrogen pseudocontinuum. – c) Monoenergetic electrons. – d) Synchrotron radiation. E_i = 9.97 eV was given in [6]. – e) Electron impact energy was 8 keV.

PES work yielded the following adiabatic and vertical ionization potentials for the outer-valence orbitals:

$E_i(5a_1; ad)$	9.868(5)	9.96(1)	(9.96)	9.96(1)	(10.13(2))	–
$E_i(5a_1; vert)$ ...	10.604(5)	10.58(1)	10.59	10.60(1)	10.59(2)	10.59(5)
$E_i(2e; ad)$	≦12.57	12.40(2)	(12.64)	12.64(2)	12.5(1)	–
$E_i(2e; vert)$ ...	13.7	13.50(5)	13.6	13.4	13.6(1)	13.59(5)
radiation	HeI	HeI, NeI, ArI	HeI	NeI	NeI	synchrotron
Ref.	[1]	[7]	[8]	[9]	[10]	[11]
remark	a)	b)	c)	d)	e)	f)

a) See also the preceding table. – b) For the vertical values, see also [12]. – c) Adiabatic values apparently taken from [9]. PE spectra of the group V (N through Sb) hydrides and halogenides were compared [8]. – d) $E_i(5a_1$; ad) = 9.98 earlier given in [13]. – e) $E_i(5a_1$; ad) was questioned [10] in view of the lower value (9.98 eV) given earlier [13]; see also [4]. – f) The angular distribution parameter β was measured [11], see also [14].

Binary (e, 2e) spectroscopy (simulating PES) yielded E_i (2e; vert) = 13.6 [15] and 13.44 eV [16] (reported as 13.4 eV in [17]).

Conventional MS work yielded data für $E_i(5a_1$; ad), which agreed within the rather large error limits with the values given above: 10.0 [18], 10.05(5) [19], 10.1(1) (and 10.15(10) for PD_3) [20], 10.2(2) (and 10.1(2) for PD_3) [21], 10.3(5) (and 10.4(3) for another extrapolation method) [22], 10.3 [23]. Two values for E_i(2e; ad) were 12.5(2) (PH_3) and 12.75(20) eV (PD_3) [20].

The **inner-valence** orbital $4a_1$ gives rise to a multiple structure (rather than a single peak), indicative of a breakdown of the one-electron picture, see PES [11, 12] and binary (e, 2e) spectroscopic [15, 16] work. As many as nine or eight features were measured by PES using synchrotron radiation [11] or by binary (e, 2e) spectroscopy [15]:

19.4	20.6	21.7	23.3	24.1	25.6	26.5	27.8	29.1	–	–	[11]1)
19.4	20.6	–	23.2	–	25.6	27.2		29.1	31.8	36.0	[15]2)

[1)] The original data [11] were rounded based on an uncertainty of at least ±0.1 eV. The angular distribution parameters β were measured and used to interprete the features [11]; see also [14, 24]. The feature at 20.6 eV was assigned to an outer-valence satellite [14]. – [2)] An s-type character of the feature at 20.6 eV (20.0 eV in the text) was observed [15].

An earlier binary (e, 2e) spectroscopic investigation had revealed three features at 19.45, 22.61 (insufficiently resolved), and 25.46 eV [16] (also reported in [12], while 22 eV was given as the centroid of the multiple final state structure in [17]). PES work using He II radiation had revealed features at 19.5 and 20.5 eV (weak) [12] and at 19.0 eV [9] (see also [25], where relative intensities for the three valence orbitals are given).

Core Orbitals

The P2p (1e and $3a_1$) ionization potentials were obtained by extrapolating the Rydberg series observed in photoabsorption (PA) spectra [26, 27] and by photoelectron spectroscopy in the X-ray region (XPS) [28 to 30]. Data for $P2p_{3/2}$, $P2p_{1/2}$, and P2p are shown below (compare also a 1980 review [31]):

$2p_{3/2}$	137.10	137.05	137.0(1.5)	136.87(5)	137.35(20)	137.3
$2p_{1/2}$	138.01	137.95	137.9(1.5)	—		
method	PA	XPS [29]	PA [27]	XPS	XPS	XPS
Ref.	[26]	[32]	[33]	[28]	[29]	[30]
remark	—	a)	b)	c)	d)	e)

a) An average value of 137.35 eV for P2p [29] was used [32] together with a spin-orbit splitting of 0.90 eV (taken from [33]). – b) The photoabsorption spectrum [27] was reinterpreted [33]. The original values 137.3(2) and 138.2(2) eV [27] had also been criticized in [34]. – c) Mg Kα radiation. – d) Mg Kα radiation [29]. The error limits were given in [35]. – e) Al Kα radiation.

The P2s ($2a_1$) ionization potential was measured (probably by XPS) to be 194.88 eV [36]. An approximate value of 189.0±0.5 eV was inferred from a PA spectrum [27].

The P1s ($1a_1$) ionization potential was first measured by XPS using Ti Kα radiation [29]. This value (2150.5±0.5 eV; see also [31]) was repeatedly corrected (2150.88(20) [35], 2150.84 [37]) and is now given to be 2150.87 eV [38]. E_i = 2150.5 eV was also obtained from the X-ray line P $K\alpha_1$ at 2013.6 eV and the ionization limit for $P2p_{3/2}$ (taken from [28]) [39].

Theoretical Work

Ionization potentials of only the outermost orbital $5a_1$ were calculated in [2, 40 to 43], of only $5a_1$ and 2e in [44], of the three valence orbitals in [45 to 49], and of all occupied orbitals in [50].

Orbital energies (ε_i) were calculated for only $5a_1$ in [40, 51 to 53], for only $5a_1$ and 2e in [54], for the three valence orbitals in [47, 49, 55 to 59], and for all occupied orbitals in [50, 60 to 68]. Orbital energies were also calculated for the unoccupied orbitals 3e [47, 52, 59, 64, 67], $6a_1$, 4e, $7a_1$ [47, 64, 67], and 5e [67].

References:

[1] Maripuu, R.; Reineck, I.; Ågren, H.; Nian-Zu, Wu; Rong, Ji Ming; Veenhuizen, H.; Al-Shamma, S. H.; Karlsson, L.; Siegbahn, K. (Mol. Phys. **48** [1983] 1255/67, 1258).

[2] Berkowitz, J.; Curtiss, L. A.; Gibson, S. T.; Greene, J. P.; Hillhouse, G. L.; Pople, J. A. (J. Chem. Phys. **84** [1986] 375/84, 380).

[3] McAllister, T.; Lossing, F. P. (J. Phys. Chem. **73** [1969] 2996/8).
[4] Xia, T. J.; Chien, T. S.; Wu, C. Y. R.; Judge, D. L. (J. Quant. Spectrosc. Radiat. Transfer **45** [1991] 77/91, 83/4).
[5] Zarate, E. B.; Cooper, G.; Brion, C. E. (Chem. Phys. **148** [1990] 277/88, 284).
[6] Xia, T. J.; Wu, C. Y. R.; Judge, D. L. (Phys. Scr. **41** [1990] 870/3).
[7] Maier, J. P.; Turner, D. W. (J. Chem. Soc. Faraday Trans. II **68** [1972] 711/9).
[8] Grodzicki, M.; Walther, H.; Elbel, S. (Z. Naturforsch. **39b** [1984] 1319/30, 1320).
[9] Potts, A. W.; Price, W. C. (Proc. R. Soc. [London] A **326** [1971/72] 181/97, 190).
[10] Branton, G. R.; Frost, D. C.; McDowell, C. A.; Stenhouse, I. A. (Chem. Phys. Lett. **5** [1970] 1/2).

[11] Cauletti, C.; Piancastelli, M. N.; Adam, M. Y. (J. Mol. Struct. **174** [1988] 135/40).
[12] Domcke, W.; Cederbaum, L. S.; Schirmer, J.; Niessen, von, W.; Maier, J. P. (J. Electron Spectrosc. Relat. Phenom. **14** [1978] 59/72, 66).
[13] Price, W. C.; Passmore, T. R. (Discuss. Faraday Soc. No. 35 [1963] 232).
[14] Cauletti, C.; Adam, M. Y.; Piancastelli, M. N. (J. Electron Spectrosc. Relat. Phenom. **48** [1989] 379/87).
[15] Clark, S. A. C.; Brion, C. E.; Davidson, E. R.; Boyle, C. (Chem. Phys. **136** [1989] 55/66, 59).
[16] Hamnett, A.; Hood, S. T.; Brion, C. E. (J. Electron Spectrosc. Relat. Phenom. **11** [1977] 263/74, 271).
[17] Cook, J. P. D.; Brion, C. E.; Hamnett, A. (Chem. Phys. **45** [1980] 1/13, 10).
[18] Morrison, J. D.; Traeger, J. C. (Int. J. Mass Spectrom. Ion Phys. **11** [1973] 277/88, 283).
[19] Fehlner T. P.; Callen, R. B. (Adv. Chem. Ser. **72** [1968] 181/90, 184/5).
[20] Märk, T. D.; Egger, F. (J. Chem. Phys. **67** [1977] 2629/35).

[21] Wada, Y.; Kiser, R. W. (Inorg. Chem. **3** [1964] 174/7).
[22] Fischler, J.; Halmann, M. (J. Chem. Soc. **1964** 31/6).
[23] Kiser, R. W.; Gallegos, E. J. (J. Phys. Chem. **66** [1962] 947/8).
[24] Adam, M. Y.; Cauletti, C.; Piancastelli, M. N.; Svensson, A. (J. Electron Spectrosc. Relat. Phenom. **50** [1990] 219/27).
[25] Dechant, P.; Schweig, A.; Thiel, W. (Angew. Chem. **85** [1973] 358/9; Angew. Chem. Int. Ed. Engl. **12** [1973] 308/9).
[26] Ishiguro, E.; Iwata, S.; Mikuni, A.; Suzuki, Y.; Kanamori, H.; Sasaki, T. (J. Phys. B **20** [1987] 4725/39, 4733).
[27] Hayes, W.; Brown, F. C. (Phys. Rev. [3] A **6** [1972] 21/30, 27).
[28] Perry, W. B.; Schaaf, T. F.; Jolly, W. L. (J. Am. Chem. Soc. **97** [1975] 4899/905).
[29] Cavell, R. G.; Sodhi, R.N.S. (J. Electron Spectrosc. Relat. Phenom. **15** [1979] 145/50).
[30] Ashe, A. J.; Bahl, M. K.; Bomben, K. D.; Chan, W.-T.; Gimzewski, J. K.; Sitton, P. G.; Thomas, T. D. (J. Am. Chem. Soc. **101** [1979] 1764/7).

[31] Bakke, A. A.; Chen, H.-W.; Jolly, W. L. (J. Electron Spectrosc. Relat. Phenom. **20** [1980] 333/66, 359).
[32] Sodhi, R. N. S.; Brion, C. E. (J. Electron Spectrosc. Relat. Phenom. **37** [1985] 97/123, 103, 114).
[33] Schwarz, W. H. E. (Chem. Phys. **11** [1975] 217/28, 221, 223).
[34] Robin, M. B. (Higher Excited States of Polyatomic Molecules, Vol. I, Academic, New York 1974, pp. 231/6).
[35] Sodhi, R. N. S.; Cavell, R. G. (J. Electron Spectrosc. Relat. Phenom. **32** [1983] 283/312, 294).
[36] Sodhi, R. N. S.; Cavell, R. G. (unpublished results from [32]).

[37] Cavell, R. G.; Sodhi, R. N. S. (J. Electron Spectrosc. Relat. Phenom. **41** [1986] 25/35, 27).
[38] Cavell, R. G.; Sodhi, R. N. S. (J. Electron Spectrosc. Relat. Phenom. **43** [1987] 215/23, 220).
[39] Dolenko, G. N.; Krupoder, S. A.; Mazalov, L. N. (Zh. Strukt. Khim. **20** [1979] 334/6; J. Struct. Chem. [Engl. Transl.] **20** [1979] 279/81).
[40] Eyermann, C. J.; Jolly, W. L. (J. Phys. Chem. **87** [1983] 3080/2).

[41] Müller, J.; Ågren, H.; Canuto, S. (J. Chem. Phys. **76** [1982] 5060/8).
[42] Graf, P.; Mehler, E. L. (Int. J. Quantum Chem. Quantum Biol. Symp. No. 8 [1981] 49/61, 59).
[43] Noack, W.-E. (Int. J. Quantum Chem. **17** [1980] 1125/41, 1135).
[44] Hatano, Y. (Chem. Phys. Lett. **56** [1978] 314/7).
[45] Lisini, A.; Brosolo, M.; Decleva, P.; Fronzoni, G. (J. Mol. Struct. **253** [1992] 333/48, 341 [THEOCHEM **85**]).
[46] Lisini, A.; Decleva, P.; Fronzoni, G. (J. Mol. Struct. **228** [1991] 97/116, 109 [THEOCHEM **74**]).
[47] Xiao, S.-X.; Trogler, W. C.; Ellis, D. E.; Berkovitch-Yellin, Z. (J. Am. Chem. Soc. **105** [1983] 7033/7).
[48] Sukhorukov, V. L.; Demekhin, V. F. (Koord. Khim. **9** [1983] 158/67; Sov. J. Coord. Chem. [Engl. Transl.] **9** [1983] 98/106).
[49] Guest, M. F.; Saunders, V. R. (Mol. Phys. **29** [1975] 873/84, 878).
[50] Chong, D. P.; Herring, F. G.; McWilliams, D. (J. Chem. Phys. **61** [1974] 3567/70).

[51] Glidewell, C.; Thomson, C. (J. Comput. Chem. **4** [1983] 9/14).
[52] Hillier, I. H.; Saunders, V. R. (J. Chem. Soc. D **1970** 316/8).
[53] Cook, D. B.; Palmieri, P. (Chem. Phys. Lett. **3** [1969] 219/22).
[54] Hillier, I. H.; Saunders, V. R. (Trans. Faraday Soc. **66** [1970] 2401/7).
[55] Gáspár, R.; Gáspár, R., Jr. (Can. J. Chem. **63** [1985] 1922/4).
[56] Fantucci, P.; Polezzo, S. (Mol. Phys. **37** [1979] 831/41, 836).
[57] Moldoveanu, S. (Rev. Roum. Chim. **23** [1978] 453/65, 463).
[58] Barthelat, J. C.; Durand, P.; Serafini, A. (Mol. Phys. **33** [1977] 159/80, 170).
[59] Hillier, I. H.; Saunders, V. R. (J. Chem. Soc. A **1970** 2475/7).
[60] Hinchliffe, A. (J. Mol. Struct. **105** [1983] 335/41 [THEOCHEM **14**]).

[61] Wallmeier, H.; Kutzelnigg, W. (J. Am. Chem. Soc. **101** [1979] 2804/14, 2812).
[62] Ahlrichs, R.; Keil, F.; Lischka, H.; Kutzelnigg, W.; Staemmler, V. (J. Chem. Phys. **63** [1975] 455/63).
[63] Norman, J. G., Jr. (J. Chem. Phys. **61** [1974] 4630/5).
[64] Petke, J. D.; Whitten, J. L. (J. Chem. Phys. **59** [1973] 4855/66, 4861).
[65] Lehn, J. M.; Munsch, B. (Mol. Phys. **23** [1972] 91/107, 95).
[66] Robert, J.-B.; Marsmann, H.; Schaad, L. J.; Van Wazer, J. R. (Phosphorus Relat. Group V Elem. **2** [1972] 11/8).
[67] Boyd, D. B.; Lipscomb, W. N. (J. Chem. Phys. **46** [1967] 910/9).
[68] Moccia, R. (J. Chem. Phys. **40** [1964] 2176/85, 2183).

1.3.1.4.1.4 Dipole Moment μ in D

An **absolute** value of the electric dipole moment $|\mu|=0.57397(20)$ was measured for the vibronic **ground state** of PH_3 from an rf spectrum, which was observed with a molecular-beam electric-resonance (MBER) spectrometer in the rotational state J=K=2 at high electric field strengths (~98 and ~147.5 V/cm). This $|\mu|$ was also in accordance with another spectrum for J=K=1 ($|\mu|=0.57395\pm0.0003$ was given in the abstract) [1]. A small correction factor (1.00009) to $|\mu|$ was obtained from laser Stark spectroscopic work on several lines of the v_2 and v_4 fundamental bands [2]. The similar value $|\mu|=0.5743(3)$ was measured in the sub-mm range spectrum for the rotational transition J=2←1, K=1 [3]. $|\mu|=0.5796(12)$ for the rotational state J=14, K=12 (but possibly to be interchanged with 0.5768(12) D given for the vibrationally excited state $v_2=1$; see below) was obtained by Doppler-free optical double-resonance spectroscopy using a single-frequency laser and modulation side bands [4]. $|\mu|=0.58\pm0.01$ is a selected value [5], based on older data (see "Phosphor" C, 1965, p. 20).

The dipole moment was also measured for **vibrationally excited** states (for a possible rotational-state dependence, see discussions in the original papers). Absolute values $|\mu|$ are given for several vibrational quantum numbers v_n (n=2, 3, 4). For $v_2=1$: 0.57420(27) [2], 0.5740(3) [3], 0.5740(2) [6], 0.574(3) [7], and 0.5768(12) [4] (possibly to be interchanged with 0.5796(12) given above for $v_2=0$ [4]). For $v_2=2$: 0.5738(5) [8]. For $v_3=1$ and 2 (both for PD_3): 0.56228(10) and 0.54900(21) [9]. For $v_4=1$: 0.57904(32) [2], 0.5791(13) [10], and 0.5784(1) [6].

The **direction** of μ is along the C_3 symmetry axis (z axis) of PH_3 or PD_3. Thus, $\mu_z=-0.58$ is given in [11, 12]; the absolute value was taken from [5], and a negative end of μ at P was inferred from theoretical calculations [13]. Similarly, $\mu_z=-0.578$ in [14] was based on an older absolute value [15] and on a polarity $P^-H_3^+$ apparently found from their own calculations [14]. The components μ_x, μ_y, and μ_z of the isotopically substituted PHDT molecule were given to be −0.34, −0.43, and +0.20 (y and z along the principal axes with the largest and the smallest moment of inertia) [11, 12].

For **derivatives** of μ with respect to the normal coordinates, see the work on IR intensities (p. 188).

A P–H **bond moment** $\mu_b=0.30$ D, directed in the P^+H^- sense, was derived from theoretically calculated values for the total dipole moment $\mu=-0.66$ (see below) and for a lone-pair contribution of −1.15 D (due to the $5a_1$ orbital having mainly P3p character; see p. 142). The resultant of the three bond moments (0.49 D) was thus directed oppositely to the total dipole moment [14]. Lone-pair moments of ~0.4 or ~0.2 D (and small bond moments) were derived from the experimental μ value and its variation with the symmetry coordinate S_2 (for two different signs of $\partial\mu/\partial S_2$) [37]. A lone-pair moment of only 0.2 D was used in an investigation of the influence of the inductive effect on the bond moments in a series of phosphanes [16]. Bond moments were also derived assuming a vanishing lone-pair moment (and using an experimental value $|\mu|=0.58$): $\mu_b=0.3574$ D [17] and 0.36 D [18, 19].

Theoretical calculations yielded strongly different absolute magnitudes and also different signs of μ. A few theoretical values which differ from the experimental result $|\mu|=0.574$ D (or 0.226 au) by not more than ±0.1 D are listed below:

$\|\mu\|$ in D	$\|\mu\|$ in au	method	Ref.
0.563		density functional	[20]
	0.2264	many-body perturbation	[21]
0.624		configuration interaction	[22]
	0.2353	configuration interaction	[23]
0.653		MP2 perturbation	[24]

Various sizes of basis sets, especially an inclusion of d functions, and/or various methods of calculation were tested in each of the following papers [14, 25 to 35]; see also theoretical work listed in the section dealing with atomic charges, p. 142. For a calculation of bond and lone-pair moments, see also [36].

References:

[1] Davies, P. B.; Neumann, R. M.; Wofsy, S. C.; Klemperer, W. (J. Chem. Phys. **55** [1971] 3564/8).
[2] Takagi, K.; Itoh, K.; Miura, E.; Tanimura, S. (J. Opt. Soc. Am. B **4** [1987] 1145/57, 1152, 1156/7).
[3] Krupnov, A. F.; Melnikov, A. A.; Skvortsov, V. A. (Opt. Spektrosk. **46** [1979] 1012/3; Opt. Spectrosc. [Engl. Transl.] **46** [1979] 569/70).
[4] Orr, J.; Oka, T. (Appl. Phys. **21** [1980] 293/306, 302).
[5] Nelson, R. D.; Lide, D. R.; Maryott, A. A. (NSRDS-NBS-10 [1967] 1/49, 13).
[6] Di Lonardo, G.; Trombetti, A. (Chem. Phys. Lett. **76** [1980] 307/10).
[7] Shimizu, F. (Chem. Phys. Lett. **17** [1972] 620/2).
[8] Takagi, K. (Chem. Phys. Lett. **112** [1984] 302/5).
[9] Tanaka, K.; Ito, H.; Tanaka, T. (J. Mol. Spectrosc. **115** [1986] 383/92, 389/90).
[10] Scappini, F.; Schwarz, R. (Chem. Phys. Lett. **80** [1981] 350/1).

[11] Salzman, W. R. (Chem. Phys. Lett. **167** [1990] 417/20).
[12] Salzman, W. R. (Chem. Phys. **94** [1991] 5263/9).
[13] Borfield, M. (private communication from [11, 12]).
[14] Lehn, J. M.; Munsch, B. (Mol. Phys. **23** [1972] 91/107, 96).
[15] Burrus, C. A. (J. Chem. Phys. **28** [1958] 427/9).
[16] Raevskii, O. A.; Khalitov, F. G. (Izv. Akad. Nauk SSSR Ser. Khim. **1970** 2368/70; Bull. Acad. Sci. USSR Div. Chem. Sci. [Engl. Transl.] **1970** 2222/4).
[17] Addepalli, V. B.; Satyavathi, N.; Rao, N. R. (Indian J. Pure Appl. Phys. **15** [1977] 370/2).
[18] Weaver, J. R.; Parry, R. W. (Inorg. Chem. **5** [1966] 718/23).
[19] Morse, J. G.; Parry, R. W. (J. Chem. Phys. **46** [1967] 4159/60).
[20] Chong, D. P. (Chin. J. Phys. [Taipei] **30** [1992] 115/28 from C. A. **116** [1992] No. 242058), reported according to Chong, D. P.; Papoušek, D. (J. Mol. Spectrosc. **155** [1992] 167/76, 170).

[21] Sadlej, A. J. (Theor. Chim. Acta **79** [1991] 123/40, 129).
[22] Clark, S. A. C.; Brion, C. E.; Davidson, E. R.; Boyle, C. M. (Chem. Phys. **136** [1989] 55/66, 58).
[23] Feller, D.; Boyle, C. M.; Davidson, E. R. (J. Chem. Phys. **86** [1987] 3424/40, 3436).
[24] Latajka, Z.; Scheiner, S. (J. Chem. Phys. **81** [1984] 2713/6).
[25] Kraka, E.; Gauss, J.; Cremer, D. (J. Mol. Struct. **234** [1991] 95/126, 115, 118 [THEOCHEM **80**]).
[26] Novak, I. (Z. Phys. Chem. [Munich] **167** [1990] 251/4).
[27] Spackmann, M. A. (J. Phys. Chem. **93** [1989] 7594/603, 7597).
[28] Ikuta, S.; Kebarle, P. (Can. J. Chem. **61** [1983] 97/102).
[29] Gordon, M. S.; Binkley, J. S.; Pople, J. A.; Pietro, W. J.; Hehre, W. J. (J. Am. Chem. Soc. **104** [1982] 2797/803).
[30] Graf, P.; Mehler, E. L. (Int. J. Quantum Chem. Quantum Biol. Symp. No. 8 [1981] 49/61, 59).

[31] Wallmeier, H.; Kutzelnigg, W. (J. Am. Chem. Soc. **101** [1979] 2804/14, 2809).
[32] Scott, J. M.; Sutcliffe, B. T. (Theor. Chim. Acta **41** [1976] 141/8).

[33] Burton, P. G.; Carlsen, N. R.; Magnusson, E. A. (Mol. Phys. **32** [1976] 1687/94).
[34] Robert, J.-B.; Marsmann, H.; Schaad, L. J.; Van Wazer, J. R. (Phosphorus Relat. Group V Elem. **2** [1972] 11/8).
[35] Santry, D. P.; Segal, G. A. (J. Chem. Phys. **47** [1967] 158/74, 164/5).
[36] Schmiedekamp, A.; Skaarup, S.; Pulay, P.; Boggs, J. E. (J. Chem. Phys. **66** [1977] 5769/76).
[37] Scrocco, M.; de Luca, B. (Ric. Sci. **37** [1967] 250/7).

1.3.1.4.1.5 Quadrupole Moment Θ in 10^{-26} esu·cm². Octupole Moment

Data for the **quadrupole moment** were based on molecular Zeeman effect measurements and on theoretical calculations. The measurements (on PH_2D and PHD_2) [1], yielding the molecular g values (see p. 160) and the magnetic susceptibility anisotropy $\Delta\chi$ (see p. 156), were reanalyzed in [2]: $\Theta_{\parallel} = -2.3(12)$ or $-3.9(12)$ was obtained for the component being parallel to the PH_3 symmetry axis ($\Theta_{\parallel} = -2\Theta_{\perp}$ for an oblate top) and for two possible choices of the signs of $g_{\parallel}$ and $g_{\perp}$. The first value is preferred in view of its agreement with a value of $\Theta_{\parallel} = -2.4$ from an SCF calculation (see below) [2]. $\Theta_{\parallel} = -2.1 \pm 1.0$ was favored over $\Theta_{\parallel} = -4.1 \pm 1.0$ in the original work, mainly by a comparison with the related NH_3 molecule [1]. It was also quoted in a review [3] and more recently ($\Theta_{\parallel} = -7.0 \times 10^{-40}$ C·m²) in [4].

Ab initio calculations of Θ (referred to the center of mass) used SCF [2, 4 to 6] or CI [7, 8] methods. Response densities were determined and used in [9]. Explicit SCF data for $-\Theta_{\parallel}$ were 2.43 [2], 2.5006 (1.8593 au) [6], 2.53 (8.4556×10^{-40} C·m²) [5], and 2.82 (9.402×10^{-40} C·m²; another basis [10] gave 5.79×10^{-40} C·m²) [4]. The two CI results were 2.0230 [7] and 2.25 (1.6761 au) [8].

Components of the **octupole moment** were theoretically calculated [6].

References:

[1] Kukolich, S. G.; Flygare, W. H. (Chem. Phys. Lett. **7** [1970] 43/6).
[2] Combariza, J.; Salzman, W. R.; Kukolich, S. G. (Chem. Phys. Lett. **167** [1990] 607/8).
[3] Flygare, W. H.; Benson, R. C. (Mol. Phys. **20** [1971] 225/50, 241).
[4] Novak, I. (Z. Phys. Chem. [Munich] **167** [1990] 251/4).
[5] Hinchliffe, A. (J. Mol. Struct. **105** [1983] 335/41 [THEOCHEM **14**]).
[6] Rothenberg, S.; Young, R. H.; Schaefer, H. F., III (J. Am. Chem. Soc. **92** [1970] 3243/50).
[7] Petke, J. D.; Whitten, J. L. (J. Chem. Phys. **59** [1973] 4855/66, 4862).
[8] Feller, D.; Boyle, C. M.; Davidson, E. R. (J. Chem. Phys. **86** [1987] 3424/40, 3436).
[9] Kraka, E.; Gauss, J.; Cremer, D. (J. Mol. Struct. **234** [1991] 95/126, 115 [THEOCHEM **80**]).
[10] Kikuchi, O.; Wang, H.; Nakano, T.; Morihashi, K. (J. Mol. Struct. **205** [1990] 301/15 [THEOCHEM **64**]).

1.3.1.4.1.6 Electric Polarizability α in Å³. Magnetic Susceptibility χ in 10^{-6} cm³/mol

The static average electric dipole **polarizability** $\alpha = 4.84$ was derived [1] from older data [2] for the dielectric constant, the molar refraction, and the dipole moment. A vibrational contribution $\alpha_{vib} = 0.086$ (0.096×10^{-40} C²·m²·J⁻¹; formerly called "atomic polarization") was obtained

[3] from IR-intensity data taken from [4]. This contribution was subtracted from α given above to yield a value of 4.75 Å³ (5.28×10^{-40} $C^2 \cdot m^2 \cdot J^{-1}$) to be compared with the results of theoretical calculations [5].

Polarizabilities α (without the vibrational contribution) and the components $\alpha_{\parallel}$ (parallel) and $\alpha_{\perp}$ (perpendicular to the molecular symmetry axis) from three theoretical calculations are given below:

method	α	$\alpha_{\parallel}$	$\alpha_{\perp}$	Ref.	remark
SCF	4.29	4.27	4.30	[5]	a)
MBPT	4.49	4.64	4.41	[6]	a)
MP2	4.50	4.62	4.43	[5]	c)

a) 4.773, 4.749, and 4.784×10^{-40} $C^2 \cdot m^2 \cdot J^{-1}$. – b) Many-body perturbation theory; 30.29, 31.32, and 29.78 au. – c) 4.999, 5.142, and 4.927×10^{-40} $C^2 \cdot m^2 \cdot J^{-1}$.

Experimental and a few theoretical values for the average magnetic **susceptibility** χ, its components $\chi_{\parallel}$ (parallel) and $\chi_{\perp}$ (perpendicular to the molecular symmetry axis), the anisotropy $\Delta\chi = \chi_{\perp} - \chi_{\parallel}$, and the dia- and paramagnetic contributions χ^d and χ^p are given in the table below. The experimental data (in the first row) are based on an earlier measurement of χ for PH_3 [7] (see also "Phosphor" C, 1965, p. 30) and on later observations of the Zeeman effect in rotational transitions of PH_2D and PHD_2 [8]. The theoretical data (all for the experimental geometry) were often based on coupled Hartree-Fock (CHF) calculations (see the remarks):

$-\chi$	$-\chi_{\parallel}$	$-\chi_{\perp}$	$\Delta\chi$	$-\chi^d$	χ^p	method	Ref.	remark
26.2(8) [7]	24.4(13)	27.1(11)	−2.7(8)	42.4	16.2	Zeeman	[8, 9]	a)
27.0(2)	—	—	−3.1	—	—	IGLO	[10]	b)
27.3	25.2	28.4	−3.2	43.9	16.6	CHF	[11]	c)
27.56	25.45	28.82	−3.4	—	17.21	CHF	[12]	d)
26.98	—	—	—	44.45	17.47	CHF	[13]	e)
26.2 [7]	—	—	—	44.00	17.8	SCF	[14]	f)

a) Δχ from a reanalysis [9] of the original work [8] (see also [15]), where a different sign was given. χ^d and χ^p derived from the components $\chi^d_{\parallel}$ and $\chi^d_{\perp}$ (−42.9(14) and −42.1(11)) and $\chi^p_{\parallel}$ and $\chi^p_{\perp}$ (18.56(3) and 15.04(2)) given in [8]. – b) The individual gauge for localized orbitals (IGLO) method and various basis sets were used. The results apparently converged to the tabulated values. – c) Gauge-invariant Gaussian functions were used. The original data (slightly different values were obtained for a second basis set) were given in au. The converted values in the table are rounded. Also given: $\chi^d_{\parallel} = -44.2$, $\chi^d_{\perp} = -43.8$, $\chi^p_{\parallel} = 19.0$, $\chi^p_{\perp} = 15.4$. – d) Δχ was reported in [10]. The original data [12] for χ, $\chi_{\parallel}$, and $\chi_{\perp}$ (for the larger of two basis sets) are not fully consistent. Also given: $\chi^p_{\parallel} = 19.08$ and $\chi^p_{\perp} = 16.28$ (and nondiagonal components). The origin of the vector potential was at P. – e) Origin at P. Values of χ, χ^d, and χ^p for the origin at H were −28.47, −137.90, and 109.43. – f) χ^p was based on the calculated χ^d and on the measured χ.

χ^d and χ^p were also calculated for different origins by a variational method yielding the susceptibilities of localized P–H bonds [16] and for the origin at the center of mass by an ab initio sum-over-states (SOS) configuration interaction method [17]. $\chi^d = -40.93$ was obtained from a united-atom approximation, and χ^p was estimated to be 16.35 (or 7.70) [18].

References:

[1] Miller, T. M. (in: Lide, D. R.; CRC Handbook of Chemistry and Physics, 73rd Ed., CRC Press, Ann Arbor 1992/93, pp. 10-194/10-210, 10-201).
[2] Maryott, A. A.; Buckley, F. (NBS-C-537 [1953] 1/29).
[3] Bishop, D. M.; Cheung, L. M. (J. Phys. Chem. Ref. Data **11** [1982] 119/33, 124).
[4] McKean, D. C.; Schatz, P. N. (J. Chem. Phys. **24** [1956] 316/25).
[5] Spackman, M. A. (J. Phys. Chem. **93** [1989] 7594/603, 7597).
[6] Sadlej, A. J. (Theor. Chim. Acta **79** [1991] 123/40, 129).
[7] Barter, C.; Meisenheimer, R. G.; Stevenson, D. P. (J. Phys. Chem. **64** [1960] 1312/6).
[8] Kukolich, S. G.; Flygare, W. H. (Chem. Phys. Lett. **7** [1970] 43/6).
[9] Combariza, J.; Salzman, W. R.; Kukolich, S. G. (Chem. Phys. Lett. **167** [1990] 607/8).
[10] Fleischer, U.; Schindler, M.; Kutzelnigg, W. (J. Chem. Phys. **86** [1987] 6337/47, 6337/8).

[11] Lazzeretti, P.; Zanasi, R. (J. Chem. Phys. **72** [1980] 6768/76).
[12] Höller, R.; Lischka, H. (Mol. Phys. **41** [1980] 1041/50, 1045).
[13] Keil, F.; Ahlrichs, R. (J. Chem. Phys. **71** [1979] 2671/5).
[14] Rothenberg, S.; Young, R. H.; Schaefer, H. F., III (J. Am. Chem. Soc. **92** [1970] 3243/50).
[15] Flygare, W. H.; Benson, R. C. (Mol. Phys. **20** [1971] 225/50, 241).
[16] Aminova, R. M.; Sadykova, A. Yu. (Teor. Eksp. Khim. **22** [1986] 487/91; Theor. Exp. Chem. [Engl. Transl.] **22** [1986] 465/8).
[17] Galasso, V. (Theor. Chim. Acta **63** [1983] 35/41).
[18] Banyard, K. E.; Hake, R. B. (J. Chem. Phys. **43** [1965] 2684/9).

1.3.1.4.1.7 **Nuclear Quadrupole Coupling Constants** eqQ in kHz

Deuterium

A coupling constant eqQ = 7.8 ± 0.6 along the molecular **symmetry axis** was obtained from a ^{2}D Fourier transform NMR spectrum of PD_3 in a nematic solvent [1]. After correcting for a supposed increase of the DPD bond angle by $\Delta\alpha = 0.7°$ vs. the gas-phase value (as was observed for PH_3 [2]), a value of eqQ = 6.5 ± 0.6 was obtained [2]. A still smaller theoretical value (although the theoretical value is usually larger than the experimental one) of eqQ = 5.7 was based [1] on SCF calculations of the components of the electric field gradient q in its principal axis system [3] (which was assumed to lie along the P–D bond; see also [4]). The true uncertainty of eqQ was probably also larger than given above [1].

The coupling constant along the **P–D bond** was assumed to be slightly larger than that in the related $C_6H_5PD_2$ molecule, for which eqQ = 115 ± 2 had been obtained from NMR measurements in liquid-crystal solutions. Axial symmetry of the field gradient, i.e., a vanishing asymmetry parameter $\eta = (q_{\beta\beta} - q_{\gamma\gamma})/q_{\alpha\alpha}$ was supposed (α parallel, β and γ perpendicular to the P–D bond) [5]. This eqQ value was then used for PD_3 in a nuclear spin relaxation study [6]. Ab initio SCF calculations yielded eqQ = 122.4 and $\eta = 0.082$ [7] or eqQ = 127.9 [8]. A later CI calculation gave q = 0.1977 au [9], which converts into a rounded value of eqQ = 133 (using a conversion factor of 672.0 kHz/au for D as given in [7, 8]). This q value was compared [9] to $q_{\alpha\alpha}$ from an earlier SCF calculation [3] of the elements of the q tensor in its principal axis system, yielding $q_{\alpha\alpha} = 0.1026$, $q_{\beta\beta} = 0.0868$, $q_{\gamma\gamma} = -0.1893$ au, and $\eta = (q_{\beta\beta} - q_{\alpha\alpha})/q_{\gamma\gamma} = 0.0834$ (the convention $|q_{\alpha\alpha}| \geqq |q_{\beta\beta}| \geqq |q_{\gamma\gamma}|$ was not obeyed). For derivations of eqQ with respect to the P–D bond length, see [7, 8].

Microwave spectroscopy of rotational transitions in PH_2D gave an upper limit of 20 kHz to the components $eq_{aa}Q+eq_{bb}Q$ and $eq_{aa}Q+eq_{cc}Q$ (a, b, c = principal inertial axes) [10].

Phosphorus

The components of the field gradient tensor in its principal axis system were found to be $q_{\alpha\alpha}=1.2581$, $q_{\beta\beta}=q_{\gamma\gamma}=-0.6288$ au, thus $\eta=0$, from an SCF calculation [3], and $q_{\alpha\alpha}=1.6858$ [11] and −1.3001 au [9] from two CI calculations.

References:

[1] Zumbulyadis, N.; Dailey, B. P. (J. Chem. Phys. **60** [1974] 4223/5).
[2] Zumbulyadis, N.; Dailey, B. P. (Mol. Phys. **27** [1974] 633/40).
[3] Rothenberg, S.; Young, R. H.; Schaefer, H. F., III (J. Am. Chem. Soc. **92** [1970] 3243/50).
[4] Kukolich, S. G. (Mol. Phys. **29** [1975] 249/55).
[5] Fung, B. M.; Wei, I. Y. (J. Am. Chem. Soc. **92** [1970] 1497/501).
[6] Sawyer, D. W.; Powles, J. G. (Mol. Phys. **21** [1971] 83/95, 84).
[7] Huber, H. (J. Chem. Phys. **83** [1985] 4591/8).
[8] Huber, H. (J. Mol. Struct. **121** [1985] 207/11 [THEOCHEM **22**]).
[9] Feller, D.; Boyle, C. M.; Davidson, E. R. (J. Chem. Phys. **86** [1987] 3424/40, 3436).
[10] Kukolich, S.; Schaum, L.; Murray, A. (J. Mol. Spectrosc. **94** [1982] 393/8).

[11] Petke, J. D.; Whitten, J. L. (J. Chem. Phys. **59** [1973] 4855/66, 4862).

1.3.1.4.1.8 Magnetic Shielding Constants σ

This section covers the magnetic shielding constants of the ^{31}P and ^{1}H nuclei in phosphane. The variation of the shielding constants with external conditions (phase, temperature, pressure) is treated in the context of chemical shift measurements; see Section 1.3.1.4.3.1, p. 182.

^{31}P Magnetic Shielding Constants

The following table lists the average shielding constant σ of ^{31}P in gaseous PH_3, referenced to the naked nucleus, and the shielding anisotropy $\Delta\sigma=\sigma_{\parallel}-\sigma_{\perp}$ (parallel and perpendicular to the C_3 axis of PH_3). The experimental values σ ($=\sigma^p+\sigma^d$) at 300 K [1, 2, 3] were obtained from the ^{31}P spin-rotation interaction constants (see p. 160), measured by molecular-beam electric-resonance spectroscopy of PH_3 in the vibrational ground state [3] (paramagnetic contribution σ^p), and the calculated diamagnetic constants σ^d.

	experimentally determined			calculated
σ in ppm	594.45±0.63	599.93	594.40	577 to 634
Δσ in ppm	−55.98	−64.6	−55.5±5	−26.6 to −40
remark	a)	b)	c)	d)
Ref.	[1]	[2]	[3]	[2, 4 to 12, 22]

a) The paramagnetic term $\sigma^p=-386.56\pm0.63$ ppm was obtained after improving the spin-rotation constants [3] using more accurate rotational constants of PH_3 [13, 14]. The ab initio calculated diamagnetic contribution $\sigma^d=981.01$ ppm was taken from [15]. – b) The spin-rotation constants for PH_3 in the vibrational ground-state [3] were recalculated in line with the PH_3 equilibrium geometry. σ^d [15] was used as above. – c) $\sigma^p=-366.43$ ppm was derived from the measured spin-rotation constants. The calculation of σ^d according [16] yielded 960.83 ppm [3].

The anisotropy $\Delta\sigma$ was determined from the ^{31}P NMR spectrum of PH_3 in a liquid-crystal solvent [17]. $\sigma = 594 \pm 10$ ppm and $\Delta\sigma = -50 + 15$ ppm [18] were reported [6]. – [d] The shielding constants were obtained by ab initio calculations of the paramagnetic and diamagnetic contributions, applying the conventional coupled Hartree-Fock method [8 to 10], local origin methods (GIAO [4, 7, 22], IGLO [6], LORG [3, 5]), a pseudo-potential method [11], and the sum-over-states CI procedure [12]; for details, see the papers.

^1H Magnetic Shielding Constants

From the ^1H NMR spectrum of PH_3 in benzene [19], the ^1H magnetic shielding constant $\sigma = 28.28$ ppm was derived [3, 20]. From σ and the paramagnetic contribution $\sigma^p = -98.13$ ppm (the sum of the ^1H spin-rotation term $\sigma^{sr} = 10.14$ ppm and the term $\sigma^{nuc} = -108.27$ ppm), the diamagnetic part $\sigma^d = 126.41$ ppm was obtained [3]. Ab initio calculations gave average shielding constants between 28 and 31 ppm [6 to 10, 12].

The analysis of the ^1H NMR spectrum of PH_3 (at 28°C) in a nematic solvent (*N*-(*p*-ethoxybenzylidene)-*p*′-*n*-butylaniline) led to the ^1H shielding anisotropy $\Delta\sigma = -7.5 \pm 0.5$ ppm [17] which agrees very well with $\Delta\sigma = -7.5$ ppm [18], derived with the ^1H spin-rotation constants [3]. $\Delta\sigma = -13.5$ ppm followed from an investigation of the ^1H NMR spectrum of PH_3 in a nematic solvent (60 mol% butyl-(*p*-ethoxyphenoxycarbonyl)-phenylcarbonate and 40 mol% *p*-capronyloxy-*p*′-ethoxyazobenzene) [21].

Ab initio calculated anisotropies are $\Delta\sigma = -7.7$ [8], -7.5 [9], and 10.6 ppm [6].

References:

[1] Jameson, J. C.; De Dios, A.; Jameson, A. K. (Chem. Phys. Lett. **167** [1990] 575/82).
[2] Jameson, J. C.; De Dios, A.; Jameson, A. K. (J. Chem. Phys. **95** [1991] 9042/53).
[3] Davies, P. B.; Neuman, R. M.; Wofsy, S. C.; Klemperer, W. (J. Chem. Phys. **55** [1971] 3564/8).
[4] Chesnut, D. B.; Rusiloski, B. E. (Chem. Phys. **157** [1991] 105/10).
[5] Bouman, T. D.; Hansen, A. E. (Chem. Phys. Lett. **175** [1990] 292/9).
[6] Fleischer, U.; Schindler, M.; Kutzelnigg, W. (J. Chem. Phys. **86** [1987] 6337/47).
[7] Chesnut, D. B.; Foley, C. K. (J. Chem. Phys. **85** [1986] 2814/20).
[8] Lazzeretti, P.; Zanasi, R. (J. Chem. Phys. **72** [1980] 6768/76).
[9] Höller, R.; Lischka, H. (Mol. Phys. **41** [1980] 1041/50).
[10] Keil, F.; Ahlrichs, R. (J. Chem. Phys. **71** [1979] 2671/5).

[11] Ridard, J.; Levy, B.; Millie, P. (Mol. Phys. **36** [1978] 1025/35).
[12] Galasso, V. (Theor. Chim. Acta **63** [1983] 35/41).
[13] Belov, S. P.; Burenin, A. V.; Gershtein, L. I.; Krupnov, A. F.; Markov, V. N.; Maslovsky, A. V.; Shapin, S. M. (J. Mol. Spectrosc. **86** [1981] 184/92).
[14] Belov, S. P.; Burenin, A. V.; Polyansky, O. L.; Shapin, S. L. (J. Mol. Spectrosc. **90** [1981] 579/89).
[15] Rothenberg, S.; Young, R. H.; Schaefer, H. F., III (J. Am. Chem. Soc. **92** [1970] 3243/50).
[16] Flygare, W. H.; Goodisman, J. (J. Chem. Phys. **49** [1968] 3122/5).
[17] Zumbulyadis, N.; Dailey, B. P. (Mol. Phys. **27** [1974] 633/40).
[18] Gierke, T. D.; Flygare, W. H. (J. Am. Chem. Soc. **94** [1972] 7277/83).
[19] Ebsworth, E. A. V.; Sheldrick, G. M. (Trans. Faraday Soc. **63** [1967] 1071/6).
[20] Wofsy, S. C.; Muenter, J. S.; Klemperer, W. (J. Chem. Phys. **55** [1971] 2014/9).

[21] Spiesecke, H. (Z. Naturforsch. **25a** [1970] 650/2).
[22] Chesnut, D. B. (Chem. Phys. **110** [1986] 415/20).

1.3.1.4.1.9 Rotational g Factor. Nuclear Spin-Rotation Coupling Constant C in kHz

The components $g_{\parallel}$ (parallel) and $g_{\perp}$ (perpendicular to the molecular symmetry axis C_3) are $g_{\parallel}=-0.011(1)$ and $g_{\perp}=0.0325(9)$ [1], see also [2]; previously, $g_{\parallel}=-0.0113$ and $g_{\perp}=0.0325$ [3]. The absolute values and the relative signs of both components were based on a reanalysis [1] of Zeeman effect measurements on rotational transitions of PH_2D and PHD_2 [4]; the true sign was determined by comparing the quadrupole moment values either derived from the g factors or from an ab initio MO calculation [1]. (The "experimental" quadrupole moments were, however, not very different for the two possible choices of the signs (see p. 155), and thus left a certain ambiguity; accordingly (?), the signs of $g_{\parallel}$ and $g_{\perp}$ were reversed in the abstract [1].) The original data $g_{\parallel}=-0.033\pm0.001$ and $g_{\perp}=0.017\pm0.001$ [4] had been interchanged with respect to the axes in a review [5].

Diagonal and nondiagonal elements of the g tensor of the PHDT molecule were based on the data above for PH_3 [2, 3].

The components $C_{\parallel}$ (parallel) and $C_{\perp}$ (perpendicular to C_3) of the nuclear spin-rotation coupling constant were derived for both ^{31}P and ^{1}H from an rf spectrum of PH_3, taken with a molecular-beam electric-resonance spectrometer [6]:

	^{31}P	^{1}H
$C_{\parallel}$	-116.38 ± 0.32	7.69 ± 0.19 *)
$C_{\perp}$	-114.90 ± 0.13	8.01 ± 0.08

*) 7.67 in the abstract [6].

Rovibrational corrections were applied to the data above for ^{31}P (which are valid for the zero-point vibrational state) to yield the equilibrium values $C_{\parallel,e}(^{31}P)\approx C_{\parallel}(^{31}P)-1.30$ and $C_{\perp,e}(^{31}P)\approx C_{\perp}(^{31}P)+1.45$ [7].

Effective values $C^2_{eff}(^{31}P)=4800\pm100$ kHz² and $C^2_{eff}(^{1}H)=25.0\pm0.5$ kHz² were obtained from the spin-lattice relaxation times T_1, measured for ^{31}P and ^{1}H in gaseous PH_3 (see p. 183). $C^2_{eff}=k_1C^2_a+k_2C_aC_d+k_3C^2_d$ appeared in an expression for T_1 with coefficients k_i given in the original paper and with $C_a=1/3\,(C_{\parallel}+2C_{\perp})$ (isotropic), $C_d=C_{\perp}-C_{\parallel}$ (anisotropic spin rotation interaction). The molecular beam data for $C_{\parallel}$ and $C_{\perp}$ from above [6] were converted into $C^2_{eff}(^{31}P)=13900\pm50$ kHz² and $C^2_{eff}(^{1}H)=65.5\pm0.2$ kHz² [8]. An earlier estimate from proton relaxation data was $6.1<|C_{eff}(^{1}H)|<8.5$ kHz [9]. $|C_{eff}(^{31}P)|=102$ kHz (from NMR shielding calculations [10]) and $|C_{eff}(^{1}H)|=7.8$ kHz (from the above inequality [9]) were used in another spin relaxation study [11]. ^{31}P NMR shielding calculations, which yielded $C_a=-146$, $C_{\parallel}=-124$, and $C_{\perp}=-160$ [12], were questioned in [11].

The PH_2D microwave spectrum for two rotational transitions showed a hyperfine structure due to ^{31}P spin-rotation interaction with $(C_{aa}+C_{bb})/2=(C_{aa}+C_{cc})/2=-98\pm3$ kHz (a, b, c = principal inertial axes) [13].

References:

[1] Combariza, J.; Salzman, W. R.; Kukolich, S. G. (Chem. Phys. Lett. **167** [1990] 607/8).
[2] Salzman, W. R. (J. Chem. Phys. **94** [1991] 5263/9).
[3] Salzman, W. R. (Chem. Phys. Lett. **167** [1990] 417/20).
[4] Kukolich, S. G.; Flygare, W. H. (Chem. Phys. Lett. **7** [1970] 43/6).
[5] Flygare, W. H.; Benson, R. C. (Mol. Phys. **20** [1971] 225/50, 241).
[6] Davies, P. B.; Neumann, R. M.; Wofsy, S. C.; Klemperer, W. (J. Chem. Phys. **55** [1971] 3564/8).

[7] Jameson, C. J.; de Dios, A. C.; Jameson, A. K. (J. Chem. Phys. **95** [1991] 9042/53, 9047).
[8] Armstrong, R. L.; Courtney, J. A. (Can. J. Phys. **50** [1972] 1262/72, 1266/7, 1269/70).
[9] Armstrong, R. L.; Courtney, J. A. (J. Chem. Phys. **51** [1969] 457/8).
[10] Sawyer, D. W.; Deverell, C. (private communication from [11]).

[11] Sawyer, D. W.; Powles, J. G. (Mol. Phys. **21** [1971] 83/95, 84).
[12] Deverell, C. (Mol. Phys. **18** [1970] 319/25).
[13] Kukolich, S.; Schaum, L.; Murray, A. (J. Mol. Spectrosc. **94** [1982] 393/8).

1.3.1.4.1.10 Rotational Constants. Rotation-Vibration Constants. Centrifugal Distortion Constants

Very accurate rotational constants (B, C), rotation-vibration constants (α, β), and centrifugal distortion constants (D, H, K, L) of PH_3 and its isotopomers in the electronic ground state $\tilde{X}\ ^1A_1$ have been derived from high-resolution rotational and rotation-vibration spectra.

1.3.1.4.1.10.1 Vibrational Ground State

The molecule $\mathbf{PH_3}$ (C_{3v} symmetry) is an oblate symmetric top (C<B=A) which is nearly spherical, as shown by the relatively small difference (about $-0.53\ cm^{-1}$, see Table 6, p. 162) of the rotational constants C (refers to rotation around the C_3 axis) and B (perpendicular to C_3). Since the permanent electric dipole moment is pointed parallel to the C_3 axis, only pure rotational transitions with the selection rule $\Delta K=0$ are allowed (K is the quantum number of the component about the C_3 axis of the total angular momentum J). Their analysis leads to the parameters B, D_J, D_{JK}, and H_{JK}. From the "perturbation-allowed" transitions $\Delta K=\pm 3n$ (n=1, 2, ...), which become weakly allowed by centrifugal distortion effects (inducing a small dipole moment of about 8×10^{-5} D perpendicular to the C_3 axis [1, 2, 3]), the K-related constants (C, D_K, H_K) were obtained; see, e.g. [1, 3, 4].

Well-agreeing sets of constants from three detailed analyses of the spectrum [4, 5, 6] are presented in Table 6, p. 162. The distortion constants are defined by the effective rotational Hamiltonian for a nondegenerate vibrational state of a symmetric-top C_{3v} molecule, including terms up to the eighth power in the angular momentum (according to [7]). An effective rotational Hamiltonian in form of a Padé operator was derived [8 to 11]. Optimum versions of a rational expansion of the effective rotational Hamiltonian for C_{3v} molecules were developed and some of them critically discussed [12 to 17]. For an ab initio calculation of centrifugal distortion effects for phosphane, see [18].

The constants in Table 6 were determined [4, 5, 6] by using the very precise microwave and submillimeter-wave data [1, 8, 19 to 22] (see Table 14, p. 185) and by combining these data with data of rotational transitions in the far IR [4] and with ground-state combination differences from high-resolution measurements of the ν_2 band of PH_3 [5]. Similarly detailed rotational analyses are given in [4, 12].

Some earlier rotation analyses led to sets of well-agreeing but less accurate rotation and distortion constants [1, 20 to 32].

Table 6

Rotational and Centrifugal Distortion Constants of PH_3 in the Vibrational Ground State.

constant	in cm^{-1} [4]	in cm^{-1} [5]	in MHz [6]
rotational constants			
B_0	4.452417639(59)	4.45241610(58)	133480.1264(65)
C_0	3.919025393(243)	3.91902563	117489.436(10)
quartic centrifugal distortion constants			
$D_J \times 10^4$	1.312928(35)	1.312081(192)	3937.89(99)
$D_{JK} \times 10^4$	−1.723619(29)	−1.725001(109)	−5172.39(12)
$D_K \times 10^4$	1.412250(170)	1.414369(337)	4238.95(55)
sextic centrifugal distortion constants*)			
$H_J \times 10^8$	1.0865(130)	1.2807(145)	488(46)
$H_{JK} \times 10^8$	−2.21930(243)	−4.7948(45)	−1442.04(72)
$H_{KJ} \times 10^8$	1.5481(92)	5.9668(277)	1765.0(13)
$H_K \times 10^8$	0.5094(344)	−1.5749(674)	485(11)

*) Octic centrifugal distortion constants 10^{11} $L_J = -0.12328(142)$, 10^{11} $L_{JK} = 1.1643(130)$, 10^{11} $L_{KKJ} = -1.8727(311)$, $L_{JJK} = L_K = 0$ (constrained) [4], see also [5, 6].

Constants for deuterated phosphanes are presented in Table 7. The constants of $\mathbf{PH_2D}$ (prolate asymmetric top) and $\mathbf{PHD_2}$ (oblate asymmetric top) result from an analysis of their rotational spectra (see Table 15, p. 187) through Watson's [25] asymmetric rotor Hamiltonian in its S reduction [2]. Linear combinations of rotational constants of PH_2D (A − C, 2B − A − C) were obtained by fitting them to Q-branch transitions [34].

The values for $\mathbf{PD_3}$ were obtained [2, 33] by a reanalysis of earlier PD_3 rotational spectra [1, 20, 22]. The Hamiltonian above was used with the constants $h_1 = h_2 = d_1 = d_2 = 0$ because of symmetry for the limiting case of asymmetric rotor [2].

Table 7

Rotational and Centrifugal Distortion Constants of Deuterated Phosphanes in the Vibrational Ground State.

	PD_3 [33]	PD_3 [2]	PH_2D [2]	PHD_2 [2]
rotational constants (in MHz)				
A_0	—	—	129843.14(4)	93917.79(2)
B_0	69471.145(23)	69471.09(11)	89279.34(2)	81910.92(2)
C_0	58973.960(556)	58974.42(11)	83250.52(2)	64841.80(3)
quartic centrifugal distortion constants (in kHz)				
D_J	1031.4(50)	1020.4(68)	2726.04*)	1612.75*)
D_{JK}	−1311.48(87)	−1312.28(46)	−3475.8(23)	−794.3(64)
D_K	928.76(9036)	1026.9(22)	1873.54(56)	−352(19)
d_1	0	0	−226.7(1.7)	−233.4(3)
d_2	0	0	449.1(6)	−326.20(17)

Table 7 (continued)

	PD_3 [33]	PD_3 [2]	PH_2D [2]	PHD_2 [2]
sextic centrifugal distortion constants (in Hz)				
H_J	588(306)	—	—	—
H_{JK}	−163.98(36)	−171(19)	−346(18)	221(31)
H_{KJ}	106(50)	140(23)	116(20)	—
H_K	−3956(3596)	—	67(20)	−1290(150)
h_1	0	0	—	−52(3)
h_2	0	0	−71(4)	−4.2(8)
h_3	9.21(14)	4.63(3)	—	—

*) D_J could not be determined independently since the only observed R-branch transitions for both molecules were $J=1 \leftarrow 0$. Therefore D_J was fixed at the value calculated from the harmonic force field.

For $\mathbf{PT_3}$ the constants $D_J=0.491$, $D_{JK}=-0.611$, and $D_K=0.457$ MHz were calculated from the harmonic force field [2]. $D_J=0.431$, $D_{JK}=-0.500$, and $D_K=0.377$ MHz were obtained by applying the concept of kinetic constants [35].

References:

[1] Chu, F. Y.; Oka, T. (J. Chem. Phys. **60** [1974] 4612/8).
[2] McRae, G. A.; Gerry, M. C. L.; Cohen, E. A. (J. Mol. Spectrosc. **116** [1986] 58/70).
[3] Oka, T. (Mol. Spectrosc. Mod. Res. **2** [1976] 229/53).
[4] Fusina, L.; Carlotti, M. (J. Mol. Spectrosc. **130** [1988] 371/81).
[5] Přádná, S.; Papoušek, D.; Kauppinen, J.; Belov, S. P.; Krupnov, A. F.; Scappini, F.; Di Lonardo, G. (Collect. Czech. Chem. Commun. **50** [1985] 2480/92).
[6] Tarrago, G.; Dang-Nhu, M. (J. Mol. Spectrosc. **111** [1985] 425/39).
[7] Watson, J. K. G. (Vib. Spectra Struct. **6** [1977] 1/89).
[8] Belov, S. P.; Burenin, A. V.; Polyansky, O. L.; Shapin, S. M. (J. Mol. Spectrosc. **90** [1981] 579/89).
[9] Burenin, A. V.; Polyanskii, O. L.; Shchapin, S. M. (Opt. Spektrosk. **54** [1983] 436/41; Opt. Spectrosc. [Engl. Transl.] **54** [1983] 256/9).
[10] Burenin, A. V.; Polyanskii, O. L.; Shchapin, S. M. (Opt. Spektrosk. **53** [1982] 666/72; Opt. Spectrosc. [Engl. Transl.] **53** [1982] 395/8).

[11] Burenin, A. V.; Fevral'skikh, T. M.; Karyakin, E. N.; Polyansky, O. L.; Shchapin, S. M. (J. Mol. Spectrosc. **100** [1983] 182/92).
[12] Burenin, A. V. (Mol. Phys. **75** [1992] 305/9).
[13] Burenin, A. V. (Opt. Spektrosk. **70** [1991] 57/60; Opt. Spectrosc. [Engl. Transl.] **70** [1991] 31/3).
[14] Burenin, A. V. (J. Mol. Spectrosc. **142** [1990] 117/21).
[15] Sarka, K. (J. Mol. Spectrosc. **151** [1992] 534/5).
[16] Sarka, K. (J. Mol. Spectrosc. **134** [1989] 354/61).
[17] Sarka, K. (J. Mol. Spectrosc. **133** [1989] 461/6).
[18] Taleb-Bendiab, A.; Lohr, L. L. (J. Mol. Spectrosc. **132** [1988] 413/21).
[19] Davies, P. B.; Neumann, R. M.; Wofsy, S. C.; Klemperer, W. (J. Chem. Phys. **55** [1971] 3564/8).
[20] Helms, D. A.; Gordy, W. (J. Mol. Spectrosc. **66** [1977] 206/18).

[21] Belov, S. P.; Burenin, A. V.; Gershtein, L. I.; Krupnov, A. F.; Markov, V. N.; Maslovsky, A. V.; Shapin, S. M. (J. Mol. Spectrosc. **86** [1981] 184/92).
[22] Helminger, P.; Gordy, W. (Phys. Rev. [2] **188** [1969] 100/8).
[23] Krupnov, A. F.; Mel'nikov, A. A.; Skvortsov, V. A. (Opt. Spektrosk. **46** [1979] 1012/3; Opt. Spectrosc. [Engl. Transl.] **46** [1979] 569/70).
[24] Andreev, B. A.; Belov, S. P.; Burenin, A. V.; Gershtein, L. I.; Krupnov, A. F.; Maslovskii, A. V.; Shchapin, S. M. (Opt. Spektrosk. **44** [1978] 620/1; Opt. Spectrosc. [Engl. Transl.] **44** [1978] 363/4).
[25] Yin, P. K. L.; Narahari Rao, K. (J. Mol. Spectrosc. **51** [1974] 199/207).
[26] Burrus, C. A. (J. Chem. Phys. **28** [1958] 427/9).
[27] Hoffman, J. M.; Nielsen, H. H.; Narahari Rao, K. (Z. Elektrochem. **64** [1960] 606/16).
[28] Maki, A. G.; Sams, R. L.; Olson, W. B. (J. Chem. Phys. **58** [1973] 4502/12).
[29] Bernard, P.; Oka, T. (J. Mol. Spectrosc. **75** [1979] 181/96).
[30] Goldman, A.; Cook, G. R.; Bonomo, F. S. (J. Quant. Spectrosc. Radiat. Transfer **24** [1980] 211/8).

[31] Tarrago, G.; Dang-Nhu, M.; Goldman, A. (J. Mol. Spectrosc. **88** [1981] 311/22).
[32] Goldman, A. (NASA-CP-2223 [1980/82] 635/55; C.A. **98** [1963] No. 116538).
[33] Burenin, A. V.; Shchapin, S. M. (Opt. Spektrosk. **52** [1982] 375/6; Opt. Spectrosc. [Engl. Transl.] **52** [1982] 225).
[34] Kukolich, S.; Schaum, L.; Murray, A. (J. Mol. Spectrosc. **94** [1982] 393/8).
[35] Thirugnanasambandam, P.; Gnanasekaran, S. (Bull. Soc. Chim. Belg. **86** [1977] 11/5).

1.3.1.4.1.10.2 Vibrationally Excited States

A simultaneous analysis of the five strongly coupled $\mathbf{PH_3}$ IR vibrational bands ν_1, ν_3, $2\nu_2$, $\nu_2+\nu_4$, and $2\nu_4$ (see p. 189), taking into account all vibrational-rotational interactions among the bands (pentad model), and the adoption of the vibrational ground state constants of PH_3 in [1] (see Table 6, p. 162) gave the following vibration-rotation constants α_i and β_i (in cm^{-1}) (variation of the rotational constants B and C and of the distortion constant D, respectively, with the vibrational modes i=1 to 4). For the adjustment of these upper-state parameters, 3766 relatively unblended out of about 4400 transitions were retained for the analysis [2].

$10^2\alpha_1^B$	$10^2\alpha_1^C$	$10^2\alpha_3^B$	$10^2\alpha_3^C$	$10^3\alpha_2^B$	$10^2\alpha_2^C$	$10^2\alpha_4^B$
4.528(9)	3.52(2)	4.435(6)	1.712(9)	3.79(7)	−2.452(6)	−1.554(4)

$10^2\alpha_4^C$	$10^6\beta_1^J$	$10^5\beta_1^{JK}$	$10^5\beta_1^K$	$10^6\beta_3^J$	$10^6\beta_3^{JK}$	$10^5\beta_3^K$
2.381(4)	−1.9(6)	1.3(2)	−1.5(2)	5.2(4)	9.7(1)	−1.5(1)

$10^6\beta_2^J$	$10^5\beta_2^{JK}$	$10^5\beta_2^K$	$10^6\beta_4^J$	$10^6\beta_4^{JK}$	$10^6\beta_4^K$
3.3(4)	−2.9(2)	3.2(2)	3.2(2)	2.0(9)	−6.8(8)

Fitting the changes of the rotational constants ΔX (X=B, C) obtained from the ν_2, $2\nu_2-\nu_2$ [3], $4\nu_2-\nu_2$, and $3\nu_2$ bands of PH_3 [4] to an expansion of X in a power series of v according to $X_2=X_0-\alpha_2^X v_2+\gamma_{22}^X v_2^2+\gamma_{222}^X v_2^3+\ldots$ yielded the following values of α and γ (in cm^{-1}) which reproduce ΔX with a deviation less than 0.0016 (X=B) and 0.00014 cm^{-1} (X=C) [3].

α_2^B	$10^2\gamma_{22}^B$	$10^3\gamma_{222}^B$	$10^2\alpha_2^C$	$10^4\gamma_{22}^C$	γ_{222}^C
0.2042(46)	2.75(32)	−1.89(56)	2.610(33)	6.74(81)	0 (fixed)

The constants B, C, D, H for the vibrational states $v_2=1$ and 2 were obtained [3] by analysing the ν_2 [5] and $2\nu_2-\nu_2$ bands of PH_3 [3].

Rotational and distortion constants (up to the fourth order) for PH_3 in the vibrational states $v_i=1$ (i=1 to 4) were derived in less extended studies of all four fundamental bands ν_i, analysing separately the two pairs ν_1, ν_4 [6] and ν_2, ν_4 [7] each pair being characterized by strong Coriolis coupling between the vibrations [6, 7].

The rotational analysis of the four **PD_3** fundamental bands, taking into account the ν_1/ν_3 and ν_2/ν_4 Coriolis coupling, gave a set of constants which reproduce the measured spectrum with a standard deviation of 0.007 cm^{-1}. The constants up to the fourth order are presented in Table 8; for higher-order constants, see the paper [12].

Table 8
Rotational and Centrifugal Distortion Constants of PD_3 in Vibrationally Excited States [12].

	ν_1 state	ν_3 state	ν_2 state	ν_4 state
ν (cm^{-1})	1682.1091	1693.3352	728.2825(16)	803.1279(15)
B (cm^{-1})	2.307638	2.298139	2.313476(35)	2.323027(46)
C (cm^{-1})	1.953363	1.961762	1.978026(43)	1.959726(45)
D_J (10^{-5} cm^{-1})	3.514	3.2954	2.9151(172)	3.7787(259)
D_{JK} (10^{-5} cm^{-1})	−4.783	−4.363	−4.704(54)	−4.348(68)
D_K (10^{-5} cm^{-1})	3.673	3.476	4.378(45)	3.122(51)

References:

[1] Tarrago, G.; Dang-Nhu, M. (J. Mol. Spectrosc. **111** [1985] 425/39).
[2] Tarrago, G.; Lacome, N.; Lévy, A.; Guelachvili, G.; Bézard, B.; Drossart, P. (J. Mol. Spectrosc. **154** [1992] 30/42).
[3] Hashinami, S.; Mito, H.; Matushima, F.; Takagi, K. (J. Mol. Spectrosc. **132** [1988] 1/12).
[4] Maki, A. G.; Sams, R. L.; Olson, W. B. (J. Chem. Phys. **58** [1973] 4502/12).
[5] Takagi, K.; Itoh, K.; Miura, E.; Tanimura, S. (J. Opt. Soc. Am. B **4** [1987] 1145/57).
[6] Baldacci, A.; Malathy Devi, V.; Narahari Rao, K. (J. Mol. Spectrosc. **81** [1980] 179/206).
[7] Tarrago, G.; Dang-Nhu, M.; Goldman, A. (J. Mol. Spectrosc. **88** [1981] 311/22).
[8] Hoffman, J. M.; Nielsen, H. H.; Narahari Rao, K. (Z. Elektrochem. **64** [1960] 606/16).
[9] Yin, P. K. L.; Narahari Rao, K. (J. Mol. Spectrosc. **51** [1974] 199/207).
[10] Andreev, B. A.; Belov, S. P.; Burenin, A. V.; Gershtein, L. I.; Krupnov, A. F.; Maslovskii, A. V.; Shchapin, S. M. (Opt. Spektrosk. **44** [1978] 620/1; Opt. Spectrosc. [Engl. Transl.] **44** [1978] 363/4).

[11] Tipton, T.; Choe, J.-I.; Kukolich, S. G. (J. Phys. Chem. **90** [1986] 1534/7).
[12] Kijima, K.; Tanaka, T. (J. Mol. Spectrosc. **89** [1981] 62/75).
[13] Bernard, P.; Oka, T. (J. Mol. Spectrosc. **75** [1979] 181/96).

1.3.1.4.1.11 Geometric Structure. Inertia Defect

Bond Distance. Bond Angle

Spectroscopically derived geometric parameters of PH_3 and PD_3 in the **electronic ground state** $\tilde{X}\ ^1A_1$ are presented in Table 9.

Table 9
Bond Distance r and Bond Angle α of PH_3 and PD_3 (C_{3v} symmetry).

type	parameter	PH_3	PD_3	remark	Ref.
effective	r_o in Å	1.420	1.4176	a)	[1]
	α_o	93.345°	93.359°	a)	[1]
average	r_z in Å	1.42774(9)	1.42373(13)	b)	[2]
		1.42731±0.0001	1.42287±0.0001	a)	[3]
		1.42699±0.0002	1.42265±0.0001	a)	[1]
	α_z	93.286(12)°	93.332(18)°	b)	[2]
		93.280°±0.02°	93.301°±0.01°	a)	[3]
		93.2287°±0.005°	93.2567°±0.004°	a)	[1]
equilibrium	r_e in Å	1.4135(20)		b)	[2]
		1.41175±0.0005		c)	[4]
		1.4115±0.0005		a)	[3]
		1.41159±0.0006		a)	[1]
	α_e	93.45(9)°		b)	[2]
		93.421°±0.06°		c)	[4]
		93.36°±0.08°		a)	[3]
		93.328°±0.02°		a)	[1]

a) From an analysis of the rotation spectrum [1, 3]. – r_0(P–H)=1.42002±0.00006 Å and α_o=93.3454°±0.0043° were derived [6] from microwave [5] and IR data [6] and cited in [7]. – b) The r_z structure is based on the rotational constants of PH_3 [8], PH_2D, and PHD_2 [2] and the reanalyzed [2] rotational constants of PD_3 [1]. The r_e structure was estimated [2] from the r_z parameters using the "diatomic approximation" [9] for the bond length or bond angle in a polyatomic molecule. – c) Based on PD_3 rotational constants.

A bond length of 1.419±0.001 Å and an angle of 94.3°±0.04° were derived from the ^1H NMR spectrum of PH_3 in a liquid crystal solvent [18].

Ab initio calculations of PH_3 in the D_{3h} transition state (see the section on the inversion barrier on p. 170) gave optimized bond lengths between 1.37 and 1.38 Å; see for example [15, 16, 17]. For further information, consult the bibliography on quantum-chemical calculations on p. 176.

The parameters r(P–H)=1.4176 Å and α=93.359° were derived (assuming equal H–H and H–D distances and equal P–H and P–D distances) from linear combinations of the rotational constants obtained by the analysis of rotational transitions in the Q branch of PH_2D [10].

Electronically Excited States. Only calculated values were reported. CI calculations yielded for PH_3 in the $\tilde{A}\ ^1A_1$ state $r_e = 2.694$ au and $\beta = 14.6°$ (angle between the P–H bond and the plane perpendicular to the C_3 axis) and in the $\tilde{a}\ ^3A_2''$ state (D_{3h} symmetry; see p. 145) $r_e = 2.667$ au [11].

Inertia Defect

Adopting the normal coordinate approach proposed in [12], the inertia defects $\Delta(PH_3) = 0.132$ amu·Å² and $\Delta(PD_3) = 0.176$ amu·Å² were derived [13] from the force field and fundamental frequencies in [14].

References:

[1] Helms, D. A.; Gordy, W. (J. Mol. Spectrosc. **66** [1977] 206/18).
[2] McRae, G. A.; Gerry, M. C. L.; Cohen, E. A. (J. Mol. Spectrosc. **116** [1986] 58/70).
[3] Chu, F. Y.; Oka, T. (J. Chem. Phys. **60** [1974] 4612/8).
[4] Kijima, K.; Tanaka, T. (J. Mol. Spectrosc. **89** [1981] 62/75).
[5] Helminger, P.; Gordy, W. (Phys. Rev. [2] **188** [1969] 100/8).
[6] Maki, A. G.; Sams, R. L.; Olson, W. B. (J. Chem. Phys. **58** [1973] 4502/12).
[7] Callomon, J. H.; Hirota, E.; Kuchitsu, K.; Lafferty, W. J.; Maki, A. G.; Pote, C.; Buck, I.; Starck, B. (Landolt-Börnstein New Ser. Group II **7** [1976] 1/395, 90).
[8] Belov, S. P.; Burenin, A. V.; Gershtein, L. I.; Krupnov, A. F.; Markov, V. N.; Maslovsky, A. V.; Shapin, S. M. (J. Mol. Spectrosc. **86** [1981] 184/92).
[9] Oka, T.; Morino, Y. (J. Mol. Spectrosc. **8** [1962] 300/14).
[10] Kukolich, S.; Schaum, L.; Murray, A. (J. Mol. Spectrosc. **94** [1982] 393/8).

[11] Müller, J.; Ågren, H. (J. Chem. Phys. **76** [1982] 5060/9).
[12] Herschbach, D. R.; Laurie, C. W. (J. Chem. Phys. **37** [1962] 1668/86, **40** [1964] 3142/53).
[13] Lalitha, M.; Srinivasamoorthy, R.; Savariray, G. A. (Indian J. Pure Appl. Phys. **19** [1981] 330/4).
[14] Shimanouchi, T.; Nakagawa, I.; Hiraishi, J.; Ishii, M. (J. Mol. Spectrosc. **19** [1966] 78/107).
[15] Schmiedekamp, A.; Skaarup, S.; Pulay, P.; Boggs, J. E. (J. Chem. Phys. **66** [1977] 5769/76).
[16] Dixon, D. A.; Arduengo, A. J. (J. Am. Chem. Soc. **109** [1987] 338/41).
[17] Marynick, D. S.; Dixon, D. A. (J. Phys. Chem. **86** [1982] 914/7).
[18] Zumbulyadis, N.; Dailey, B. P. (Mol. Phys. **27** [1974] 633/40).

1.3.1.4.1.12 Molecular Vibrations

Fundamental Frequencies, Harmonic Frequencies, Anharmonicity Constants

PH_3, PD_3, PT_3. The six normal modes of the PH_3 (PD_3) molecule form two totally symmetrical vibrations, $\nu_1(A_1)$ and $\nu_2(A_1)$, and two doubly degenerate vibrations, $\nu_3(E)$ and $\nu_4(E)$. Force field calculations (see p. 172) showed that approximately the higher frequency mode in each of the two symmetry species is a bond stretch and the lower an angle deformation.

The following fundamental frequencies (in cm⁻¹) of PH_3 and PD_3 in the **electronic ground state** $\tilde{X}\ ^1A_1$ in the gas phase were spectroscopically determined (for the fundamental frequencies of PH_3 and PD_3 in solid phases, see p. 193). They are given in a table on the next page.

	$\nu_1(A_1)$	$\nu_2(A_1)$	$\nu_3(E)$	$\nu_4(E)$	remark	Ref.
PH_3	2321.124(3)	—	2326.503(9)	—	a)	[1]
	2321.1314(27)	—	2326.8766(23)	—	b)	[2]
	—	992.1301(23)	—	1118.3131(15)	b)	[3]
	—	992.13454(68)	—	—	c)	[4]
PD_3	1682.1091(31)	728.2825(16)	1693.3352(13)	803.1279(15)		[5]

a) Obtained by simultaneously fitting the IR absorption bands ν_1, ν_3, $2\nu_2$, $\nu_2+\nu_4$, and $2\nu_4$, considering the strong vibrational-rotational coupling (Coriolis coupling, Fermi resonance) among all five bands [1]. – b) From an analysis of the band pairs ν_1/ν_3 [2] and ν_2/ν_4 [3] considering the Coriolis coupling between both vibrations within each pair [2]. Earlier, less extended studies [6, 7] gave the well-agreeing wave numbers $\nu_2=992.13$ (±0.02 [6], ±0.03 [7]) cm^{-1} and $\nu_4=1118.30$ (±0.01 [6], ±0.03 cm^{-1} [7]). – c) From laser Stark spectroscopy of the ν_2 fundamental [8] and the hot band $2\nu_2-\nu_2$ [4].

The harmonic frequencies ω_i were obtained from the experimentally determined fundamental frequencies [2, 3, 5] by using approximate anharmonicity corrections [9] and from ab initio calculations, e.g. [10, 11] (for further calculations, see p. 170).

ω_i (in cm^{-1})	PH_3 [9]	PH_3 [10]	PD_3 [9]	PD_3 c) [10]
ω_1	2405.4 a)	2411	1725.9 a)	1724
ω_2	1012.4 b)	1017	739.1 b)	743
ω_3	2411.4 a)	2409	1737.6 a)	1733
ω_4	1141.1 b)	1139	814.8 b)	813

a) Applying the Dennison rule and assuming the anharmonicity constants $x_1=x_3=0.035$ (anharmonicity of the P–H stretching in PHD_2 [12]). – b) Application of the Dennison rule and assuming $x_2=x_4=0.020$. – c) SCF MO/6-31G** calculation.

A set of ν_i values (1398, 623, 1401, 668 cm^{-1}) and ω_i values (1443, 643, 1424, 679 cm^{-1}) for PT_3 were reported [13].

For PH_3 in the **electronically excited states** $\tilde{B}$ and $\tilde{C}$ (in the older numbering, see p. 145) $\nu_2=495.8$ cm^{-1} and $x_{22}=7.8$ cm^{-1} (state $\tilde{B}$) and $\nu_2=420$ cm^{-1} (state $\tilde{C}$) were reported [14] based on an analysis of the $\tilde{B}\leftarrow\tilde{X}$ (1590 to 1490 Å) and $\tilde{C}\leftarrow\tilde{X}$ transitions (1500 to 1360 Å) [15].

Ab initio (CI) calculations gave for PH_3 in the $\tilde{A}$ $^1A_1$ state $\omega_1=0.201$ eV and $\omega_2=0.097$ eV [16].

PH_2D, PHD_2. The frequency of the P–H stretching vibration in PHD_2, $\nu=2323.81$ cm^{-1}, was determined by Fourier transform spectroscopy and used together with the overtone $2\nu=4563.26$ cm^{-1} to derive ω(P–H)$=2408.17$ cm^{-1}, treating the P–H unit as a diatomic molecule [12]. – For the fundamental frequencies of the species in solid phases, see p. 192.

Vibrational Amplitudes

Mean vibrational amplitudes of PH_3 and isotopomers were derived [17] (with minor adjustments) from the force constants [18] (this force field [18] was later corrected [9]); see Table 10.

Table 10

Mean Vibrational Amplitudes (in Å) of PH_3 and Isotopomers (Y,Y′ = H, D, T) [17].

molecule	T in K	P–H	P–D	P–T	Y...Y	Y...Y′
PH_3	0	0.0868	–	–	0.1560	–
	298	0.0869	–	–	0.1568	–
PH_2D	0	0.0868	0.0737	–	0.1559	0.1449
	298	0.0868	0.0737	–	0.1567	0.1462
PHD_2	0	0.0868	0.0736	–	0.1315	0.1448
	298	0.0868	0.0737	–	0.1341	0.1462
PD_3	0	–	0.0736	–	0.1314	–
	298	–	0.0737	–	0.1341	–
PT_3	0	–	–	0.0670	0.1189	–
	298	–	–	0.0672	0.1236	–

Mean amplitudes for PH_3, PD_3, and PT_3 were calculated by applying the concept of kinetic constants [19].

Coriolis Coupling Constants ζ

The Coriolis coupling constants in Table 11 were derived from the effect of the zero point vibration on the rotational constants through vibration-rotation interaction [20] and from an harmonic force field [21] (symmetry of vibration A_1, E; z‖C_3 axis).

Table 11

Coriolis Coupling Constants ζ of PX_3 (X = H, D, T).

sym.	$E_a \times E_b$			$E_a \times E_a$	$A_1 \times E_a$				Ref.
	ζ^z_{33}	ζ^z_{44}	ζ^z_{34}	$\zeta^y_{3a\,4a}$	ζ^y_{13}	ζ^y_{14}	ζ^y_{23}	ζ^y_{24}	
PH_3	−0.002	−0.429	0.754	−0.465	0.004	0.003	0.707	0.535	[20]
	0.024	−0.429	–	–	0.004	–	–	0.522	[21]
PD_3	0.038	−0.449	0.756	−0.453	−0.030	0.022	0.721	0.524	[20]
	0.065	−0.451	–	–	0.059	–	–	0.512	[21]
PT_3	0.103	−0.495	–	–	0.062	–	–	0.501	[21]

The simultaneous analysis of the five PH_3 absorption bands ν_1, ν_3, $2\nu_2$, $\nu_2+\nu_4$, and $2\nu_4$ yielded (upper indices omitted in the following) $|\zeta_{24}| \cong 0.5352(4)$ and $|\zeta_{13}| \cong 0.0042(2)$ [1] in excellent agreement with the values [20] in Table 11. The agreement is much poorer for $\zeta_{33} = 0.0098(4)$ and $\zeta_{44} = -0.4527(2)$ [1]. $\zeta_{33} = 0.0167$ was derived by a simultaneous analysis of the ν_1 and ν_3 bands of PH_3 [2]. Coupling constants $\zeta_{44} = -0.448$ [3], -0.453 ± 0.001 [6], −0.45620(11) [7], −0.441 [22] from less extended analyses of the PH_3 spectra were reported.

The analysis of the four fundamentals of PD_3 led to $\zeta_{33} = 0.056$, $\zeta_{44} = -0.463$, and $|\zeta_{13}| =$ 0.062. The sum rule $\zeta_{33} + \zeta_{44} = B_e/2C_e - 1$ is adequately fulfilled (observed: −0.407, calculated: −0.410) [5].

For Coriolis coupling constants calculated from force fields, see [5, 9, 13, 18, 21, 23].

References:

[1] Tarrago, G.; Lacome, N.; Lévy, A.; Guelachvili, G.; Bézard, B.; Drossart, P. (J. Mol. Spectrosc. **154** [1992] 30/42).
[2] Baldacci, A.; Devi, V. M.; Narahari Rao, K.; Tarrago, G. (J. Mol. Spectrosc. **81** [1980] 179/206).
[3] Tarrago, G.; Dang-Nhu, M.; Goldman, A. (J. Mol. Spectrosc. **88** [1981] 311/22).
[4] Hashinami, S.; Mito, H.; Matsushima, F.; Takagi, K. (J. Mol. Spectrosc. **132** [1988] 1/12).
[5] Kijima, K.; Tanaka, T. (J. Mol. Spectrosc. **89** [1981] 62/75).
[6] Yin, P. K. L.; Narahari Rao, K. (J. Mol. Spectrosc. **51** [1974] 199/207).
[7] Goldman, A.; Cook, G. R.; Bonomo, F. S. (J. Quant. Spectrosc. Radiat. Transfer **24** [1980] 211/8).
[8] Takagi, K.; Itoh, K.; Miura, E.; Tanimura, S. (J. Opt. Soc. Am. B **4** [1987] 1145/57).
[9] Duncan, J. L.; McKean, D. C. (J. Mol. Spectrosc. **107** [1984] 301/5).
[10] Breidung, J.; Thiel, W. (J. Phys. Chem. **92** [1988] 5597/602).

[11] Breidung, J.; Thiel, W.; Komornicki, A. (J. Phys. Chem. **92** [1988] 5603/11).
[12] McKean, D. C.; Torto, I.; Morrisson, A. R. (J. Phys. Chem. **86** [1982] 307/9).
[13] De Alti, G.; Costa, G.; Galasso, V. (Spectrochim. Acta **20** [1964] 965/75).
[14] Herzberg, G. (Molecular Spectra and Molecular Structure III. Electronic Spectra and Electronic Structure of Polyatomic Molecules, Van Nostrand-Reinhold, New York - Cincinnati - Toronto - London - Melbourne 1966, p. 610).
[15] Humphries, C. M.; Walsh, A. D.; Warsop, P. A. (Discuss. Faraday Soc. No. 35 [1963] 148/57).
[16] Müller, J.; Ågren, H. (J. Chem. Phys. **76** [1982] 5060/8).
[17] Cyvin, S. J. (Molecular Vibrations and Mean Square Amplitudes, Universitetsforlaget-Elsevier, Oslo - Amsterdam 1968, pp. 212, 214).
[18] Duncan, J. L.; Mills, I. M. (Spectrochim. Acta **20** [1964] 523/46).
[19] Thirugnanasambandam, P.; Mohan, S. (Indian J. Phys. **49** [1975] 808/17).
[20] Chu, F. Y.; Oka, T. (J. Chem. Phys. **60** [1974] 4612/8).

[21] McRae, G. A.; Gerry, M. C. L.; Cohen, E. A. (J. Mol. Spectrosc. **116** [1986] 58/70).
[22] Hoffman, J. M.; Nielsen, H. H.; Narahari Rao, K. (Z. Elektrochem. **64** [1960] 606/16).
[23] Thirugnanasambandam, P.; Gnanasekaran, S. (Bull. Soc. Chim. Belg. **86** [1977] 11/5).

1.3.1.4.1.13 Molecular Inversion. Force Field

Molecular Inversion

The displacement of the P atom of the pyramidal PH_3 molecule through the H_3 plane (related to the ν_2 umbrella vibration) leads to an inverted configuration. The potential energy as a function of the distance of the P atom from the H_3 plane is characterized by a double minimum (inversion barrier B).

Ab initio MO calculations, including contributions of the electron correlation by a second-order Møller-Plesset (MP2) perturbation approach, predict PH_3 (C_{3v}) to invert by a vertex inversion process through a planar trigonal transition state (D_{3h}) rather than by an edge inversion process through a planar, T-shaped transition state (C_{2v}) [1, 2]; for a topological electron density analysis of tricoordinate phosphorus inversion processes, see [3].

Inversion Barrier. Numerous ab initio calculations of the inversion barrier height B (difference of the total molecular energy of PH_3 at C_{3v} and D_{3h} equilibrium geometry) at different levels of computation have been performed. The following table shows values of the barrier for PH_3 in the **electronic ground state** $\tilde{X}\ ^1A_1$, obtained with differently extended basis sets by ab initio SCF calculations with (B_{correl}) and without (B_{SCF}) considering the electron correlation effects (PTCI second-order perturbation configuration interaction, CEPA coupled electron pair approximation, MPn n^{th}-order Møller-Plesset perturbation).

barrier	method						
	MP2	CI	CI	PTCI	CEPA	MP4	MP3
B_{correl} in kJ/mol	142.7	143.1	143.9	122.5	145.8	149.8	145.6
B_{SCF} in kJ/mol	—	—	153.6	157.2	158.1	158.2	162.4
Ref.	[2]	[4]	[5]	[6]	[7]	[8]	[9]

Inversion barriers were derived from the inversion potential functions (see below) as follows: for PH_3 11018 cm^{-1} [13], 11274 cm^{-1} [14], and 11700 cm^{-1} [15]; for PD_3 11416 cm^{-1} [14].

The inversion of the pyramidal phosphorus has been measured for substituted phosphetanes [16]. Based on these experiments, a lower limit of 146 kJ/mol was estimated for the inversion barrier of PH_3 [17].

For further calculations of the barrier with various basis sets at SCF and higher levels, see [5, 10 to 12, 18, 31, 59 to 62]; see also the bibliography of ab initio calculations on p. 176.

The influence of energy effects, associated with the interaction between the highest occupied MO (HOMO) and the lowest unoccupied MO (LUMO), on the inversion barrier was investigated for PH_3 and other molecules by simple MO theory and SCF calculations [19 to 21].

The origin of the inversion barrier was discussed in the framework of the theory of atoms in molecules [4].

An ab initio (SCF) calculated low barrier of inversion of 21 kJ/mol was reported for PH_3 in the **electronically excited state** $\tilde{A}\ ^1A_1$ [23]. A CNDO/2 + V_{n-1} potential calculation resulted in barriers of 114 and 135 kJ/mol for PH_3 in the excited states ^1E and ^3E, respectively [64].

Inversion Splitting. The inversion splitting Δ_i of PH_3 in the vibrational ground state is extremely small and all efforts to detect it by direct measurement failed [24 to 26]. An attempt to measure Δ_i with a molecular-beam electric-resonance spectrometer revealed that the inversion splitting must be lower than the resolution of the spectrometer (1 kHz) [26]. Similarly, from a high-resolution IR study of the $4\nu_2$ band of PH_3 followed $\Delta_i < 0.02\ cm^{-1}$ [25]. Much lower upper limits of $\Delta_i < 0.6(801)$ Hz [27], 3.8(400) Hz [28], and 8.4(418) Hz [29] were obtained from an evaluation of the modulation of the K splitting of rotational levels by the inversion contribution.

The calculation of the inversion splitting of PH_3 on the basis of inversion potential functions (see below) gave for PH_3 in the $v_2 = 0$ vibrational state $\Delta_i \approx 10^{-16}\ cm^{-1}$ and in the $v_2 = 7$ state $\Delta_i \approx 10^{-2}\ cm^{-1}$ [13]; see also [14, 32].

Inversion Potential Function. In a large-scale study, 93 points on the potential energy surface of the ground electronic state $\tilde{X}\ ^1A_1$ of PH_3 were determined by ab initio SCF + second-order Møller-Plesset perturbation calculations (MP2). A polynomial potential function (earlier applied to NH_3 [30]) was fitted to these points [13]. The calculated vibration-rotation energies are in reasonable agreement with the experiment [13]. For calculated inversion splittings, see above.

Further studies treated only sections of the full potential surface. The results of ab initio MO calculations [17, 31], the inversion barrier and the shape of the double-minimum potential, were fitted by suitable analytic expressions and used in calculations of the relevant vibrational energies and inversion splittings [17, 32].

An effective potential function was determined using a three-parameter (inversion coordinate, barrier height, harmonic force constant) reduced potential curve on the basis of MO calculations (ab initio MP2, CNDO/2) [15]. Another one was constructed as the sum of the pure inversion potential, V_0, and a vibrational modification, V_{vib}. The potential V_0 was described by a two-parameter (barrier height, height of the PH_3 pyramid) reduced potential [14]. The vibrational modification of the inversion barrier is about 273 cm^{-1} for PH_3 [32].

Force Field

The quadratic valence force field of pyramidal XY_3-type molecules contains six independent constants. In symmetry coordinates, these are the main diagonal constants $F_{11}(A_1)$, $F_{22}(A_1)$, $F_{33}(E)$, and $F_{44}(E)$ and the two interaction constants $F_{12}(A_1)$ and $F_{34}(E)$. Table 12 shows the relation between the symmetry force constants and the force constants in internal coordinates; these are the stretching (f_r) and bending (f_α) force constants and the four interaction force constants $f_{rr'}$, $f_{r\alpha}$, $f_{r\alpha'}$, and $f_{\alpha\alpha'}$. Of the two interaction constants F_{12} and F_{34}, the latter was adequately expressed by the Coriolis coupling constants ζ_3 and ζ_4 of the doubly degenerate vibrations. The evaluation of F_{12} was difficult due to the absence of phosphorus isotopic frequency data and the high sensitivity of F_{12} to centrifugal distortion [33]. A previous empirical force field with a negative F_{12} value [34] (see also [35]) was revised on the basis of new experimental data to have a positive F_{12} value [33] following a corresponding prediction from high-level ab initio calculations [36, 37].

Table 12 lists two empirically derived (A, B) and two quantum-chemically calculated (C, D) sets of constants. The constants of the harmonic force field A (reproducing the harmonic frequencies ω_i; see p. 168) were derived from experimentally determined fundamental frequencies, Coriolis coupling constants, and centrifugal distortion constants of PH_3 and PD_3 [33]. Improving earlier force fields [34, 35, 38, 56], the effective force field B (reproducing the observed fundamental frequencies; see p. 168) was obtained using the fundamental frequencies of PH_3, PD_3, and PT_3, quartic distortion constants of PH_3, PD_3, PH_2D, and PHD_2, and Coriolis coupling constants of PH_3 and PD_3 [39]. The ab initio (CEPA, coupled electron pair approximation) calculated force field C [36] is regarded to be the best available [40]. The ab initio SCF calculated force field D [40] is in excellent agreement with the empirical force field A.

Table 12
Symmetry Force Constants of Phosphane.
For explanation, see the text.

force constant	force field				dimension
	A [33]	B [39]	C [36]	D*) [40]	
$F_{11}=f_r+2f_{rr'}$	3.341 (33)	3.194 (20)	3.557	3.367	mdyn/Å
$F_{12}=f_{r\alpha'}+2f_{r\alpha}$	0.100 (21)	0.156 (10)	0.088	0.152	mdyn/rad
$F_{22}=f_\alpha+2f_{\alpha\alpha'}$	0.612 (7)	0.594 (4)	0.690	0.641	mdyn·Å·rad^{-2}
$F_{33}=f_r-f_{rr'}$	3.339 (33)	3.165 (18)	3.553	3.327	mdyn/Å
$F_{34}=f_{r\alpha'}-f_{r\alpha}$	−0.048 (8)	−0.036 (10)	−0.036	0.044	mdyn/rad
$F_{44}=f_\alpha-f_{\alpha\alpha'}$	0.730 (10)	0.719 (4)	0.783	0.708	mdyn·Å·rad^{-2}

*) Scaling of the angle-bending coordinates with a unit bond length of 1 Å yielded constants in mdyn/Å.

Several theoretically and empirically derived and estimated force constants of the quadratic valence force field were reported [41 to 55, 58, 63].

The 14 independent parameters of the cubic force field of PH_3 were determined by ab initio SCF calculations [57].

References:

[1] Dixon, D. A.; Arduengo, A. J.; Fukunaga, T. (J. Am. Chem. Soc. **108** [1986] 2461/2).
[2] Dixon, D. A.; Arduengo, A. J. (J. Am. Chem. Soc. **109** [1987] 338/41).
[3] Edgecombe, K. E. (J. Mol. Struct. **226** [1991] 157/79 [THEOCHEM **72**]).
[4] Bader, R. F. W.; Cheeseman, J. R.; Laidig, K. E.; Wiberg, K. B.; Breneman, C. (J. Am. Chem. Soc. **112** [1990] 6530/6).
[5] Marynick, D. S.; Dixon, D. A. (J. Phys. Chem. **86** [1982] 914/7).
[6] Maripuu, R.; Reineck, I.; Ågren, H.; Nian-Zu, Wu; Rong, Ji Ming; Veenhuizen, H.; Al-Shamma, S. H.; Karlsson, L.; Siegbahn, K. (Mol. Phys. **48** [1983] 1255/67).
[7] Ahlrichs, R.; Keil, F.; Lischka, H.; Kutzelnigg, W.; Staemmler, V. (J. Chem. Phys. **63** [1975] 455/63).
[8] Reed, A. E.; Schleyer, P. v. R. (Chem. Phys. Lett. **133** [1987] 553/61).
[9] Latajka, Z.; Scheiner, S. (J. Chem. Phys. **81** [1984] 2713/6).
[10] Dixon, D. A.; Marynick, D. S. (J. Am. Chem. Soc. **99** [1977] 6101/3).

[11] Marynick, D. S.; Dixon, D. A. (J. Chem. Phys. **69** [1978] 498/500).
[12] Jolly, C. A.; Chan, F.; Marynick, D. S. (Chem. Phys. Lett. **174** [1990] 320/4).
[13] Civiš, S.; Čársky, P.; Špirko, V. (J. Mol. Spectrosc. **118** [1986] 88/95).
[14] Špirko, V.; Civiš, S.; Ebert, M.; Danielis, V. (J. Mol. Spectrosc. **119** [1986] 426/32).
[15] Špirko, V.; Civiš, S.; Beran, S.; Čársky, P.; Fabian, J. (Collect. Czech. Chem. Commun. **50** [1985] 1519/36).
[16] Cremer, S. E.; Chorvat, R. J.; Chang, C. H.; Davis, D. W. (Tetrahedron Lett. **1968** 5799/802).
[17] Schoeller, W. W. (J. Mol. Struct. **137** [1986] 341/5).
[18] Bernardi, F.; Bottoni, A.; Olivucci, M.; Tonachini, G. (J. Mol. Struct. **133** [1985] 243/61 [THEOCHEM **26**]).
[19] Lambert, J. (Top. Stereochem. **6** [1971] 19/105).
[20] Epiotis, N. D.; Cherry, W. R.; Shaik, S.; Yates, R.; Bernardi, F. (Top. Curr. Chem. **70** [1977] 140/3).

[21] Cherry, W.; Epiotis, N. (J. Am. Chem. Soc. **98** [1976] 1135/40).
[22] Cherry, W.; Epiotis, N.; Borden, W. T. (Acc. Chem. Res. **10** [1977] 167/73).
[23] Müller, J.; Ågren, H. (J. Chem. Phys. **76** [1982] 5060/8).
[24] Helminger, P.; Gordy, W. (Phys. Rev. [2] **188** [1969] 100/8).
[25] Maki, A. G.; Sams, R. L.; Olson, W. B. (J. Chem. Phys. **58** [1973] 4502/12).
[26] Davies, P. B.; Neumann, R. M.; Wofsy, S. C.; Klemperer, W. (J. Chem. Phys. **55** [1971] 3564/8).
[27] Fusina, L.; Carlotti, M. (J. Mol. Spectrosc. **130** [1988] 371/81).
[28] Belov, S. P.; Burenin, A. V.; Polyansky, O. L.; Shapin, S. L. (J. Mol. Spectrosc. **90** [1981] 579/89).
[29] Belov, S. P.; Burenin, A. V.; Gershtein, L. I.; Krupnov, A. F.; Markov, V. N.; Maslovsky, A. V.; Shapin, S. M. (J. Mol. Spectrosc. **86** [1981] 184/92).
[30] Špirko, V. (J. Mol. Spectrosc. **101** [1984] 30/47).

[31] Lehn, J. M.; Munsch, B. (Mol. Phys. **23** [1972] 91/107).
[32] Špirko, V.; Papoušek, D. (Mol. Phys. **36** [1978] 791/6).
[33] Duncan, J. L.; McKean, D. C. (J. Mol. Spectrosc. **107** [1984] 301/5).
[34] Duncan, J. L.; Mills, I. M. (Spectrochim. Acta **20** [1964] 523/46).
[35] Shimanouchi, T.; Nakagawa, I.; Hiraishi, J. (J. Mol. Spectrosc. **19** [1966] 78/107).
[36] Kutzelnigg, W.; Wallmeier, H.; Wasilewski, J. (Theor. Chim. Acta **51** [1979] 261/73).
[37] Schlegel, H. B.; Wolfe, S.; Bernardi, F. (J. Chem. Phys. **63** [1975] 3632/8).
[38] Kuchitsu, K. (J. Mol. Spectrosc. **7** [1961] 399/409).
[39] McRae, G. A.; Gerry, M. C. L.; Cohen, E. A. (J. Mol. Spectrosc. **116** [1986] 58/70).
[40] Breidung, J.; Thiel, W. (J. Phys. Chem. **92** [1988] 5597/602).

[41] Thyagarajan, G.; Sundaram, S.; Cleveland, F. F. (J. Mol. Spectrosc. **5** [1960] 307/18).
[42] Sundaram, S.; Suszek, F.; Cleveland, F. F. (J. Chem. Phys. **32** [1960] 251/4).
[43] Sundaram, S.; Cleveland, F. F. (J. Mol. Spectrosc. **5** [1960] 61/4).
[44] King, S.-T.; Overend, J. (J. Phys. Chem. **73** [1969] 406/12).
[45] Pariseau, M.; Wu, E.; Overend, J. (J. Chem. Phys. **39** [1963] 217/23).
[46] Arnold, T. H.; Swanson, B. I.; Yamaguchi, Y.; Nelson, D. J. (J. Mol. Spectrosc. **78** [1979] 267/76).
[47] Mohan, S.; Ravikumar, K. G.; Gunasekaran, S. (Acta Cienc. Indica Phys. **10** [1984] 39/43).
[48] Purnachandra Rao, B.; Ramamurthy, V. (Indian Chem. J. **12** [1977] 20/1).
[49] Vallamattam, A. J.; Babu Joseph, K.; Pillai, M. G. K. (Indian J. Pure Appl. Phys. **15** [1977] 49/51).
[50] Thirugnanasambandam, P.; Mohan, S. (Indian J. Phys. **49** [1975] 808/17).

[51] Ramaswamy, K.; Sridharan, T. (Indian J. Pure Appl. Phys. **13** [1975] 98/100).
[52] Thyagarajan, G.; Subhedar, M. K. (Indian J. Pure Appl. Phys. **12** [1974] 309/11).
[53] Padmaja, K. M.; Aruldhas, G. (Indian J. Pure Appl. Phys. **12** [1974] 658/60).
[54] Ponomarev, Yu. I. (Opt. Spektrosk. **35** [1973] 828/31; Opt. Spectrosc. [Engl. Transl.] **35** [1973] 482/3).
[55] Whitmer, J. C. (J. Chem. Phys. **56** [1972] 1050/7).
[56] De Alti, G.; Costa, G.; Galasso, V. (Spectrochim. Acta **20** [1964] 965/75).
[57] Breidung, J.; Schneider, W.; Thiel, W.; Schaefer, H. F., III (J. Mol. Spectrosc. **140** [1990] 226/36).
[58] Díaz Fleming, G.; Cyvin, S. J.; Cyvin, B. N. (Spectrosc. Lett. **20** [1987] 881/97).
[59] Zahradník, R.; Havlas, Z.; Hess, B. A., Jr.; Hobze, P. (Collect. Czech. Chem. Commun. **55** [1990] 869/89).
[60] Marynick, D. S.; Dixon, D. A. (Faraday Discuss. Chem. Soc. No. 62 [1977] 47/8).

[61] Moccia, R. (J. Chem. Phys. **40** [1964] 2176/85).
[62] Scott, J. M.; Sutcliffe, B. T. (Theor. Chim. Acta **41** [1976] 141/8).
[63] Walker, W. (J. Chem. Phys. **59** [1973] 1537).
[64] Banerjee, M.; Bhattacharyya, S. P. (Proc. Indian Acad. Sci. Chem. Sci. **89** [1980] 549/59).

1.3.1.4.1.14 Bond Dissociation Energy $D(H_2P–H)$, Atomization Energy ΔH_{at}, in kJ/mol

The most recent determination of the dissociation energy yielded 345 kJ/mol. Other values obtained with various methods are scattered over a relatively wide range:

	D_0	D_{298}	D_0	D_{298}	D_0	D_0	D_0
D in kJ/mol	345.0±1.9	351±8	356±4	378	≤339±17	331	326±6
Ref.	[1]	[2]	[3]	[4]	[5]	[6]	[7]
remark	a)	b)	c)	d)	e)	f)	g)

a) Evaluated in a photoionization mass spectrometric study from the difference in the thresholds for PH_2^+ from PH_3 and PH_2^+ from PH_2 [1]. This value is only 4 kJ/mol higher than the almost coincident values calculated with the 4th-order Møller-Plesset perturbation procedure [1, 8] and with the generalized valence bond model [9]. – b) From the electron impact ionization of PH_3 and PH_2. – c) From the fluorescence excitation of PH_3 photolysis fragments. – d) From the appearance potential of PH_2^+ in electron impact studies of PH_3 and the ionization potential of PH_2. – e) From the appearance potential of PH_2^- from PH_3 (2.2, 2.3 eV) and the electron affinity of the PH_2 radical (1.25 eV, see p. 62) [5]; for earlier results (D≤326 kJ/mol), see [10]. – f) From the highest populated rotational-vibrational level of HF, which is produced in a hydrogen abstraction reaction of PH_3 with F atoms in a flowing afterglow experiment [6]; for earlier results, see [11]. – g) Literature value based on the upper limit D≤326 kJ/mol.

D_0 = 345 kJ/mol for the P–H bond in PHD_2 was derived from the experimental P–H stretching frequency by treating the P–H unit as a diatomic molecule [12].

The atomization energy ΔH_{at} = 950.2±1.6 kJ/mol at 0 K was reported [1] (see also [13]); it is based on the measured enthalpy of formation, $\Delta_f H^\circ_{298}$ = 5.4 kJ/mol [14] (see the thermodynamic data of formation on p. 179). Atomization energies for about 65 molecules including PH_3 were calculated with density functional models [15].

References:

[1] Berkowitz, J.; Curtiss, L. A.; Gibson, S. T.; Green, J. P.; Hillhouse, G. L.; Pople, J. A. (J. Chem. Phys. **84** [1986] 375/84).
[2] McAllister, T.; Lossing, F. P. (J. Phys. Chem. **73** [1969] 2996/8).
[3] Di Stefano, G.; Lenzi, M.; Margani, A.; Mele, A.; Nguyen Xuan, C. (J. Photochem. **7** [1977] 335/44).
[4] Saalfeld, F. E.; Svec, H. J. (Inorg. Chem. **3** [1964] 1442/3).
[5] Smyth, K. C.; Brauman, J. I. (J. Chem. Phys. **56** [1972] 1132/42).
[6] Manocha, A. S.; Setser, D. W.; Wickramaaratchi, M. A. (Chem. Phys. **76** [1983] 129/46).
[7] Glushko, V. P.; Gurvich, L. V.; Bergman, G. A.; Veits, I. V.; Medvedev, V. A.; Khachkuruzov, G. A.; Yungman, V. S. (Termodinamicheskie Svoistva Individual'nykh Veshchestv, Izdatel'stvo Nauka, Moscow 1978, Vol. I, Pt. 1, p. 369).
[8] Pople, J. A.; Luke, B. T.; Frisch, M. J.; Binkley, J. S. (J. Phys. Chem. **89** [1985] 2198/203).
[9] Goddard, W. A., II; Harding, L. B. (Annu. Rev. Phys. Chem. **29** [1978] 363/96).
[10] Brauman, J. I.; Eyler, J. R.; Blair, L. K.; White, M. J.; Comisarov, M. B.; Smyth, K. C. (J. Am. Chem. Soc. **93** [1971] 6360/2).

[11] Duewer, W. H.; Setser, D. W. (J. Chem. Phys. **58** [1973] 2310/20).
[12] McKean, D. C.; Torto, I.; Morrisson, A. R. (J. Phys. Chem. **86** [1982] 307/9).
[13] Curtiss, L. A.; Jones, C.; Trucks, G. W.; Raghavachari, K.; Pople, J. A. (J. Chem. Phys. **93** [1990] 2537/45).
[14] Gunn, S. R.; Green, L. G. (J. Phys. Chem. **65** [1961] 779/83).
[15] Clementi, E.; Chakravorty, S. J. (J. Chem. Phys. **93** [1990] 2591/602).

1.3.1.4.1.15 Quantum-Chemical Calculations

The phosphane molecule is the subject of numerous quantum-chemical calculations. The methods applied range from simple LCAO MO approaches to high-quality calculations, such as Hartree-Fock (SCF MO), configuration interaction (CI), perturbation theory (MP, MBPT), and, more recently, density function procedures.

The sections on the molecular properties of the phosphane quote relevant calculations which supplement experimental results or where experimental results are not available or experiments had yielded diagreeing or uncertain results. Otherwise, the reader is referred to the following bibliography:

Richards, W. G.; Walker, T. E. H.; Hinkley, R. K.; A Bibliography of *ab initio* Molecular Wave Functions, Clarendon Press, Oxford, 1971.

Richards, W. G.; Walker, T. E. H.; Farnell, L.; Scott, P. R.; Bibliography of *ab initio* Molecular Wave Functions. Supplement for 1970—1974, Clarendon Press, Oxford, 1974.

Richards, W. G.; Scott, P. R.; Colburn, E. A.; Marchington, A. F.; Bibliography of *ab initio* Molecular Wave Functions. Supplement for 1974—1977, Clarendon Press, Oxford, 1978.

Richards, W. G.; Scott, P. R.; Sackwild, V.; Robins, S. A.; A Bibliography of *ab initio* Molecular Wave Functions. Supplement for 1978—1980, Clarendon Press, Oxford, 1981.

Ohno, K.; Morokuma, K.; Quantum Chemistry Literature Data Base-Bibliography of *ab initio* Calculations for 1978—1980, Elsevier, Amsterdam, 1982; Annual Supplements appeared in the following volumes of the Journal of Molecular Structure: **91** [1982], **106** [1983], **119** [1984], **134** [1985], **148** [1986], **154** [1987], **182** [1988], **203** [1989], **211** [1990], **252** [1991], **298** [1992].

1.3.1.4.2 Thermal, Crystallographic, and Mechanical Properties

Earlier investigations of the physical properties of phosphane are described in "Phosphor" C, 1965, pp. 24/30; thermodynamic data of formation are given on pp. 10/1 and solubility data on p. 50 of that volume.

Phase Transition. Temperatures. The boiling point of PH_3 at normal pressure, 185.5 K, was extrapolated from vapor pressure measurements; see "Phosphor" C, 1965, p. 26. PD_3 boils at 188 K [1]. The other phase transitions of PH_3 under its own pressure were identified by measuring the heat capacity. Liquid PH_3 (and PD_3 [1]) solidifies at the triple point temperature of 139.4 K, forming the plastic crystalline phase α. Cooling to 88.3 K results in conversion to the crystalline phase β. The conversion of the β phase to the stable crystalline phase γ requires several days and has a transition temperature of 49.4 K. The temperature for the fast transition between the β phase and the metastable crystalline δ phase is 30.3 K; details are described in "Phosphor" C, 1965, pp. 24, 29. An IR investigation showed the conversion of the δ phase into the more stable γ phase to be an extremely slow process [2].

The phase transition of PH_3 and PD_3 were investigated more recently by vibrational spectroscopy of crystalline films. The resulting transition temperatures in K are as follows [3]:

transition	PH_3(Raman)	PH_3(IR)	PD_3(Raman)	PD_3(IR)
$\gamma\rightarrow\beta$	47.5 to 53.5	49.5 to 53.5	56	56
$\beta\leftrightarrow\delta$	25.0 to 29.5	28.5 to 28.7	—	33

The $\beta \leftrightarrow \gamma$ phase transition of PD_3 was found to proceed in the temperature range 50 to 60 K with a hysteresis typical for a first-order transition. The low-temperature γ phase was stable down to 6 K [4]. The $\beta \rightarrow \gamma$ conversion is quite fast for PD_3 in contrast to that of PH_3 [3] as described in the preceeding paragraph. The temperature for the transition of PD_3 between the crystalline phase β and the plastic crystalline phase α of 93.5 K is significantly higher than the corresponding temperature of 88.2 K for PH_3 [5].

Indications of an additional phase transition of PH_3 at 10 K [6] were found to be erroneous [2, 3].

Structural Features of the Crystalline Phases. Many properties of phosphane in the plastic crystalline α phase resemble those in the liquid and are described in the following section.

Recent X-ray investigations of solid phosphane do not seem to exist; for older results, see "Phosphor" C, 1965, p. 25. However, some conclusions concerning crystalline phases were drawn from vibrational spectra. PH_3 probably occupies sites with C_{3v} or C_3 symmetry in the β phase [7]. The line widths observed for the β and δ phases indicate some degree of disorder which might be due to rotational motions [6]. The β phase of PD_3 seems to be a disordered version of the δ phase [3]. The γ phase of PD_3 probably contains only two inequivalent C_s sites [3, 4] in contrast to higher numbers discussed in [2]; possible unit cells were discussed [3]. The nearly constant frequency ratio of the IR peaks of PH_3 and PD_3 suggests an identical crystal structure of their δ phases [2].

Properties of the Liquid and Plastic Crystalline Phases. Early investigations led to the conclusion that the molecules in liquid phosphane only very weakly interact with one another; see "Phosphor" C, 1965, p. 25. The absence of hydrogen bonding was confirmed by measurements demonstrating the rapid rotation of the molecules in the liquid and the plastic crystalline phase α and their rapid translation in the liquid phase by the small values of the reorientational correlation times τ_Θ and the angular momentum correlation times τ_J The correlation times were obtained for PH_3 [8, 9], PHD_2 [10], and PD_3 [1] from analyzing Raman scattering and for PH_3 [5, 11, 12] and PD_3 [5, 12] from determining the spin-lattice relaxation times and the self-diffusion coefficient [11]. Values for the correlation times as a function of the reciprocal temperature are shown in Fig. 8, p. 178 [5]. The transition from the liquid to the plastic crystalline phase has only a small effect on the reorientational motion, but slows the translational motion considerably [12]. The continuity in τ_J of PD_3 suggests that the efficiency of the collision process is unchanged, whereas the discontinuity in the reorientational correlation times at the melting point indicates a change in the molecular field of PD_3 [5].

Reorientational correlation times τ_Θ of liquid and plastic crystalline PH_3 [9], PHD_2 [10], and PD_3 [1] in the range 0.2 to 0.5 ps were also obtained from the line width of stretching vibrations in Raman spectra by using a memory-function approach. The values agree with results from NMR [5] and Raman spectra [8] which are given in **Fig. 8**, p. 178.

The small apparent activation energies derived from the temperature dependence of the correlation times confirm that the hindrance of molecular motions is low; activation energies for the reorientation are in the range 2.1 to 2.5 kJ/mol [2, 8, 12] in the liquid and about 1.3 kJ/mol in the solid [5]. A larger activation energy of 4.6 kJ/mol was found for the translation in the liquid [12].

The graphs of the experimentally derived correlation times of liquid and plastic crystalline PH_3 [5, 11] and PD_3 [5] (which are identical for both isotopomers) were compared with the curves calculated for a spherical top molecule in the two limiting cases of the classical extended rotational diffusion model. The experimental graphs were found to lie between the curves for the J-diffusion limit (where magnitude and orientation of the angular momentum of the molecule is randomized by collisions) and the M-diffusion limit (where only the orientation of the angular momentum is randomized). However, the molecular orientation in the liquid phase seems to be more characteristic of the M-diffusion model than of the J-diffusion model which in most cases provides a better description of the diffusion in liquids [5, 11].

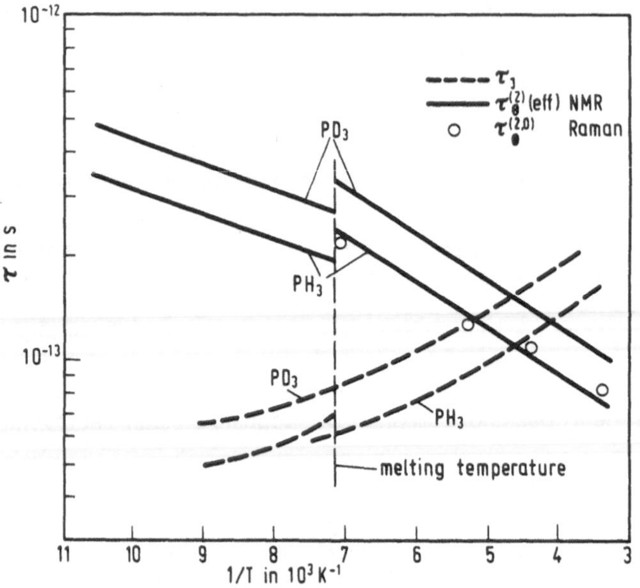

Fig. 8. (Effective) reorientational correlation times τ_Θ and angular momentum correlation times τ_J from NMR [5] and Raman spectra [8] as a function of the reciprocal temperature in the plastic crystalline and liquid phases of PH_3 and PD_3 (from [5]).

The reduced correlation times of liquid PH_3 and PD_3 are almost identical. This shows that the molecular reorientation is dominated by inertial effects. These can be expected to be nearly isotropic, because PH_3 and PD_3 are approximately spherical top molecules [5, 11]. Also, a comparison of the τ_Θ values from relaxation measurements in [12] (reflecting both spinning and tumbling motions) with those from Raman spectra (indicative solely of the tumbling motion) show that the rates of spinning and tumbling motions for both PH_3 and PD_3 are nearly equal [8].

Density. Molar Volume. The density of PH_3 from the triple (melting) point of 139 K to the boiling point of 186 K was measured by the hydrostatic weighing method. The values obtained fit the equation $d=(0.594\pm0.003)-[(171\pm3)\times10^{-5}](T-273)$ with d in g/cm and T in K. The molar volume at the boiling point is $V_m=45.72$ cm³/mol [13]. The data agree well with previous results given in "Phosphor" C, 1965, pp. 26/7.

Surface Tension. Viscosity. The surface tension σ in dyn/cm of liquid PH_3 between 139 and 186 K was measured by the capillary rise method. The experimental data can be expressed by $\sigma=(49\pm4)-(0.15\pm0.02)T$ with T in K [14].

The viscosity η (in P) of liquid PH_3 was measured between 155 K and the boiling point (given to be 187 K) by the capillary flow method. The values were fitted by $\ln\eta=7.115-4912.4\times T^{-1}+455270\times T^{-2}$ with T in K [15]. The viscosity of gaseous PH_3 was measured at 193 to 273 K by the same method; the experimental values are given by $\eta=3.648\times10^{-7}\cdot T^{1.013}$. The data were used to calculate the constant $\varepsilon/k=270$ K (k is the Boltzmann constant) and the effective diameter $\sigma=3.897$ Å of the Lennard-Jones (6-12) potential [16]. The data of the surface tension and the viscosity of gaseous PH_3 agree with literature values [14, 16]; earlier results are given in "Phosphor" C, 1965, p. 28.

Self-Diffusion. The self-diffusion coefficient D of liquid PH_3 and PD_3 in sealed tubes was determined by spin echo measurements of the nuclei ^1H, ^2H, and ^{31}P from 139 K to ambient temperature. Numerical values for both phosphanes are given by $D(cm^2/s) = 5.18\times10^{-4} \exp(-413/T)$ at temperatures up to 200 K. Above this temperature, D rises faster with temperature to reach 1.5×10^{-4} cm²/s at 293 K. Attempts to correlate D and the viscosity η failed except at the lowest temperatures [11]. The self-diffusion constant of plastic crystalline PH_3 was estimated for a vacancy diffusion mechanism on the basis of the spin-lattice relaxation time. The increase from $D = 2\times10^{-12}$ to 1×10^{-9} cm²/s between 103 and 138 K is typical for a plastic crystal. A diffusion activation energy of 19 kJ/mol was estimated [12].

Vapor Pressure. The vapor pressure over solid and liquid PH_3 was measured by a differential method. In the temperature range 122.5 to 183.5 K, the curve of the vapor pressure follows $\log p = -1277.475/T - 5.85652661 \log T - 0.0020449708\,T + 23.397806$ with p in Torr and T in K [17]. These pressure data differ slightly from the older results which were quoted in "Phosphor" C, 1965, p. 26. Vapor pressure data of PH_3 could also be fitted to the Antoine equation to give $\log p = 4.2303 - 154.6895/(T - 85.06)$ with p in Torr and T in K [18].

The measured vapor pressures of mixtures of PH_3 with SiH_4 [18, 19] and AsH_3 [20, 21] were used to calculate physicochemical parameters of the systems. The relative vapor pressure of infinitely diluted PH_3 at the triple points and boiling points of the solvents CH_4, SiH_4, and AsH_3 was calculated by a statistical method [22].

Solubility. The solubility of PH_3 in water at about 276 K is 1960 ± 28 μg/L (0.000196%) [23]; a value of 0.000160 to 0.000166% was found at room temperature [24]; older data are considerably higher; see "Phosphor" C, 1965, p. 50. Graphs of the technical solubility coefficient of PH_3 in aqueous solutions of H_2SO_4, NaOH, and NaCl as a function of temperature and concentration are given in [25].

The solubility of PH_3 in nonaqueous solvents is usually considerably higher than in water. The solubilities of PH_3 in CS_2 and a series of organic solvents at ambient temperature are compiled in [26]; in most cases, two to eight volume parts of gaseous PH_3 are dissolved per volume part of the solvent. Earlier results are given in "Phosphor" C, 1965, p. 50.

Thermodynamic Data of Formation. The standard enthalpy of formation of gaseous PH_3, $\Delta_f H^\circ_{298} = 5.4 \pm 1.7$ kJ/mol, was calculated from the heats of the explosive decomposition of PH_3/SbH_3 mixtures with white phosphorus as the reference state and is the recommended value; different experimental values are given in "Phosphor" C, 1965, pp. 10/1. The experimentally determined value yields $\Delta_f G^\circ_{298} = 13.4$ and $\Delta_f H^\circ_0 = 13.39$ kJ/mol [27]. Calculations from molecular constants with standard methods yielded thermodynamic data of formation and the equilibrium constant for the formation of PH_3 as an ideal gas; the reference states used were crystalline α white phosphorus (0 to 195 K), crystalline β white phosphorus (195 to 317 K), the melt (317 to 1180 K), and an ideal P_2 gas at higher temperatures. Selected results are as follows [28]:

T in K	0	100	200	298	400	600
$\Delta_f H^\circ$ in kJ/mol	30.811	28.751	25.790	22.886	19.164	14.408
$\Delta_f G^\circ$ in kJ/mol	30.811	28.326	28.803	30.893	34.277	42.937
$\log K_p$	—	−14.796	−7.523	−5.412	−4.476	−3.738

T in K	800	1000	1500	2000	4000	6000
$\Delta_f H^\circ$ in kJ/mol	11.279	9.356	−53.925	−50.766	−39.951	−36.668
$\Delta_f G^\circ$ in kJ/mol	52.958	63.620	108.272	266.807	369.850	573.735
$\log K_p$	−3.458	−3.323	−3.770	−4.228	−4.830	−4.995

In an earlier calculation, the reference states were crystalline red (V) phosphorus from 0 to 704 K and ideal P_2 gas from 704 to 6000 K [29]. Ab initio calculations at the MP4 level gave $\Delta_f H^\circ_{298}$ values of 9.1 [30] and 16.7 kJ/mol [31]; for a MINDO/3 calculation, see [32].

Thermodynamic data for the formation of PH_3 in aqueous solution (1 mol/kg) at 1 bar and 298.15 K were calculated to be $\Delta_f H^\circ = -9.50$ and $\Delta_f G^\circ = 25.36$ kJ/mol; the standard enthalpy under these conditions is 120.1 $J \cdot K^{-1} \cdot mol^{-1}$ [27].

Thermodynamic Functions. Thermodynamic functions of gaseous PH_3 at 1 atm between 0 and 6000 K were calculated from molecular constants by standard methods. Selected results in $J \cdot K^{-1} \cdot mol^{-1}$ are as follows [28]:

T in K	0	100	200	298	400	600
C°_p	0	33.259	33.933	37.102	41.782	50.905
S°	0	173.027	196.180	210.243	221.787	240.489
$-(G^\circ - H^\circ_{298})/T$	∞	241.134	213.516	210.243	211.755	218.323

T in K	800	1000	1500	2000	4000	6000
C°_p	58.511	64.300	72.822	76.837	81.427	82.368
S°	256.218	269.929	297.837	319.404	374.567	407.798
$-(G^\circ - H^\circ_{298})/T$	225.879	233.348	250.411	265.068	307.480	335.744

Similar data for 298 K were given in [27]. Least squares polynomials were fitted to partition functions of polyatomic molecules including phosphane in the temperature range 1000 to 6000 K and pressure range 0.1 to 10^6 Pa [33]. Standard entropies and heat capacities at 298 K for XY_n-type compounds with n = 3, 4, 5 and including PH_3 were predicted by an empirical relation making use of the normal boiling points [34]. Constants for the polynomial of the heat capacity of PH_3 as a function of temperature from 143 to 186 K were given; $\Delta H = 1.131$ kJ/mol at the triple point temperature of 139.35 K was reported [35].

Thermodynamic functions of gaseous PH_3 and the deuterated isotopomers between 1000 and 6000 K and at 1 atm were calculated from molecular constants; selected values are as follows [36].

T in K	$-(G^\circ - H^\circ_0) \cdot R^{-1} \cdot T^{-1}$				S° in $J \cdot K^{-1} \cdot mol^{-1}$			
	PH_3	PH_2D	PHD_2	PD_3	PH_3	PH_2D	PHD_2	PD_3
1000	26.863	28.568	29.156	28.625	270.197	286.210	293.062	290.758
1400	28.884	30.665	31.333	30.888	293.261	309.973	317.506	315.893
2000	31.307	33.167	33.917	33.558	320.134	337.345	345.376	344.262
4000	36.783	38.771	39.651	39.426	376.855	394.781	403.503	403.046
6000	40.387	42.443	43.393	43.232	412.591	431.240	440.578	440.661

Earlier results for PH_3 and isotopically substituted molecules are described in "Phosphor" C, 1965, pp. 28/30.

References:

[1] Wilde, R. E.; Cohen, S. S. (J. Chem. Phys. **70** [1979] 4557/61).
[2] Francia, M. D.; Nixon, E. R. (J. Chem. Phys. **58** [1973] 1061/5).
[3] Huang, T.-H.; Decius, J. C.; Nibler, J. W. (J. Phys. Chem. Solids **38** [1977] 897/904).
[4] Wilde, R. E.; Covington, B. (J. Phys. Chem. Solids **36** [1975] 1225/7).

[5] Boden, N.; Folland, R. (Chem. Phys. Lett. **32** [1975] 127/32).
[6] Hardin, A. H.; Harvey, K. B. (Can. J. Chem. **42** [1964] 84/9).
[7] Heinemann, A. (Ber. Bunsen-Ges. Phys. Chem. **68** [1964] 280/6).
[8] Schwartz, M.; Wang, C. H. (Chem. Phys. Lett. **25** [1974] 26/33; AD-766802-3 [1973] 1/24; C.A. **80** [1974] No. 54219).
[9] Wilde, R. E.; Chang, T.-C. (J. Chem. Phys. **72** [1980] 1293/6).
[10] Wilde, R. E. (Chem. Phys. Lett. **67** [1979] 555/7).

[11] Krynicki, K.; Sawyer, D. W.; Powles, J. G. (Magn. Reson. Relat. Phenom. Proc. 18th Congr. AMPERE, Nottingham, Engl., 1974 [1975], Vol. 2, pp. 511/2; C.A. **83** [1975] No. 197955).
[12] Sawyer, D. W.; Powles, J. G. (Mol. Phys. **21** [1971] 83/95).
[13] Zorin, A. D.; Runovskaya, I. V.; Lyakhmanov, S. B.; Yudanova, L. V. (Zh. Neorg. Khim. **12** [1967] 2529/34; Russ. J. Inorg. Chem. [Engl. Transl.] **12** [1967] 1335/8).
[14] Devyatykh, G. G.; Zorin, A. D.; Runovskaya, I. V. (Dokl. Akad. Nauk SSSR **188** [1969] 1082/3; Dokl. Phys. Chem. [Engl. Transl.] **184/189** [1969] 663/4).
[15] Runovskaya, I. V.; Zorin, A. D.; Devyatykh, G. G. (Zh. Neorg. Khim. **15** [1970] 2581/2; Russ. J. Inorg. Chem. [Engl. Transl.] **15** [1970] 1338/9).
[16] Vlasov, S. M.; Devyatykh, G. G. (Zh. Neorg. Khim. **11** [1966] 2681/4; Russ. J. Inorg. Chem. [Engl. Transl.] **11** [1966] 1439/41).
[17] Zorin, A. D.; Krasnova, S. G. (Izv. Vyssh. Uchebn. Zaved. Khim. Khim. Tekhnol. **10** [1967] 1097/100; C.A. **68** [1968] No. 98953).
[18] Krasnova, S. G.; Zorin, A. D.; Yudanova, L. V. (Zh. Fiz. Khim. **39** [1965] 2440/4; Russ. J. Phys. Chem. [Engl. Transl.] **39** [1965] 1302/4).
[19] Devyatykh, G. G.; Zorin, A. D.; Balabanov, V. V.; Stepanov, V. M.; Gur'yanov, A. N.; Kedyarkin, V. M. (Izv. Akad. Nauk SSSR Ser. Khim. **1970** 1960/5; Bull. Acad. Sci. USSR Div. Chem. Sci. [Engl. Transl.] **1970** 1835/9).
[20] Devyatykh, G. G.; Zorin, A. D.; Postnikova, T. K.; Umilin, V. A. (Zh. Neorg. Khim. **14** [1969] 1626/30; Russ. J. Inorg. Chem. [Engl. Transl.] **14** [1969] 851/4).

[21] Kut'in, A. M.; Frolov, I. A.; Zaburdyaev, V. S.; Kulakov, S. I. (Zh Fiz. Khim. **49** [1975] 297/300; Russ. J. Phys. Chem. [Engl. Transl.] **49** [1975] 177/8).
[22] Devyatykh, G. G.; Vlasov, S. M. (Zh. Fiz. Khim. **39** [1965] 1171/5; Russ. J. Phys. Chem. [Engl. Transl.] **39** [1965] 620/2).
[23] Berck, B. (J. Agric. Food Chem. **16** [1968] 419/25; C.A. **69** [1968] No. 1857).
[24] Rauscher, H.; Mayr, G.; Hild, K. (unpublished report, Degesch, Frankfurt am Main 1965 from [23]).
[25] Kruis, A. (Landolt-Börnstein 6th Ed. Vol. 4 Pt. 4 C **1** [1976] 33, 333, 338, 340).
[26] Fogg, P. G. T. (Solubility Data Ser. **21** [1985] 281/301; C.A. **104** [1986] No. 136823).
[27] Wagman, D. D.; Evans, W. H.; Parker, V. B.; Schumm, R. H.; Halow, I.; Bailey, S. M.; Churney, K. L.; Nuttall, R. L. (J. Phys. Chem. Ref. Data **11** Suppl. No. 2 [1982] 2-73).
[28] Chase, M. W., Jr.; Davies, C. A.; Downey, J. R., Jr.; Frurip, D. J.; McDonald, R. A.; Syverud, A. N. (J. Phys. Chem. Ref. Data **14** Suppl. No. 1 [1985] 1298, 1735).
[29] Stull, D. R.; Prophet, H. (Natl. Stand. Ref. Data Ser. U.S. Natl. Bur. Stand. No. 37 [1971]).
[30] Wong, M. W.; Gill, P. M. W.; Nobes, R. H.; Radom, L. (J. Phys. Chem. **92** [1988] 4875/80).

[31] Pople, J. A.; Luke, B. T.; Frisch, M. J.; Binkley, J. S. (J. Phys. Chem. **89** [1985] 2198/203).
[32] De Santo, J. T.; Mosbo, J. A.; Storhoff, B. N.; Bock, P. L.; Bloss, R. E. (Inorg. Chem. **19** [1980] 3086/92).
[33] Irwin, A. W. (Astron. Astrophys. Suppl. Ser. **74** [1988] 145/60 from C.A. **109** [1988] No. 100938).

[34] Stølevik, R. (Acta Chem. Scand. **43** [1989] 758/62).
[35] Zábranský, M.; Růžička, V.; Barcal, P. (Chem. Prum. **38** [1988] 67/73).
[36] Shabur, V. N.; Katashinskii, A. S.; Titova, I. S.; Koval'chuk, D. S. (Teplofiz. Vys. Temp. **7** [1969] 369/71; High Temp. [Engl. Transl.] **7** [1969] 338/9).

1.3.1.4.3 Spectra

1.3.1.4.3.1 Nuclear Magnetic Resonance

Earlier NMR investigations of phosphane were reported in "Phosphor" C, 1965, p. 17.

On account of the spin I=1/2 for the nuclei ^{31}P and ^{1}H, the ^{31}P NMR spectrum of PH_3 consists of a 1:3:3:1 quartet and the ^{1}H NMR spectrum of a 1:1 doublet [1, 2]. Deuterium with I=1 brings about a 1:3:6:7:6:3:1 septet in the ^{31}P NMR spectrum of PD_3 [3]. Measured ^{31}P and ^{1}H chemical shifts and spin-spin coupling constants of PH_3 are presented in Table 13.

Table 13
^{31}P and ^{1}H Chemical Shifts δ (in ppm) and Spin-Spin Coupling Constants J (P–H) (in Hz) of PH_3.

state or solvent	gas (300 K)	gas (—)	liquid (300 K)	liquid (180 K)	P_2H_4 (183 K)	THF (300 K)	$CDCl_3$ (300 K)
$-\delta(^{31}P)$	266.10	254.2	241.0	234.76	227.47	245.4±0.1	379.3
$\delta(^{1}H)$	—	—	1.73	1.832	1.968	—	1.81
J(P–H)	—	187	182.2±0.3	186.4	189.8	189±3	188.0, 189.2
remark	a)	b)	c)	d)	d)	e)	f)
Ref.	[4]	[3, 8]	[7]	[9]	[9]	[10]	[11]

solvent g)	C_6H_6	$(CH_3)_3N$	CH_3CN	CCl_4	NH_3	$CHCl_3$	CS_2
$\delta(^{1}H)$ [1]	1.548	1.732	1.786	1.792	1.812	1.842	1.872
J(P–H) [1]	186.6	183.9	189.0	184.9	189.9	189.2	185.6

a) Obtained by extrapolating the ^{31}P resonance frequency of PH_3 in Ar gas (up to 30 amagat) to zero density of Ar, referenced to 85% H_3PO_4 (corrected for bulk susceptibility).

b) Proton-decoupled ^{31}P spectrum, referenced to 85% aqueous H_3PO_4 at 291 K. The linear increase of the ^{31}P NMR shift for PH_3 in various solvents (CCl_4, benzene, toluene, and cyclohexane) with increasing temperature (30 to 90°C) is illustrated in the paper [8].

c) J(P–H) was taken from [12].

d) Low concentration of PH_3 in P_2H_4. ^{31}P shifts referenced to 85% aqueous H_3PO_4 at 303 K. Internal ^{1}H NMR standard: $Si(CH_3)_4$.

e) 1-Molar solution of PH_3 in tetrahydrofuran. External standard: 85% aqueous H_3PO_4.

f) Saturated solution of PH_3 in $CDCl_3$. Standards are internal $Si(CH_3)_4$ and external $PO(CH_3)_3$.

g) Infinitely diluted solutions of PH_3 at 294 K. Shifts referenced to $Si(CH_3)_4$. Variations of $\delta(^{1}H)$ and J(P–H) with temperature (294 to 249 K) were given. Additional coupling constants J(P–H) ranging from 182 to 189 Hz were measured for PH_3 in various isotropic and liquid-crystal solvents [13].

The ^{31}P NMR resonance of gaseous PH_3 was found to be shifted significantly from the resonance signal in liquid phosphane towards higher fields. From the gas-phase **chemical shift** $\delta = -266.10$ ppm for PH_3 (see Table 13) at the zero-pressure limit at 300 K, referenced to 85% aqueous H_3PO_4, and the absolute shielding constant of PH_3, 594.45 ± 0.63 ppm (see p. 158), an absolute shielding scale for ^{31}P was established with a shielding constant of 328.35 ppm for 85% H_3PO_4 (corrected for bulk susceptibility) [4]. (The earlier ^{31}P scale [5] was based on the chemical shift $\delta = -238 \pm 1$ ppm [6] for liquid PH_3 at 183 K.)

An investigation of the isotope effect on the ^{31}P shielding in liquid PH_nD_{3-n} at 227 K yielded $\sigma(PH_3) - \sigma(PD_3) = -2.54$ ppm [14] in good agreement with -2.5 ± 0.1 ppm for gaseous phosphane [15]. The shifts $\sigma(PH_3) - \sigma(PH_2D) = -0.804$ ppm, $\sigma(PH_2D) - \sigma(PHD_2) = -0.845$ ppm, and $\sigma(PHD_2) - \sigma(PD_3) = -0.88$ ppm showed the isotope effect to satisfy the additivity rule [14]. The deuterium-induced isotope effect was shown to be dominated by contributions due to bond stretching; for a detailed discussion, see [15].

The effect of the deuterium isotope on the ^{31}P shielding constant in the isoelectronic series PH_2^-, PH_3, PH_4^+, and the deuterated species is largest for the negative ion and essentially zero for the phosphonium ion [16].

The variation of the chemical shift of gaseous PH_3 with temperature and gas density (pure PH_3 [15, 17], PH_3 in excess Ar [4]) was monitored between 270 and 400 K and in the range of approximately 7 to 30 amagat (1 amagat $= 2.68 \times 10^{19}$ molecules/cm^3) [4, 15, 17, 18]. Compared to the value extrapolated to zero gas density, the resonance signal shifts downfield by 4.2 ppm on increasing the PH_3 density to 50 amagat [17]. When PH_3 is replaced by the light noble gas Ar as collision partner the downfield shift of PH_3 becomes smaller (1.1 ppm) and the extrapolation to zero density becomes more accurate [4]. The experimental results were evaluated in terms of a virial expansion of the shielding constant, $\sigma(T, \rho) = \sigma_0(T) + \sigma_1(T)\rho$. The term $\sigma_0(T)$ denotes the shielding in the isolated PH_3 molecule and $\sigma_1(T)$ is the second virial coefficient of shielding in low-density samples. $\sigma_1(T)$ was fitted to the power series $\sigma_1(T) = 77.8 - 0.250\,(T-300) + 2.26 \times 10^{-3}\,(T-300)^2 - 1.71 \times 10^{-5}\,(T-300)^3$ ppb·amagat^{-1}, revising an earlier [17] $\sigma_1(T)$ power series [15]. σ_0 was found to decrease with increasing temperature [4, 15] (by about 0.1 ppm between 270 and 400 K [15]), superseding the earlier positive temperature coefficient [17, 18, 19].

A positive sign of the **spin-spin coupling constant** J(P–H) (see Table 13) was established by the ^1H NMR spectrum of PH_3 taken in a liquid-crystal solvent (*N*-(*p*-ethoxybenzylidene)-*p*′-*n*-butylaniline, EBBA) [3], supporting the prediction $J(P–H) > 0$ from semiempirical calculations [20]. Ab initio [21, 22, 23] and semiempirical [24, 25, 26] calculations of J(P–H) were reported. A qualitative perturbation model [27] and a nonempirical double perturbation method [28] were applied to calculate J(P–H) in PH_3.

The coupling constant $J(P–D) = 29.0 \pm 0.1$ Hz was determined from the ^{31}P NMR spectrum of PD_3 in the isotropic phase CCl_4 and the dipolar coupling constant in the liquid-crystal solvent EBBA (the doublet splitting in the ^2H NMR spectrum led to the deuterium nuclear quadrupole coupling constant; see p. 157) [3].

$J(H–H) = -13.3$ Hz was obtained from the geminal coupling constant $J(H–D) = -2.03$ Hz, derived from the ^1H NMR spectrum of liquid PH_2D at room temperature [12]. The negative sign was established by multiple resonance experiments [29] and ab initio calculations [21, 22, 23, 30].

The change in the P–H spin-spin coupling of phosphane brought about by D–H isotopic substitution was examined [14, 31, 32].

The spin lattice **relaxation times** T_1 of ^{31}P and ^1H were measured on gaseous samples of PH_3 at densities between 0.03 and 4 amagat. The spin-rotation interaction provides the dominant relaxation mechanism. The effective ^{31}P and ^{1}H spin-rotation constants (see p. 160) were derived from the density dependence of T_1 [33]; see [34] concerning previous work.

The ^{31}P and ^{2}H spin-lattice relaxation times of PH_3 and PD_3 were measured as a function of temperature in the course of studies on the molecular motion in the liquid and plastic crystalline phases of phosphane (see p. 177) [35, 36, 37].

References:

[1] Ebsworth, E. A. V.; Sheldrick, G. M. (Trans. Faraday Soc. **63** [1967] 1071/6).
[2] Fluck, E.; Binder, H. (Z. Naturforsch. **22b** [1967] 805/8).
[3] Zumbulyadis, N.; Dailey, B. P. (J. Chem. Phys. **60** [1974] 4223/5).
[4] Jameson, J. C.; De Dios, A.; Jameson, A. K. (Chem. Phys. Lett. **167** [1990] 575/82).
[5] Appleman, B. R.; Dailey, B. P. (Adv. Magn. Reson. **7** [1974] 231/320, 280).
[6] van Wazer, J. R.; Callis, C. F.; Shoolery, J. N.; Jones, R. C. (J. Am. Chem. Soc. **78** [1956] 5715/26, 5719).
[7] Fluck, E.; Bürger, H.; Goetze, U. (Z. Naturforsch. **22b** [1967] 912/5).
[8] Zumbulyadis, N.; Dailey, B. P. (Mol. Phys. **27** [1974] 633/40).
[9] Junkes, P.; Baudler, M.; Dobbers, J.; Rackwitz, D. (Z. Naturforsch. **27b** [1972] 1451/6).
[10] Batchelor, R.; Birchall, T. (J. Am. Chem. Soc. **104** [1982] 674/9).

[11] Vincent, E.; Verdonck, L.; van der Kelen, G. P. (Spectrochim. Acta A **36** [1980] 699/704).
[12] Lynden-Bell, R. M. (Trans. Faraday Soc. **57** [1961] 888/92).
[13] Wasser, R.; Lounila, J.; Diehl, P. (Mol. Cryst. Liq. Cryst. **141** [1986] 51/67).
[14] Jameson, A. K.; Jameson, C. J. (J. Magn. Reson. **32** [1978] 355/6).
[15] Jameson, J. C.; De Dios, A.; Jameson, A. K. (J. Chem. Phys. **95** [1991] 9042/53).
[16] Wasylishen, R. E.; Burford, N. (Can. J. Chem. **65** [1987] 2707/12).
[17] Jameson, C. J.; Jameson, A. K.; Parker, H. (J. Chem. Phys. **68** [1978] 2868/72).
[18] Jameson, C. J.; Jameson, A. K. (J. Chem. Phys. **69** [1978] 615/21).
[19] Osten, H.-J.; Jameson, C. J. (J. Chem. Phys. **82** [1985] 4595/606).
[20] Cowley, A. H.; White, D. W. (J. Am. Chem. Soc. **91** [1969] 1917/21).

[21] Galasso, V. (Theor. Chim. Acta **63** [1983] 35/41).
[22] Guest, M. F.; Overill, R. E. (Chem. Phys. Lett. **73** [1980] 612/5).
[23] Lazzeretti, P.; Rossi, E.; Taddei, F.; Zanasi, R. (J. Chem. Phys. **77** [1982] 408/14).
[24] Beer, M. D.; Grinter, R. (J. Magn. Reson. **26** [1977] 421/3).
[25] Barbieri, G.; Benassi, R.; Taddei, F. (Gazz. Chim. Ital. **105** [1975] 807/26).
[26] Albright, T. A. (Org. Magn. Reson. **8** [1976] 489/99).
[27] Shustorovich, E. (Inorg. Chem. **18** [1979] 2108/11).
[28] Giessner-Prettre, C.; Pullman, B. (J. Theor. Biol. **72** [1978] 751/9).
[29] Manatt, S. L.; Cohen, E. A.; Cowley, A. H. (J. Am. Chem. Soc. **91** [1969] 5919/20).
[30] Kowalewski, J.; Laaksonen, A.; Saunders, V. R. (J. Chem. Phys. **74** [1981] 2412/5).

[31] Jameson, C. J.; Osten, H.-J. (J. Am. Chem. Soc. **108** [1986] 2497/503).
[32] Jameson, C. J.; Osten, H.-J. (J. Am. Chem. Soc. **107** [1985] 4158/61).
[33] Armstrong, R. L.; Courtney, J. A. (Can. J. Phys. **51** [1969] 457/8).
[34] Armstrong, R. L.; Courtney, J. A. (J. Chem. Phys. **50** [1972] 1262/72).
[35] Boden, N.; Folland, R. (Chem. Phys. Lett. **32** [1975] 127/32).
[36] Sawyer, D. W.; Powles, J. H. (Mol. Phys. **21** [1971] 83/95).
[37] Krynicki, K.; Sawyer, D. W.; Powles, J. G. (Magn. Reson. Relat. Phenom. Proc. 18th Congr. AMPERE, Nottingham, Engl., 1974 [1975], Vol. 2, pp. 511/2; C.A. **83** [1975] No. 197955).

1.3.1.4.3.2 Rotation Spectra

This section deals with the rotational transitions in the vibrational ground state of the phosphanes. Transitions in vibrationally excited states are treated in the section on rotation-vibration spectra on p. 188.

PH_3 and PD_3

Rotational transitions of PH_3 and PD_3 in the vibrational ground state were observed in a wide spectral region, ranging from 60 kHz to 1068 GHz [1 to 7] and from 92 MHz to 417 GHz [2, 3, 6], respectively. A survey on experimental investigations of the rotational spectra is given in Table 14.

Table 14
Rotational Measurements for PH_3 and PD_3 in the Vibrational Ground State.
The number of observed lines is denoted by n.

	transition	n	frequency range	Ref.
	$\Delta J=+1, \Delta K=0$			
PH_3	J=0, K=0; J=1, K=0, 1	3	266.94, 533.79, 533.81 GHz	[2]
	J=1, 2; K=0, 1, 2	5	533.79 to 800.58 GHz	[3]
	J=3; K=0 to 3	4	1066.84 to 1067.21 GHz	[4]
	$J\leq22$, $K\leq19$	240	30 to 200 cm^{-1}	[7]
PD_3	J=0 to 3; K=0, 1, 2	6	138.94 to 416.75 GHz	[2]
	$\Delta J=0, \Delta K=\pm3$			
PH_3	J=6 to 17, $K=\mp1\leftarrow\pm2$	12	47392 to 43671 MHz	[1]
	J=3 to 14; $K=0\leftarrow3$	12	143702 to 134641 MHz	[5]
	J=6 to 16; $K=5\leftarrow2$	11	333840 to 310755 MHz	[4]
	J=7 to 18; $K=3\leftarrow6$	23*)	429108 to 391232 MHz	[3]
PD_3	J=13 to 20; $K=\pm1\leftarrow\mp2$	8	30805 to 29939 MHz	[1]
	J=9 to 13; $K=0\leftarrow3$	5	93499 to 92425 MHz	[5]
	J=9 to 18; $K=\pm1\leftarrow\pm4$	10	151261 to 133958 MHz	[5]

*) Including K doubling.

The allowed transitions $\Delta J=+1$, $\Delta K=0$ (permanent dipole moment along the C_3 molecular axis) with J=1, 2, 3 and K=3 were observed in the radiofrequency region [2, 3, 4] (J and $K\leq J$ are the quantum numbers of the total angular momentum and its component parallel to the C_3 axis). Over 240 rotational transitions involving quantum numbers up to J=22 and K=19 in the far IR (30 to 200 cm^{-1}) were measured with high resolution (0.002 cm^{-1}) by Fourier transform spectroscopy. In the region of the $J=13\leftarrow12$ R-branch transitions, the K structure was shown to be completely resolved, i.e., for K=3 two well-resolved lines corresponding to transitions from the A_1 and A_2 components were observed. The K=3 doublets overlapped for $J\leq9$. The A_1 and A_2 components of the K=6, 9, ... transitions overlapped for all J values. The data were analyzed together with the MW (see below) and radiofrequency data, using different formulations of the rotational Hamiltonian which included $\Delta K=\pm3$ and/or $\Delta K=\pm6$ interaction terms (see next paragraph) [7].

In the MW region rotational transitions of PH_3 according to the selection rules $\Delta J = 0$, $\Delta K = \pm 3n$ (n = 1, 2, ...) were observed which become weakly allowed, because the C_{3v} symmetry of the PH_3 equilibrium structure breaks down by centrifugal distortion, thus inducing a small dipole moment (in the order of $8 \cdot 10^{-5}$ D) perpendicular to the C_3 axis [1]. A wide variety of the $\Delta J = 0$, $\Delta K = \pm 3$ transitions were assigned up to J = 18 and K = 6 for PH_3 [1, 3 to 5]. Frequencies and intensities of higher-order transitions $\Delta K = 6$ were predicted by a perturbation-theoretical approach [8].

Seven $\Delta J = 0$ (J = 3 to 9) transitions between the A_1 and A_2 components of the K = 3 levels of PH_3 between 62.25 kHz (J = 3) and 56955.0 kHz (J = 9) were identified by molecular-beam electric-resonance spectroscopy [6].

The pressure-induced **frequency shifts** $\Delta\nu$ of rotational lines in pure PH_3 were measured for the following $\Delta J = +1$, $\Delta K = 0$ transitions in the vibrational ground state [9] (and in the $v_2 = 1$ state [10]) of PH_3; for comparison with the shifts of other symmetric top molecules, see [11].

	vibrational ground state [9]						$v_2 = 1$ [10]
J, K	0, 0	1, 0	1, 1	2, 0	2, 1	2, 2	0, 0
$\Delta\nu$ (MHz/Torr)	0.560*)	0.147	0.275	0.042	0.038	0.245	0.560

*) $\Delta\nu = 0.56 \pm 0.08$ [13].

The following table shows the self-broadened **linewidths** γ° (in 10^{-3} $cm^{-1} \cdot atm^{-1}$) and **intensities** S° (in $cm^{-2} \cdot atm^{-1}$) measured at 296 K for the six rotational transitions $\Delta J = +1$ (J = 1 to 6) of PH_3 in the vibrational ground state in the spectral range 10 to 100 cm^{-1}. The PH_3 pressure was varied between 300 and 700 Torr [12]. The γ° value for the 1←0 transition was taken from [13].

J′←J″	1←0	2←1	3←2	4←3	5←4	6←5	7←6
γ°	142.5±0.5	123.1±6.1	118.3±3.6	113.8±2.1	112.7±2.2	98.6±0.6	95.8±2.6
S°	—	0.0400	0.1717	0.5405	1.074	1.737	2.48

The H_2- and He-broadened linewidths γ° for the transitions $\Delta J = +1$ with J = 0 [13] and 1 to 7 [14] and the intensities S° for five rotational lines (J = 1 to 5) have been measured at room temperature. Pressure-induced frequency shifts and line broadening by added noble gases (He, Ne, Ar, Xe) were reported [15].

PHD_2 and PH_2D

Q-branch transitions ($\Delta J = 0$) and some R-branch transitions ($\Delta J = +1$) were identified in the rotational spectra of PHD_2 (C_s symmetry; oblate asymmetric top) and PH_2D (C_s; prolate asymmetric top). Transitions, number of lines (n), and the corresponding frequency range are given in Table 15. The rotational levels are denoted by $J_{K_aK_c}$ (in the table J, K_a, K_c); K_a and K_c are the K values for the given level for the limiting cases of prolate and oblate symmetric top, respectively [16, 17].

Table 15

Rotational Measurements for PHD_2 and PH_2D in the Vibrational Ground State. The number of observed lines is denoted by n.

		transition	n	range (GHz)	Ref.
PHD_2	$\Delta J=0$:	$(J;\ K_a,\ K_a\pm1;\ K_c,\ K_c-1,\ K_c-2) \leftarrow (J;\ K_a;\ K_c)$ with $J=1$ to 14, $K_a-K_c=-2$ to $+6$	15	9 to 264	[17]
	$\Delta J=+1$:	$1_{10},\ 1_{11} \leftarrow 0_{00}$	2		
PH_2D	$\Delta J=0$:	$(J;\ K_a,\ K_a+1;\ K_c,\ K_c-1) \leftarrow (J;\ K_a;\ K_c)$ with $J=1$ to 7, $K_a-K_c=-7$ to $+1$	9	5 to 40	[16]
		$(J,\ K_a,\ K_a+1;\ K_c,\ K_c\pm1) \leftarrow (J;\ K_a;\ K_c)$ with $J=1$ to 19, $K_a-K_c=-13$ to 0	23	5 to 257	[17]
	$\Delta J=+1$:	$1_{01} \leftarrow 0_{00}$	1		

References:

[1] Chu, F. Y.; Oka, T. (J. Chem. Phys. **60** [1974] 4612/8).
[2] Helminger, P.; Gordy, W. (Phys. Rev. [2] **188** [1969] 100/8).
[3] Belov, S. P.; Burenin, A. V.; Polyansky, O. L.; Shapin, S. M. (J. Mol. Spectrosc. **90** [1981] 579/89).
[4] Belov, S. P.; Burenin, A. V.; Gershtein, L. I.; Krupnov, A. F.; Markov, V. N.; Maslovsky, A. V.; Shapin, S. M. (J. Mol. Spectrosc. **86** [1981] 184/92).
[5] Helms, D. A.; Gordy, W. (J. Mol. Spectrosc. **66** [1977] 206/18).
[6] Davies, P. B.; Neumann, R. M.; Wofsy, S. C.; Klemperer, W. (J. Chem. Phys. **55** [1971] 3564/8).
[7] Fusina, L.; Carlotti, M. (J. Mol. Spectrosc. **130** [1988] 371/81).
[8] Ghoshal, S.; Ghosh, P. N. (J. Mol. Spectrosc. **110** [1985] 364/8).
[9] Belov, S. P.; Kazakov, V. P.; Krupnov, A. F.; Mel'nikov, A. A.; Skvortsov, V. A. (Izv. Vyssh. Uchebn. Zaved. Radiofiz. **25** [1982] 118/21; C. A. **97** [1982] No. 30547).
[10] Andreev, B. A.; Belov, S. P.; Burenin, A. V.; Gershtein, L. I.; Krupnov, A. F.; Maslovskii, A. V.; Shchapin, S. M. (Opt. Spektrosk. **44** [1978] 620/1; Opt. Spectrosc. [Engl. Transl.] **44** [1978] 363/4).

[11] Krupnov, A. F.; Belov, S. P. (Izv. Vyssh. Uchebn. Zaved. Radiofiz. **22** [1979] 901/2; Radiophys. Quantum Electron. [Engl. Transl.] **22** [1979] 628/9).
[12] Nguyen-Van-Thanh; Rossi, I.; Sergent-Rozey, M. (J. Mol. Spectrosc. **135** [1989] 410/4).
[13] Pickett, H. M.; Poynter, R. L.; Cohen, E. A. (J. Quant. Spectrosc. Radiat. Transfer **26** [1981] 197/8).
[14] Sergent-Rozey, M.; Nguyen-Van-Thanh; Rossi, I.; Lacome, N.; Levy, A. (J. Mol. Spectrosc. **131** [1988] 66/76, **133** [1989] 475 [Erratum]).
[15] Belov, S. P.; Kazakov, V. P.; Krupnov, A. F.; Markov, V. N.; Mel'nikov, A. A.; Skvortsov, V. A.; Tret'yakov, M. Yu. (J. Mol. Spectrosc. **94** [1982] 264/82).
[16] Kukolich, S.; Schaum, L.; Murray, A. (J. Mol. Spectrosc. **94** [1982] 393/8).
[17] McRae, G. A.; Gerry, M. C. L.; Cohen, E. A. (J. Mol. Spectrosc. **116** [1986] 58/70).

1.3.1.4.3.3 Rotation-Vibration Spectra

1.3.1.4.3.3.1 Gaseous PH_3

General

The rotation-vibration spectra of the symmetric-top molecule PH_3 in the gas phase show a very complex structure. The spectra are dominated by strong Coriolis coupling due to the near degeneracy, for example, of the pair of fundamentals ν_2 and ν_4 or of the group of bands ν_1, ν_3, $2\nu_2$, $\nu_2+\nu_4$, and $2\nu_4$. Several rotation-vibration interactions (vibrational l-type doubling, rotational K-type doubling, or rotational l-type doubling) lead to A_1-A_2 doublet splitting of vibrational-rotational levels, corresponding to the same rotational quantum number K. In the nondegenerate vibrational levels (e.g. $v_2=1$), the A_1-A_2 splitting occurs for rotational levels with $K=3n$ ($n=1, 2, \ldots$). In the doubly degenerate vibrational levels (e.g. $v_4=1$), the splitting occurs for $K=1, 4, 7 \ldots$ in the $+l$ level and for $K=2, 5, 8 \ldots$ in the $-l$ level; see for example [1, 2].

The Bands ν_2 and ν_4

Absorption Spectrum. The composite absorption spectrum of PH_3 in the wavelength region 9 to 10 μm arises from the fundamentals ν_2 and ν_4 and from the hot band $2\nu_2-\nu_2$ (see below). The first more rigorous analysis [3] identified three groups of absorptions: a) the range 880 to 980 cm^{-1} which shows the P branch of the ν_2 fundamental, b) the range 1150 to 1250 cm^{-1} revealing the clearly defined $^RR(J,K)$ transitions of the ν_4 band, and c) the overlapping ν_2 R-branch lines and ν_4 $^PP(J,K)$ lines in the intermediate range 980 to 1150 cm^{-1}. Rotational K-type and l-type doubling were observed [3]. Further exhaustive spectroscopic studies on both bands [4, 5], which included all transitions involving the upper-state values $J' \leq 20$, led to the identification of 1318 transitions between 818 and 1340 cm^{-1} (resolution ≤ 0.005 cm^{-1}). A refined treatment of the Coriolis coupling between ν_2 and ν_4 allowed the interpretation of the rotational structure up to high J′ values, including the large A_1-A_2 splittings observed by [6]. The wave numbers of the P-, Q-, and R-branch lines of the ν_2 band were measured by Fourier transform spectroscopy with high resolution (0.0045 cm^{-1}) [7] together with lines precisely measured with sub-Doppler saturation spectroscopy using an off-set-locked wave-guide CO_2 laser system [8] and by IR-rf double resonance [9] and are compiled in a table [7].

Twenty-four resonances of rovibrational transitions were identified in the laser Stark spectra of the ν_2 and ν_4 bands of PH_3 with CO_2 and N_2O lasers (Stark field and beam polarization parallel to each other, $\Delta M=0$) [10]. For determining the dipole moments in the $v_2=1$ and $v_4=1$ states, seven transitions of the ν_2 band and two transitions of the ν_4 band were identified by Stark measurements with $^{12}CO_2$ and $^{13}CO_2$ lasers with perpendicular polarization ($\Delta M=\pm1$) [11]. Extended laser Stark spectroscopic studies of the ν_2 and ν_4 bands were performed using all available $^{12}C^{16}O_2$, $^{13}C^{16}O_2$, $^{12}C^{18}O_2$, and N_2O laser lines with Stark fields up to 60 kV/cm. With the sub-Doppler resolution of the Lamb-dip technique, about 500 Stark resonances have been identified for 44 lines of the ν_2 band and 31 lines of the ν_4 band [13]. Two Doppler-free lines with signs opposite to those of Lamb dips were observed in a study of the ν_2 $^QP(5,3)$ transition by laser Stark spectroscopy using the 10P(18) CO_2 laser line. They were identified as infrared-infrared double resonance transitions, caused by accidental overlapping of the $^QP(5,3)$ line in the ν_2 band with the $^QP(4,3)$ line in the hot band $2\nu_2-\nu_2$ [14 to 16].

The hyperfine structure of a PH_3 rotational line, ν_2 P(7,0), was resolved with frequency-referenced saturation spectroscopy using a single CO_2 laser [22].

The **A_1-A_2 splittings** of several rovibrational levels were resolved in the ν_2 and ν_4 bands. The splitting in the $K=3$ level of the $v_2=1$ vibrational state was spectroscopically determined for J=5 to 15 [1, 6, 12, 13, 17, 21]. In the v_4=1 level the A_1-A_2 splittings of the +l, K=1 level were measured for J=1 to 4 [18, 19, 20], 1 to 17 [6]. Similarly, the splitting of the levels −l, K=2 with J=2 to 15 [1, 6, 13, 21], +l, K=4 with J=4 to 14 [1, 13, 21], and +l, K=7 with J=9 to 15, 17 [1] were reported. The deviation of the K dependence observed for the A_1-A_2 splitting from the relation (J+K)!/(J−K)! (expected within the framework of the perturbation theory) was interpreted by taking Coriolis coupling between v_2 and v_4 into account [1, 17].

Absolute **intensities** of 57 rovibrational lines in the v_2 and v_4 band were measured, and the transition dipole moments were evaluated [23].

Consult p. 186 regarding the pressure-induced frequency shift of rotational lines in the v_2=1 state.

The **multiphoton absorption** of pulsed CO_2 laser radiation was studied in the region of the v_2 and v_4 fundamentals (900 to 1100 cm^{-1}) of PH_3. At a constant phosphane pressure, the fraction of absorbed energy for lines belonging to the P and R branches of the 00^01-02^00 laser transition is higher than for those arising from the 00^01-10^00 transition (920 to 990 cm^{-1}) due to the higher density of available v_2 and v_4 rovibrational states in this region (1030 to 1080 cm^{-1}). The absorption in the whole region is characterized by low absorption cross sections between 2×10^{-20} and 4×10^{-20} $cm^2\cdot molecule^{-1}$ (50 Torr PH_3, 298 K). For a fixed laser frequency and constant incident intensity, the pressure dependence of the optical density (up to 160 Torr PH_3) can be approximated by an empirically modified Lambert-Beer law. The effect of an added nonabsorbing gas on the absorption coefficient was investigated [24].

Doppler-free two-photon absorption was observed in a study of the v_2 $^QP(5,3)$ transition by laser Stark spectroscopy using the 10 P(18) CO_2 laser line. It is caused by accidental overlapping of the $^QP(5,3)$ line in the v_2 band with the $^QP(4,3)$ line in the hot band $2v_2-v_2$ [14, 16].

Emission Spectrum. Laser emission at 935 cm^{-1} was observed in PH_3, optically pumped by a pulsed CO_2 laser (9 P(36) line) at 125 K and 0.6 Torr [25]. Optical pumping of the v_2=1 and v_4=1 levels with a pulsed CO_2 laser yielded 44 FIR lasing lines in the 83- to 223-µm region [26, 27]; see also [28].

The Hot Band $2v_2-v_2$

About 70 transitions of the hot band $2v_2-v_2$ with a band center near 980.4 cm^{-1} were identified in the PH_3 absorption spectrum in the range 9 to 10 µm (but without assigning the individual rovibrational transitions) [6].

A systematic study of the band revealed a number of weak Stark resonances which were assigned to 15 P-branch transitions (from P(2,1) at 963.22680 to P(7,6) at 920.66720 cm^{-1}) and to 4 R-branch transitions (from R(4,3) at 1020.6152 to R(5,5) at 1030.02609 cm^{-1}). The band origin was determined to be v_0=980.4389(18) [29].

See p. 188 on laser Stark spectroscopy of the accidental overlapping lines $^QP(4,3)$ of the hot band and $^QP(5,3)$ of the v_2 band.

The Bands v_1, v_3, $2v_2$, v_2+v_4, and $2v_4$

Five vibrational bands were observed in the **IR absorption** spectrum of PH_3 between 4 to 5 µm, the fundamentals $v_1(A_1)$ and $v_3(E)$ at 2321 and 2326 cm^{-1} and the three overtone/combination bands $2v_2$ (A_1), v_2+v_4 (E), $2v_4$ (A_1+E) at 1973, 2108, and 2227 cm^{-1}, respectively. Since the 4- to 5-µm region is a window in the Jovian and Saturnian atmospheres, spectroscopic monitoring of PH_3 allows the investigation of the vertical circulation and the chemistry in the deep atmosphere of these planets [30].

About 1200 lines of the fundamentals ν_1 and ν_3 were assigned in the range 2087 to 2482 cm^{-1}. Strong Coriolis coupling between both vibrations gives rise to a complicated rotational structure. The simultaneous analysis of the band pair (dyad) allowed the prediction of line positions and intensities in both bands with some reliability up to $J'=11$. At higher rotational quantum numbers, Fermi-type interaction with the neighboring $2\nu_4$ band has to be considered [31].

The lowest lying overtone ($2\nu_2$) and the combination ($\nu_2+\nu_4$) bands were recorded by Fourier transform spectroscopy. About 400 transitions in both bands were fitted with inclusion of the Coriolis coupling parameters. The band origin is $\nu_0=1972.5454$ (124) and 2108.0458 (79) cm^{-1} for the $2\nu_2$ and $\nu_2+\nu_4$ band, respectively [32]. The analysis of the $2\nu_2-\nu_2$ band gave $\nu_0=1972.57043(86)$ cm^{-1} as the origin of the ν_2 band [29]. The simultaneous analysis of the triad $2\nu_2/(\nu_2+\nu_4)/2\nu_4$, taking into account the strong Coriolis coupling between the ν_2 and ν_4 vibrational modes (see [6]), led to the assignment of most of the lines up to $J'=12$ in the $2\nu_2$ and $\nu_2+\nu_4$ bands, but failed to describe properly the rotational structure in the $2\nu_4$ band [33 to 35].

For a simultaneous analysis of the bands of the pentad $\nu_1/\nu_3/2\nu_2/(\nu_2+\nu_4)/2\nu_4$ recorded by Fourier transform spectroscopy (resolution 0.0054 cm^{-1}), 3766 relatively unblended lines out of more than 4400 transitions ($J'\leq16$) assigned in the range 1885 to 2445 cm^{-1} were selected. About 40% of the 4400 lines are perturbation-allowed transitions, getting their intensities from large vibration-rotation couplings. In addition to considering the Coriolis coupling, the Fermi coupling between the ν_1/ν_3 dyad and the $2\nu_2/(\nu_2+\nu_4)/2\nu_4$ triad was taken into account [30, 36].

Absolute line strengths and line positions for over 200 vibration-rotation transitions of PH_3 in the range 2154 to 2210 cm^{-1} were measured at 295 K with a tunable diode laser spectrometer [37]. Absolute intensities of about 1600 lines mainly of the $2\nu_2$, $\nu_2+\nu_4$ bands were measured by Fourier transform spectroscopy, and absolute band strengths were derived [30]. Four PH_3 lines in the ν_1 and ν_3 bands near 2190 cm^{-1} were pressure-broadened with H_2 gas. The pressure-broadening coefficients, which agree with the value [38] for H_2 broadening of the PH_3 pure rotational line ($J=1\leftarrow0$, $K=0$), were determined [37]. Reasonably agreeing results gave a high-resolution Fourier transform spectroscopic investigation of the pressure broadening of PH_3 by hydrogen and helium at room temperature. About 600 lines were measured, mainly in the weak transitions $\nu_2+\nu_4$, $2\nu_2$, and $2\nu_4$. The hydrogen- and helium-broadened spectra showed that for a given value of J the line widths are nearly K-independent except for K values approaching J, for which a noticeable decrease of the line width was observed [39].

The **Raman** spectrum between 2220 and 2380 cm^{-1} showed a sharp peak at about 2315 cm^{-1} (taken from the figure in the paper). The effect of temperature on the spectrum upon heating the sample from 25 to 630°C (decomposition of PH_3) was investigated [40].

The Bands $3\nu_2$, $4\nu_2$, and $4\nu_2-\nu_2$

A frequency-tunable infrared laser system, combined with Stark modulation to increase the detection sensitivity, was applied to observe the $3\nu_2$ band and the $4\nu_2-\nu_2$ hot band (recorded at about 100°C) in the 3.4-µm region and the $4\nu_2$ band in the 2.6-µm region (resolution about 0.007 cm^{-1}). The $3\nu_2$ band showed A_1-A_2 splittings for the $K'=3n$ upper-state rotational levels up to $n=3$ ($J=6$ to 16), because PH_3 is a very nearly spherical top in the $\nu_2=3$ state. The $K'=3$ splitting (for $J=5$ to 8) in the hot band was also resolved [41].

An earlier investigation (resolution 0.03 cm^{-1}) yielded the band origins $\nu_0=2940.77$ and 2903.88 cm^{-1} for the $3\nu_2$ and $4\nu_2-\nu_2$ bands, respectively. A_1-A_2 splitting of the $K=3$ levels was observed [42].

References:

[1] Papoušek, D.; Birk, H.; Magg, U.; Jones, H. (J. Mol. Spectrosc. **135** [1989] 105/18).
[2] Papoušek, D.; Aliev, M. R. (Stud. Phys. Theor. Chem. **17** [1982] 1/323, 171/87).
[3] Hoffmann, J. M.; Nielsen, H. H.; Narahari Rao, K. (Z. Elektrochem. **64** [1960] 606/16).
[4] Yin, P. K. L.; Narahari Rao, K. (J. Mol. Spectrosc. **51** [1974] 199/207).
[5] Goldman, A.; Cook, G. R.; Bonomo, F. S. (J. Quant. Spectrosc. Radiat. Transfer **24** [1980] 211/8).
[6] Tarrago, G.; Dang-Nhu, M.; Goldman, A. (J. Mol. Spectrosc. **88** [1981] 311/22).
[7] Přádná, S.; Papoušek, D.; Kauppinen, J.; Belov, S. P.; Krupnov, A. F.; Scappini, F.; Di Lonardo, G. (Collect. Czech. Chem. Commun. **50** [1985] 2480/92).
[8] Crocker, D.; Butcher, R. J. (Int. J. Infrared Millimeter Waves **3** [1982] 409/15).
[9] Tanimura, S.; Takagi, K. (J. Mol. Spectrosc. **104** [1984] 414/6).
[10] Shimizu, F. (J. Phys. Soc. Jpn. **38** [1975] 293).

[11] Di Lonardo, G.; Trombetti, A. (Chem. Phys. Lett. **76** [1980] 307/10).
[12] Carlotti, M.; Di Lonardo, G.; Trombetti, A. (J. Chem. Phys. **78** [1983] 1670/2).
[13] Takagi, K.; Itoh, K.; Miura, E.; Tanimura, S. (J. Opt. Soc. Am. B **4** [1987] 1445/57).
[14] Takagi, K. (Chem. Phys. Lett. **112** [1984] 302/5).
[15] Tanaka, K.; Harada, K.; Tanaka, T.; Takagi, K. (Chem. Phys. Lett. **119** [1985] 447/50).
[16] Takagi, K.; Tanaka, T.; Tanaka, K. (Springer Ser. Opt. Sci. **49** [1985] 157/8).
[17] Chen, Y.-T.; Oka, T. (J. Mol. Spectrosc. **133** [1989] 148/56).
[18] Scappini, F.; Schwarz, R. (Chem. Phys. Lett. **80** [1981] 350/1).
[19] Guarnieri, A.; Scappini, F.; Di Lonardo, G. (Chem. Phys. Lett. **82** [1981] 321/2).
[20] Belov, S. P.; Krupnov, A. F.; Papoušek, D.; Urban, S.; Gazzoli, G. (J. Mol. Spectrosc. **98** [1983] 265/8).

[21] Tawa, F.; Morimoto, K.; Nagasaki, H.; Matsushima, F.; Takagi, K. (J. Mol. Spectrosc. **145** [1991] 192/9).
[22] Chardonnet, C.; Butcher, R. J.; Höhe, E.; Charton, G. (Opt. Commun. **87** [1992] 233/7).
[23] Kshirsagar, R. J.; Singh, K.; D'Cunha, R.; Job, V. A.; Papoušek, D.; Ogilvie, J. F.; Fusina, L. (J. Mol. Spectrosc. **149** [1991] 152/9).
[24] Blazejowski, J.; Lampe, F. W. (J. Photochem. **29** [1985] 285/95).
[25] Rutt, H. N. (J. Phys. B At. Mol. Phys. **16** [1983] 3667/72).
[26] Malk, E. G.; Niesen, J. W.; Parsons, D. F.; Coleman, P. D. (IEEE J. Quantum Electron. **QE-14** [1978] 544/50).
[27] Coleman, P. D. (Rev. Infrared Millimeter Waves **2** [1984] 383/427).
[28] Shafik, S.; Crocker, D.; Landsberg, B. M.; Butcher, R. J. (IEEE J. Quantum Electron. **QE-17** [1981] 115/6).
[29] Hashinami, S.; Mito, H.; Matushima, F.; Takagi, K. (J. Mol. Spectrosc. **132** [1988] 1/12).
[30] Tarrago, G.; Lacome, N.; Lévy, A.; Guelachvili, G.; Bézard, B.; Drossart, P. (J. Mol. Spectrosc. **154** [1992] 30/42).

[31] Baldacci, A.; Devi, V. M.; Narahari Rao, K.; Tarrago, G. (J. Mol. Spectrosc. **81** [1980] 179/206).
[32] Tipton, T.; Choe, J.-I.; Kukolich, S. G. (J. Phys. Chem. **90** [1986] 1534/7).
[33] Tarrago, G.; Guelachvili, G.; Cadot, C. (Proc. 7th Int. Conf. High Resolut. Infrared Spectrosc., Prague 1982 from [30]).
[34] Tarrago, G.; Lacome, N.; Poussigue, G.; Guelachvili, G. (Proc. 42nd Symp. Mol. Spectrosc., Columbus, Ohio, 1987 from [30]).
[35] Lacome, N.; Lévy, G.; Tarrago, G.; Poussigue, G.; Guelachvili, G. (Proc. Atmos. Spectrosc. Appl. Workshop, Oxford 1987 from [30]).
[36] Tarrago, G. (J. Mol. Spectrosc. **139** [1990] 439/45).

[37] Lovejoy, R. W.; Schaeffer, R. D.; Frasco, D. L.; Chakerian, C. C.; Boese, R. W. (J. Mol. Spectrosc. **109** [1985] 246/55).
[38] Pickett, H. M.; Poynter, R. L.; Cohen, E. A. (J. Quant. Spectrosc. Radiat. Transfer **26** [1981] 197/8).
[39] Levy, A.; Lacome, N.; Tarrago, G. (J. Mol. Spectrosc. **157** [1993] 172/81).
[40] Abraham, P.; Bekkaoui, A.; Soulière, V.; Bouix, J.; Monteil, J. (J. Cryst. Growth **107** [1991] 26/31).

[41] Bernard, P.; Oka, T. (J. Mol. Spectrosc. **75** [1979] 181/96).
[42] Maki, A. G.; Sams, R. L.; Olson, W. B. (J. Chem. Phys. **58** [1973] 4502/12).

1.3.1.4.3.3.2 Gaseous PD_3

The fundamental bands at 1682 (ν_1), 728 (ν_2), 1693 (ν_3), and 803 (ν_4) cm^{-1} have been recorded by Fourier transform spectroscopy at a resolution of 0.06 cm^{-1}. A rotational analysis was performed, explicitly including the $\nu_1-\nu_3$ and $\nu_2-\nu_4$ Coriolis interactions. A perturbation was found in the ν_4 band and interpreted being due to a $\Delta K=3$-type interaction [1].

Three infrared-infrared double resonance signals were observed in the Stark spectrum of PD_3 for the v=15−14, P(20) line of the CO laser. They are caused by the accidentally overlapped $^RR_1(2)$ transition in the ν_3 band and the $^RR_2(3)$ transition in the $2\nu_3-\nu_3$ band [2].

References:

[1] Kijima, K.; Tanaka, T. (J. Mol. Spectrosc. **89** [1981] 62/75).
[2] Tanaka, K.; Ito, H.; Tanaka, T. (J. Mol. Spectrosc. **115** [1986] 383/92).

1.3.1.4.3.3.3 PH_3 and Isotopomers in Condensed Phases

In Ar Matrices

The IR absorption spectra of PH_3 and isotopomers in solid Ar matrices were recorded at 12 K [1, 2] (see Table 16).

Table 16
IR Absorption Bands ν_i (in cm^{-1}) of PH_3 and Isotopomers in Solid Argon.

species	$\nu_1(A_1)$	$\nu_2(A_1)$	$\nu_3(E)$	$\nu_4(E)$	Ref.
PH_3	2340	994	2345	1114	[1, 2]
PH_2D	—$^{1)}$	895	—$^{1)}$ 1701	1097$^{2)}$ 971	[1]
PHD_2	—$^{1)}$	768	1705$^{2)}$ 1698	979$^{2)}$ 914	[1]
PD_3	1698	729	1705	804	[1, 2]

$^{1)}$ Not resolved from PH_3 or PD_3. – $^{2)}$ Two bands observed for removed degeneracy in the mixed isotopic molecules.

In Liquid and Solid Phases

IR and Raman spectra of PH_3 and PD_3 films (see Table 17, p. 194) were recorded, and their splitting pattern was analyzed in terms of the phosphane modifications being present; see the chapter on the crystal structure on p. 176. The films were prepared by distilling phosphane onto a substrate precooled to the desired temperature [3 to 7]. Deposition of PH_3 at 40 K [4] or $\leqq$35 K [3] resulted in the formation of amorphous films, indicated by the broad and structureless appearance of the ν_2 band. One or more annealing cycles yielded the crystalline phase [3, 4]. Deposition between 65 and 74 K gave films that revealed the fine structure of the δ phase on first cooling [4].

The results of IR investigations [5], already reported in "Phosphor" C, 1965, p. 22, were questioned because of incorrect temperature measurement [3, 4]. IR frequencies of the ν_2 and ν_4 fundamentals were measured in dilute solutions of PD_3, PH_2D, and PHD_2 in γ-PH_3 and of PH_3 and PH_2D in γ-PD_3. Furthermore, 12 lattice vibrations (41.0 to 136.5 cm^{-1}) in the Raman spectrum of γ-PH_3 at 19 K, 14 lattice vibrations (40.3 to 136.5 cm^{-1}) in the Raman spectrum of γ-PD_3 at 18 K, and 8 lattice vibrations (53.5 to 135 cm^{-1}) in the IR spectrum of γ-PD_3 at 24 K were observed [4].

The molecular dynamics of phosphane was investigated via vibrational dephasing in liquid and crystalline phases. Vibrational correlation functions were obtained from the Raman scattering spectra of the P–H stretching vibration of liquid and solid PH_3 [8, 9] and PHD_2, present as an isotopic impurity in liquid and solid PD_3 [10]. Raman scattering data for the P–D stretching vibration in liquid and solid PD_3 are given in [11]. For details of the vibrational dephasing analysis, see [8 to 12].

References:

[1] Withnall, R.; Hawkins, M.; Andrews, L. (J. Phys. Chem. **90** [1986] 575/9).
[2] Arlinghaus, R. T.; Andrews, L. (J. Chem. Phys. **81** [1984] 4341/51).
[3] Francia, M. D.; Nixon, E. R. (J. Chem. Phys. **58** [1973] 1061/5).
[4] Huang, T.-H.; Decius, J. C.; Nibler, J. W. (J. Phys. Chem. Solids **38** [1977] 899/904).
[5] Hardin, A. H.; Harvey, K. B. (Can. J. Chem. **42** [1964] 84/9).
[6] Heinemann, A. (Ber. Bunsen-Ges. Phys. Chem. **68** [1964] 280/6).
[7] Wilde, R. E.; Covington, B. C. (J. Phys. Chem. Solids **36** [1975] 1225/7).
[8] Schwartz, M.; Wang, C. H. (Chem. Phys. Lett. **25** [1974] 26/33).
[9] Wilde, R. E.; Chang, T.-C. (J. Chem. Phys. **72** [1980] 1293/6).
[10] Wilde, R. E. (Chem. Phys. Lett. **67** [1979] 555/7).

[11] Wilde, R. E.; Cohen, S. S. (J. Chem. Phys. **70** [1979] 4557/61).
[12] Wilde, R. E. (Chem. Phys. Lett. **106** [1984] 166/9).

1.3.1.4.3.4 Photoabsorption

Excitation and ionization energies derived from photoabsorption (PA) or dipole (e,e) spectroscopy (simulating PA) and also from multiphoton ionization (MPI) work (which is not accounted for in the following) are reported in the chapters on electronically excited states (p. 144) and on ionization potentials (p. 149). The absolute photoabsorption oscillator strength df/dE and cross section σ, related via σ(in Mb) = 109.75 df/dE(in eV^{-1}), were obtained from **gas-phase** dipole (e,e) spectroscopy [1, 2] and are shown for a large range of photon energies,

Table 17

IR and Raman Spectra of Solid Phosphanes (β, δ, γ Modifications) in cm^{-1}.

mode	PH_3(β) IR at 82 K [6]	PH_3(β) IR at 35 K [3]	PH_3(δ) IR at 20 K [3]	PH_3(γ) Raman at 19 K [4]	PH_3(γ) IR at 13 K [4]	PD_3(β) IR at 35 K [3]	PD_3(δ) IR at 20 K [3]	PD_3(γ) Raman at 18 K [4]	PD_3(γ) IR at 13 K [4]
ν_2	979	980.3	970.5 to 985.3 (9 lines)	970 to 981 (3 lines)	974.1, 988	719.7	713.1 to 724.0 (6 lines)	713.1 to 721.2 (3 lines)	715.7 to 728.7 (3 lines)
ν_4	1098	1095.4	1093.7	1088 to 1106 (5 lines)	1093.8 to 1103 (4 lines)	787.0	789.0	783 to 795.6 (4 lines)	787.4 to 795.1 (3 lines)
ν_1	2268	2303.6	2298.9 to 2306.6 (7 lines)	2301.5 to 2322.6 (8 lines)	2304.2 to 2319.5 (6 lines)	1618.1	1666.4 to 1673.3 (6 lines)	1668 to 1690.5 (8 lines)	1670.6 to 1688.4 (6 lines)
ν_3	2370	2311.3	2308.2 to 2326.2 (12 lines)			1680.4	1679.8 to 1693.4 (8 lines)		

E = 5 to 220 eV (or of wavelengths λ = 250 to 6 nm), in **Fig. 9** due to [2]. The original papers give also tables of the oscillator strength df/dE for total photoabsorption, total photoionization (both are equal above 19 eV), and ionic fragmentation (i.e., for singly charged ions and PH^{2+}) at energies E = 6.0, 6.5, 7.0, ..., 130.0 eV [1] and for total, valence, P2p, and P2s photoabsorption at E = 120.0, 121.0, ..., 211.0 eV (valence-shell data above 130 eV and P2p data above 183 eV were extrapolated) [2]. Further gas-phase photoabsorption work is mentioned below and classified by the types of absorption quantities measured (σ, molar extinction coefficient ε, and others) and the energy (or wavelength) regions covered:

Graphs of σ in the vacuum-UV region are given for λ = 215 to 125 nm (at 155 K), λ = 125 to 113 nm (at 295 K) [3], and λ = 124 to 106 nm (at normal temperature; also photoionization cross section σ_i) [4]. The ratio $\sigma(PH_3)/\sigma(SiH_4)$ and $\sigma(SiH_4)$ are tabulated for λ = 200, 199, ..., 190 nm [5]. Graphs of σ in the soft X-ray region are given for E = 120 to 220 eV (enlarged in the range 130 to 139 eV) [6] and E = 130 to 280 eV [7]. Reproductions of the range 130 to 139 eV [6] are in [8, 9]. σ is also given in arbitrary units for E = 125 to 200 eV [10].

Graphs of ε (in $L \cdot mol^{-1} \cdot cm^{-1}$) are available for λ = 240 to 185 nm (42000 to 54000 cm^{-1}) [11], 240 to 200 nm [12], and 216 to 184 nm [13]. Explicit values (at normal temperature) are ε = 3400 ± 200 at 191 nm [13], 3390 ± 140 at 190 nm [14], and 4800 at 185 nm [11]. At 700°C: ε = 1100 at 203 nm and 3660 at 185 nm [15]. An absorption coefficient (in $atm^{-1} \cdot cm^{-1}$) is given for λ = 240 to 195 nm [16], and the relative absorption was measured at 0.1 Torr and a cell length of 4 cm between 230 and 120 nm [17].

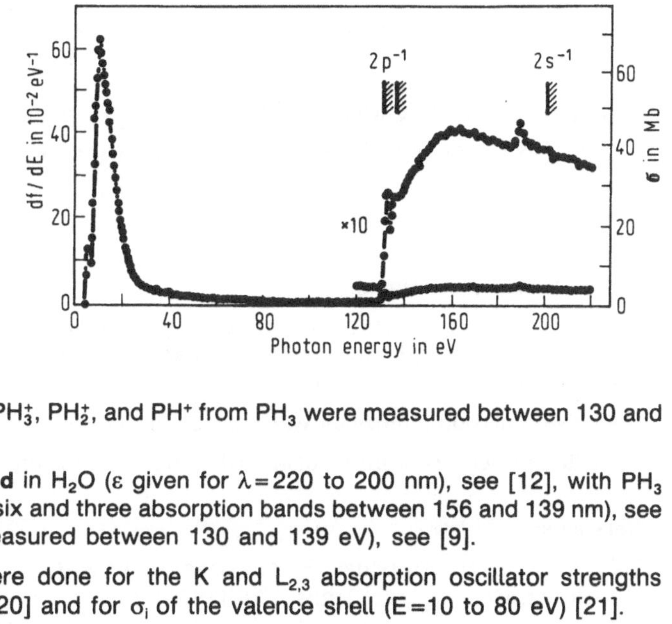

Fig. 9. Absolute photoabsorption oscillator strength and photoabsorption cross section for PH_3, covering the valence-shell and the P2p and P2s regions. The data for the inner-shell region were also increased by a factor of 10.

Relative photoion yields of PH_3^+, PH_2^+, and PH^+ from PH_3 were measured between 130 and 55 nm [18].

For work with PH_3 **dissolved** in H_2O (ε given for λ = 220 to 200 nm), see [12], with PH_3 **isolated** in Ar and Kr **matrices** (six and three absorption bands between 156 and 139 nm), see [19], and with **solid** PH_3 (σ measured between 130 and 139 eV), see [9].

Theoretical calculations were done for the K and $L_{2,3}$ absorption oscillator strengths (E around 2000 and 120 eV) [20] and for σ_i of the valence shell (E = 10 to 80 eV) [21].

References:

[1] Zarate, E. B.; Cooper, G.; Brion, C. E. (Chem. Phys. **148** [1990] 277/88, 279).
[2] Zarate, E. B.; Cooper, G.; Brion, C. E. (Chem. Phys. **148** [1990] 289/97).
[3] Xia, T. J.; Wu, C. Y. R.; Judge, D. L. (Phys. Scr. **41** [1990] 870/3).
[4] Xia, T. J.; Chien, T. S.; Wu, C. Y. R.; Judge, D. L. (J. Quant. Spectrosc. Radiat. Transfer **45** [1991] 77/91, 83/4).

[5] Clark, J. H.; Anderson, R. G. (Appl. Phys. Lett. **32** [1978] 46/9).
[6] Hayes, W.; Brown, F. C. (Phys. Rev. [3] A **6** [1972] 21/30, 27).
[7] Ishiguro, E.; Iwata, S.; Mikuni, A.; Suzuki, Y.; Kanamori, H.; Sasaki, T. (J. Phys. B **20** [1987] 4725/39, 4727).
[8] Robin, M. B. (Higher Excited States of Polyatomic Molecules, Vol. I, Academic, New York 1974, pp. 231/6).
[9] Friedrich, H.; Sonntag, B.; Raabe, P.; Butscher, W.; Schwarz, W. H. E. (Chem. Phys. Lett. **64** [1979] 360/6).
[10] Liu, Z. F.; Cutler, J. N.; Bancroft, G. M.; Tan, K. H.; Cavell, R. G.; Tse, J. S. (Chem. Phys. Lett. **172** [1990] 421/9).

[11] Halpern, A. M.; Ramachandran, B. R. (J. Mol. Spectrosc. **151** [1992] 26/32).
[12] Stevenson, D. P.; Coppinger, G. M.; Forbes, J. W. (J. Am. Chem. Soc. **83** [1961] 4350/2).
[13] Halmann, M. (J. Chem. Soc. **1963** 2853/6).
[14] Honma, N.; Sugaya, I.; Takami, K. (Jpn. J. Appl. Phys. **19** [1980] 779/80).
[15] Karlicek, R. F.; Hammarlund, B.; Ginocchio, J. (J. Appl. Phys. **60** [1986] 794/9).
[16] Kley, D.; Welge, K. H. (Z. Naturforsch. **20a** [1965] 124/31).
[17] Di Stefano, G.; Lenzi, M.; Margani, A.; Mele, A.; Nguyen Xuan, C. (J. Photochem. **7** [1977] 335/44, 339).
[18] Berkowitz, J.; Curtiss, L. A.; Gibson, S. T.; Greene, J. P.; Hillhouse, G. L.; Pople, J. A. (J. Chem. Phys. **84** [1986] 375/84, 376).
[19] Vodar, B. (AD-706847 [1970] 1/33, 12/3; C. A. **73** [1970] No. 103856).
[20] Sukhorukov, V. L.; Yavna, V. A.; Demekhin, V. F. (Izv. Akad. Nauk SSSR Ser. Fiz. **46** [1982] 763/9; Bull. Acad. Sci. USSR Phys. Ser. [Engl. Transl.] **46** No. 4 [1982] 131/6).

[21] Lavrent'ev, S. V.; Vasileva, M. E.; Petrov, I. D.; Sukhorukov, V. L. (Opt. Spektrosk. **69** [1990] 307/12; Opt. Spectrosc. [Engl. Transl.] **69** [1990] 186/9).

1.3.1.4.3.5 X-Ray Emission

The P Kα_1 line and the P Kβ band of frozen PH_3 on an Ir support (at liquid nitrogen temperature) were excited by Ag L radiation [1]. The use of th P Kβ band for identification purposes was investigated in [2].

Relative emission intensities were calculated in [3].

Satellite structures were calculated for the P Kβ [4] and P $L_{2,3}$ [4, 5] emissions.

References:

[1] Dolenko, G. N.; Krupoder, S. A.; Mazalov, L. N. (Zh. Strukt. Khim. **20** [1979] 334/6; J. Struct. Chem. [Engl. Transl.] **20** [1979] 279/81).
[2] Treiger, B. A.; Mazalov, L. N.; Dolenko, G. N.; Guzhavina, T. I. (Comput. Enhanced Spectrosc. **3** [1986] 153/63, 159).
[3] Demekhina, L. A.; Sukhorukov, V. L.; Demekhin, V. F.; Yavna, V. A. (Opt. Spektrosk. **49** [1980] 861/6; Opt. Spectrosc. [Engl. Transl.] **49** [1980] 470/3).
[4] Timonova, I. N.; Murakhtanov, V. V.; Mazalov, L. N. (Zh. Strukt. Khim. **31** No. 1 [1990] 74/83; J. Struct. Chem. [Engl. Transl.] **31** [1990] 64/72).
[5] Demekhin, V. F.; Sukhorukov, V. L.; Demekhina, L. A.; Timoshevskaya, V. V. (Opt. Spektrosk. **51** [1981] 685/90; Opt. Spectrosc. [Engl. Transl.] **51** [1981] 379/82).

1.3.1.4.3.6 Photoelectron Emission

Spectra of outer **valence electrons** were excited by NeI [1 to 3], HeI [3 to 5], ArI [3], and synchrotron [6] radiation, those of inner valence electrons by HeII [2, 4, 7] and synchrotron [6, 8] radiation. For the respective adiabatic and vertical ionization potentials, see p. 149. Intensities for HeI radiation were compared to those for HeII radiation [4] and to those for Penning ionization (by excited $He^{*}(2^3S)$ atoms) [5]. See also a review [9].

The angular distribution parameter β was measured in [6, 8]; see also a review [10].

Satellites were excited in the spectrum of inner valence electrons [7]. The satellite structure was theoretically calculated in [7, 11, 12].

Binary (e, 2e) or electron momentum spectroscopy, simulating photoelectron spectroscopy [13], was applied in the equivalent photon energy range up to 30 eV [14] and up to somewhat higher values [15].

Spectra of **core electrons** were excited by Al Kα [16] and Mg Kα [17, 18] X-rays (for P2p) and by Ti Kα [17] and Ag $L\alpha_1$ [18] X-rays (for P1s). For the respective ionization potentials, see p. 150.

References:

[1] Branton, G. R.; Frost, D. C.; McDowell, C. A.; Stenhouse, I. A. (Chem. Phys. Lett. **5** [1970] 1/2).
[2] Potts, A. W.; Price, W. C. (Proc. R. Soc. [London] A **326** [1971/72] 181/97, 189, 193).
[3] Maier, J. P.; Turner, D. W. (J. Chem. Soc. Faraday Trans. II **68** [1972] 711/9).
[4] Dechant, P.; Schweig, A.; Thiel, W. (Angew. Chem. **85** [1973] 358/9; Angew. Chem. Int. Ed. Engl. **12** [1973] 308/9).
[5] Yee, D. S. C.; Stewart, W. B.; McDowell, C. A.; Brion, C. E. (J. Electron Spectrosc. Relat. Phenom. **7** [1975] 377/83).
[6] Cauletti, C.; Piancastelli, M. N.; Adam, M. Y. (J. Mol. Struct. **174** [1988] 135/40).
[7] Domcke, W.; Cederbaum, L. S.; Schirmer, J.; Niessen, von, W.; Maier, J. P. (J. Electron Spectrosc. Relat. Phenom. **14** [1978] 59/72, 60).
[8] Cauletti, C.; Adam, M. Y.; Piancastelli, M. N. (J. Electron Spectrosc. Relat. Phenom. **48** [1989] 379/87).
[9] Bock, H. (Pure Appl. Chem. **44** [1975] 343/72, 349/50).
[10] Adam, M. Y.; Cauletti, C.; Piancastelli, M. N.; Svensson, A. (J. Electron Spectrosc. Relat. Phenom. **50** [1990] 219/27).

[11] Lisini, A.; Decleva, P.; Fronzoni, G. (J. Mol. Struct. **228** [1991] 97/116, 109 [THEOCHEM **74**]).
[12] Lisini, A.; Brosolo, M.; Decleva, P.; Fronzoni, G. (J. Mol. Struct. **253** [1992] 333/48, 341/2 [THEOCHEM **85**]).
[13] Brion, C. E. (Int. J. Quantum Chem. **29** [1986] 1397/428).
[14] Hamnett, A.; Hood, S. T.; Brion, C. E. (J. Electron Spectrosc. Relat. Phenom. **11** [1977] 263/74, 265).
[15] Clark, S. A. C.; Brion, C. E.; Davidson, E. R.; Boyle, C. (Chem. Phys. **136** [1989] 55/66, 60/2).
[16] Ashe, A. J.; Bahl, M. K.; Bomben, K. D.; Chan, W.-T.; Gimzewski, J. K.; Sitton, P. G.; Thomas, T. D. (J. Am. Chem. Soc. **101** [1979] 1764/7).
[17] Cavell, R. G.; Sodhi, R. N. (J. Electron Spectrosc. Relat. Phenom. **15** [1979] 145/50).
[18] Sodhi, R. N.; Cavell, R. G. (J. Electron Spectrosc. Relat. Phenom. **32** [1983] 283/312, 294).

1.3.1.4.3.7 Auger Electron Emission

PKLL Spectrum

Excitation by Ti Kα X-ray radiation yielded the following Auger (kinetic) energies E_{kin} and relative intensities I for the assigned final states LL (relative energies referred to $^1D_2$ were given in the original paper) [1]:

E_{kin} in eV	1726.9	1776.7	1795.3	1836.2	1841.5[1)]	1847.9
I in %	5.3	18.3	6.2	8.0	61.3	0.9
assignment	$L_1L_1(^1S_0)$	$L_1L_{2,3}(^1P_1)$	$L_1L_{2,3}(^3P)$	$L_{2,3}L_{2,3}(^1S_0)$	$L_{2,3}L_{2,3}(^1D_2)$	$L_{2,3}L_{2,3}(^3P_{0,2})$

[1)] Other values are given below.

The KLL spectrum with the final states from above (except the last one) was also theoretically derived [2].

Other experimental values for the energy of the $KL_{2,3}L_{2,3}(^1D_2)$ line were obtained by Ag Lα X-ray excitation, i.e., E_{kin}=1841.46(20) eV [3] (which was first corrected to 1841.56 [4] and then to 1841.29 eV [5]) and E_{kin}=1841.4 eV [6]. Ti Kα excitation had yielded earlier E_{kin}= 1842.0±0.5 eV [7].

$PL_{2,3}VV$ Spectrum

Excitation by 1.05 MeV He^+ ion impact yielded the following Auger (kinetic) energies E_{kin} and relative intensities I for the assigned final states VV (V = valence shell) [8]:

E_{kin} in eV	80.0	85.4	87.9	91.7
I in %	1.7	1.8	9.3	9.5
assignment	$4a_1^{-2}(^1A_1)$	$4a_1^{-1}\,2e^{-1}\,(^1E)$	$4a_1^{-1}\,2e^{-1}\,(^3E)$	$4a_1^{-1}\,5a_1^{-1}\,(^{3,1}A_1)$
E_{kin} in eV	94.6	98.1	100.6	104.3
I in %	5.7	18.3	4.2	17.7
assignment	$2e^{-2}\,(^1A_1)$	$2e^{-2}\,(^3A_2, {}^1E)$	$2e^{-1}\,5a_1^{-1}\,(^{3,1}E)$	$5a_1^{-2}\,(^1A_1)$

The $L_{2,3}VV$ spectrum was also excited by 0.4 to 2.1 MeV He^+ ions [8, 9], by 0.4 to 2.0 MeV H^+ ions [8, 10], and by 0.4 to 1.2 MeV H_2^+ ions [10].

The $L_{2,3}VV$ spectrum with the final states from above (singlets and triplets separated from each other) was also theoretically derived [2].

Other Spectra

Theoretical results are also available for the spectrum KL_1V with the final states $2a_1^{-1}\,4a_1^{-1}$ $(^{1,3}A_1)$, $2a_1^{-1}\,2e^{-1}\,(^{1,3}E)$, $2a_1^{-1}\,5a_1^{-1}\,(^{1,3}A_1)$, the spectra KVV and L_1VV with the same final states as $L_{2,3}VV$ (see the table above), and the spectra $KL_{2,3}V$ and $L_1L_{2,3}V$ with the common final states $1e^{-1}\,4a_1^{-1}\,(^{1,3}E)$, $1e^{-1}\,2e^{-1}\,(^{1,3}A_1, {}^{1,3}E)$, $3a_1^{-1}\,4a_1^{-1}\,(^{1,3}A_1)$, $3a_1^{-1}\,2e^{-1}\,(^{1,3}E)$, and $3a_1^{-1}\,5a_1^{-1}\,(^{1,3}A_1)$ [2].

References:

[1] Vayrynen, J.; Sodhi, R. N.; Cavell, R. G. (J. Chem. Phys. **79** [1983] 5329/36).
[2] Sukhorukov, V. L.; Petrov, I. D.; Sukhorukov, B. L.; Demekhina, L. A. (Khim. Fiz. **5** [1986] 175/83; C.A. **104** [1986] No. 196183).
[3] Sodhi, R. N.; Cavell, R. G. (J. Electron Spectrosc. Relat. Phenom. **32** [1983] 283/312, 294).
[4] Cavell, R. G.; Sodhi, R. N. S. (J. Electron Spectrosc. Relat. Phenom. **41** [1986] 25/35, 27).

[5] Cavell, R. G.; Sodhi, R. N. S. (J. Electron Spectrosc. Relat. Phenom. **43** [1987] 215/23).
[6] Ashe, A. J.; Bahl, M. K.; Bomben, K. D.; Chan, W.-T.; Gimzewski, J. K.; Sitton, P. G.; Thomas, T. D. (J. Am. Chem. Soc. **101** [1979] 1764/7).
[7] Cavell, R. G.; Sodhi, R. (J. Electron Spectrosc. Relat. Phenom. **15** [1979] 145/50).
[8] Ariyasinghe, W. M.; Powers, D.; Awuku, H. T. (Nucl. Instrum. Methods Phys. Res. B **56/57** [1991] 111/5).
[9] Ariyasinghe, W. M.; Powers, D. (Phys. Rev. [3] A **41** [1990] 4751/8).
[10] Ariyasinghe, W. M.; Awuku, H. T.; Powers, D. (Phys. Rev. [3] A **42** [1990] 3819/25).

1.3.1.5 Chemical Reactions

General Remarks

The principles of the chemical behavior of PH_3 towards a variety of elements and compounds to a large extent have already been covered in "Phosphor" C, 1965, pp. 31/50. Not in all cases has new information been published. Somewhat later reviews of the chemistry of PH_3 are [1, 2].

Electrochemical Reactions

The anodic oxidation of PH_3 in acidic and alkaline aqueous solutions was studied briefly. Half-cell potentials of a silver-amalgam electrode at 40°C relative to the standard hydrogen electrode were determined to be 0.11 V for 0.0016 mol/L PH_3 in 0.49 mol/L H_2SO_4 and 0.94 V for 0.0024 mol/L PH_3 in 1.00 mol/L NaOH. The potentials are slightly higher than the ones predicted for the oxidation of PH_3 to phosphorus [3]. Coulometric investigations suggested that the electrode reaction in alkaline solution proceeds according to $PH_3 + 3\,OH^- \rightarrow P + 3\,H_2O + 3\,e^-$ [4]. Earlier results are described in "Phosphor" C, 1965, pp. 30/1.

References:

[1] Fluck, E. (Fortschr. Chem. Forsch. **35** [1973] 1/64).
[2] Fluck, E.; Novobilsky, V. (Fortschr. Chem. Forsch. **13** [1969/70] 125/66).
[3] Barry, M. L.; Tobias, C. W. (Electrochem. Technol. **4** [1966] 502/6; C. A. **65** [1966] 14836).
[4] Barry, M. L. (UCRL-11573 [1964] 1/145, 105; C. A. **63** [1965] 3894).

1.3.1.5.1 Thermal, Photolytic, and Radiational Decomposition

1.3.1.5.1.1 Heterolytic Dissociation of a P–H Bond. Gas-Phase Acidity

The standard Gibbs free enthalpy change, Δ_rG°, for the heterolytic dissociation of a P–H bond according to

$$PH_3 \rightarrow PH_2^- + H^+$$

defines the **gas-phase acidity** of PH_3, whereas its deprotonation enthalpy, Δ_rH°, is quite often called the proton affinity (PA) of the PH_2^- ion. The energetics of this heterolytic cleavage of a P–H bond in the gas phase have not been determined directly.

The equilibrium constants K of a large number of proton transfer reactions, some of them involving PH_3, were determined by pulsed ICR spectroscopy, and a scale of the relative acidities of a series of acids in the gas phase at 298 K, $\delta\Delta_rG^\circ_{298}$, was deduced from the relation

$\delta\Delta_rG^\circ_{298} = -RT\ lnK$. This experimental scale of relative acidities was converted to a scale of absolute acidities by including certain compounds as anchor points. Thus, the gas-phase acidity of PH_3 was determined to be $\Delta_rG^\circ_{298} = 363 \pm 2$ kcal/mol. The entropy change for the deprotonation process was evaluated by procedures using statistical mechanics as $\Delta_rS^\circ = 24.9 \pm 2\ cal \cdot mol^{-1} \cdot K^{-1}$. From these data the deprotonation enthalpy of PH_3 at 298 K was calculated to be $\Delta_rH^\circ_{298} = PA(PH_2^-) = 370.4 \pm 2$ kcal/mol [1, 2].

A completely different method for calculating Δ_rH° uses a thermochemical cycle involving the homolytic bond dissociation energy of PH_3, $D(H_2P{-}H)$, the electron affinity of the PH_2 radical, $EA(PH_2)$, and the ionization potential of the hydrogen atom, IP (H). These thermochemical data led to $\Delta_rH^\circ = PA(PH_2^-) = 368.6 \pm 3.1$ [1, 2], 369 kcal/mol [3]. Earlier investigations of proton transfer reactions involving PH_3 and the PH_2^- ion and earlier evaluations of Δ_rH° from thermochemical cycles gave $\Delta_rH^\circ_{298} = PA(PH_2^-) = 364$ [4], 362 ± 11 kcal/mol [5].

Several ab initio MO calculations of the proton affinity of the PH_2^- anion or of the energy change of the above equation considering ΔZPE were carried out, the corrected ΔE value in some cases being denoted as gas-phase acidity. Good agreement between the computed and the most reliable experimental values was generally achieved only when the basis sets were augmented by diffuse functions and the effects of electron correlation were taken into account [6 to 11]. In two very sophisticated ab initio MO studies of two series of hydrogen compounds both including PH_3, the energetic effects of the addition of diffuse functions [9] and the correlation energy contributions were investigated in detail [7]. In particular, the effectiveness of the MP perturbation theory up to the fourth order (MP4), the linearized coupled-cluster method (LCCM), the averaged coupled-pair functional (ACPF), and the configuration interaction with all single and double excitations (CISD) were studied. An excellent agreement between the calculated electronic deprotonation energies of PH_3 and a so-called experimental electronic deprotonation energy value was achieved for nearly all of these correlation methods [7]. The results of semiempirical MNDO calculations are discussed in [10].

References:

[1] Bartmess, J. E.; Scott, J. A.; McIver, R. T., Jr. (J. Am. Chem. Soc. **101** [1979] 6046/56).
[2] Bartmess, J. E.; McIver, R. T., Jr. (in: Bowers, M. T.; Gas Phase Ion Chemistry, Vol. 2, Academic, New York - San Francisco - London 1979, pp. 87/121).
[3] Wyatt, R. H.; Holtz, D.; McMahon, T. B.; Beauchamp, J. L. (Inorg. Chem. **13** [1974] 1511/7).
[4] Braumann, J. I.; Eyler, J. R.; Blair, L. K.; White, M. J.; Comisarow, M. B.; Smyth, K. C. (J. Am. Chem. Soc. **93** [1971] 6360/2).
[5] Holtz, D.; Beauchamp, J. L.; Eyler, J. R. (J. Am. Chem. Soc. **92** [1970] 7045/55).
[6] Gronert, S. (J. Am. Chem. Soc. **113** [1991] 6041/8).
[7] Del Bene, J. E.; Shavitt, I. (J. Phys. Chem. **94** [1990] 5514/8).
[8] Pople, J. A.; Schleyer, P. von R.; Kaneti, J.; Spitznagel, G. W. (Chem. Phys. Lett. **145** [1988] 359/64).
[9] Spitznagel, G. W.; Clark, T.; Schleyer, P. von R.; Hehre, W. J. (J. Comput. Chem. **8** [1987] 1109/16).
[10] Gordon, M. S.; Davis, L. P.; Burggraf, L. W.; Damrauer, R. (J. Am. Chem. Soc. **108** [1986] 7889/93).

[11] Lohr, L. L.; Ponas, S. H. (J. Phys. Chem. **88** [1984] 2992/7).

1.3.1.5.1.2 Thermal Decomposition

The thermal decomposition reaction of PH_3 can be used to produce ultrapure phosphorus. Thus this reaction attracted much interest in recent years, for example, in experiments directed at growing layers or crystals of III-V semiconductors such as InP, GaP, $GaAs_{1-x}P_x$, etc., by vapor-phase epitaxy (cf. 1.3.1.6, pp. 293/308). An effort has not been made to systematically scan the prolific literature in that area.

Based on JANAF or equivalent thermochemical data, equilibrium constants were calculated for the gas-phase equilibria:

(1) $PH_3 \rightleftharpoons 1/4\,P_4 + 3/2\,H_2$ [1 to 3]

(2) $PH_3 \rightleftharpoons 1/2\,P_2 + 3/2\,H_2$ [1, 2]

(3) $PH_3 \rightleftharpoons P + 3/2\ H_2$ [1]

and for the heterogeneous equilibrium

(4) $PH_3(g) \rightleftharpoons P(\text{solid, white modification}) + 3/2\,H_2(g)$ [3].

(Equilibrium constants given by [3]: $\log(K_p/atm^{0.75}) = -275/T + 2.46 \log T - 4\times10^{-4}T - 3.74$ for (1) and $\log(K_p/atm^{0.5}) = (387\pm87)/T + 0.272 \log T + 1.32\times10^{-3}T$ for (4)). A comparison of the gas-phase equilibria (1) to (3) shows that, apart from H_2, P_4 is the mayor product expected thermodynamically up to ca. 1200 K. Above that temperature P_2 becomes predominant, while phosphorus atoms are less abundant than P_4 by a factor $\sim10^4$, even at 2000 K [1]. Thermodynamic data predict that PH_3 is unstable at room temperature, and disposed to decompose into gaseous [1, extrapolation of figure 3] or solid P_4 (white phosphorus) [3].

Decomposition reactions

(5) $PH_3(^1A_1) \rightarrow PH(^3\Sigma^-) + H_2(^1\Sigma_g^+)$

and (6) $PH_3(^1A_1) \rightarrow PH_2(^2B_1) + H(^2S)$

are endothermic by 207 (or 235 [47]) and 343 kJ/mol, respectively. Because the electronic ground state of PH_3 neither correlates with that of PH nor with that of PH_2, it must be assumed that both processes have an activation barrier in addition to the endothermicity [4].

Rate constants for unimolecular homogeneous PH_3 decomposition were calculated by the Rice-Ramsperger-Kassel-Marcus (RRKM) theory and by the use of estimated values for the activation energies. Rate constants at the high-pressure limit for reaction (5), $\log(k/s) = 14.18 - 11610/T$ [5] or $14.00 - 12610/T$ [4], include activation energies of 222 or 241 kJ/mol, respectively. Calculated rate constants for reaction (6) are $\log(k/s) = 15.74 - 18040/T$ with an activation energy of 345 kJ/mol. At 900 K PH formation is thus predicted to exceed PH_2 formation by a factor $\sim10^5$. Calculated fall-off pressures for both reactions which indicate the onset of second-order decomposition, are quite high, about 10^4 Torr in an H_2 bath gas [5].

On the experimental side, thermal decomposition of gaseous PH_3 appears to be a complex reaction; see "Phosphor" C, 1965, pp. 31/3, and [6 to 8]. The reaction is kinetically inhibited and, apart from two more recent studies [9 to 11], noticeable decomposition was reported to occur above 400°C. There is agreement that PH_3 decomposition shows all the characteristic features of a heterogeneous, i.e., surface-catalyzed, reaction (for possible homogeneous contributions at high temperatures, see below). Moreover it was shown recently that techniques used for sampling and analysis of the reaction mixture still have a profound influence on the results. Several conclusions drawn from the results of standard ex-situ techniques (the decomposing PH_3 gas is probed, for instance, at some distance from the hot surface by a mass spectrometer) were thus rendered obsolete [9 to 11], see Fig. 10, p. 202.

Products from PH_3 decomposition include H_2, P_4, and P_2. In the gas phase these products were identified, for instance, by mass-spectrometric techniques [1, 2, 12 to 17]. Raman techniques were used to monitor H_2 [9 to 11] and P_4 [11], laser-induced fluorescence was used for analysis of P_2 [18 to 20], UV-VIS absorption and emission for P_2 [21, 22] and P_4 [21], and gas chromatography for P_4 [23]. The decrease in PH_3 was also monitored by chemical analysis [8]. Product signals corresponding to phosphorus atoms [1, 15, 16] or to P_3 [15, 16] were occasionally observed by mass spectrometry, but the species originated from secondary reactions during ionization [1, 15, 16]. Out of the feasible intermediates, PH_2 was identified in the gas phase by Raman spectroscopy [11] and by laser-induced fluorescence [18 to 20]. PH (and tentatively PH_2) was identified by its UV-VIS emission behind reflected shock waves in PH_3/Ar mixtures [22]. A PH^+ mass signal, however, was thought to arise from a secondary process during ionization [15]; see Section 1.3.1.5.1.4, pp. 211/4.

Gaseous P_4 nucleated, and solid deposits of phosphorus were occasionally observed in colder parts of the reactor [11, 12].

Fig. 10 shows the results obtained from four experiments which have in common the fact that PH_3 was passed in an H_2 carrier gas through the hot zone of a standard flow system [17] commonly used for metalorganic vapor-phase epitaxy [9 to 11]. A large discrepancy in the degree of PH_3 decomposition was observed between results that were obtained from in-situ analysis, i.e., the laser beam used to observe the conventional Raman [11] or coherent anti-Stokes Raman scattering (CARS) [9, 10] bands of PH_3 was directed close to the hot surface, and those that were obtained by ex-situ analysis, i.e., the gas mixture was pumped from the hot surface for mass-spectrometric analysis [17]. The general trend that ex-situ results simulate a considerably higher temperature needed for a given degree of PH_3 decomposition was verified by control experiments during which CARS spectra of PH_3 were taken from mixtures pumped from the hot surface to an analysis chamber [9, 10] or Raman spectra were taken downstream from the hot zone [11]. A similar discrepancy was noted when PH_3 was pyrolyzed in N_2 and H_2 formation was followed by CARS [9].

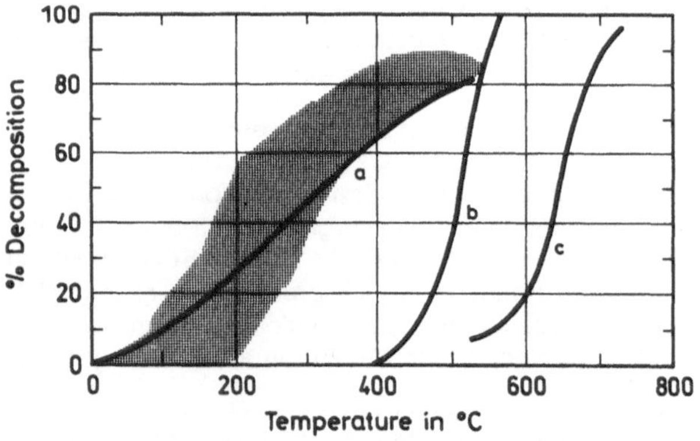

Fig. 10. Decomposition of PH_3 in an H_2 carrier gas in flow systems. In-situ results obtained over InP [9] (dotted area), graphite [11] (curve a; the scatter of results observed in [9] and [11] was similar), and ex-situ results obtained in an InP-coated (curve b) or uncoated (curve c) packed quartz tube [17] (from [11]).

In line with the classification of PH_3 thermal decomposition as an inhibited and surface-catalyzed reaction, it was observed that the extent of the decomposition and the reaction rates did not depend only on temperature, pressure, and composition of the gas mixture, but also on the material and surface area of the reactor, its geometry and prior history, on flow rates and residence times in the hot zone, etc.; see for instance [2, 8, 18, 24]. A poor reproducibility of quantitative results, see, e.g. [18], and the scatter indicated in Fig. 10 are thus an intrinsic property of the reaction. The following listing, according to the surface material at which PH_3

was decomposed is thus only intended to give a short survey on some relevant literature. Apart from the more recent studies in which PH_3 decomposition was studied at considerably lower temperatures [9 to 11], typical reaction temperatures were in the 400 to 1000°C range.

Decomposition of gaseous PH_3 was observed, for instance, on quartz [1, 2, 8, 12 to 14, 16, 17, 21, 24 to 27], glass [28], silicon [28], graphite [11], alumina [29, 30], gallium phosphide [8, 26], indium phosphide [7 to 10, 17, 22, 26, 34 to 36], magnesium oxide [31], tantalum [1, 15, 24, 29, 32], tantalum hydride [24], tungsten [24, 29, 30], molybdenum [15, 24], and gold surfaces [23, 33]. See also Section 1.3.1.5.11, pp. 287/93. Qualitative observations indicate that Mo surfaces are not very efficient in PH_3 cracking [15]. Surfaces of W, Al_2O_3, or SiO_2 catalyze PH_3 decomposition less efficiently than Ta surfaces [30, 31].

There is agreement that a change in carrier gases H_2, D_2, or N_2 has no effect on the PH_3 decomposition temperature [9, 17]. PH_3 decomposition in D_2 leads only to H_2 (minor HD traces were of unclear origin) [17].

The experimental yield of H_2 is lower than expected for complete PH_3 decomposition [9]. Up to about 250°C, H_2 partial pressures followed those expected for decomposition of PH_3 into PH_2. Up to about 550°C, H_2 partial pressures increased but did not reach those expected for decomposition into P_4 [11, figure 8]. The apparently lower PH_3 decomposition derived from ex-situ observations indicates equilibrium reactions which lead to reformation of PH_3 during ex-situ analysis [9 to 11]. Because P_4 is thermodynamically more stable than PH_3, PH or PH_2 must be the sources of PH_3 reformation [11].

A linear relation $\ln(p/p_0)$ vs. τ for the change in PH_3 partial pressures in flow systems (τ = residence time in the hot zone) indicated that decomposition was first order in PH_3 [17]. Most published activation energies, however, simply rely on the assumption of a first-order reaction. The following activation energies were measured for the individual surfaces: quartz 150 kJ/mol above 670 K [27], 153 kJ/mol above 850 K [14], 245 kJ/mol [18], 18 kJ/mol below 1070 K and 290 kJ/mol above 1170 K [8]; glass 185 kJ/mol [28]; silicon 231 kJ/mol [28]; InP 150 kJ/mol [17], ca. 50 kJ/mol at lower temperatures and less than 21 kJ/mol at higher temperatures [9] (see Fig. 10). The scatter in these data illustrates the unsettled debate on the mechanism of surface-catalyzed PH_3 decomposition.

Recent in-situ experiments could not be represented by a first-order rate expression [9]. A mechanism, however, which would account for both the generally lower activation energies and the opposite temperature trend compared to what was observed before, was only stated to be typical for NH_3 decomposition [9] but not specified for PH_3. An early mechanism which ascribed the lower activation energies up to 800°C to PH_3 adsorption and the high apparent activation energies above 900°C to homogeneous PH_3 decomposition [8] was modified with respect to the low temperature region: $PH_3(g) \rightleftharpoons PH_3(ads)$; $PH_3(ads) \rightleftharpoons PH_2(ads) + H(ads)$ slow step E_{app} = 150 kJ/mol on SiO_2 or InP; $PH_2(ads) \rightleftharpoons P(ads) + 2H(ads)$; $2H(ads) \rightleftharpoons H_2(g)$; ... [17], see also [27]. However, this modification was criticized as being not compatible with the extensive equilibrium reactions observed in low temperature in-situ experiments [9]. Another modification included the assumption of the surface reaction $PH_3(ads) \rightleftharpoons PH(ads) + H_2(g)$ [14]. The effect of PH_3 decomposition on the homoepitaxial growth of InP crystals (InP(100) surfaces cut 2° or 6° off the $\langle 110 \rangle$ direction) [35, 36] was also discussed in terms of a model including two temperature regimes, but the assumption of homogeneous decomposition at high temperatures was abandoned: $PH_3(g) \rightleftharpoons PH(ads) + H_2(g)$ and $2PH(ads) \rightleftharpoons P_2(g) + H_2(g)$ (which would be thermodynamically feasible in the gas phase, $\Delta H = -328$ kJ/mol [4]) above 970 K and $PH_3(g) \rightleftharpoons PH_3(ads)$; $PH_3(ads) \rightleftharpoons PH(ads) + H_2(g)$; $PH_3(ads) + PH(ads) \rightarrow 2PH_2(g)$ or $P_2(g) + 2H_2(g)$ below 870 K [36].

Dissociative PH_3 adsorption on clean polycrystalline Fe [37] on Pt(111) [38] and Rh(100) surfaces [39 to 41] was noticed at 100 K and even at 25 K on Ni(100) [39]. At 100 K, no isotopic scrambling took place with preadsorbed D_2 [37, 38, 40, 41]. Temperature-programmed desorption from Rh(100) and Pt(100) produced PH_3(g) and significant amounts of H_2(g) above 300 K [38, 41]. Significant decomposition was also observed when PH_3 was desorbed from clean polycrystalline Fe at 170 K; from oxidized Fe surfaces, however, only molecular PH_3 desorption occurred [37]. In contrast to the decomposition product H_2, molecular PH_3 which desorbed from Rh(100) showed no isotopic scrambling with preadsorbed D_2 [40]. See also Section 1.3.1.5.11, pp. 287/93.

The thermal decomposition of PH_3/AsH_3 mixtures is described in Section 1.3.1.5.6, p. 249.

Pyrolysis of PH_3 in a D_2 carrier gas which was saturated with $(CH_3)_3In$ occurred at significantly lower temperatures as compared to the pyrolysis of pure PH_3 on an InP-coated quartz surface. The temperature difference amounts to about 200°C and depends on the $(CH_3)_3In$ concentration. In addition to the pyrolysis products of PH_3, CH_4 was observed. This is in contrast to the pyrolysis of pure $(CH_3)_3In$ in D_2 which results in CH_3D [17, 34]. Activation energies that were measured above 400°C for PH_3 pyrolysis on an InP surface and in the presence of $(CH_3)_3In$ are similar and amount to ca. 150 kJ/mol. Below 400°C, however, the activation energy was reduced to ca. 76 kJ/mol when $(CH_3)_3In$ was present. Thereby a change in mechanisms is indicated (decomposition via an adduct; see also Section 1.3.1.5.8, p. 259) [27]. The addition of up to 0.1 mbar $(CH_3)_3In$ to 100 mbar of a 95:5 H_2/PH_3 mixture had no effect on PH_3 decomposition [9, 10]. The decomposition of PH_3/R_3In ($R=CH_3$, C_2H_5) mixtures found some application in growing ultrapure InP by metalorganic vapor-phase epitaxy, e.g. [42 to 44]; see also Section 1.3.1.6, pp. 294/5.

The regime of homogeneous, thermal PH_3 decomposition in the gas phase may be reached under the special condition of a photosensitized reaction. Experimentally, SiF_4 was used as the photosensitizer which was excited by a pulsed CO_2 laser (~1025 cm^{-1}) at fluences of 0.2 to 0.6 J/cm^2. The degree of PH_3 decomposition in PH_3/SiF_4 mixtures (up to 16 Torr PH_3, varying $PH_3:SiF_4$ ratios, additions of He or N_2) was measured by mass spectrometric observation of the H_2 product after irradiation. An initial model which was used to quantitatively describe the dynamics of the system included multi-quantum absorption by SiF_4 (up to the PH_3 dissociation energy) followed by energy transfer to PH_3 [45]. A theoretical reexamination of the system was done by a model which included energy absorption by SiF_4, energy redistribution in the gas mixture as described by the generation of a shock wave (which reaches temperatures up to 3000 K), and homogeneous $PH_3 \rightarrow PH + H_2$ decomposition with rate constants calculated by the RRKM theory (see p. 201) [4], see also [46]. Information from an early shock-tube study performed in PH_3/Ar mixtures [22] was too incomplete for a comparison with the aforementioned model. The SF_6-photosensitized pyrolysis of PH_3 and PH_3/SiH_4 mixtures, which were irradiated by a CO_2 laser, was explained by a combination of homogeneous and heterogeneous (at formed aerosol particles) decomposition processes [47].

There is no danger that PH_3 packaged in common commercial cylinders is prone to explosive decomposition [46].

References:

[1] Chow, R.; Chai, Y. G. (J. Vac. Sci. Technol. [2] A **1** [1983] 49/54).
[2] Ban, V. S.; Ettenberg, M. (J. Phys. Chem. Solids **34** [1973] 1119/29).
[3] Devyatykh, G. G.; Yushin, A. S. (Zh. Fiz. Khim. **38** [1964] 957/62; Russ. J. Phys. Chem. [Engl. Transl.] **38** [1964] 517/20).
[4] Błażejowski, J.; Rak, J.; Lampe, F. W. (J. Photochem. Photobiol. A **52** [1990] 347/62).
[5] Buchan, N. I.; Jasinski, J. M. (J. Cryst. Growth **106** [1990] 227/38).
[6] Fluck, E. (Fortschr. Chem. Forsch. **35** [1973] 1/64, 21/2).

[7] Goodridge, I. H.; Hasdell, N. B. (Inst. Phys. Conf. Ser. No. 74 [1985] 205/10; C.A. **103** [1985] No. 224658).
[8] Larsen, C. A.; Stringfellow, G. B. (J. Cryst. Growth **75** [1986] 247/54).
[9] Lückerath, R.; Tommack, P.; Hertling, A.; Koss, H. J.; Balk, P.; Jensen, K. F.; Richter, W. (J. Cryst. Growth **93** [1988] 151/8).
[10] Lückerath, R.; Richter, W.; Jensen, K. F. (NATO ASI Ser. B **198** [1989] 157/67; C.A. **113** [1990] No. 29957).

[11] Abraham, P.; Bekkaoui, A.; Souliere, V.; Bouix, J.; Monteil, Y. (J. Cryst. Growth **107** [1991] 26/31).
[12] Ban, V. S. (J. Electrochem. Soc. **118** [1971] 1473/8).
[13] Tietjen, J. J.; Ban, V. S.; Enstrom, R. E.; Richman, D. (J. Vac. Sci. Technol. **8** [1971] S5/S11).
[14] Anderson, T. J.; Quinlan, K. P. (RADC-TR-88-31 [1988] 1/250, 21/2; C.A. **111** [1989] No. 124049).
[15] Baillargeon, J. N.; Cheng, K. Y.; Jackson, S. L.; Stillman, G. E. (J. Appl. Phys. **69** [1991] 8025/30).
[16] Huet, D.; Lambert, M.; Bonnevie, D.; Dufresne, D. (J. Vac. Sci. Technol. [2] B **3** [1985] 823/9).
[17] Larsen, C. A.; Buchan, N. I.; Stringfellow, G. B. (J. Cryst. Growth **85** [1987] 148/53).
[18] Donnelly, V. M.; Karlicek, R. F. (J. Appl. Phys. **53** [1982] 6399/407).
[19] Karlicek, R. F.; Donnelly, V. M.; Johnston, W. D., Jr. (Proc. SPIE Int. Soc. Opt. Eng. **323** [1982] 62/6; C.A. **97** [1982] No. 172612).
[20] Donnelly, V. M.; Karlicek, R. F. (Proc. SPIE Int. Soc. Opt. Eng. **476** [1984] 102/9; C.A. **101** [1984] No. 238003).

[21] Karlicek, R. F.; Hammarlund, B.; Ginocchio, J. (J. Appl. Phys. **60** [1986] 794/9).
[22] Guenebaut, H.; Pascat, B. (C. R. Hebd. Sceances Acad. Sci. **255** [1962] 1741/3).
[23] Sowinski, E. J.; Suffet, I. H. (Am. Ind. Hyg. Assoc. J. **38** [1977] 363/7; C.A. **88** [1978] No. 27192).
[24] Karlicek, R. F., Jr.; Mitcham, D.; Ginocchio, J. S.; Hammarlund, B. (J. Electrochem. Soc. **134** [1987] 470/4).
[25] Frolov, I. A. (Zh. Prikl. Khim. [Leningrad] **51** [1978] 721/3; J. Appl. Chem. USSR [Engl. Transl.] **51** [1978] 709/11).
[26] Stringfellow, G. B. (J. Cryst. Growth **68** [1984] 111/22).
[27] Buchan, N. I.; Larsen, C.A.; Stringfellow, G. B. (J. Cryst. Growth **92** [1988] 605/15).
[28] Devyatykh, G. G.; Kedyarkin, V. M.; Zorin, A. D. (Zh. Neorg. Khim. **14** [1969] 2011/4; Russ. J. Inorg. Chem. [Engl. Transl.] **14** [1969] 1055/7).
[29] Panish, M. B.; Hamm, R. A. (J. Cryst. Growth **78** [1986] 445/52).
[30] Panish, M. B.; Temkin, H.; Sumski, S. (J. Vac. Sci. Technol. [2] B **3** [1985] 657/65).

[31] Apoyan, A. K.; Artsruni, G. K.; Arutyunyan, G. A.; Azatyan, V. V. (Arm. Khim. Zh. **43** [1990] 8/14 from C.A. **114** [1991] No. 215482).
[32] Panish, M. B.; Sumski, S. (J. Appl. Phys. **55** [1984] 3571/6).
[33] Sowinski, E. J.; Suffet, I. H. (Adv. Environ. Sci. Technol. **8** [1977] 167/92, 186/8; C.A. **88** [1978] No. 65430).
[34] Buchan, N. I.; Larsen, C.A.; Stringfellow, G. B. (Appl. Phys. Lett. **51** [1987] 1024/6).
[35] Harrous, M.; Chaput, L.; Bendraoui, A.; Cadoret, M.; Pariset, C.; Cadoret, R. (J. Cryst. Growth **92** [1988] 423/31).
[36] Harrous, M.; Laporte, J. L.; Cadoret, M.; Pariset, C.; Cadoret, R. (J. Cryst. Growth **83** [1987] 279/85).
[37] Hegde, R. I.; White, J. M. (J. Phys. Chem. **90** [1986] 2159/63).

[38] Mitchell, G. E.; Henderson, M. A.; White, J. M. (Surf. Sci **191** [1987] 425/48).
[39] Greenlief, C. M.; Hegde, R. I.; White, J. M. (J. Phys. Chem. **89** [1985] 5681/5).
[40] Hegde, R. I.; White, J. M. (Surf. Sci. **157** [1985] 17/28).

[41] Hegde, R. I.; Tobin, J.; White, J. M. (J. Vac. Sci. Technol. [2] A **3** [1985] 339/45).
[42] Chen, C. H.; Kitamura, M.; Cohen, R. M.; Stringfellow, G. B. (Appl. Phys. Lett. **49** [1986] 963/5).
[43] Zhu, L. D.; Chan, K. T.; Ballantyne, J. M. (Appl. Phys. Lett. **47** [1985] 47/8).
[44] Di Forte-Poisson, M. A.; Brylinksi, C.; Duchemin, J. P. (Appl. Phys. Lett. **46** [1985] 476/8).
[45] Blażejowski, J.; Lampe, F. W. (J. Phys. Chem. **88** [1984] 1666/70).
[46] Blażejowski, J.; Lampe, F. W. (Spectrochim. Acta A **46** [1990] 627/30).
[47] Adamova, Yu. A.; Kaganyuk, D. S.; Pimenov, H. P.; Skachkov, A. N.; Stolyarova, G. I.; Shmerling, G. V. (Khim. Fiz. **8** [1989] 1354/61; C.A. **112** [1990] No. 105876).
[48] Horiguchi, S.; Urano, Y.; Kondo, S. (Koatsu Gasu **28** [1991] 351/63 from C.A. **115** [1991] No. 74561).

1.3.1.5.1.3 Photolysis

1.3.1.5.1.3.1 UV Photolysis

The UV photolysis of PH_3 requires irradiation with wavelengths $\lambda \leqq 230$ nm; see the description of the UV spectrum (Section 1.3.1.4.3.4, p. 193). The present section describes the overall products of the UV photolysis of PH_3, their quantum yields, and the primary pathways of PH_2, PH, and H formation. The application of the UV photolysis of PH_3 for the formation of PH_2 and PH is described in Section 1.2.1.1, p. 47, and Section 1.1.1.1, p. 2. Subsequent reactions of these intermediates are described in the respective sections on PH_2, PH, and PH_3 in the present volume. The ionization of PH_3 by UV irradiation is described in Section 1.3.1.4.1.3, p. 149. Earlier results on the photolysis of PH_3 which, however, are outdated to a great extent, are summarized in "Phosphor" C, 1965, pp. 33/4. Results up to 1978 were included in a review [1]. The UV photolysis of PH_3 in the presence of additional reactants is described together with other reactions of PH_3 (Section 1.3.1.5.6, pp. 244/55) because these investigations usually lack details on the formation of primary fragments of PH_3. The formation of phosphorus via photolysis of PH_3 was also discussed in relation to the photochemistry of the atmospheres of Jupiter and Saturn, see, e.g. [2 to 7].

UV Photolysis in the Gas Phase

Stable Products. The UV photolysis of gaseous PH_3 at ambient temperature initially yields H_2 and P_2H_4 in equimolar quantities [6, 8] and a solid deposit on the walls of the photolysis cell [6, 9]. The concentration of H_2 increases continuously during the photolysis, whereas the concentration of P_2H_4 peaks after a relatively short time and then decreases to a small fraction of the maximum value [9, 10]. At 157 K, a considerable quantity of P_2H_4 can be isolated because condensation of the product minimizes its decomposition during photolysis [11]. The search for other gaseous products like P_4 or higher phosphanes by mass spectrometry failed [6]. The deposition of a solid layer sets in after an induction period [9, 11] and increases monotonically during the photolysis [6, 9]. The pressure increase at the end of the photolysis at ambient temperature indicates an overall reaction to form H_2 and a solid layer of elemental phosphorus, but only a small quantity of P_2H_4 [12]. The color of the formed deposit is reddish [11 to 13] to yellowish [6, 7]. The mass balance and qualitative physical investigations indicate a composition of phosphorus allotropes and a small quantity of polyphosphanes [6, 7, 11]. The influence of added N_2 or SF_6 on the rate of the reaction and the distribution of the products is small [9].

Quantum Yields. The following quantum yields for a small turnover were determined for the photolysis at 147 nm and a total pressure of 40 Torr using the diluents He for PH_3 and Ne for PD_3; values for PD_3 are given in parentheses [6]:

$p(PH_3)$ in Torr	1.4×10^{15} quanta/s $\Phi(-PH_3)$	$\Phi(P_2H_4)$	$\Phi(H_2)$	3.5×10^{15} quanta/s $\Phi(-PH_3)$	$\Phi(P_2H_4)$	$\Phi(H_2)$
0.08	2.0	0.18	—	1.2	0.10	0.93
0.16	2.9	0.39	2.4	1.8 (1.2)	0.22 (0.040)	1.7 (0.93)
0.32	4.0	0.60	4.1	3.1 (1.7)	0.34 (0.10)	3.1 (1.9)
0.40	4.9	0.74	4.9	—	—	3.5
0.48	4.7	0.71	—	3.4 (1.9)	0.54 (0.10)	3.8 (2.4)

As indicated above, all quantum yields increase with increasing PH_3 pressure and decrease with increasing light intensity. The high quantum yields for the PH_3 destruction suggest that the photodecomposition involves a short chain reaction. The quantum yields for the photolysis of PD_3 are smaller than the corresponding PH_3 values [6].

At 206.2 nm irradiation of 90 Torr of PH_3 at 298 K and photon fluxes of $(2.0\text{ to }3.2)\times10^{14}$ quanta/s the following initial quantum yields were determined: $\Phi(P_2H_4)=0.92\pm0.17$, $\Phi(H_2)=0.93\pm0.07$ [11]. Note that the quantum yield for H_2 formation replaces a too low value of 0.43 measured earlier [9]. Values for PH_3 destruction, formation of P_2H_4, and formation of the solid deposit (ascribed to P_4), $\Phi(-PH_3)=1.78\pm0.18$ $\Phi(P_2H_4)=0.80\pm0.08$, $\Phi(4P_4)=0.04\pm0.16$ were confirmed or revised only slightly [11] with respect to the original results in [9]. When the PH_3 pressure was reduced to 26 and 11 Torr, $\Phi(P_2H_4)$ dropped to 0.49 ± 0.03 and 0.40 ± 0.02, respectively. At 11 Torr PH_3, the initial H_2 yield was 0.74 ± 0.08. P_2H_4 yields measured around 90 Torr PH_3 pressure were found to be independent of temperature down to 157 K [11].

Approximate values of quantum yields $\Phi(-PH_3)=2$ and $\Phi(H_2)=\Phi(P_2H_4)=1$ are compatible with a simplified reaction scheme [6, 11]:

(1) $PH_3+h\nu\rightarrow PH_2+H$

(2) $PH_3+H\rightarrow PH_2+H_2$

(3) $2PH_2+M\rightarrow P_2H_4+M$

(4) $PH_3+h\nu\rightarrow PH+H_2$

The importance of H_2 formation by the reaction (4) in addition to reaction (2), at least at 147 nm, is shown by the photolysis of PH_3/PD_3 mixtures where formation of H_2 and D_2 predominates relative to the formation of HD. The origin of P_2H_4 from recombination of PH_2 radicals (reaction (3)) is confirmed by the formation of major quantities of $P_2H_2D_2$ during the photolysis of PH_3/PD_3 and rules out extensive P_2H_4 formation via insertion of PH into PH_3 [6]. The complete mechanism of the photolysis is clearly very complicated.

Formation of Intermediates. Quantum Yields. The photolysis at 193.3 nm induces excitation of the first electronically excited state of PH_3, designated as $\tilde{A}\,^1A_1$, which probably has a relatively short lifetime [14]. Photolysis in the vacuum UV range ($\lambda<180$ nm) proceeds principally via excitation of PH_3 to the higher electronically excited states designated as $\tilde{B}\,^1E$, $\tilde{C}\,^1A_1'$ [15], and $\tilde{D}\,^1A_2''$ states [6] (see p. 145 for a discussion of electronic states and their revised numbering). The shape of the $PH_3(\tilde{A})$ potential energy surface (PES) indicates that dissociation could be very fast, and it is believed that the coupling to the triplet surface is not strong enough to let the singlet-triplet transitions become competitive with dissociation. The distribution of the excess energy among the formed PH_2 and H indicates that the dissociation of electronically excited PH_3 does not occur via a unimolecular reaction mechanism on a PES which correlates without a barrier to $PH_2(\tilde{X})$ and $H(^2S)$ [16]. The experimental observation agrees with a calculated small barrier, found in an ab initio calculation at the CI level [17]. The PH_2 internal energy distribution, as obtained from the measured H atom kinetic energy distribution, agrees reasonably well with the calculated statistical distribution of energy among the PH_2 fundamentals [14]. The possible energetic states of formed PH_2 and H were discussed in [16].

The major decay processes during UV photolysis lead to the neutral product channels even beyond the ionization threshold of $\lambda=124$ nm. The quantum yield for the formation of neutral species decreases from 1.0 near the ionization threshold to 0.7 at $\lambda=106$ nm [18].

$PH_2(\tilde{A}\,^2A_1)$ is a primary product of the photolysis (at least in the far-UV region) under collisionless conditions. However, its low quantum yield indicates that the main primary product is $PH_2(\tilde{X}\,^2B_1)$ [19]. The direct observation of nascent $PH_2(\tilde{X})$ by laser-induced fluorescence (LIF) spectroscopy was obstructed by the fluorescence of $PH_2(\tilde{A})$ [14]. Thresholds of these and the other species which were investigated in detail are as follows:

species	threshold	reaction of formation, comment
$PH_2(\tilde{X}\,^2B_1)$	>193 nm	deduced as main product from the low $PH_2(\tilde{A})$ yield [9]
$PH_2(\tilde{A}\,^2A_1)$	208±1 nm	$PH_3 \rightarrow PH_2(\tilde{A}) + H(^2S)$, spin-forbidden; quantum yield 0.014±0.006; the radical vanishes at 119 nm [15, 19]
$PH(A\,^3\Pi_i)$	159±2 nm	$PH_3 \rightarrow PH(A) + H_2$, spin-forbidden; low yield [15, 20]
	121 nm	$PH_3 \rightarrow PH(A) + 2H(^2S)$, spin-allowed; moderate yield [15]
$PH(b\,^1\Sigma^+)$	147±2 nm	$PH_3 \rightarrow PH(b) + 2H(^2S)$, spin-allowed; very low yield [21]

The bands of PH(A) are of very low intensity when PH_3 is photolyzed in the 159 to 121 nm range but increase very sharply at <121 nm suggesting formation by a spin-allowed reaction [15]. The maximum emission of the spin-allowed primary product PH(b) is weaker than those of PH(A) of $PH_2(\tilde{A})$ by a factor of about 100; no explanation was proposed [21].

The formation of $PH(A\,^3\Pi_i)$ with populated spin sublevels i=0, 1, 2 during photolysis at 193 nm under collisionless conditions results from a two-photon process, probably via $PH_3 + h\nu \rightarrow PH_2(\tilde{X}) + H$ followed by $PH_2(\tilde{X}) + h\nu \rightarrow PH + H$ [19]. Early investigations of the flash photolysis of PH_3 extending to the far-UV range (210 to 180 nm) led to the identification of vibrationally excited PH_2, ground-state PH [12, 22], atomic phosphorus in the $^2D^0$ and $^2P^0$ states [23], and P_2 in the ground state [12] as unstable intermediates. The observation of a P_4 aerosol and of gaseous P_2H_4 after flash photolysis of ~1 Torr of PH_3 in ~200 Torr of H_2 [9] contradicts earlier results [12].

Quantum yields Φ for the primary reactions of the photolysis of PH_3/PD_3 mixtures at 147 nm were calculated with a simplified kinetic model of the formation of the hydrogen isotopomers using mass spectrometric data (* indicates electronic excitation) [6]:

reaction	Φ (Y=H)	Φ (Y=D)
$PY_3 = h\nu\,(147\ nm) \rightarrow PY^* + Y_2$	0.8	0.35
$\rightarrow PY_2^* + Y$	0.1	0.4
$\rightarrow PY + 2Y$	0.1	0.25

Photodissociation of PH_3 at 193 nm into PH_2 and H atoms leaves an energy of ~22000 cm^{-1} to be distributed among the products degrees of freedom. The H atom kinetic energy distribution was measured by sub-Doppler spectroscopy of the Lyman-α transition. The distribution peaks at rather low energies around 3500 cm^{-1}. The corresponding distribution of the PH_2

internal energy (obtained by conservation of energy and momentum) peaks at ~18000 cm^{-1} with a mean value of ~14000 cm^{-1}. The available energy would allow for $PH_2(\tilde{A}\,^2A_1)$ formation which requires about 19000 cm^{-1}. In this case the H atoms would be very cold. However, in agreement with other data [19], vibrationally highly excited $PH_2(\tilde{X}\,^2B_1)$ is probably the main product. The failure to observe the nascent rotational-vibrational excitation of $PH_2(\tilde{X})$ by laser-induced fluorescence was attributed to virtually unoccupied low vibrational levels [14, 16]; for preliminary results see [16, 24].

Vibrational excitation of the $PH_2(\tilde{A})$ bending level ν_2 up to v=6 [19] or 7 [15] was found at 193 nm irradiation, but the vibrational ground state appeared to be the most populated one [19]; see also [14, p. 885] for a comment on the distribution. While PH(A) was observed only in its vibrational ground state [15], results on PH(b) are insufficient for a definite statement and formation in the vibrational ground state was supposed [21].

The photolysis of PH_3 was claimed to yield PH_2 in the ground state mostly by secondary processes; kinetics were investigated by LIF [25].

UV Photolysis in Solid Phases

Far-UV photolysis of frozen PH_3 at 77 K yields H_2 and elemental phosphorus just as the photolysis in the gas phase [7].

There is no decomposition of PH_3 in an Xe matrix upon irradiation at $\lambda \geqq 250$ nm and temperatures between 4.2 and 100 K [26]. Photolysis at 184.9 nm of PH_3 in an Ar matrix at 4.2 K was used to demonstrate the formation of phosphorus atoms; analysis was by EPR spectroscopy [27, 28]. The same result was described for PH_3 and PD_3 under identical conditions upon irradiation with a low-pressure Hg lamp [29]. Far-UV photolysis at 77 K was applied to PH_3 absorbed at ambient temperature by a zeolite (cancrinite, $Na_6Al_6Si_6O_{24} \cdot CaCO_3 \cdot 2H_2O$). The hydrogen atoms formed and roughly equivalent quantities of PH_2 and atomic phosphorus were identified by the EPR spectrum; the products are stabilized by the host lattice [30].

When a mixture of PH_3 and Ar (1:200) is condensed at 14 K with simultaneous photolysis at λ=121.6 nm, absorption spectra of PH_2 and PH can be observed in the formed solid. Absorptions of P_2 were thought to result from secondary photodecomposition [31, 32]. An analogous behavior was found for PD_3 [32]. The vacuum-UV photolysis of PH_3 in Kr (1:5) at 4.2 K and the investigation by EPR spectroscopy led to the identification of PH_2, atomic H, atomic P, and a signal originally thought to be PH_4 [33]. However, the assignment of the last species is erroneous [34]; see also [35].

References:

[1] Ashfold, M. N. R.; Macpherson, M. T.; Simons, J. P. (Top. Curr. Chem. **86** [1979] 1/90, 27/9).
[2] Wagener, R.; Caldwell, J.; Owen, T. (Icarus **66** [1986] 188/91; C. A. **104** [1986] No. 233582).
[3] Hildebrand, R. H.; Loewenstein, R. F.; Harper, D. A.; Orton, G. S.; Keene, J.; Whitcomb, S. E. (Icarus **64** [1985] 64/87; C. A. **104** [1986] No. 37318).
[4] Drossart, P.; Encrenaz, T.; Tokunaga, A. T. (Icarus **60** [1984] 613/20; C. A. **102** [1985] No. 69791).
[5] Prinn, R. G.; Lewis, J. S. (Science **190** [1975] 274/6).
[6] Blazejowski, J.; Lampe, F. W. (J. Phys. Chem. **85** [1981] 1856/64).
[7] Noy, N.; Podolak, M.; Bar-Nun, A. (J. Geophys. Res. C **86** [1981] 11985/8; C. A. **96** [1982] No. 43568).
[8] Zakhar'in, V. I.; Nadtochenko, V. A.; Sarkisov, O. M.; Teitel'boim, M. A. (Dokl. Akad. Nauk SSSR **263** [1982] 127/30; Dokl. Phys. Chem. [Engl. Transl.] **262/267** [1982] 168/70).

[9] Ferris, J. P.; Benson, R. (J. Am. Chem. Soc. **103** [1981] 1922/7).
[10] Ferris, J. P.; Benson, R. (Nature **285** [1980] 156/7).

[11] Ferris, J. P.; Bossard, A.; Khwaja, H. (J. Am. Chem. Soc. **106** [1984] 318/24).
[12] Norrish, R. W. G.; Oldershaw, G. A. (Proc. R. Soc. [London] A **262** [1961] 1/9).
[13] Vera Ruiz, H. G.; Rowland, F. S. (Geophys. Res. Lett. **5** [1978] 407/10; C. A. **89** [1978] No. 92778).
[14] Baugh, D.; Koplitz, B.; Xu, Z.; Wittig, C. (J. Chem. Phys. **88** [1988] 879/87).
[15] di Stefano, G.; Lenzi, M.; Margani, A.; Mele, A.; Nguyen Xuan, C. (J. Photochem. **7** [1977] 335/44).
[16] Koplitz, B.; Xu, Z.; Baugh, D.; Buelow, S.; Häusler, D.; Rice, J.; Reisler, H.; Qian, C. X. W.; Noble, M.; Wittig, C. (Faraday Discuss. Chem. Soc. No. 82 [1986] 125/48).
[17] Müller, J.; Ågren, H.; Canuto, S. (J. Chem. Phys. **76** [1982] 5060/8).
[18] Xia, T. J.; Chien, T. S.; Wu, C. Y. R.; Judge, D. L. (J. Quant. Spectrosc. Radiat. Transfer **45** [1991] 77/91; C. A. **114** [1991] No. 111069).
[19] Sam, C. L.; Yardley, J. T. (J. Chem. Phys. **69** [1978] 4621/7).
[20] Becker, K. H.; Welge, K. H. (Z. Naturforsch. **19a** [1964] 1006/15).

[21] di Stefano, G.; Lenzi, M.; Margani, A.; Nguyen Xuan, C. (J. Chem. Phys. **68** [1978] 959/63; Extend. Abstr. 5th Int. Conf. Vac. Ultraviolet Radiat. Phys., Montpellier, Fr., 1977, Vol. 1, pp. 149/52; C. A. **90** [1979] No. 31829).
[22] Kley, D.; Welge, K. H. (Z. Naturforsch. **20a** [1965] 124/31).
[23] Basco, N.; Yee, K. K. (Nature **216** [1967] 998/9).
[24] Xu, Z.; Koplitz, B.; Buelow, S.; Baugh, D.; Wittig, C. (Chem. Phys. Lett. **127** [1986] 534/40).
[25] Stephens, K. M. (Diss. Univ. Alabama 1990 from Diss. Abstr. Int. B **52** [1991] 865).
[26] Shimokoshi, K.; Nakamura, K.; Sato, S. (Mol. Phys. **53** [1984] 1239/49).
[27] Cochran, E. L. (4th Int. Symp. Free Radical Stab. Trapped Radicals Low Temp., Washington, D.C., 1959, pp. D-I-1/D-I-6; C. A. **57** [1962] 1787).
[28] Cochran, E. L.; Adrian, F. J. (Prepr. Pap. 5th Int. Symp. Free Radicals, Uppsala 1961, pp. 12-1/12-7; C. A. **59** [1963] 10905).
[29] Adrian, F. J.; Cochran, E. L.; Bowers, V. A. (Adv. Chem. Ser. No. 36 [1962] 50/67).
[30] Raghunathan, P.; Sur, S. K. (Proc. Indian Acad. Sci. Chem. Sci. **92** [1983] 597/604; C. A. **100** [1984] No. 148374).

[31] Larzillière, M.; Jacox, M. E. (J. Mol. Spectrosc. **79** [1980] 132/50).
[32] Larzillière, M.; Jacox, M. E. (NBS Spec. Publ. [U.S.] No. 561-1 [1979] 529/43; C. A. **92** [1980] No. 31431).
[33] McDowell, C. A.; Mitchell, K. A. R.; Raghunathan, P. (J. Chem. Phys. **57** [1972] 1699/703).
[34] Colussi, A. J.; Morton, J. R.; Preston, K. F. (J. Chem. Phys. **62** [1975] 2004/6).
[35] Griller, D.; Roberts, B. P. (J. Chem. Soc. Perkin Trans. II **1973** 1339/42).

1.3.1.5.1.3.2 IR Photolysis

Gaseous PH_3, either neat or in the presence of photosensitizers, may decompose under certain conditions upon exposure to laser IR radiation. The IR-photosensitized thermolysis of PH_3 is described in Section 1.3.1.5.1.2, p. 204. The reactions in PH_3/SiH_4 mixtures under the influence of IR radiation in the absence and presence of photosensitizers are described in Section 1.3.1.5.6, pp. 249/51.

The fundamental frequencies of the ν_2 and ν_4 modes of PH_3 lie in the emission range of the CO_2 laser. However, the absorption by PH_3 is too weak to directly result in decomposition by an unfocused CO_2 laser with fluences up to 1 J/cm^2 at all PH_3 pressures and in the presence of nonabsorbing gases [1, 2]. PH_3 decomposition is, however, observed in the presence of photosensitizers [1, 3] or in laser beams with high fluences [4, 5].

The decompositions of PH_3 by pulsed IR laser beams exhibit energy thresholds and yield H_2 and solid $(PH_x)_n$ with $x<1$. The range of x differs for experiments with a focused beam from those with a parallel beam in the presence of photosensitizers. The product yield is much higher when a focused beam is used [4].

In SiF_4-photosensitized IR decomposition the PH_3 molecules are excited via collisional transfer of vibration energy from photochemically excited SiF_4 [1]. The dissociation probably follows the channel $PH_3 + nh\nu \rightarrow PH(\tilde{X}\ ^3\Sigma^-) + H_2(\tilde{X}\ ^1\Sigma_g^+)$. No H atoms could be trapped with added C_2H_4 [6]. The SF_6-photosensitized decomposition of PH_3 is similar [3]. In contrast to the UV photolysis (cf. 1.3.1.5.1.3.1, pp. 206/10), P_2H_4 was not formed [1].

The decomposition of gaseous PH_3 in a focused CO_2 laser beam is not an IR multiphoton decomposition, but due to a dielectric breakdown. The decomposition arises from the interaction of the generated free electrons with PH_3. The decomposition is accompanied by a visible chemiluminescence stemming from formed PH_2, PH, H_2, H, and probably H_2^+ [4].

References:

[1] Blazejowski, J.; Lampe, F. W. (J. Phys. Chem. **88** [1984] 1666/70).
[2] Blazejowski, J.; Lampe, F. W. (J. Photochem. **29** [1985] 285/95).
[3] Adamova, Yu. A.; Kaganyuk, D. S.; Pimenov, V. P.; Skachkov, A. N.; Stolyarova, G. I.; Shmerling, G. V. (Khim. Fiz. **8** [1989] 1354/61; C.A. **112** [1990] No. 105876).
[4] Blazejowski, J.; Lampe, F. W. (J. Appl. Phys. **59** [1986] 2283/92).
[5] Bordé, C.; Henry, A.; Henry, L. (C. R. Seances Acad. Sci. B **263** [1966] 619/20).
[6] Blazejowski, J.; Lampe, F. W. (J. Photochem. **24** [1984] 235/8).

1.3.1.5.1.4 Interactions with Electrons. Behavior in Discharges and Afterglows

Interactions with Electrons

Neutral PH molecules in electronically excited states resulted from the bombardment of PH_3 with a beam of 60-eV electrons [1] or pulses of 20-keV electrons [2].

The formation of **negative ions** was systematically investigated by mass spectrometric techniques for electron energies up to 10 eV, a range characteristic for resonance electron capture [3], and for electron energies up to 75 eV which also include those energies needed for ion pair formation [4]. Appearance potentials which were obtained from measured ionization efficiencies by linear extrapolation and tentative reaction paths $PH_3 + e^- \rightarrow$ products, as characterized by the negative ion and the accompanying products, are compiled in a table on p. 212 [4].

Measurements up to 10 eV carried out by [3] agree within ± 0.1 eV. The only difference is a weak and broad feature in the H^- ion efficiency curve observed at 3 to 4 eV [3] which was missing in the other study [4]. This feature either originates from a background or a secondary reaction [3] or corresponds to $PH_3 + e^- \rightarrow H^- + PH_2$ for which an appearance potential of 3.0 eV has been estimated [4]. The H^- appearance potentials given below replace earlier data [5]. Total cross sections for the formation of PH_2^-, PH^-, and P^- (all formation channels up to 10 eV) of 4×10^{-18}, 2×10^{-18}, and 3×10^{-19} cm^2 and corresponding relative abundances PH_2^- (100%),

PH^- (51.5%), and P^- (7%) have been reported [3]. The cross sections for the formation of negative ions are small as compared to those for the corresponding positive ions (see below). This is in agreement with the small relative abundances of negative ions reported in "Phosphor" C, 1965, p. 24. Further information on the formation of PH_2^- and PH^- by electron impact is found in Section 1.2.4.1, p. 102, and Section 1.1.4, p. 45, respectively.

ion	AP in eV	other products	ion	AP in eV	other products
H^-	see text	PH_2	PH^-	2.2 ± 0.2	H_2
	5.4 ± 0.1	$P+H_2$		6.3 ± 0.1	2H
	7.5	PH+H		8.2 ± 0.2 [1)]	2H
	16.3 ± 0.1	$P^++H_2+e^-$		19.8 ± 0.1	H^++H+e^-
PH_2^-	2.2 ± 0.1	H	P^-	5.8 ± 0.2	H_2+H
	5.2 ± 0.1 [1)]	H		8.4 ± 0.2 [1)]	H_2+H
	8 [1)]	H(?)		13.4 ± 0.2 [1)]	3H(?)
	15.8 ± 0.2	H^++e^-		22.6 ± 0.1	H^++2H+e^-

[1)] Reaction channel leads to excited products.

The formation of **positive ions** was systematically studied by mass spectrometric techniques for electron energies up to 180 eV. Their appearance potentials were derived from the energy-dependent ionization cross sections by linear extrapolation. The data for the individual ions originating from PH_3 are as follows [6]:

ion	$PH_3^+(^2A_1)$	$PH_3^+(^2E)$	PH_2^+	PH^+		P^+	
AP in eV	10.1 ± 0.1	12.5 ± 0.2	14.2 ± 0.2	13.3 ± 0.2	17.2 ± 0.2	18.8 ± 0.3	23.5 ± 0.3
ion	H_2^+		H^+	PH_3^{2+}	PH_2^{2+}	PH^{2+}	P^{2+}
AP in eV	14.8 ± 0.2	24.8 ± 0.5	25.3 ± 0.5	30.0 ± 1.0	35.5 ± 0.5	33.0 ± 1.0	45.6 ± 0.5

For PH^+ a second appearance potential of 16.4 ± 0.4 eV was reported [7]. The positive ions that originate from electron impact on PD_3 have the following appearance potentials [6]:

ion	$PD_3^+(^2A_1)$	$PD_3^+(^2E)$	PD_2^+	PD^+		P^+	
AP in eV	10.15 ± 0.1	12.75 ± 0.2	14.3 ± 0.2	13.3 ± 0.2	17.7 ± 0.3	19.5 ± 0.3	22.0 ± 0.3
ion	D_2^+		D^+	PD_3^+	PD_2^+	PD^{2+}	P^{2+}
AP in eV	14.9 ± 0.2	24.8 ± 0.5	25.3 ± 0.5	29.8 ± 0.5	34.9 ± 0.5	34.4 ± 0.5	46.7 ± 0.5

D_3^+ was detected in addition to the abovementioned ions. The appearance potentials of ions derived from PD_3 are very similar to those for respective ions derived from PH_3. A tendency toward slightly higher values can be anticipated taking the difference in zero-point energies into account [6].

Appearance potentials that had been measured before with variations in experimental techniques and methods of data evaluation mostly agree. Reported values for singly ionized ions derived from PH_3 fall into the following ranges: 9.97 to 11.5 eV for $PH_3^+(^2A_1)$ [7 to 13] (see also Section 1.3.2.1, p. 309); 13.2 to 14.4 eV for PH_2^+ [8 to 12] (see also Section 1.2.2.1,

p. 93), a second value of 14.4 eV [7] was corrected to 13.4 eV [14]; 12.4 to 13.6 eV for PH^+ [7 to 11] (see also Section 1.1.2, p. 33); 15.9 to 17.2 eV for the first P^+ appearance potential [8 to 11]; 20 and 20.8 eV for the second P^+ appearance potential [7, 8]. Appearance potentials reported earlier for doubly ionized ions, in particular for PH_3^{2+} and PH^{2+} [8], are much lower. Reasons for this discrepancy were discussed [6]. Appearance potentials for $PD_3^+(^2A_1)$, PD^+, and P^+ (first potential) [10] are close to those given above.

The first and third [6], first and second experimental P^+ appearance potentials [7, 8, 10] most probably indicate reaction channels that lead to accompanying H_2 or 2H products, respectively; see, however, the discussion in [9].

The appearance potential of PH^+ around 13.3 eV was ascribed to channels $PH_3 + e^- \rightarrow PH^+ + H^- + H + e^-$ [5] (which is not compatible with the H^- appearance potentials mentioned above) or $PH_3 + e^- \rightarrow PH^+ + H_2 + 2e^-$ [7, 10]. The second PH^+ appearance potential observed only once at 16.4 eV was tentatively ascribed to $PH_3 + e^- \rightarrow PH^+ + 2H + 2e^-$ [7].

Absolute ionization cross sections for the individual singly and doubly ionized ions arising from electron impact on PH_3 are depicted in **Fig. 11**.

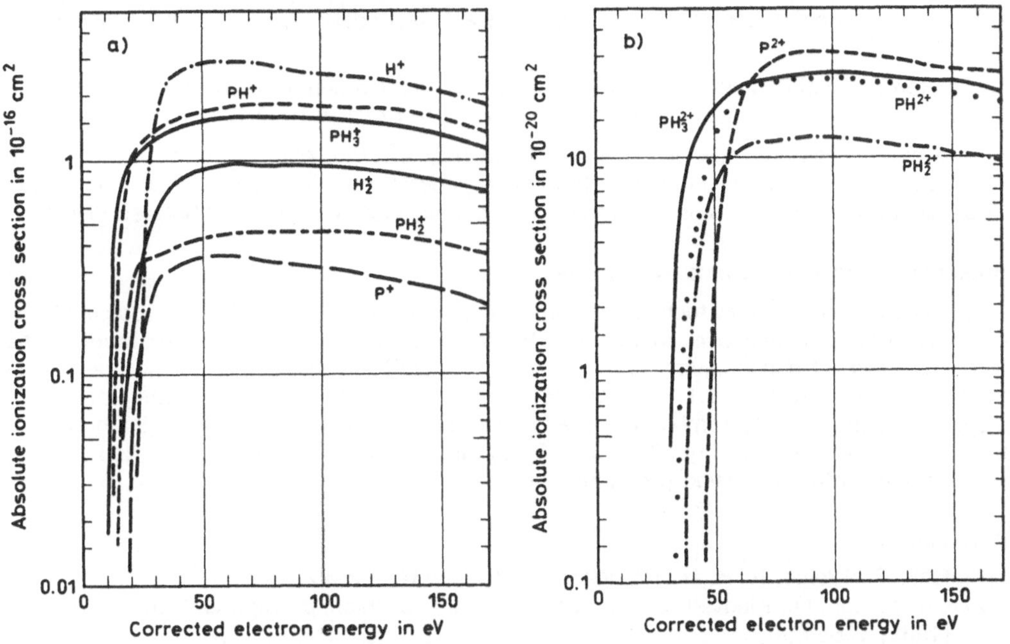

Fig. 11. Electron impact ionization of PH_3. Absolute cross sections for the formation of individual a) singly and b) doubly ionized ions as function of the electron energy from [6].

The cross sections for singly ionized ions steeply increase from the threshold to a broad maximum in the 60 to 80 eV region and then slowly decrease again toward higher electron energies. The cross sections for doubly ionized ions show similar patterns. The maxima, however, are shifted toward higher energies (ca. 90 to 100 eV) and their absolute values are lower by a factor of about 1000, see Fig. 11. A cross section of ca. 5×10^{-21} cm^2 at an electron energy of 90.5 eV was estimated for D_3^+ formation from PD_3. An insufficient isotopic purity of the sample prevented the measurement of the absolute cross sections for ion formation from PD_3 [6].

Total dissociative ionization cross sections for PH_3 that were measured in the 20 to 100 eV range, e.g., 5×10^{-16}, 11×10^{-16}, and 9×10^{-16} cm^2 at energies of 20, 50, and 100 eV, respectively [15], are about three times higher than the sum of the individual contributions for positive ions [6]. Because positive ion formation dominates all other inelastic channels, their total cross section can be compared to theoretical values which were calculated for inelastic and elastic electron scattering at PH_3 in the 10 to 5000 eV range. Theoretical cross sections for the sum of all inelastic channels (excitation, ionization, and dissociation processes) fairly agree with total experimental cross sections for positive ion formation, e.g., at 50 and 100 eV theoretical values are about 8×10^{-16} and 6.3×10^{-16} cm^2 [16] while experimental data are about 4×10^{-16} and 4.2×10^{-16} cm^2 [6].

The general patterns of the energy-dependent cross section shown in Fig. 11, p. 213, agree with the observation that no major dependences in the abundances of positive ions are to be expected when mass spectra of PH_3 are recorded at electron energies in the 40 to 80 eV range [17, figure 1]. Relative abundances of the major species, e.g., PH^+ (100), PH_3^+ (81.7), P^+ (34.1), and PH_2^+ (25.0) at 50 eV [17], are roughly established by several mass spectra taken in the 50 to 70 eV range [7, 10, 11, 17, 18]. Relative abundances measured at 50 eV for minor species, e.g., PH_3^{2+} (0.04), PH_2^{2+} (0.03), PH^{2+} (0.05), P^{2+} (0.02), H_2^+ (0.5), H^+ (0.3) (reference PH^+ (100)) [17], are naturally more prone to impurity contributions and secondary reactions, see the discussion in [6].

The formation of PH_4^+ upon electron impact on PH_3 is the result of a consecutive reaction of the primary molecular ion, $PH_3^+ + PH_3 \rightarrow PH_4^+ + PH_2$, see Section 1.4.2.1, p. 315.

Ionization cross sections of gaseous PH_3 at high electron energies of 0.1 to 2.7 MeV (electrons originated from β-decay) fitted into a general formula originally given by Bethe. Coefficients for several atoms and molecules, among them PH_3, were tabulated [19].

Behavior in Discharges and Afterglows

Discharge and afterglow techniques applied to pure gaseous PH_3 and to gaseous mixtures containing PH_3 have been used to produce PH_3 fragments, e.g., for spectroscopic studies. The formation of PH_2, PH in various electronic states, and of PH^+ is described in the corresponding Sections 1.2.1.1, pp. 47/9, 1.1.1.4.2, pp. 26/9, and 1.1.2, pp. 33/42. P_2 which was produced in a spark discharge in PH_3/Ar mixtures, see for example [20], and P_2H_4 which originated from a low-pressure electric discharge (2.6 to 13.8 kV) through PH_3 (with a maximum yield at 5 kV) [21] are two further identified products.

References:

[1] Fink, E.; Welge, K. H. (Z. Naturforsch. **19a** [1964] 1193/201).
[2] Gustafsson, O.; Kindvall, G.; Larsson, M.; Senekowitsch, J.; Sigray, P. (Mol. Phys. **56** [1985] 1369/80).
[3] Ebinghaus, H.; Kraus, K.; Müller-Duysing, W.; Neuert, H. (Z. Naturforsch. **19a** [1964] 732/6).
[4] Halmann, M.; Platzner, I. (J. Phys. Chem. **73** [1969] 4376/8).
[5] Halmann, M.; Klein, Y. (Adv. Mass Spectrom. **3** [1966] 267/72, 270/1).
[6] Märk, T. D.; Egger, F. (J. Chem. Phys. **67** [1977] 2629/35).
[7] Saalfeld, F. E.; Svec, H. J. (Inorg. Chem. **2** [1963] 46/50).
[8] Fischler, J.; Halmann, M. (J. Chem. Soc. **1964** 31/6).
[9] Morrison, J. D.; Traeger, J. C. (Int. J. Mass Spectrom. Ion Phys. **11** [1973] 277/88).
[10] Wada, Y.; Kiser, R. W. (Inorg. Chem. **3** [1964] 174/7).

[11] Fehlner, T. P.; Callen, R. B. (Adv. Chem. Ser. No. 72 [1968] 181/90).
[12] McAllister, T.; Lossing, F. P. (J. Phys. Chem. **73** [1969] 2996/8).

[13] Kiser, R. W.; Gallegos, E. J. (J. Phys. Chem. **66** [1962] 947/8).
[14] Saalfeld, F. E.; Svec, H. J. (Inorg. Chem. **3** [1964] 1442/3).
[15] Lloret, A.; de Rosny, G. (Int. J. Mass Spectrom. Ion Processes **62** [1984] 89/98).
[16] Jain, A.; Baluja, K. L. (Phys. Rev. [3] A **45** [1992] 202/18).
[17] Larin, N. V.; Devyatykh, G. G.; Agafonov, I. L. (Zh. Neorg. Khim. **9** [1964] 205/7; Russ. J. Inorg. Chem. [Engl. Transl.] **9** [1964] 110/1).
[18] Halmann, M. (J. Chem. Soc. **1962** 3270/2).
[19] Rieke, F. F.; Prepejchal, W. (Phys. Rev. [3] A **6** [1972] 1507/19, 1513).
[20] Howe, J. D.; Puyuelo, P.; Ashfold, M. N. R.; Western, C. M. (J. Chem. Soc. Faraday Trans. **89** 1993 2337/43).

[21] Albrand, J. P.; Anderson, S. P.; Goldwhite, H.; Huff, L. (Inorg. Chem. **14** [1975] 570/3).

1.3.1.5.1.5 Interactions with Neutrons

Thermal neutrons induce the nuclear reaction $^{31}P(n,\gamma)^{32}P$ at the phosphorus atom of PH_3, see "Phosphor" B, 1964, pp. 132/6. Based on scavenging experiments with butadiene, an upper limit for the bond retention in PH_3 during this process was set at about 4% [1]. This value agrees with a theoretical one of 3.2% which was calculated with the assumption that the γ cascade after neutron capture can be represented by three equal quanta with random angular correlation [2]. PH_3 itself is an effective scavenger for the ^{32}P atoms which recoil from the nuclear reaction. Absolute $^{32}PH_3$ yields up to almost 80% were reported [3] (for values in the 50 to 60% range, see [1, 4, 5]). The remaining ^{32}P activity was found at the walls of the reaction vessels, probably in the form of phosphorus oxoacids. This was due to the fact that total ^{32}P activities achieved under neutron irradiation were much below the level of oxygen and/or water impurities, even in pure PH_3 [2, 4]; see also [5]. $^{32}PH_3$ yields were independent of the PH_3 pressure in the 50 to 800 Torr range [1]. Scavenging of hot ^{32}P atoms by PH_3 is most probably a multistep process initiated by $PH_3 + {}^{32}P \rightarrow {}^{32}PH + PH_2$; however, no P_2H_4 product was found which might arise from intermediate PH_2 [1, 3, 6]. A speculation that $^{32}PH_3$ is formed by reactions of hot and thermalized ^{32}P atoms in a 60 : 40 ratio [3] must be viewed with scepticism [6].

Noble gases He, Ne, Ar were used to moderate the recoiling ^{32}P atoms. With 98% neon in the gas mixture the $^{32}PH_3$ yield was reduced to about 1/4 of the value found in unmoderated systems [6]. This result is in some contrast with earlier observations where the $^{32}PH_3$ yield was found to be insensitive to moderation or scavenging by O_2, H_2O, NH_3 [4], NO [1], H_2S [3], CH_4, CCl_4 [4], ethene [3], 1-butene [1], and cyclohexene [4]. ^{32}P-labeled methylphosphanes were observed in the presence of an excess of methane [2]; see also [4]. Butadiene proved to be an effective scavenger [1].

Fast neutrons (from $^{9}Be(d,n)^{10}Be$ with 6 MeV deuterons) induce the nuclear reaction $^{31}P(n,p)^{31}Si$ at the phosphorus atom of PH_3. ^{31}Si atoms recoil with an energy of about 6×10^5 eV. This energy assures that all bonds in the PH_3 precursor molecule are broken [7]. Irradiation of pure PH_3 gives $^{31}SiH_4$ which contains 60 to 100% of the ^{31}Si yield in volatile products [7, 8] but only 2% of the total ^{31}Si yield [8]. $^{31}SiH_2$ probably plays a major role as a reactive intermediate, as demonstrated by the products formed in irradiated gas samples of pure PH_3 and mixtures of PH_3 with several scavenger gases (among them silanes, butadiene, NO); see "Silicon" Supp. Vol. B 1, 1982, pp. 30, 39, and [9, 10].

When $PH_3/^{3}He$ mixtures are irradiated by neutrons, the additional nuclear reaction $^{3}He(n,p)T$ has to be taken into account. The tritium atoms formed react with PH_3 to give HT and PH_2T [11].

References:

[1] Gennaro, G. P.; Tang, Y.-N. (J. Inorg. Nucl. Chem. **36** [1974] 259/62).
[2] Halmann, M.; Kugel, L. (J. Inorg. Nucl. Chem. **25** [1963] 1343/50).
[3] Stewart, G. W.; Hower, C. O. (J. Inorg. Nucl. Chem. **34** [1972] 39/45).
[4] Halmann, M.; Kugel, L. (J. Chem. Soc. **1964** 4025/8).
[5] Halmann, M. (Chem. Rev. **64** [1964] 689/702).
[6] Zeck, O. F.; Ferrieri, R. A.; Copp, C. A.; Gennaro, G. P.; Tang, Y.-N. (J. Inorg. Nucl. Chem. **41** [1979] 785/9).
[7] Gaspar, P. P.; Pate, B. D.; Eckelman, W. (J. Am. Chem. Soc. **88** [1966] 3878/9).
[8] Gaspar, P. P.; Bock, S. A.; Eckelman, W. C. (J. Am. Chem. Soc. **90** [1968] 6914/22).
[9] Tominaga, T.; Tachikawa, E. (Modern Hot-Atom Chemistry and its Applications, Springer, Berlin 1981, pp. 73/5).
[10] Tang, Y.-N. (in: Abramovitch, R. A.; Reactive Intermediates. Reactions of Silicon Atoms and Silylenes, Vol. 2, Plenum, New York 1982, pp. 297/366, 299/301, 303/5, 343, 351).

[11] Castiglioni, M.; Volpe, P. (Radiochim. Acta **34** [1983] 165/8).

1.3.1.5.1.6 Gamma Irradiation

Decomposition of gaseous PH_3 by irradiation with ^{60}Co γ-rays yields hydrogen and red phosphorus with only traces of P_2 and P_2H_4 being detected. The yield of H_2 was proportional to the dose between 0.4×10^{20} and 5×10^{20} eV and also proportional to the initial PH_3 pressure between 50 Torr and 1 atm. Experimental G values (in molecules per 100 eV) of $G(H_2)=11.3$ and $G(-PH_3)=7.5$ are in line with an overall stoichiometry $PH_3 \rightsquigarrow P(solid, red) + 3/2\,H_2$. Only at high doses up to 9×10^{20} eV did a slight decrease of $G(H_2)$ occur [1]. In cases where gaseous mixtures of PH_3 and NH_3 were irradiated, PH_3 acted as a radical scavenger, e.g., $H(NH_2) + PH_3 \rightarrow PH_2 + H_2(NH_3)$ [2]. Similarly, PH_3 acted as a scavenger in irradiated PH_3/CH_3I mixtures [3].

Irradiation of $PH_3(PD_3)$ in low-temperature Kr or Xe matrices yielded $PH_2(PD_2)$ (see also Section 1.2.1.1, p. 50) along with atomic species H(D) and P [4]; see also [5]. PH_4 was detected after irradiation of PH_3 in an Xe matrix at 77 K [5]. Formation of $PH_4(PD_4)$ was confirmed by γ-irradiation of a neopentane/5% $PH_3(PD_3)$ mixture [8]. Analogously, PH_3F [5, 9] or PD_3F molecules [5] were detected at 133 K after γ-irradiation of an SF_6 matrix (at 77 K) which contained 1% $PH_3(PD_3)$ [5] or 5% PH_3 [9]. The PH_3^+ radical cation is the major product of γ-irradiation of a frozen, solid sulution of PH_3 in concentrated aqueous H_2SO_4 [6, 7]. The formation of PH_2 (PD_2 and traces of PHD in concentrated D_2SO_4 in D_2O) under the above condition was also observed [6].

References:

[1] Buchanan, J. W.; Hanrahan, R. J. (Radiat. Res. **42** [1970] 244/54).
[2] Buchanan, J. W.; Hanrahan, R. J. (Radiat. Res. **44** [1970] 296/304).
[3] Buchanan, J. W.; Hanrahan, R. J. (Radiat. Res. **44** [1970] 305/12).
[4] Morehouse, R. L.; Christiansen, J. J.; Gordy, W. (J. Chem. Phys. **45** [1966] 1747/51).
[5] Sogabe, K. (J. Sci. Hiroshima Univ. A **39** [1975] 11/25).
[6] Fullam, B. W.; Mishra, S. P.; Symons, M. C. R. (J. Chem. Soc. Dalton Trans. **1974** 2145/8).
[7] Begum, A.; Lyons, A. R.; Symons, M. C. R. (J. Chem. Soc. A **1971** 2290/3).
[8] Colussi, A. J.; Morton, J. R.; Preston, K. F. (J. Chem. Phys. **62** [1975] 2004/6).
[9] Colussi, A. J.; Morton, J. R.; Preston, K. F. (J. Chem. Phys. **79** [1975] 1855/8).

1.3.1.5.2 Reactions with Gaseous Ions

The reactions of PH_3 with ions in the gas phase generally were investigated by mass spectrometric methods. The experimental rate constants were occasionally compared to maximum rate constants which were calculated by using the average dipole orientation (ADO) theory [1 to 3].

With Cations

Proton Affinity. Gas Phase Basicity. H_2^+. The proton affinity (PA) of PH_3 is given by the negative enthalpy change, $-\Delta_rH^\circ$, for the reaction $PH_3 + H^+ \rightarrow PH_4^+$, whereas the corresponding negative Gibbs free enthalpy change, $-\Delta_rG^\circ$, is called the gas-phase basicity (GB) of PH_3. Direct experimental investigations of the protonation reaction with respect to the determination of PA seem to be missing, but data for PH_3 were determined mass spectrometrically relative to the PA and GB values of other molecules, using mostly ion cyclotron resonance (ICR) spectroscopy at various pressures of the gas mixtures.

The selected $PA(PH_3)$, 188.6 kcal/mol at 300 K in [4], is based on the experimental results in [5, 6]. An absolute value of $PA(PH_3)$ of 187.4 kcal/mol was obtained from $\Delta_rH^\circ_{320} = -14.9$ kcal/mol for the reaction $PH_4^+ + NH_3 \rightarrow PH_3 + NH_4^+$ and a reference value, $PA(NH_3) = 202.3 \pm 2$ kcal/mol. A temperature of 300 K was given originally [6], but was found to have been actually somewhat higher [7]. A PA value of 185.6 kcal/mol at 320 K relative to the reference value $PA(NH_3) = 203.6$ kcal/mol [8] was derived from experiments with PH_3/C_3H_6 mixtures in [5]. A $PA(PH_3)$ of 191.1 kcal/mol was calculated from $GB(PH_3) = 182.5$ kcal/mol (relative to $GB(NH_3) = 196.4$ kcal/mol), correcting for symmetry entropy changes on protonation [9]. Earlier bracketing experiments with the mixtures PH_3/CH_3CHO and $PH_3/(CH_3)_2CO$ yielded $PA(PH_3) = 185 \pm 4$ [10 to 12] and 186 ± 1 kcal/mol [13], however, using the reference value $PA(NH_3) = 207 \pm 3$ kcal/mol. A selected value, $GB(PH_3) = 180.2$ kcal/mol at 300 K, was calculated [4] from the experimental results in [5, 6].

Calculations of the proton affinity by thermodynamic cycles in [13, 14] lead to values in the range of the above given experimental data. More sophisticated ab initio MO calculations predict a proton affinity at 298 K in the range of 188.1 kcal/mol at the SCF-CI level [15] to 191.7 kcal/mol at the MP4/6-31G**//HF/6-31G* level [16]. Both values were corrected for zero-point energy, whereas only the first value includes a correction for the difference in the specific heats of PH_3 and PH_4^+. In a very sophisticated ab initio Mo study, theoretical methods for the determination of the protonation energies of a series of hydrogen compounds including PH_3 were compared. These methods included the MP perturbation theory up to the fourth order (MP4), the linearized coupled-cluster method (LCCM), the averaged coupled-pair functional (ACPF), and the configuration interaction with all single and double excitations (CISD). The agreement between the computed electronic protonation energies and a so-called experimental electronic protonation energy value was excellent for all of the correlation methods [17]. Other ab initio calculations are described in [18 to 24]; for semiempirical calculations, see [25 to 28].

Qualitative mass spectrometric investigations at increasing ion source gas pressures showed that the proton affinity of PH_3 is smaller than that of NH_3 [10, 12], c-C_6H_{10} [5], $(CH_3)_2CO$ [10], ferrocene [29], and $Co(CO)_3NO$ [30], i.e., the molecules deprotonate PH_4^+ to PH_3, and larger than that of H_2O [10, 12], AsH_3 [31], CH_4, C_2H_4, and CH_3CHO [10]; in the latter cases the protonated molecules are deprotonated by PH_3.

The bombardment of gaseous PH_3 by H^+ ions (0.4 to 2.0 MeV) and H_2^+ ions (0.4 to 1.2 MeV) was used to measure the L-shell Auger electron spectrum of phosphorus and to calculate the ionization cross section of the L-shell from the Auger electron yields [32].

Noble Gas Ions. Electron transfer is the dominant channel in the reactions of PH_3 with the cations of the noble gases. The rate constants k at ambient temperature and the products were investigated by the ICR method [2]:

reaction	k in $cm^3 \cdot molecule^{-1} \cdot s^{-1}$	products[a) (mole fraction)
$PH_3 + He^+$	2.6×10^{-9}	not identified
$PH_3 + Ne^+$	1.2×10^{-9}	PH^+ (0.8), P^+ (0.2)
$PH_3 + Ar^+$	0.4×10^{-9}	PH_2^+ (0.8), PH^+ (0.2)
$PH_3 + Kr^+$	1.3×10^{-9}	probably PH_2^+ and PH^+
$PH_3 + Xe^+$	1.0×10^{-9}	PH_3^+ (0.85)[b), PH^+ (0.15)

a) Identification is restricted to cations; for possible formation reactions, see the original paper. – b) Only this cation observed below the IP of Xe^+ ($^2P_{1/2}$).

The rate constants were found not to be correlated with the exothermicity of the reactions. However, their Franck-Condon factors offer a general guide to reactivity [2].

The bombardment of gaseous PH_3 by He^+ ions (0.4 to 2.0 MeV) was used to measure the L-shell Auger electron spectrum of phosphorus and to calculate the ionization cross section of the L-shell from the Auger electron yields [33].

Halogen Ions. The formation of the tetrahedral van der Waals complexes PH_3X^+ from PH_3 and X^+ with X = F, Cl, Br was investigated in an ab initio SCF MO calculation at the STO-3G* level. The complexation energy was predicted to decrease from PH_3F^+ to PH_3Br^+, and the complexes with F^+ and Cl^+ were found to be more stable than PH_4^+; however, the numerical values of the complexation energies were considered to be overestimated [34].

H_3O^+. NH_4^+. The hypothetical formation of van der Waals complexes between PH_3 and H_3O^+ was computed to be endothermal for three structural isomers in an ab initio MO calculation at the MP4/6-31G* level. Both constituents are linked by a single hydrogen bridge between H_3O^+ and PH_3 in the most favorable isomer, by two hydrogen bridges in a less favorable isomer, and by three hydrogen bridges in the isomer of highest energy [35]. On the other hand, the formation of singly hydrogen-bridged complexes between H_3O^+ or NH_4^+ and PH_3 was predicted to be exothermal for both cations at the RHF/4-31G level and to be more favorable for H_3O^+ [36]. The exothermal formation of a singly hydrogen-bridged complex between PH_3 and NH_4^+ was confirmed at the MP4/6-31G* level [35].

P^+. P_2^+. The reactions of PH_3 with P^+ and P_2^+ were investigated in studies related to the decomposition of PH_3 in an electron beam; see Section 1.3.1.5.1.4, p. 212. The rate constant for the reaction of P^+ with PH_3 of $k = 1.2 \times 10^{-9}$ $cm^3 \cdot molecule^{-1} \cdot s^{-1}$ at 300 K was determined in a selected-ion flow tube [37]. A rate constant of $k = 1.5 \times 10^{-9}$ $cm^3 \cdot molecule^{-1} \cdot s^{-1}$ resulted from a trap-mode ICR investigation [1]. An investigation in a drift-tube ion source gave $k = 2.0 \times 10^{-9}$ $cm^3 \cdot molecule^{-1} \cdot s^{-1}$ at 330 and 430 K [38]. The main products of the reaction were found to be P_2H^+ with a yield of 62% [37] or 85% [1] and H_2; this reaction channel has an estimated enthalpy change of $\Delta_rH^\circ_{298} = -80 \pm 8$ kcal/mol [47]. Charge transfer led to 26% [37] or 6% [1] of PH_3^+, and a third reaction channel gave 12% [37] or 9% [1] of $P_2H_2^+$. An early investigation by the drift-mode ICR method is described in [10]. The reaction of PH_3 with the radical cation P_2^+ leads to the formation of the cation $P_3H_2^+$ and the radical cation P_3H^+ [10].

$P_xH_y^+$. The reactions of PH_3 with PH_n^+ and PH_n^- ($n \geq 1$) are treated in the present volume in the corresponding chapters on these ions. A mixture of P_3^+ and $P_3H_2^+$ resulted from the interaction of PH_3 and P_2H^+ with a rate constant of 4.1×10^{-10} $cm^3 \cdot molecule^{-1} \cdot s^{-1}$ at 430 K, determined in a drift tube [10]. A rate constant of $k=3\times10^{-10}$ $cm^3\cdot molecule^{-1}\cdot s^{-1}$ was measured under the same conditions for the reaction of PH_3 with $P_2H_3^+$ [38].

The solvation of cations from the decomposition of PH_3 in an electron beam by other PH_3 molecules leads to equilibria between several solvated cations. The equilibrium constants for four series were determined in a drift tube at 196 K and at elevated source pressures. Solvation reactions according to $PH_3+[PH_4(PH_3)_{n-1}]^+ \rightleftharpoons [PH_4(PH_3)_n]^+$ were found for n up to 5. Analogous series were observed for $[P_2H_5(PH_3)_n]^+$ with n=1 to 4, $[P_3H_6(PH_3)_n]^+$ with n=1 to 3, and for $[P_4H_7(PH_3)_n]^+$ with n=1, 2. Thermodynamic data for these solvation reactions were calculated from the changes in the equilibrium constants with temperature [38]:

$[P_xH_y(PH_3)_n]^+$	temperature in K	n	$-\Delta_{solv}G$ in kcal/mol	$-\Delta_{solv}H$ in kcal/mol	$-\Delta_{solv}S$ in $cal\cdot K^{-1}\cdot mol^{-1}$
$[PH_4(PH_3)_n]^+$	298	2	2.5	9.2	22.3
		3	1.8	7.3	18.4
		4	1.7	6.2	15.0
		5	1.6	5.5	13.2
$[P_2H_5(PH_3)_n]^+$	283	1	3.55	9	19.3
	233	1	4.5	9	19.3
	215	2	3.66	9	—
		3	3.33	9	—
$[P_3H_6(PH_3)_n]^+$	247	1	2.39	10.8	34
	233	1	2.85	10.8	34
	215	2	3.27	—	—

The low values for the Gibbs free energy and the enthalpy of solvation are attributed to the small dipole moment and the large radius of the PH_3 molecule. The cations $[P_2H_5PH_3]^+$ and $[P_3H_6PH_3]^+$ can be observed at temperatures as high as 300 K [38].

CO^+. CO_2^+. The rate constant for the reaction of PH_3 with CO^+ at ambient temperature was determined to be $k=1.9\times10^{-9}$ $cm^3\cdot molecule^{-1}\cdot s^{-1}$. The products PH_3^+ (17%), PH_2^+ (23%), PH^+ (53%), and COH^+ (7%) were identified. The corresponding reaction with CO_2^+ has $k=1.3\times10^{-9}$ $cm^3\cdot molecule^{-1}\cdot s^{-1}$ at ambient temperature. The products consisted of PH_3^+ (25%), PH_2^+ (10%), and PH^+ (65%) [2].

Organic Cations. The reactions of PH_3 with cations of organic compounds were investigated by the ICR method. The reaction with CH_3^+ at ambient temperature leads to the formation of PCH_2^+ (63%) with a rate constant of $k=7.0\times10^{-10}$ $cm^3\cdot molecule^{-1}\cdot s^{-1}$ and of PCH_4^+ (37%) with a rate constant of $k=4.1\times10^{-10}$ $cm^3\cdot molecule^{-1}\cdot s^{-1}$ [10]. PH_3 and s-$C_3H_7^+$ yield PH_4^+ (43%), the rate constant being $k=(5.2\pm0.5)\times10^{-10}$ $cm^3\cdot molecule^{-1}\cdot s^{-1}$ at 350 K [5]. In the corresponding reaction with c-$C_3H_6^+$ at ambient temperature, PH_4^+ (33%) was formed with $k=4\times10^{-10}$ $cm^3\cdot molecule^{-1}\cdot s^{-1}$; the second product was found to be $H_2CPH_3^+$ (67%) [39].

The interaction between PH_3 and the radical cation of ethylene oxide gives the radical cation $CH_2PH_3^+$, whereas protonated ethylene oxide yields $C_2H_6P^+$ which is probably a cyclic species [40]. The equilibrium constants of the proton exchange reactions $PH_3+BH^+ \rightleftharpoons PH_4^+ + B$ were determined and yielded ΔG° values in kcal/mol of 0.1 for $B=C_2H_5OH$, 0.6 for $B=CH_3CN$, 1.0 for $B=HCOOCH_3$, and −1.5 for $B=CH_3CHO$. A temperature of 300 K was given originally [6], but was found actually to be 320 K [7]. The cyclic onium ions $C_2H_4X^+$ with X=Cl, Br react with PH_3 forming the cyclic cation $C_2H_4PH_2^+$. The analogous reaction of acyclic CH_3CHX^+ yields

$CH_3CHPH_2^+$ [41]. Proton transfer to PH_3 was found to be the exclusive reaction of CF_2H^+, whereas CCl_2H^+ did not react with PH_3 [3]. The reaction of PH_3 with $C_3H_6Li^+$ leads to the formation of PH_3Li^+ via transfer of Li^+ [30].

Transition Metal Carbonyl and Nitrosyl Ions. Substitution of one CO ligand was observed in reactions of PH_3 with $CoCO^+$ and $Co(CO)_2^+$. Reactions of PH_3 with $Co(NO)(CO)_n^+$ (n=1, 2, 3) proceeded by substitution of CO exclusively and resulted in mixtures of cationic nitrosyl complexes with up to three PH_3 ligands. The ICR investigation indicated the rate constant of $Co(NO)(CO)_3^+$ to be about 80% of that of $Co(NO)(CO)_2^+$. PH_3/H_2O and PH_3/AsH_3 mixtures and $Co(NO)(CO)_2^+$ react initially by substitution of one CO ligand by PH_3. The intermediate Co(NO)-$CO(PH_3)^+$ then reacts with PH_3, H_2O, or AsH_3 by substitution of the second CO ligand. The displacement of coordinated H_2O and AsH_3 by PH_3 was also observed [30]. An ICR investigation demonstrated the displacement of NO in $C_5H_5NiNO^+$ by PH_3 with formation of $C_5H_5NiPH_3^+$ [42].

With Anions

The deprotonation of PH_3 by OH^-, NH_2^- [10, 43], CH_3O^-, $C_2H_5O^-$, n-$C_3H_7O^-$ [10], and SiH_3^- [43] was observed by ICR spectroscopy. The reactions were used to bracket the gas-phase acidity of PH_3 (cf. 1.3.1.5.1.1, p. 199) [10, 43] and show that PH_3 has the highest relative acidity with respect to the corresponding neutral molecules [43]. Some rate constants of reactions with anions were determined at ambient temperature in flowing afterglow systems. The rate constants for the reactions of PH_3 with both OH^- and NH_2^- at 298 K were found to be $k = 2.0 \times 10^{-9}$ $cm^3 \cdot molecule^{-1} \cdot s^{-1}$ [44]. Hydrogen atom transfer from PH_3 to the radical anion $(CF_3)_2C^-$ yields PH_2 and $(CF_3)_2CH^-$ with $k = (3.7 \pm 0.6) \times 10^{-11}$ $cm^3 \cdot molecule^{-1} \cdot s^{-1}$ [45]. The reaction $PH_3 + H^- \rightarrow PH_4^-$ was predicted to have an enthalpy $\Delta_r H = 1.4$ kcal/mol by an ab initio MO calculation with inclusion of configuration interaction but omitting the zero-point vibrational energies [23].

For the reaction of PH_3 with $Mn(CO)_3^-$ a rate constant of $k = (6.8 \pm 0.2) \times 10^{-10}$ $cm^3 \cdot molecule^{-1} \cdot s^{-1}$ was found, the products being $Mn(CO)_3PH^-$ (98%) and $Mn(CO)_3(H)PH_2^-$ (2%). The rate constant of the analogous reaction of PH_3 with the radical anion $Fe(CO)_2^-$ was determined to be $k = (5.1 \pm 0.3) \times 10^{-10}$ $cm^3 \cdot molecule^{-1} \cdot s^{-1}$ yielding the radical anions $Fe(CO)(H)PH_2^-$ (68%) and $Fe(CO)_2PH^-$ (38%) [46].

References:

[1] Thorne, L. R.; Anicich, V. G.; Huntress, W. T. (Chem. Phys. Lett. **98** [1983] 162/6).
[2] Chau, M.; Bowers, M. T. (Chem. Phys. Lett. **44** [1976] 490/4).
[3] Lias, S. G.; Ausloos, P. (Int. J. Mass Spectrom. Ion Phys. **22** [1976] 135/45).
[4] Lias, S. G.; Liebman, J. F.; Levin, R. D. (J. Phys. Chem. Ref. Data **13** [1984] 695/808, 759).
[5] Lias, S. G.; Shold, D. M.; Ausloos, P. (J. Am. Chem. Soc. **102** [1980] 2540/8).
[6] Wolf, J. F.; Staley, R. H.; Koppel, I.; Taagepera, M.; McIver, R. T., Jr.; Beauchamp, J. L.; Taft, R. W. (J. Am. Chem. Soc. **99** [1977] 5417/29).
[7] Taft, R. W. (private communication, ref. 3f in [5]).
[8] Lias, S. G. (private communication, ref. 26 in [15]).
[9] Aue, D. H.; Bowers, M. T. (in: Bowers, M. T.; Gas Phase Ion Chemistry, Vol. 2, Academic, New York 1979, pp. 1/51, 17).
[10] Holtz, D.; Beauchamp, J. L.; Eyler, J. R. (J. Am. Chem. Soc. **92** [1970] 7045/55).

[11] Eyler, J. R. (Inorg. Chem. **9** [1970] 981/2).
[12] Holtz, D.; Beauchamp, J. L. (J. Am. Chem. Soc. **91** [1969] 5913/5).
[13] Haney, M. A.; Franklin, J. L. (J. Phys. Chem. **73** [1969] 4328/31).

[14] Waddington, T. C. (Trans. Faraday Soc. **61** [1965] 2652/5).
[15] Marynick, D. S.; Scanlon, K.; Eades, R. A.; Dixon, D. A. (J. Phys. Chem. **85** [1981] 3364/6).
[16] Lohr, L. L.; Schlegel, H. B.; Morokuma, K. (J. Phys. Chem. **88** [1984] 1981/7).
[17] Del Bene, J. E.; Shavitt, I. (J. Phys. Chem. **94** [1990] 5514/8).
[18] Müller, B.; Schüler, M.; Reinhold, J. (Chem. Phys. Lett. **172** [1990] 478/82).
[19] Korkin, A. A.; Tsvetkov, E. N. (Zh. Neorg. Khim. **34** [1989] 290/4; Russ. J. Inorg. Chem. [Engl. Transl.] **34** [1989] 161/4; Bull. Soc. Chim. Fr. **1988** 335/8).
[20] Hendewerk, M. L.; Frey, R.; Dixon, D. A. (J. Phys. Chem. **87** [1983] 2026/32).

[21] Latajka, Z.; Scheiner, S. (J. Chem. Phys. **81** [1984] 2713/6).
[22] Del Bene, J. E.; Frisch, M. J.; Pople, J. A. (J. Phys. Chem. **89** [1985] 3669/74).
[23] Trinquier, G.; Daudey, J.-P.; Caruana, G.; Madaule, Y. (J. Am. Chem. Soc. **106** [1984] 4794/9).
[24] Aue, D. H.; Webb, H. M.; Davidson, W. R.; Vidal, M.; Bowers, M. T.; Goldwhite, H.; Vertal, L. E.; Douglas, J. E.; Kollman, P. A.; Kenyon, G. I. (J. Am. Chem. Soc. **102** [1980] 5151/7).
[25] Friedemann, R.; Gründler, W.; Issleib, K. (Tetrahedron **26** [1970] 2861/7).
[26] Graffeuil, M.; Labarre, J.-F.; Lappert, M. F.; Leibovici, C.; Stelzer, O. (J. Chim. Phys. Phys.-Chim. Biol. **72** [1975] 799/802).
[27] Banyard, K. E.; Hake, R. B. (J. Chem. Phys. **43** [1965] 2684/9).
[28] Hake, R. B.; Banyard, K. E. (J. Chem. Phys. **45** [1966] 3199/203).
[29] Foster, M. S.; Beauchamp, J. L. (J. Am. Chem. Soc. **97** [1975] 4814/7).
[30] Weddle, G. H.; Allison, J.; Ridge, D. P. (J. Am. Chem. Soc. **99** [1977] 105/9).

[31] Wyatt, R. H.; Holtz, D.; McMahon, T. B.; Beauchamp, J. L. (Inorg. Chem. **13** [1974] 1511/7).
[32] Ariyasinghe, W. M.; Awuku, H. T.; Powers, D. (Phys. Rev. [3] A **42** [1990] 3819/25).
[33] Ariyasinghe, W. M.; Powers, D. (Phys. Rev. [3] A **41** [1990] 4751/8).
[34] Alcamí, M.; Mó, O.; Yáñez, M.; Abboud, J.-L. M. (J. Phys. Org. Chem. **4** [1991] 177/91).
[35] Evleth, E. M.; Hamou-Tahra, Z. D.; Kassab, E. (J. Phys. Chem. **95** [1991] 1213/20).
[36] Desmeules, P. J.; Allan, L. C. (J. Chem. Phys. **72** [1980] 4731/48).
[37] Smith, D.; McIntosh, B. J.; Adams, N. G. (J. Chem. Phys. **90** [1989] 6213/9).
[38] Long, J. W.; Franklin, J. L. (J. Am. Chem. Soc. **96** [1974] 2320/7).
[39] Lias, S. G.; Buckley, T. J. (Int. J. Mass Spectrom. Ion Processes **56** [1984] 123/37).
[40] Staley, R. H.; Corderman, R. R.; Foster, M. S.; Beauchamp, J. L. (J. Am. Chem. Soc. **96** [1974] 1260/1).

[41] Berman, D. W.; Anicich, V.; Beauchamp, J. L. (J. Am. Chem. Soc. **101** [1979] 1239/48).
[42] Corderman, R. R.; Beauchamp, J. L. (J. Am. Chem. Soc. **98** [1976] 3998/4000).
[43] Brauman, J. I.; Eyler, J. R.; Blair, L. K.; White, M. J.; Comisarow, M. B.; Smyth, K. C. (J. Am. Chem. Soc. **93** [1971] 6360/2).
[44] Anderson, D. R.; Bierbaum, V. M.; DePuy, C. H. (J. Am. Chem. Soc. **105** [1983] 4244/8).
[45] McDonald, R. N.; Chowdhury, A. K.; McGhee, W. D. (J. Am. Chem. Soc. **106** [1984] 4112/6).
[46] McDonald, R. N.; Chowdhury, A. K.; Jones, M. T. (J. Am. Chem. Soc. **108** [1986] 3105/7).
[47] O'Hair, R. A. J.; Krempp, M.; Damrauer, R.; DePuy, C. H. (Inorg. Chem. **31** [1992] 2092/6).

1.3.1.5.3 Isotope Exchange

The results of earlier investigations are described in "Phosphor" C, 1965, p. 31. The exchange by ^{32}P (or T) after neutron irradiation of undiluted PH_3 or its mixture with ^3He is described in Section 1.3.1.5.1.5, pp. 215/6. Some exchange of hydrogen by deuterium was found when PH_3 was coadsorbed with D_2; see Section 1.3.1.5.11, pp. 287/93.

The isotopic fractionation factor for the isotope exchange reaction $PH_2D(g) + H_2O(g) \rightleftharpoons PH_3(g) + HDO(g)$ at 298 K was derived from the known experimental equilibrium constant to be $\alpha_{H/D} = 2.28$, with $\alpha_{H/D}$ being defined as $([D]/[H])_{PH_3}/([D]/[H])_{H_2O}$. The result calculated from the reduced partition function of PH_2D, $\alpha_{H/D} = 2.38$, is in reasonably good agreement. The computed values of $\alpha_{H/D}$ range from 9.98 at 100 K to 1.28 at 1000 K [1]. The phosphane–water system was considered for the production of heavy water, but the isotope exchange reactions were found to suffer from slow reaction rates [2].

A value of $\alpha_{H/T} = (\alpha_{H/D})^r = 3.26$ at 298 K for the exchange of tritium instead of deuterium in the exchange reaction quoted above was obtained from the experimental $\alpha_{H/D}$ with r=1.435 derived from reduced partition functions; a purely calculated value of $\alpha_{H/T} = 3.47$ was obtained from reduced partition functions. The calculated values of $\alpha_{H/T}$ range from 27.18 at 100 K to 1.41 at 1000 K. For the exchange with liquid water, $\alpha_{H/T} = 3.57$ was deduced from the gas-phase value $\alpha_{H/T} = 3.26$ [1].

Rapid H/D exchange occurs in PH_3/DX mixtures to give $PH_2D \cdot HX$, $PH_2D \cdot DX$ (X=F [3] or ^{79}Br [4]), $PHD_2 \cdot HF$, and $PHD_2 \cdot DF$ [3]. Tritium distribution coefficients at 293 K in gaseous systems of PH_3 and the hydrogen-containing compounds HX with X=H, F, Cl, Br, I, OH, SH, NH_2, CH_3, Li, Na, K, Rb, and Cs were calculated from β-factors which were derived from mostly estimated fundamental vibrations via a statistical summation method [5].

References:

[1] Weston, R. E., Jr. (Z. Naturforsch. **28a** [1973] 177/84).
[2] Barr, F. T.; Drews, W. P. (Chem. Eng. Prog. **56** No. 3 [1960] 49/56; C. A. **1960** 19043).
[3] Legon, A. C.; Willoughby, L. C. (Chem. Phys. **74** [1983] 127/36).
[4] Willoughby, L. C.; Legon, A. C. (J. Phys. Chem. **87** [1983] 2085/90).
[5] Varshavskii, Ya. M.; Vaisberg, S. E. (Dokl. Akad. Nauk SSSR **140** [1961] 1361/3; Dokl. Phys. Chem. [Engl. Transl.] **136/141** [1961] 789/91).

1.3.1.5.4 Oxidation Reactions Applicable for the Removal of PH_3 from Gases or Solutions

The oxidation reactions presented in this section in most cases are used to remove less than 1 vol% of PH_3 from gases like N_2, C_2H_2, or their mixtures. Earlier results are described in "Phosphor" B, 1964, pp. 58/9. For the removal of PH_3 by physical or other chemical methods, see Section 1.3.1.3.13, pp. 138/41. The oxidation of diluted PH_3 by metal complexes and free or bound oxygen, mostly for the removal from gases, was reviewed. Kinetics and mechanisms were discussed for most of these reactions [1].

With Oxygen and Ozone. Air is used in most oxidation reactions. The oxidation by excess O_2 at 770 to 870 K removes PH_3 from CO [2]. The combustion of PH_3 in waste gases from microelectronic processing with oxygen yields P_4O_{10} [3, 4]. Versatile burners are described in [4 to 9]. The combustion in reactors is covered in [3, 10 to 12]. Oxidation by an oxygen plasma [13, 14] generated in a corona discharge [15] and by dynamic flame combustion [16] were also proposed. The oxidation of PH_3 in air is catalyzed by activated carbon treated with HI [17], by

$CuCl_2$ on a solid support [18], and at elevated temperature by contact with Cu, Ag, Au, or Pt on a solid support [19]. The catalytic oxidation of PH_3 by excess O_2 using activated carbon can be applied for the purification of C_2H_2 [20]. The catalytic oxidation of PH_3 with formation of H_3PO_4 in acidic solutions of Cu^{II} halides [21, 22], [23, pp. 156/61] and Fe^{III} halides [23, pp. 232/41], [24, 25] was studied, in both cases with added halide ions. The oxidation in the presence of a Cu^{II}-containing catalytic mixture is faster if the catalyst is supported on silica gel; the presence of moisture in this case is required [26 to 28]. The reactivity of a Cu^{II} catalytic solution containing $HgCl_2$ was also investigated [29]. The initial reaction of PH_3 with O_3 in air in a column yields P_4O_{10}, and the remaining PH_3 is converted to P_4O_6 after adsorption on activated carbon containing an oxidation catalyst [30].

With Halogens and Halogen Compounds. Gaseous halogens X_2, mostly Cl_2, can be used to oxidize PH_3 in waste gases of microelectronic fabrication to PX_3 [31, 32]. Scrubbing with aqueous solutions, containing Cl_2, its disproportionation product ClO^-, or both, was applied in [34 to 36]. Cl_2 in alkaline solution removes PH_3 from C_2H_2 with formation of H_3PO_4 [37]. An aqueous solution containing I_2, KI, and HI was used successfully in a pilot plant [180]. Ion exchangers in the polyiodide form oxidize PH_3 yielding I^- [38, 39]. The oxidation to PO_4^{3-} by aqueous ClO^- can remove PH_3 from air [40], C_2H_2 [41], and alkaline waste water [42]. The reaction of PH_3/H_2S mixtures with hypochlorite solutions is described in [43, 44]. Other solutions suitable for scrubbing PH_3 from waste gases contain $HBrO_3$, BrO_3^- [45], HIO_3 [45, 46], IO_3^- [45], HIO_4 [46], or I_2 at different pH [47]. The oxidation of low concentrations of PH_3 by acidic solutions of XO_3^- with X = Cl, Br, I begins after an induction period during which X_2 is formed and is accelerated autocatalytically by formed X^-; the oxidation product is H_3PO_4 [48 to 50]. The oxidation by BrO_3^- is accelerated by the addition of Br^- and I^-, whereas the effect of added Cl^- is small [51 to 53]. The induction period of the reactions with BrO_3^- [49, 54] and IO_3^- [50, 54, 55] decreases, and the reaction rates increase in the presence of added Br^- and I^- and the catalysts MoO_4^{2-} [49, 50, 54] and WO_4^{2-} [54, 55]. Details of the reaction of PH_3 with IO_3^- are also given in [50, 56]; oxalate decelerates the oxidation [57]. The oxidation of PH_3 by acidic solutions of IO_4^- is autocatalytic and yields H_3PO_4 and I^- [58]; addition of Cl^-, Br^-, or I^- accelerates this reaction [59]. The ions MoO_4^{2-} and WO_4^{2-} again act catalytically [54, 60]; MoO_4^{2-} is deactivated by oxalate [61].

With Compounds of the Main Groups 6 and 5. For a thermodynamic analysis of the reactions of PH_3 with elemental Se and Te, see Section 1.3.1.5.5, p. 239. Bubbling impure C_2H_2 through concentrated H_2SO_4 containing SO_3 removes PH_3 oxidatively, the reduction product being SO_2 [62]. PH_3 at low concentrations can be oxidized by an acidic solution of H_2SeO_3 with formation of H_3PO_4. The rate of the reaction increases upon addition of I^- [63, 64]. A similar reactivity is observed when the oxidizing mixture is on a silica gel carrier in the presence of moisture [65]. Oxidation of PH_3 can also be achieved with an acidic solution of TeO_3^{2-}. The reaction is accelerated upon addition of Cl^-, Br^-, or I^- [66, pp. 75/84]. PH_3 can be removed from waste gases by reaction with $H_2S_2O_8$ [46] or with H_2O_2 in the presence of Ag^+ ions [67]. The removal of PH_3 from C_2H_2/N_2 mixtures by reaction with acidic solutions of H_2O_2 yields H_3PO_4 along with lower acids. The reaction rate increases by adding Br^-, I^- [68, 69], or Cl^- and decreases when H_3PO_4 or CH_3COOH is added [70]. The acceleration of this oxidation by Cu^{II} salts [23, table 2, no. 45 to 53], [71] and Fe^{III} salts [23, table 2, no. 47 to 53], [70, 72] was investigated. These reactions are favored similarly by the addition of halides but slowed by the addition of H_3PO_4, sulfate, and acetate [70, 71]. The oxidation of a PH_3/H_2S mixture by acidic H_2O_2 in the presence of halides is described in [43, pp. 79/80]. The oxidation by H_2O_2 can also be used for the purification of waste water [42].

The oxidation of PH_3 in C_2H_2 to H_3PO_3 and H_3PO_4 by aqueous HNO_3 is improved in the presence of small quantities of $AgNO_3$ [73], $CaCl_2$, and NaI [74] or $CuCl_2$, $HgCl_2$, and Cl^- [29].

The oxidation of PH_3 by NO_2^- in concentrated H_2SO_4 yields H_3PO_4 and NO which is partly reduced to N_2O by PH_3. The rate of this reaction also increases with increasing concentration of Cl^-, Br^-, or I^- [66, pp. 91/7].

With Ce^{IV} and V^V Compounds. The oxidation of PH_3 in C_2H_2/N_2 mixtures by acidic Ce^{IV} solutions with formation of H_3PO_4 and Ce^{III} takes place at an acceptable rate only in the presence of KBr [75] or KI [75, 76]. The oxidation of PH_3 in low concentration by oxygen complexes of V^V supported on silica gel in the presence of KI and $HClO_4$ was investigated in [77].

With Cr^{VI} and Mo^{VI} Compounds. The oxidative removal of PH_3 from C_2H_2 by contact with solid supports covered with acidified CrO_3 [78 to 80] and from waste gases by contact with acidified K_2CrO_4 [81] was described. The reactions of PH_3 with the system Cr^{VI}– H_2SO_4–KI in aqueous solutions and on solid supports were investigated [82]. The oxidation of PH_3 by an acidic solution of dichromate is slow yielding H_3PO_4 and Cr^{III}. The reaction can be accelerated by Br^- [83, 84], I^- [84, 85], and Fe^{3+} and decelerated by Na_2SO_4 and H_3PO_4 [84]. Scrubbing with acidic $Na_2Cr_2O_7$ solution in the presence of Cl^- and $CuSO_4$ can also be used [86]. The oxidation of PH_3/H_2S mixtures by dichromate is described in [43, pp. 78/9]. Oxidation of PH_3 in N_2 takes place upon contact with MoO_3 mixed with Co_2O_3 in a column at 470 K [87].

With Mn^{IV} and Mn^{VII} Compounds. PH_3 in waste gases can be oxidized by $KMnO_4$ [88 to 90] in the presence of an acid [81] on porous carriers and also by scrubbing the gas with aqueous, acidic $KMnO_4$ solutions [36, 46, 91]. The addition of KI to the solution was found to be useful for the removal from C_2H_2 [92]; the presence of Br^- or Cl^- in the solution accelerates the reaction to a lesser degree [66, pp. 64/71], [93]. The oxidation of PH_3/H_2S mixtures by permanganate is described in [43, pp. 78/9]. Solid MnO_2 on alumina at 490 K oxidizes PH_3 in waste gases [94]; aqueous suspensions of MnO_2 can be used to oxidize PH_3 absorbed in concentrated H_2SO_4 [67].

With Fe^{III} Compounds. The scrubbing of PH_3 from gases like C_2H_2 succeeds with acidic solutions of Fe^{III} yielding H_3PO_4 and Fe^{II} [23, pp. 174/90, 323/6], [95 to 97]. The oxidation is accelerated by addition of acids except H_3PO_4 [23, pp. 181/6], by NCS^-, CN^-, and by halides, especially I^- [23, pp. 177, 182, 187, 190], [98 to 100]. Addition of sulfate [23, pp. 174/82], [95] or acetate [97] slows the reaction; see also [101]. Numerous reactions of PH_3 with $FeCl_3$ solutions containing the additives HCl and NaI [102], $ZnCl_2$ [103], $HgCl_2$ [104], or HCl, $HgCl_2$ and $CuCl_2$ [105, 106] were described. Solutions of $Fe_2(SO_4)_3$ [107] and a complex of Fe^{III} with triethanolamine and ethylene diamine tetraacetate were also applied [108]. The oxidation of PH_3 by Fe^{III} chloro complexes on coal ash [109], on activated carbon in the presence of acid [110] or of $CuCl_2$ and palladium chloro complexes [111], and on silica gel in the presence of I^- and acid [112] was studied.

With Compounds of Ni^{II}, Pd^{II}, and Pt^{IV}. The removal of PH_3 from inert gases with nickel oxide on activated carbon [113] and with Pd^{II} chloro complexes on kieselguhr [114] was described. Acidic solutions of H_2PtCl_6 oxidize PH_3 to H_3PO_4 with formation of Pt^{II} which activates PH_3 [115]. The rate of the reaction increases with increasing acidity of the solution and with increasing concentration of Cl^-, Br^-, I^- [23, pp. 305/7], or SCN^- [116].

With Cu^{II} Compounds. The PH_3 in waste gases of the microelectronic fabrication is decomposed on contact with solid CuO at 490 K [94, 117]. The reaction proceeds at 370 K or less in contact with CuO and the promoter MnO_2 on a solid support [118, 119]; other transition metal oxides were added in [120]. Pellets of CuO and MnO_2 containing Ag compounds [121], oxides of Al, Si, or Zn [122], or other metal oxides [123 to 125] can be applied at ambient temperature or below. Oxidizing pellets of CuO containing hydroxides of Li or alkaline earth metals [126], or oxides of Sn, Pb, Sb, Bi [127], La, Ti, Zr [128], V [129], Nb, Ta [130], Mo, W

[129], Fe, Co, or Ni [131] were also claimed to be useful. The removal of PH_3 from C_2H_2 by $CuCl_2$ on coal ash involves oxidative adsorption [132]; see also [133]. A layer of $CuCl_2$ on activated carbon is also effective [134]; the system also removes PH_3 from air when Br^- is added in order to increase the activity of the catalyst [135].

The oxidation of PH_3 in C_2H_2 by solutions of CuX_2 (X=Cl, Br, I, NCS) yields H_3PO_4 at low partial pressures of PH_3 [23, pp. 43/50, 92/145, 320/1], [136 to 140] and additional H_3PO_3 at higher partial pressures [141]. The Cu^I compounds thus formed act autocatalytically [23, p. 95], [98, 100, 137]. Halide ions and KNCS increase the reaction rate whereas a decrease is found if $ZnCl_2$, $CdCl_2$, $BiCl_3$, or CH_3COO^- are present [23, pp. 102, 116, 126, 133, 138], [142]. The reactions are assumed to involve the formation of complexes as intermediates [23, 98, 116, 137]; examples are given in [23, pp. 138, 140], [100, 143] and their electronic properties were calculated in [101, 143 to 150]. Scrubbing solutions are based on $CuCl_2$ in aqueous HCl and a small amount of $HgCl_2$ as an additive [104, 151, 152]. Other halides of Cu and Hg were also used [29, 98, 153]. The reaction is accelerated by NaCl [154], NH_4Br, or NaI [155]. Aqueous $CuBr_2$ containing LiBr and NH_4Br can also be applied [156]. The oxidation of both PH_3 and H_2S in C_2H_2 by solutions containing Cu^{II} halides or Cu^{II}/Hg^{II} halide mixtures are described in [43, pp. 80/92, 97/107], [157 to 159].

Scrubbing with Cu^{II} solutions can remove PH_3 from the waste gases of smelting furnaces [160] and phosphorus production [107, 161 to 164]. Addition of $Fe_2(SO_4)_3$ accelerates the scrubbing [165]. The removal of PH_3 from waste gases by $CuSO_4$ on silica gel [166], $CuSO_4 \cdot H_2O$ on a porous support [167], or by oxidation in a propanol solution of $Cu(NO_3)_2$ [168] was described. The oxidative adsorption of PH_3 to $Cu(NO_3)_2$ on an activated carbon cloth was investigated in [169].

With Compounds of Ag^I and Au^{III}. The removal of PH_3 from an inert gas by $AgNO_3$ on solid supports involves adsorption and oxidation [169, 170] and was proposed for the purification of air [171]. The oxidation of PH_3 by an acidic solution of $AuCl_3$ yields H_3PO_4 and AuCl. The reaction is accelerated by the as-formed Au^I [98] and by adding HCl, LiCl [172], NaBr, or NaI [116, 172].

With $HgCl_2$. $HgCl_2$ on activated carbon chemisorbs PH_3 from C_2H_2/PH_3 mixtures [173]. Aqueous $HgCl_2$ can also be used. The rate of this reaction decreases upon addition of HCl, Cl^-, Br^-, NCS^-, SO_3^{2-}, or CH_3COO^- [29]. The reaction of a PH_3/H_2S mixture with solutions of $HgCl_2$ is described in [43, 174 to 176].

With Organic Oxidants. The oxidation of PH_3 in C_2H_2/N_2 mixtures by p-benzoquinone in acidic solution yielding H_3PO_4 and hydroquinone proceeds at a high rate only in the presence of catalytic amounts of I^- [177]. Other catalysts are LiBr and $CuBr_2$ [178]. PH_3-contaminated gases can be purified by treating them with an aqueous solution of CH_2O in the presence of $CdCl_2$ or $NiCl_2$ as catalysts [179].

References:

[1] Dorfman, Ya. A.; Yukht, I. M.; Levina, L. V.; Polimbetova, G. S.; Petrova, T. V.; Emel'yanova, V. S. (Usp. Khim. **60** [1991] 1190/228; Russ. Chem. Rev. [Engl. Transl.] **60** [1991] 605/27).

[2] Munday, T. F.; Goldstein, D.; Walden, J. (U.S. 4185079 [1978/80] 7 pp. from C.A. **92** [1980] No. 113145).

[3] Berth, M.; Laporte, G. (Fr. Demande 2591509 [1985/87] 5 pp. from C.A. **107** [1987] No. 222480).

[4] Yoneda, N.; Kudo, H.; Iwamoto, N.; Nakamura, M.; Munekazu, K.; Kojima, C.; Kaneko, K.; Mori, Y.; Ishikawa, H. (Jpn. Kokai Tokkyo Koho 02-126014 [1988/90] 11 pp. from C.A. **113** [1990] No. 236958).

[5] Hashimoto, H.; Kudo, H.; Iwamoto, N.; Hoshino, N.; Kawai, H.; Ishikawa, H.; Mori, Y. (Jpn. Kokai Tokkyo Koho 03-160207 [1989/91] 5 pp. from C.A. **116** [1992] No. 90465).
[6] Hashimoto, H.; Kudo, H.; Yoneda, N.; Iwamoto, N.; Kawai, H.; Ishikawa, H.; Mori, Y. (Jpn. Kokai Tokkyo Koho 03-055415 [1989/91] 4 pp. from C.A. **115** [1991] No. 14620).
[7] Yoneda, N.; Kudo, H.; Iwamoto, N.; Nakamura, M.; Munekazu, K.; Kojima, C.; Kaneko, K.; Mori, Y.; Ishikawa, H. (Jpn. Kokai Tokkyo Koho 02-103311 [1988/90] 8 pp. from C.A. **113** [1990] No. 196952).
[8] Konagaya, Y.; Tanaka, T.; Takaine, M.; Tsubouchi, T. (U.S. 4801437 [1986/89] 12 pp. from C.A. **110** [1989] No. 218288).
[9] Deai, K.; Yamazaki, T. (Jpn. Kokai Tokkyo Koho 63-87519 [1986/88] 3 pp. from C.A. **109** [1988] No. 60752).
[10] Elliot, B.; Balma, F.; Johnson, F. (Solid State Technol. **33** No. 1 [1990] 89/92 from C.A. **114** [1991] No. 87748).

[11] Vickery, E. C. (Res. Dev. **31** No. 5 [1989] 107, 109/10 from C.A. **111** [1989] No. 102086).
[12] Kroedel, G.; Fabian, L.; Moeller, R.; Stelzer, H. (Ger. [East] 273009 [1988/89] 5 pp. from C.A. **112** [1990] No. 185079).
[13] Tsukune, A.; Nishimura, M.; Koyama, K. (Jpn. Kokai Tokkyo Koho 63-288026 [1987/88] 3 pp. from C.A. **110** [1989] No. 164573).
[14] Itoga, M.; Sato, J.; Tanikawa, E.; Fujiwara, T. (Jpn. Kokai 76-129868 [1975/76] 3 pp. from C.A. **87** [1977] No. 43504).
[15] Fabian, L.; Moeller, R.; Tiller, H. J.; Berg, D.; Lenke, D. (Ger. [East] 277843 [1988/90] 3 pp. from C.A. **113** [1990] No. 217262).
[16] Thomson, M. (Solid State Technol. **31** No. 2 [1988] 93/7 from C.A. **108** [1988] No. 100396).
[17] Dorfman, Ya. A.; Yukht, I. M.; Nadyrova, G. M.; Polimbetova, G. S.; Aibasov, E. Zh.; Shokorova, L. A.; Kurakbaeva, R. Kh.; Kozlovskii, V. A. (Zh. Prikl. Khim. [Leningrad] **64** [1991] 490/5; J. Appl. Chem. USSR [Engl. Transl.] **64** [1991] 438/42).
[18] Rakitskaya, T. L.; Abramova, N. N.; Red'ko, T. D. (Kinet. Katal. **30** [1989] 1084/8; Kinet. Catal. [Engl. Transl.] **30** [1989] 947/51).
[19] Svara, J.; Thuemmler, U. (Ger. Offen. 3822777 [1988/90] 7 pp. from C.A. **112** [1990] No. 144924).
[20] Saito, G.; Nakamura, T. (Jpn. Kokai Tokkyo Koho 02-045428 [1988/90] 3 pp. from C.A. **113** [1990] No. 61725).

[21] Rakitskaya, T. L.; Abramova, N. N.; Poklad, N. S.; Red'ko, T. D. (Kinet. Katal. **28** [1987] 872/5; Kinet. Catal. [Engl. Transl.] **28** [1987] 762/5).
[22] Sokol'skii, D. V.; Dorfman, Ya. A.; Ernestova, L. S. (Zh. Prikl. Khim. [Leningrad] **46** [1973] 1127/9; J. Appl. Chem. USSR [Engl. Transl.] **46** [1973] 1194/6).
[23] Sokol'skii, D. V.; Dorfman, Ya. A. (Ligand Catalysis of Oxidation-Reduction Reactions in Aqueous Solutions, Nauka Kaz. SSR, Alma-Ata 1972, pp. 1/336; C.A. **78** [1973] No. 128923).
[24] Sokol'skii, D. V.; Dorfman, Ya. A.; Rakitskaya, T. L. (Dokl. Akad. Nauk SSSR **203** [1972] 155/8; Dokl. Phys. Chem. [Engl. Transl.] **202/207** [1972] 193/6).
[25] Sokol'skii, D. V.; Dorfman, Ya. A.; Rakitskaya, T. L.; Protopopova, G. D.; Sapova, R. G.; Rogova, Z. I.; Pole, G. P.; Kaidarova, R. K. (Dokl. 4th Vses. Konf. Khim. Atsetilena, Alma-Ata 1972, Vol. 3, pp. 360/2; C.A. **79** [1973] No. 77958).
[26] Rakitskaya, T. L.; Abramova, N. N.; Poklad, N. S.; Red'ko, T. D. (Kinet. Katal. **28** [1987] 872/5; Kinet. Catal. [Engl. Transl.] **28** [1987] 762/5).

[27] Rakitskaya, T. L.; Abramova, N. N. (Izv. Vyssh. Uchebn. Zaved. Khim. Khim. Tekhnol. **26** No. 11 [1983] 1334/8; C. A. **100** [1984] No. 40395).
[28] Rakitskaya, T. L.; Kosheleva, N. N.; Shkol'nikova, O. L. (Koord. Khim. **7** [1981] 355/8; C. A. **94** [1981] No. 163270).
[29] Sokol'skii, D. V.; Dorfman, Ya. A.; Kazantseva, I. A.; Shalabaeva, G. B. (Tr. Inst. Org. Katal. Elektrokhim. Akad. Nauk Kaz. SSR **4** [1973] 54/79; C. A. **80** [1974] No. 113144).
[30] Nishimura, S.; Watanabe, M. (Jpn. Kokai 76-11068 [1974/76] 3 pp. from C. A. **85** [1976] No. 51218).

[31] Suzuki, K.; Kumagai, Y. (Jpn. Kokai Tokkyo Koho 63-162025 [88-162025] [1986/88] 4 pp. from C. A. **109** [1988] No. 236143).
[32] Yamashita, Y.; Takemori, T.; Yamada, Y. (Jpn. Kokai Tokkyo Koho 02-035912 [1988/90] 4 pp. from C. A. **113** [1990] No. 117799).
[33] Takemori, T.; Yamada, Y.; Kumagai, Y.; Konagaya, Y. (Jpn. Kokai Tokkyo Koho 63-162026 [88-162026] [1986/88] 4 pp. from C. A. **109** [1988] No. 236144).
[34] Taura, T. (Jpn. Kokai Tokkyo Koho 01-262930 [1988/89] 8 pp. from C. A. **112** [1990] No. 164256).
[35] Yamashita, Y.; Takemori, T. (Jpn. Kokai Tokkyo Koho 63-49227 [88-49227] [1986/88] 4 pp. from C. A. **109** [1988] No. 11087).
[36] Herman, T.; Soden, S. (AIP Conf. Proc. No. 166 [1988] 99/108 from C. A. **109** [1988] No. 236085).
[37] Sokol'skii, D. V.; Dorfman, Ya. A.; Pole, G. P. (Zh. Prikl. Khim. [Leningrad] **44** [1971] 2571/3; J. Appl. Chem. USSR [Engl. Transl.] **44** [1971] 2647/9).
[38] Vulikh, A. I.; Dubinina, E. G. (Okislitel'no-Vosstanov. Vysokomol. Soedin. Tr. Vses. Nauchno-Tekh. Konf. Sint. Issled. Svoistv Primen. Okislitel'no-Vosstanov. Vysokomol. Soedin., Leningrad 1966 [1967], pp. 91/8 from C. A. **70** [1969] No. 89171).
[39] Vulikh, A. I.; Dubinina, E. G.; Kim, F. I.; Fedoseeva, Yu. V.; Shukalkina, L. I. (Khim. Prom. [Moscow] **43** [1967] 337/9 from C. A. **67** [1967] No. 108105).
[40] Matsumoto, R.; Ishikawa, H.; Inamoto, I. (Jpn. Kokai 75-95104 [1973/75] 3 pp. from C. A. **84** [1976] No. 64835).

[41] Sokol'skii, D. V.; Dorfman, Ya. A.; Solov'eva, L. S.; Kaidarova, R. K.; Pole, G. P. (Zh. Prikl. Khim. [Leningrad] **44** [1971] 1239/45; J. Appl. Chem. USSR [Engl. Transl.] **44** [1971] 1259/64).
[42] Nakayama, Y.; Wada, N.; Hirao, K. (Jpn. Kokai Tokkyo Koho 79-28447 [1977/79] 3 pp. from C. A. **91** [1979] No. 26827).
[43] Sokol'skii, D. V.; Dorfman, Ya. A.; Kazanzeva, I. A.; Emel'yanova, V. S.; Pole, G. P.; Rakitskaya, T. L.; Utegenova, G. S. (Tr. Inst. Khim. Nauk Akad. Nauk Kaz. SSR **30** [1970] 74/108; C. A. **74** [1971] No. 6833).
[44] Sokol'skii, D. V.; Dorfman, Ya. A.; Pole, G. P. (Tr. Inst. Khim. Nauk Akad. Nauk Kaz. SSR **30** [1970] 46/53; C. A. **73** [1970] No. 126461).
[45] Fabian, W.; Roehle, H.; Wolfram, P. (Ger. Offen. 3342816 [1983] 21 pp.; C. A. **103** [1985] No. 92316).
[46] Fabian, W. (Ger. Offen. 3839153 [1988/90] 9 pp. from C. A. **114** [1991] No. 108173).
[47] Sokol'skii, D. V.; Dorfman, Ya. A.; Pole, G. P. (Dokl. Akad. Nauk SSSR **201** [1971] 875/6; Dokl. Chem. [Engl. Transl.] **196/201** [1971] 1004/5).
[48] Dorfman, Ya. A.; Rakitskaya, T. L.; Kaidarova, R. K. (Tr. Inst. Org. Katal. Elektrokhim. Akad. Nauk Kaz. SSR **10** [1975] 19/27; C. A. **84** [1976] No. 50320).
[49] Dorfman, Ya. A.; Rakitskaya, T. L.; Kaidarova, R. K. (Kinet. Katal. **17** [1976] 1197/201; Kinet. Catal. [Engl. Transl.] **17** [1976] 1034/8).

[50] Dorfman, Ya. A.; Rakitskaya, T. L.; Kaidarova, R. K. (Zh. Obshch. Khim. **46** [1976] 2643/6; J. Gen. Chem. USSR [Engl. Transl.] **46** [1976] 2527/30).

[51] Sokol'skii, D. V.; Dorfman, Ya. A.; Rakitskaya, T. L.; Kaidarova, R. K. (Dokl. Akad. Nauk SSSR **209** [1973] 1354/6; Dokl. Chem. [Engl. Transl.] **208/213** [1973] 340/2).
[52] Sokol'skii, D. V.; Dorfman, Ya. A.; Rakitskaya, T. L. (Zh. Fiz. Khim. **47** [1973] 59/61; Russ. J. Phys. Chem. [Engl. Transl.] **47** [1973] 31/2).
[53] Sokol'skii, D. V.; Dorfman, Ya. A.; Rakitskaya, T. L. (Kinet. Katal. **14** [1973] 1406/8; Kinet. Catal. [Engl. Transl.] **14** [1973] 1245/7).
[54] Dorfman, Ya. A.; Rakitskaya, T. L.; Kaidarova, R. K. (Tr. Inst. Org. Katal. Elektrokhim. Akad. Nauk Kaz. SSR **13** [1976] 220/34; C.A. **86** [1977] No. 71434).
[55] Dorfman, Ya. A.; Rakitskaya, T. L.; Kaidarova, R. K. (Zh. Prikl. Khim. [Leningrad] **48** [1975] 2742/5; J. Appl. Chem. USSR [Engl. Transl.] **48** [1975] 2823/5).
[56] Sokol'skii, D. V.; Dorfman, Ya. A.; Rakitskaya, T. L.; Kaidarova, R. K. (Elektrokhimiya **9** [1973] 1806/8; Sov. Electrochem. [Engl. Transl.] **9** [1973] 1697/9).
[57] Dorfman, Ya. A.; Rakitskaya, T. L.; Kaidarova, R. K. (Zh. Fiz. Khim. **51** [1977] 1434/7; Russ. J. Phys. Chem. [Engl. Transl.] **51** [1977] 843/5).
[58] Dorfman, Ya. A.; Rakitskaya, T. L.; Kaidarova, R. K. (Elektrokhimiya **12** [1976] 4799/81; Sov. Electrochem. [Engl. Transl.] **12** [1976] 471/3).
[59] Dorfman, Ya. A.; Rakitskaya, T. L.; Kaidarova, R. K. (Zh. Obshch. Khim. **46** [1976] 1668/71; J. Gen. Chem. USSR [Engl. Transl.] **46** [1976] 1624/77).
[60] Dorfman, Ya. A.; Rakitskaya, T. L.; Kaidarova, R. K. (Zh. Fiz. Khim. **51** [1977] 1103/5; Russ. J. Phys. Chem. [Engl. Transl.] **51** [1977] 649/51).

[61] Dorfman, Ya. A.; Rakitskaya, T. L.; Kaidarova, R. K. (Zh. Fiz. Khim. **52** [1978] 2129/30; Russ. J. Phys. Chem. [Engl. Transl.] **52** [1978] 1233).
[62] Kadic, K. (Czech. 112239 [1963/64] 2 pp. from C.A. **62** [1965] 10268).
[63] Dorfman, Ya. A.; Rakitskaya, T. L.; Amanzholova, A. S. (Zh. Prikl. Khim. [Leningrad] **52** [1979] 2385; J. Appl. Chem. USSR [Engl. Transl.] **52** [1979] 2262).
[64] Dorfman, Ya. A.; Rakitskaya, T. L.; Amanzholova, A. S. (VINITI No. 3914-78 [1978] 1/10).
[65] Rakitskaya, T. L.; Pshenitsa, M. P.; Koroznikova, L. I. (Izv. Vyssh. Uchebn. Zaved. Khim. Khim. Tekhnol. **29** [1986] 21/6; C.A. **105** [1986] No. 214536).
[66] Dorfman, Ya. A.; Emel'yanova, V. S.; Amanzholova, A. S.; Kamaletdinova, A. K.; Polimbetova, G. S.; Shokorova, L. A.; Shalabaeva, G. B.; Khaleeva, G. I.; Khasanova, G. N.; Supieva, Kh. T. (Tr. Inst. Org. Katal. Elektrokhim. Akad. Nauk Kaz. SSR **22** [1980] 64/129; C.A. **97** [1982] No. 99040).
[67] Symossek, F.; Lau, H. R. (Ger. [East] 279683 [1989/90] 3 pp. from C.A. **114** [1991] No. 9063).
[68] Sokol'skii, D. V.; Dorfman, Ya. A.; Rakitskaya, T. L. (Zh. Fiz. Khim. **45** [1971] 2771/4; Russ. J. Phys. Chem [Engl. Transl.] **45** [1971] 1571/3).
[69] Sokol'skii, D. V.; Dorfman, Ya. A.; Rakitskaya, T. L. (Gomogennyi Katal. Mater. 1st Koord. Soveshch., Alma-Ata-Frunze 1969 [1970], pp. 79/89; C.A. **76** [1972] No. 145334).
[70] Dorfman, Ya. A.; Rakitskaya, T. L.; Sapova, R. G. (Tr. Inst. Org. Katal. Elektrokhim. Akad. Nauk Kaz. SSR **8** [1974] 59/87; C.A. **82** [1975] No. 160744).

[71] Dorfman, Ya. A.; Rakitskaya, T. L.; Sapova, R. G. (Tr. Inst. Org. Katal. Elektrokhim. Akad. Nauk Kaz. SSR **10** [1975] 3/18; C.A. **84** [1976] No. 65680).
[72] Sokol'skii, D. V.; Dorfman, Ya. A.; Rakitskaya, T. L.; Sapova, R. G. (Zh. Fiz. Khim. **48** [1974] 2758/61; Russ. J. Phys. Chem. [Engl. Transl.] **48** [1974] 1625/6).

[73] Kal'manovich, S. P.; Strizhevskii, I. I. (Tr. Vses. Nauchno-Issled. Inst. Avtog. Obrab. Met. No. 9 [1963] 124/35 from C.A. **60** [1964] 6734).
[74] Sokol'skii, D. V.; Dorfman, Ya. A.; Shalabaeva, G. B. (U.S.S.R. 440057 [1971/76] from C.A. **84** [1976] No. 179635).
[75] Sokol'skii, D. V.; Dorfman, Ya. A.; Rogoza, Z. I. (Zh. Fiz. Khim. **48** [1974] 585/7; Russ. J. Phys. Chem. [Engl. Transl.] **48** [1974] 340/1).
[76] Sokol'skii, D. V.; Dorfman, Ya. A.; Rogoza, Z. I. (Zh. Fiz. Khim. **46** [1972] 998/9; Russ. J. Phys. Chem. [Engl. Transl.] **46** [1972] 578/9).
[77] Rakitskaya, T. L.; Koroznikova, L. I. (Kinet. Katal. **32** [1991] 98/102; Kinet. Catal. [Engl. Transl.] **32** [1991] 86/90).
[78] Rybkin, A. P.; Chuklenkova, G. P.; Kozhichkina, T. N.; Zolotova, L. F.; Shurubtsov, V. N. (U.S.S.R. 1467080 [1986/89] from C.A. **111** [1989] No. 96637).
[79] Rybkin, A. P.; Neshumova, S. P.; Chuklenkova, G. P.; Kozhichkina, T. N.; Dzyubanov, I. Ya. (U.S.S.R. 1518358 [1987/89] from C.A. **112** [1990] No. 178048).
[80] Dorfman, Ya. A.; Polimbetova, G. S.; Shindler, Yu. M.; Sharopin, A. N.; Syrovatskii, E. I.; Kuznetsov, L. M. (U.S.S.R. 596274 [1975/78] from C.A. **89** [1978] No. 5893).

[81] Sonobe, S.; Fukui, M. (Jpn. Kokai Tokkyo Koho 63-72338 [88-72338] [1986/88] 5 pp. from C.A. **109** [1988] No. 42931).
[82] Dorfman, Ya. A.; Polimbetova, G. S.; Kel'man, I. V.; Doroshkevich, D. M.; Petrova, T. V.; Mansurov, B. A.; Bikmukhametova, A. K.; Yukht, I. M. (Koord. Khim. **15** [1989] 77/81 from C.A. **110** [1989] No. 180124).
[83] Sokol'skii, D. V.; Dorfman, Ya. A.; Utegenova, G. S. (Zh. Fiz. Khim. **46** [1972] 507/8; Russ. J. Phys. Chem. [Engl. Transl.] **46** [1972] 293/4).
[84] Dorfman, Ya. A.; Polimbetova, G. S. (Tr. Inst. Org. Katal. Elektrokhim. Akad. Nauk Kaz. SSR **13** [1976] 235/49; C.A. **86** [1977] No. 60950).
[85] Sokol'skii, D. V.; Dorfman, Ya. A.; Utegenova, G. S. (Dokl. Akad. Nauk SSSR **199** [1971] 1082/3; Dokl. Chem. [Engl. Transl.] **196/201** [1971] 681/2).
[86] Sokol'skii, D. V.; Dorfman, Ya. A.; Utegenova, G. S. (U.S.S.R. 618124 [1972/78] from C.A. **90** [1979] No. 5878).
[87] Nippon Pionics K. K. (Jpn. Kokai Tokkyo Koho 85-68034 [1983/85] 4 pp. from C.A. **103** [1985] No. 92307).
[88] Nakayasu, T.; Morikawa, Y.; Goto, Y. (Jpn. Kokai Tokkyo Koho 61-293549 [86-293549] [1985/86] 5 pp. from C.A. **107** [1987] No. 12191).
[89] Kobayashi, Y.; Ichimura, S.; Yamakawa, T. (Jpn. Kokai Tokkyo Koho 62-95119 [87-95119] [1985/87] 6 pp. from C.A. **107** [1987] No. 140169).
[90] Nikolaikin, N. I.; Nikolaikina, N. E.; Chekhov, O. S. (Khim. Neft. Mashinostr. **1988** No. 10, p. 33 from C.A. **110** [1989] No. 62910).

[91] Fabian, L.; Kleinert, M.; Stelzer, H.; Petermann, K. (Ger. [East] 149470 [1980/81] 7 pp. from C.A. **96** [1982] No. 11107).
[92] Dorfman, Ya. A.; Polimbetova, G. S.; Solov'eva, L. S. (U.S.S.R. 521915 [1975/76] from C.A. **85** [1976] No. 142594).
[93] Dorfman, Ya. A.; Polimbetova, G. S.; Mansurov, B. A.; Bikmukhametova, A. K.; Doroshkevich, D. M. (Koord. Khim. **14** [1988] 1219/23 from C.A. **109** [1988] No. 238061).
[94] Kikuchi, Y. (Jpn. Kokai Tokkyo Koho 63-200820 [1987/88] 4 pp. from C.A. **110** [1989] No. 28546).
[95] Sokol'skii, D. V.; Dorfman, Ya. A.; Rakitskaya, T. L. (Dokl. Akad. Nauk SSSR **199** [1971] 618/21; Dokl. Chem. [Engl. Transl.] **196/201** [1971] 626/9).
[96] Sokol'skii, D. V.; Dorfman, Ya. A.; Rakitskaya, T. L. (Kinet. Katal. **14** [1973] 507/10; Kinet. Catal. [Engl. Transl.] **14** [1973] 431/3).

[97] Sokol'skii, D. V.; Dorfman, Ya. A.; Protopopova, G. D. (Zh. Fiz. Khim. **47** [1973] 1982/6; Russ. J. Phys. Chem. [Engl. Transl.] **47** [1973] 1116/8).
[98] Golodov, V. A. (J. Res. Inst. Catal. Hokkaido Univ. **29** [1981] 49/60; C.A. **96** [1982] No. 41579).
[99] Sokol'skii, D. V.; Dorfman, Ya. A.; Rakitskaya, T. L. (Dokl. Akad. Nauk SSSR **201** [1971] 123/5; Dokl. Chem. [Engl. Transl.] **196/201** [1971] 913/5).
[100] Dorfman, Ya. A.; Polimbetova, G. S.; Kel'man, I. V.; Doroshkevich, D. M.; Petrova, T. V.; Zil'bert, I. G.; Saltykov, Yu. P.; Yukht, I. M. (Tr. Inst. Org. Katal. Elektrokhim. Akad. Nauk Kaz. SSR **23** [1984] 142/63; C.A. **102** [1985] No. 12817).

[101] Dorfman, Ya. A.; Kel'man, I. V.; Doroshkevich, D. M. (Teor. Eksp. Khim. **19** [1983] 300/4; Theor. Exp. Chem. [Engl. Transl.] **19** [1983] 274/8).
[102] Dorfman, Ya. A.; Protopopova, G. D. (U.S.S.R. 539593 [1974/76] from C.A. **86** [1977] No. 160527; U.S.S.R. 474348 [1974/75] from C.A. **83** [1975] No. 131117).
[103] Tleukulov, O. M.; Batkaev, R. I.; Moldabekov, Sh. M. (U.S.S.R. 978898 [1980/82] from C.A. **98** [1983] No. 95059).
[104] Veda, I.; Shinotuni, H. (Yoshitsu Gakkaishi **29** [1960] 333/5 from C.A. **1960** 12425).
[105] Dorfman, Ya. A.; Kazantseva, I. A. (Katal. Reakts. Zhidk. Faze Mater. 4th Vses. Konf., Alma-Ata 1974, Vol. 3, pp. 642/9; C.A. **83** [1975] No. 157199).
[106] Dorfman, Ya. A.; Kazantseva, I. A. (U.S.S.R. 570386 [1974/77] from C.A. **87** [1977] No. 183966).
[107] Al'zhanov, S.; Ershov, V. A.; Moldabekov, Sh. M. (Deposited Doc. VINITI-4803-83 [1983] 1/13; C.A. **101** [1984] No. 198967).
[108] Drahorad, J.; Kucera, J. (Czech. 117293 [1963/66] 2 pp. from C.A. **65** [1966] 16549).
[109] Evtikov, N. I. (U.S.S.R. 940809 [1979/82] from C.A. **98** [1983] No. 56447).
[110] Rakitskaya, T. L.; Kostyukova, I. S.; Red'ko, T. D. (Kinet. Katal. **28** [1987] 1501/3; Kinet. Catal. [Engl. Transl.] **28** [1987] 1299/301).

[111] Rakitskaya, T. L.; Ennan, A. A.; Abramova, N. N.; Gaziev, G. A. (U.S.S.R. 1028350 [1981/83] from C.A. **99** [1983] No. 124889).
[112] Rakitskaya, T. L.; Novitsyuk, E. D. (Zh. Prikl. Khim. [Leningrad] **63** [1990] 652/5; J. Appl. Chem. USSR [Engl. Transl.] **63** [1990] 619/21).
[113] Yunusov, U. I.; Topolova, N. A. (U.S.S.R. 1159625 [1983/85] from C.A. **103** [1985] No. 107111).
[114] Rakitskaya, T. L.; Abramova, N. N.; Paina, V. Ya. (Izv. Vyssh. Uchebn. Zaved. Khim. Khim. Tekhnol. **33** No. 3 [1990] 19/22 from C.A. **113** [1990] No. 139647).
[115] Sokol'skii, D. V.; Dorfman, Ya. A.; Emel'yanova, V. S. (Dokl. Akad. Nauk SSSR **198** [1971] 1110/2; Dokl. Chem. [Engl. Transl.] **196/201** [1971] 515/7).
[116] Sokol'skii, D. V.; Dorfman, Ya. A.; Emel'yanova, V. S. (Dokl. Akad. Nauk SSSR **201** [1971] 1125/7; Dokl. Chem. [Engl. Transl.] **196/201** [1971] 1036/8).
[117] Skuta, D.; Tesarik, P.; Smrcek, J. (Czech. 259456 [1986/89] 4 pp. from C.A. **112** [1990] No. 59106).
[118] Serayssol, J. M. (Eur. Appl. 335792 [1989] 7 pp. from C.A. **112** [1990] No. 83283).
[119] Inaba, M.; Tomiyama, Y. (Jpn. Kokai Tokkyo Koho 01-015135 [1987/89] 6 pp. from C.A. **111** [1989] No. 200891).
[120] Yoneda, N.; Iwamoto, N.; Nakamura, M. (Jpn. Kokai Tokkyo Koho 02-126936 [1988/90] 6 pp. from C.A. **113** [1990] No. 236956).

[121] Kitahara, K.; Shimada, T.; Akita, N.; Hiramoto, T.; Sasaki, K. (Eur. Appl. 309099 [1988/89] 25 pp. from C.A. **111** [1989] No. 63274).

[122] Kitahara, K.; Shimada, T.; Akita, N.; Hiramoto, T.; Sasaki, K. (Eur. Appl. 294142 [1987/88] 14 pp. from C.A. **110** [1989] No. 120497).
[123] Delobel, O.; Louise, J.; Cornut, P. (Eur. Appl. 419356 [1990/91] 8 pp. from C.A. **115** [1991] No. 98427).
[124] Cornut, P.; Delobel, O.; Leondaridis, P. (Fr. 2654362 [1989/91] 16 pp. from C.A. **116** [1992] No. 118903).
[125] Yoneda, N.; Iwamoto, N.; Nakamura, M. (Jpn. Kokai Tokkyo Koho 02-139033 [1988/90] 4 pp. from C.A. **113** [1990] No. 137816).
[126] Kitahara, K.; Shimada, T. (Jpn. Kokai Tokkyo Koho 62-286520 [87-286520] [1986/87] 7 pp. from C.A. **108** [1988] No. 118178).
[127] Kitahara, K.; Shimada, T. (Jpn. Kokai Tokkyo Koho 62-286523 [87-286523] [1986/87] 6 pp. from C.A. **108** [1988] No. 155806).
[128] Kitahara, K.; Shimada, T. (Jpn. Kokai Tokkyo Koho 62-286521 [87-286521] [1986/87] 6 pp. from C.A. **108** [1988] No. 118204).
[129] Kitahara, K.; Shimada, T. (Jpn. Kokai Tokkyo Koho 62-286524 [87-286524] [1986/87] 6 pp. from C.A. **108** [1988] No. 155807).
[130] Kitahara, K.; Shimada, T. (Jpn. Kokai Tokkyo Koho 62-286525 [87-286525] [1986/87] 6 pp. from C.A. **108** [1988] No. 155808).

[131] Kitahara, K.; Shimada, T. (Jpn. Kokai Tokkyo Koho 62-286522 [87-286522] [1986/87] 6 pp. from C.A. **108** [1988] No. 155805).
[132] Goreev, A. M.; Kolotushkin, V. V.; Samosudov, Yu. I. (Khim. Neft. Mashinostr. **1988** No. 4, pp. 13/4 from C.A. **108** [1988] No. 223424).
[133] Goreev, A. M.; Kolotushkin, V. V.; Kuz'mina, R. P.; Samosudov, Yu. I.; Suleimanova, L. N.; Shukailo, V. I. (U.S.S.R. 1389827 [1985/88] from C.A. **109** [1988] No. 95484).
[134] Rakitskaya, T. L.; Litvinskaya, V. V.; Abramova, N. N.; Red'ko, T. D.; Popova, N. A. (Zh. Prikl. Khim. [Leningrad] **60** [1987] 1415/7; J. Appl. Chem. USSR [Engl. Transl.] **60** [1987] 1340/2).
[135] Rakitskaya, T. L.; Ennan, A. A.; Abramova, N. N.; Pimenov, S. D.; Kleptsova, A. P. (U.S.S.R. 1156729 [1982/85] from C.A. **103** [1985] No. 75653).
[136] Sokol'skii, D. V.; Dorfman, Ya. A.; Rakitskaya, T. L. (Gomogennyi Katal. Mater. 1st Koord. Soveshch., Alma-Ata-Frunze 1969 [1970], pp. 259/79; C.A. **77** [1972] No. 10079).
[137] Sokol'skii, D. V.; Dorfman, Ya. A.; Prazdnikova, Z. F. (Zh. Org. Khim. **7** [1971] 2048/52; J. Org. Chem. USSR [Engl. Transl.] **7** [1971] 2127/31).
[138] Sokol'skii, D. V.; Dorfman, Ya. A.; Prazdnikova, Z. F.; Rodnikova, I. V. (Tr. Inst. Org. Katal. Elektrokhim. Akad. Nauk Kaz. SSR **4** [1973] 92/110; C.A. **80** [1974] No. 699941).
[139] Sokol'skii, D. V.; Dorfman, Ya. A.; Prazdnikova, Z. F.; Rodnikova, I. V. (Dokl. Akad. Nauk SSSR **197** [1971] 365/8; Dokl. Chem. [Engl. Transl.] **196/201** [1971] 227/30).
[140] Sokol'skii, D. V.; Dorfman, Ya. A.; Prazdnikova, Z. F.; Rodnikova, I. V. (Zh. Org. Khim. **9** [1973] 1569/73; J. Org. Chem. USSR [Engl. Transl.] **9** [1973] 1594/7).

[141] Sokol'skii, D. V.; Dorfman, Ya. A.; Emel'yanova, V. S. (Zh. Obshch. Khim. **41** [1971] 1918/21; J. Gen. Chem. USSR [Engl. Transl.] **41** [1971] 1931/4).
[142] Sokol'skii, D. V.; Dorfman, Ya. A.; Ernestova, L. S. (Zh. Prikl. Khim. [Leningrad] **45** [1972] 1344/6; J. Appl. Chem. USSR [Engl. Transl.] **45** [1972] 1388/90).
[143] Dorfman, Ya. A.; Kel'man, I. V.; Doroshkevich, D. M. (Zh. Neorg. Khim. **30** [1985] 132/6; Russ. J. Inorg. Chem. [Engl. Transl.] **30** [1985] 74/6).
[144] Dorfman, Ya. A.; Kel'man, I. V. (Teor. Eksp. Khim. **21** [1985] 718/22; Theor. Exp. Chem. [Engl. Transl.] **21** [1985] 683/6).

[145] Dorfman, Ya. A.; Kel'man, I. V. (Teor. Eksp. Khim. **20** [1984] 485/8; Theor. Exp. Chem. [Engl. Transl.] **20** [1984] 455/7).
[146] Dorfman, Ya. A.; Kel'man, I. V.; Doroshkevich, D. M. (Teor. Eksp. Khim. **18** [1982] 621/3; Theor. Exp. Chem. [Engl. Transl.] **18** [1982] 573/5).
[147] Bersuker, I. B.; Budnikov, S. S.; Doroshkevich, D. M.; Dorfman, Ya. A.; Kel'man, I. V. (Teor. Eksp. Khim. **17** [1981] 535/7; Theor. Exp. Chem. [Engl. Transl.] **17** [1981] 417/9).
[148] Dorfman, Ya. A.; Kel'man, I. V.; Doroshkevich, D. M. (Koord. Khim. **10** [1984] 320/4; C.A. **100** [1984] No. 216367).
[149] Dorfman, Ya. A. (Vestn. Akad. Nauk Kaz. SSR **1978** No. 2, pp. 3/16; C.A. **88** [1978] No. 159022).
[150] Dorfman, Ya. A.; Kel'man, I. V. (Teor. Eksp. Khim. **16** [1980] 692/5; Theor. Exp. Chem. [Engl. Transl.] **16** [1980] 496/8).

[151] Sokol'skii, D. V.; Dorfman, Ya. A.; Kazantseva, I. A. (U.S. 3974085 [1972/76] 3 pp. from C.A. **86** [1977] No. 54963; Ger. Offen. 2213483 [1972/73] 9 pp. from C.A. **79** [1973] No. 145943).
[152] Long, G. L.; Boss, C. B. (Anal. Chem. **53** [1981] 2363/5).
[153] Sokol'skii, D. V.; Dorfman, Ya. A.; Kazantseva, I. A. (Dokl. Akad. Nauk SSSR **213** [1973] 103/4; Dokl. Chem. [Engl. Transl.] **208/213** [1973] 846/7).
[154] Boguslavskii, E. A.; Lapshina, N. M.; Érler, L. N. (Zh. Prikl. Khim. [Leningrad] **51** [1978] 1783/6; J. Appl. Chem. USSR [Engl. Transl.] **51** [1978] 1686/8).
[155] Dorfman, Ya. A.; Ernestova, L. S. (U.S.S.R. 474349 [1974] from C.A. **83** [1975] No. 131118).
[156] Sokol'skii, D. V.; Dorfman, Ya. A.; Evtikov, N. I. (U.S.S.R. 381373 [1971/73] from C.A. **79** [1973] No. 136459).
[157] Dorfman, Ya. A.; Evtikov, N. I. (Tr. Inst. Khim. Nauk Akad. Nauk Kaz. SSR **30** [1970] 37/45; C.A. **74** [1971] No. 25504).
[158] Sokol'skii, D. V.; Dorfman, Ya. A.; Emel'yanova, V. S.; Shindler, Yu. M. (Tr. Inst. Khim. Nauk Akad. Nauk Kaz. SSR **30** [1970] 18/27; C.A. **74** [1971] No. 25490).
[159] Sokol'skii, D. V.; Dorfman, Ya. A.; Prazdnikova, Z. F.; Ernestova, L. S. (Tr. Inst. Khim. Nauk Akad. Nauk Kaz. SSR **30** [1970] 66/73; C.A. **73** [1970] No. 134331).
[160] Koverya, V. M.; Belyakov, B. P.; Derkach, O. N.; Los, L. M.; Marinova, N. V.; Monin, V. Ya.; Chernogorenko, V. B.; Mel'nikov, B. I. (Khim. Tekhnol. [Kiev] **1984** No. 3, pp. 51/3; C.A. **101** [1984] No. 93612).

[161] Lynchak, K. A.; Chernogorenko, V. B.; Koverya, V. M. (Poluch. Svoistva Primen. Fosfidov **1977** 93/6; C.A. **88** [1978] No. 76002).
[162] Chernogorenko, V. B.; Alzhanov, T. M.; Lynchak, K. A.; Muchnik, S. V.; Ishkhanov, E. S.; Sergienko, V. Y.; Sapyan, V. G.; Koverya, V. M.; Pobortsev, M. E.; Markovsky, E. A.; Dmitrenko, V. V.; Bykov, V. I.; Kipchakbaev, A. D.; Vopilov, A. N. (U.S. 4192853 [1978/80] 9 pp.; C.A. **93** [1980] No. 74857).
[163] Chernogorenko, V. B.; Lynchak, K. A.; Koverya, V. M.; Dmitrenko, V. V.; Bykov, V. I. (U.S.S.R. 709142 [1976/80] from C.A. **92** [1980] No. 185174).
[164] Bykov, V. I.; Koverya, V. M. (Khim. Prom-st. Ser. Fosfornaya Prom-st. **1980** No. 2, p. 24; C.A. **94** [1981] No. 35537).
[165] Al'zhanov, S.; Moldabekov, Sh. M.; Ershov, V. A.; Popov, A. P. (Deposited Doc. VINITI-4802-83 [1983] 1/11; C.A. **101** [1984] No. 194512).
[166] Monderkamp, V.; Bierhals, J. (Gas Aktuell No. 37 [1989] 26/31 from C.A. **111** [1989] No. 159459).
[167] Goekcek, C. (U.S. 5024823 [1988/91] 3 pp. from C.A. **115** [1991] No. 74673; Ger. Offen. 3802496 [1988/89] 3 pp. from C.A. **112** [1990] No. 41729).

[168] Yagi, T.; Watanabe, S. (Jpn. Kokai Tokkyo Koho 01-148331 [1987/89] 6 pp. from C.A. **111** [1989] No. 139758).
[169] Hall, P. G.; Gittins, P. M.; Winn, J. M.; Robertson, J. (Carbon **23** [1985] 353/71; C.A. **103** [1985] No. 93382).
[170] Greifer, B.; Taylor, J. K. (PB-251220 [1976] 1/32; C.A. **87** [1977] No. 122138).

[171] Alder, J. F.; McCallum, J. J. (PCT Int. Appl. 89-04290 [1988/89] 17 pp. from C.A. **111** [1989] No. 137037).
[172] Sokol'skii, D. V.; Dorfman, Ya. A.; Emel'yanova, V. S. (Kinet. Katal. **14** [1973] 1573/7; Kinet. Catal. [Engl. Transl.] **14** [1973] 1386/9).
[173] Smirnov, V. M. (Khim. Prom-st. Ser. Kislorodn. Prom-st. **1981** No. 3, pp. 5/7 from C.A. **96** [1982] No. 124920).
[174] Sokol'skii, D. V.; Dorfman, Ya. A.; Kazantseva, I. A.; Utegenova, G. S. (Khim. Atsetilena Tr. 3rd Vses. Konf., Dushanbe, Tadzh. SSR, 1968 [1972], pp. 225/8; C.A. **79** [1973] No. 23920).
[175] Sokol'skii, D. V.; Dorfman, Ya. A.; Kazantseva, I. A.; Utegenova, G. S. (Zh. Fiz. Khim. **44** [1970] 2263/7; Russ. J. Phys. Chem. [Engl. Transl.] **44** [1970] 1281/3).
[176] Sokol'skii, D. V.; Dorfman, Ya. A.; Kazantseva, I. A. (Tr. Inst. Khim. Nauk Akad. Nauk Kaz. SSR **30** [1970] 28/36; C.A. **74** [1971] No. 2551).
[177] Sokol'skii, D. V.; Dorfman, Ya. A.; Evtikov, N. I. (Zh. Org. Khim. **8** [1972] 2566/9; J. Org. Chem. USSR [Engl. Transl.] **8** [1972] 2615/8).
[178] Sokol'skii, D. V.; Dorfman, Ya. A.; Evtikov, N. I. (Zh. Org. Khim. **9** [1973] 735/7; J. Org. Chem. USSR [Engl. Transl.] **9** [1973] 757/8).
[179] Valetdinov, R. K.; Sharifullin, A. Sh.; Gafarova, A. F.; Volozhin, L. M.; Abdullina, N. A.; Yunusov, U. I. (U.S.S.R. 1588431 [1988/90] from C.A. **114** [1991] No. 11508).
[180] Nikandrov, G. A.; Vulikh, A. I.; Kim, F. I.; Shukalkina, L. I. (Khim. Prom-st. [Moscow] **45** [1969] 903/4; Sov. Chem. Ind. [Engl. Transl.] **45** No. 12 [1969] 26/8).

1.3.1.5.5 Reactions with Elements

The results of earlier investigations are described in "Phospor" C, 1965, pp. 35/40. The sorption of PH_3 on the surfaces of a series of metallic and nonmetallic elements is covered in Section 1.3.1.5.11, pp. 287/93. The formation of films and layers on various substrates by reacting PH_3 with elements is described in Section 1.3.1.6, pp. 293/308.

With Nonmetals

Helium. Argon. The collision of PH_3 with electronically excited **He** (2^1S, 2^3S) yields PH_3^+ [1]. An ab initio MO calculation of the potential energy surface between PH_3 and **Ar** suggested the formation of a van der Waals complex $H_3P\cdots Ar$ with an internuclear distance $r(P\cdots Ar) = 3.75$ Å. An interaction energy of about −1.7 kJ/mol was calculated [2]. This value represents a lower limit because the basis set used for Ar was not optimal; the actual interaction can be assumed to be stronger by 13 to 18% [3].

Hydrogen. The reaction of PH_3 with atomic **H** proceeds with an intense, greenish yellow chemiluminescence and yields H_2 and red phosphorus as stable products; see "Phosphor" C, 1965, p. 35. The unstable intermediates PH_2 [4 to 6], PH [4, 7], P [8], and probably P_2 [9] were identified spectroscopically; the corresponding deuterated species were obtained with the reactant PD_3. Bands of PHD were found in reactions with PH_3/PD_3 mixtures [5, 6]. The initial hydrogen abstraction from PH_3 by H yields PH_2 and can be used for preparative purposes

[8, 10]; see also Section 1.2.1.1, p. 47. The reaction with excess H then leads to the sequential formation of PH and P [8, 11]; their rates of formation could be optimized by controlling the relative flows of the reactants [8]. The role of the primary reaction with H atoms during the UV photolysis of PH_3 is discussed in Section 1.3.1.5.1.3.1, pp. 206/9. The reaction $PH_3 + H \rightarrow PH_2 + H_2$ at room temperature is exothermic by $\Delta H° = -131$ kJ/mol [11]. The rate constants in the temperature range 209 to 495 K were measured by flash photolysis of PH_3 in He, followed by time-resolved detection of H via resonance fluorescence; their numerical values in $cm^3 \cdot molecule^{-1} \cdot s^{-1}$ are given by $k = (4.52 \pm 0.39) \times 10^{-11} \exp[-(1470 \pm 50)/1.987\ T]$. The room temperature rate constant $k_{298} = (3.45 \pm 0.46) \times 10^{-12}$ [12] was confirmed by the results from experiments using PH_3 and H atoms generated in jets of H_2/He mixtures by UHF discharges giving $k_{298} = (2.7 \pm 1.0) \times 10^{-12}$ [11]. The mean of the two results, $k = (3.1 \pm 1.0) \times 10^{-12}\ cm^3 \cdot molecule^{-1} \cdot s^{-1}$, was proposed as the recommended rate constant at 298 K [13].

A barrier height of 51 kJ/mol for the identity reaction $H_3P + H \rightarrow H + H_2PH$ was estimated from the empirical Morse function of the H_2P–H bond and geometrical parameters of the activated complex [14]. The reaction of recoil tritium atoms with PH_3 yields PH_2T and HT, see Section 1.3.1.5.1.5, p. 215.

The formation of PH_4 from PH_3 and H in low-temperature matrices upon γ irradiation is described in Section 1.4.1, p. 312. An ab initio MO calculation at the UHF/4-31G level predicted that PH_3 and H individually are energetically more favorable than PH_4 by 135 kJ/mol [15]; see also Section 1.4.1, p. 313. The differences between the total energies of the reactants and the products for the formation of PH_5 from PH_3 and H_2 (+188 kJ/mol) or 2H (−246 kJ/mol) were obtained from ab initio MO calculations including configuration interactions but neglecting the differences in zero-point vibrational energies [16]; see also Chapter 1.5, p. 321.

Halogens. The reaction of PH_3 with **F** atoms was investigated in flow reactors. The green emission observed in this reaction [17] is less intense than the one resulting from the reaction with F_2 [18] (see below). The products are: HF, which is vibrationally and rotationally excited [17, 19], PH_2, PH, P, P_2 as well as PF, PF_2 [20], and PF_3 [21]. Fluorinated products probably result from secondary reactions ($P + F_2$, $P_2 + F$, etc.) [22]. The reaction of PH_3 with F atoms is a convenient source of PH_2 [10, 23, 24]; see also Section 1.2.1.1, p. 48.

The reaction $PH_3 + F \rightarrow PH_2 + HF$ at ambient temperature has a standard enthalpy change of −246 [19], −249 kJ/mol [25]. The rate constant of $k = (1.2 \pm 0.2) \times 10^{-10}\ cm^3 \cdot molecule^{-1} \cdot s^{-1}$ at 300 K was determined relative to the reaction rate of CH_4 and F [19]. An earlier value, measured relative to the reaction rate of NO with F and Ar [25], was considered to be less reliable [19]. The nascent vibrational and rotational excitation of the HF formed typically was as follows [19]:

vibrational level	v=0	v=1	v=2	v=3	v=4	v=5
relative distribution	0.06	0.10	0.15	0.29	0.29	0.11
highest J level	—	>25	24	23	19	12

The vibrational excitation represents about 54% of the available energy of the reaction [19].

The formation of PH_3F during γ irradiation of PH_3 in an SF_6 low-temperature matrix presumably results from the addition of F atoms to PH_3 and is described in Section 1.3.1.5.1.6, p. 216.

The reaction of PH_3 with $\mathbf{F_2}$ can be carried out in a concentric burner [26] or by mixing the diluted reactants in a flow tube [18]. The reaction is accompanied by a bright emission with a continuum in the region of 320 to 600 nm [18, 26]. Cocondensation of Ar-diluted PH_3 and a small excess of F_2 in Ar at 16 K yielded PH_2F as the dominant product, with PHF_2 and PH_3F_2 as other major products, as well as HF, $(HF)_2$, $H_3P \cdot HF$, and a very small amount of PF_3. The analogous deuterated products were found for the reaction of PD_3. The low concentration of

PF_3 indicated that secondary reactions of the products with F_2 seemed to be of little importance. The products were identified by their IR bands. Initial formation of PH_3F_2 is probably followed by the prevailing loss of HF or the less important loss of H_2 [27].

The hydrogen abstraction $PH_3 + \mathbf{Cl} \rightarrow PH_2 + HCl$ has an enthalpy change of $\Delta_rH^\circ_0 = -97.5$ kJ/mol; a value of $\Delta_rH^\circ_0 = -91.2$ kJ/mol was obtained for the corresponding reaction of PD_3 [28]. The reaction proceeds at 298 K with a rate constant of $k = (2.35 \pm 0.25) \times 10^{-10}$ $cm^3 \cdot molecule^{-1} \cdot s^{-1}$. It was measured for the competitive reactions of thermal ^{38}Cl atoms (generated by neutron irradiation of, and moderated by multiple collisions with, $CClF_3$) with PH_3 and C_2H_3Br in 4000 Torr of $CClF_3$ via determination by radio gas chromatography of the $C_2H_3{}^{38}Cl$ formed [29]. A similar rate constant of $(2.6 \pm 0.9) \times 10^{-10}$ $cm^3 \cdot molecule^{-1} \cdot s^{-1}$ at room temperature was found relative to the reaction of Cl in He with HI in a flowing afterglow apparatus by observing the summed relative intensities of the IR emissions resulting from the HCl formed [28].

The initial vibrational distribution of HCl(DCl) formed from $PH_3(PD_3)$ was determined as follows [28]:

vibrational level	v=0	v=1	v=2	v=3	v=4
relative distribution[a)] for HCl	0.09[c)]	0.49	0.39	0.03	—
relative distribution[b)] for DCl	0.04[c)]	0.19	0.46	0.21	0.10

[a)] Experimental uncertainty ±10%. – [b)] Experimental uncertainty ±20%. – [c)] Extrapolated value.

Secondary processes in the reaction of PH_3 with Cl lead to more highly excited HCl and to the formation of PCl in the A $^3\Pi$ state, as evidenced by a weak, blue-green chemiluminescence [28]. An earlier investigation of the reaction of PH_3 with Cl in a flow apparatus showed a second-order behavior in PH_3 with a rate constant of $k = (2.2 \pm 1.0) \times 10^{-23}$ $cm^6 \cdot molecule^{-2} \cdot s^{-1}$ at ambient temperature. This dependence differed from the first-order behavior of a series of other main group 4 and 5 hydrides [30].

The reaction of gaseous PH_3 with $\mathbf{Cl_2}$ (molar ratio ~1:2) was investigated at total pressures below 10 Torr under static conditions and in a jet. The products, PCl_3 and HCl, form below a threshold of ~305 K and a total pressure of <6 Torr in a moderately exothermic unbranched chain reaction with an apparent activation energy of 42 kJ/mol. Above the threshold, this chain reaction is supplemented by branching reactions. Their onset after a short time is indicated by a blue chemiluminescence (shortwave limit ~400 nm) and a simultaneous temperature jump. The high velocity of the chain reaction, its observation even at 196 K in the dark, and its immediate start in a jet seem to rule out the formation of Cl atoms as a first step. The primary reaction $PH_3 + Cl_2 \rightarrow [PH_3Cl_2] \rightarrow PH_2Cl + HCl$ with $\Delta_rH = -178$ kJ/mol was proposed [31]. The application of the reaction in the oxidative removal of PH_3 from gases is described in Section 1.3.1.5.4, p. 223.

Twin-jet cocondensation of PH_3 and Cl_2 in Ar at 14 K resulted in the formation of $H_3P \cdots Cl_2$, which according to the IR spectrum is a van der Waals complex rather than a rearrangement product. The formation of the 1:1 complex was strongly favored; some evidence of higher aggregates was noted only at high Cl_2 concentration [32]. Stabilization energies of $H_3P \cdots Cl_2$ were calculated by the semiempirical CNDO/2 method assuming geometries with the Cl–Cl bond perpendicular to the C_3 axis of PH_3. Even though the values obtained of 608 to 830 kJ/mol are probably considerably overestimated, they suggest that the formation of such a complex is energetically favorable [31].

The oxidation of gaseous PH_3 by a neutral aqueous suspension of $\mathbf{I_2}$ follows the equation $PH_3 + 2\,I_2 + 2\,H_2O \rightarrow H_3PO_2 + 4\,HI$ (see "Phosphor" C, 1965, p. 39) only at the beginning of the reaction. As the pH value of the solution decreases due to the acids formed, oxidation of H_3PO_2 sets in and the reaction approaches the stoichiometry $PH_3 + 3I_2 + 3H_2O \rightarrow H_3PO_3 + 6HI$ [33]. Nearly quantitative formation of H_3PO_2 resulted from PH_3 oxidation by I_2 in weakly alkaline $NaHCO_3$ solution [34]. A mixture of H_3PO_2 and H_3PO_3 in a ratio of 20:1 results in aqueous solution when the pH value decrease is limited to 5.5 by suspended $BaCO_3$. A product ratio of 1:60 is obtained in 1.5 M HCl [33]. For the use of I_2 in the oxidative removal of PH_3, see Section 1.3.1.5.4, p. 223.

The reaction of gaseous PH_3 with I_2 in CS_2 yielded the main products P_2I_4 and HI; a small quantity of PHI_2 was identified in the solution. The equilibria established between PH_3 (generated in situ from PH_4I in nonaqueous solvents) and I_2 were investigated in closed vessels under the pressure of the HI formed. Products at a reactant ratio of 1:2 or with less I_2 were PI_3, a small quantity of PHI_2, and very little P_2I_4 and PH_2I. Only PI_3 was found at a reactant ratio of 1:3 [35].

Oxygen. The highly exothermic, chemiluminescent gas-phase reaction of PH_3 with **O atoms** in flow tubes is usually carried out by using NO or N_2O as the source of O atoms. The compounds are decomposed by flash photolysis or a discharge, followed by rapid relaxation of the $O(^1D)$ formed via collisions with the diluents He or Ar to $O(^3P)$; a general discussion of the method is given in [13]. A complex chain reaction ensues; the secondary reactions lead to the overall consumption of about two O atoms per PH_3 in mixtures containing excess O and of 2 PH_3 per O with excess PH_3 [36]. Initially an aerosol forms [37, 38], and finally a deposit consisting mainly of red phosphorus and H_3PO_3 [38].

A mass spectrometric investigation of the products resulting from the reaction of PH_3 with O indicated that the main reaction channel ($\geqq$90%) consists of an addition–decomposition process in accordance with the equation $PH_3 + O \rightarrow H_2PO + H$, probably via H_3PO in its triplet state as a very short-lived intermediate. The minor importance of the abstraction reaction $PH_3 + O \rightarrow PH_2 + OH$ ($\Delta_rH^\circ_{298} = -280$ kJ/mol) was demonstrated by the virtual absence of the PH_2^+ signal in a mass spectrometric investigation of the reaction [36] and by the absence of the bands of OH as a primary product in a chemiluminescence study [39]. The minor importance of this reaction channel also agrees with the small rate constant for the formation of OH at ambient temperature ($k = (2.8 \pm 1.2) \times 10^{-12}$ $cm^3 \cdot molecule^{-1} \cdot s^{-1}$ [11]) in comparison to the overall rate constant for the reaction of PH_3 with O (discussed in the next paragraph) [13]. The estimated temperature dependence of the rate constant for the abstraction channel suggested a measurable contribution to the experimental overall rate constant only at T>1000 K [40]. However, the abstraction product PH_2 was obtained in satisfactory yield by reacting PH_3 with O_2 which had been exposed to a dc discharge [41]. The intermediate formation of PH_2 was discussed for the reaction of PH_3 with approximately equimolar quantities of $O(^3P)$ and $N(^4S)$ yielding $PO(X\ ^2\Pi)$ [102].

The rate constant of the reaction of PH_3 with $O(^3P)$ was determined in flow systems. A value of $k = (4.75 \pm 1.09) \times 10^{-11}$ $cm^3 \cdot molecule^{-1} \cdot s^{-1}$, obtained by measuring the resonance fluorescence of O in excess PH_3, remained unchanged in the temperature range 208 to 423 K and did not depend on either the total pressure or the concentrations of PH_3, the oxygen precursor N_2O, and O. The independence of the O atom concentration indicates the true isolation of the reaction under study from secondary reactions [40]. The cited rate constant agrees with an earlier value which was obtained by the same method at 298 K [13] and also with a value of $k = (4.6 \pm 0.8) \times 10^{-11}$ $cm^3 \cdot molecule^{-1} \cdot s^{-1}$ which was determined at 293 K with excess O by a mass spectrometric method [36]. An early but differing rate constant ($k = 4.2 \times 10^{-12}$ $cm^3 \cdot molecule^{-1} \cdot s^{-1}$ at ambient temperature) [11] probably resulted from an insufficient resolution of the resonant fluorescent spectrometer used, which led to the simultaneous measurement of the reaction of PH_3 with generated H atoms [13]. An estimation of the rate constant based on a kinetic model for the reaction of PH_3 with O_2 [42] differed by two orders of magnitude from the experimental value. An approximate classical trajectory calculation yielded a rate constant which was independent of temperature as observed experimentally, but the numerical value was too large by a factor of 10 [43].

The strong chemiluminescence of the reaction between PH_3 and O atoms was studied in connection with the search for chemical lasers operating in the visible range [38, 44]. The intensity of the continuous emission increases with increasing reaction time [39, 45] and with increasing content of moisture in the gases used [36, 46]. The emission observed with excess O atoms is whitish green (originally described as whitish to yellow-white [46]) and extends down to ~360 nm; the blue-white emission observed with excess PH_3 extends down to ~285 nm and has an increased intensity in the UV range [44, 46]. The upper limit of the emission lies above 990 nm [38]. Spectra of the emission are displayed in [38, 44, 45]. Superimposed bands were assigned to the intermediates PH_2, PH, OH, PO(B $^2\Pi$, A $^2\Sigma^+$) [37, 38], P [11], and HPO [38]. The bands of PH appear earlier than those of PO [37], which are observed earlier than the bands of OH [37, 39]. The reaction of PH_3 with an excess of O(^{3}P) and N(^{4}S) in equal concentration (from titrating half of the present N(^{4}S) atoms with NO) greatly enhances the formation of PO(X $^2\Pi$) [102]. The intermediates H_3PO, H_2PO, HPO, HPO_2, PO, and PO_2 were identified using mass spectrometry [44].

The attempted identification of emitting species by absorption spectroscopy failed; the chemiluminescence was attributed tentatively to the transition $PO_2(^2B_1) \rightarrow PO_2(^2A_1)$ or to the decomposition of the excimer $(PO)_2^*$ into 2 PO(X $^2\Pi$) [38]. However, the presence of two or more emitting species was deduced from the broadness of the emission range and the influence of the reactant ratio on the color of the chemiluminescence. The consecutive reactions of O atoms with HPO and H_2PO were proposed as the source of the blue component of the chemiluminescence; the origin of the green component was attributed to the reaction $PO + O \rightarrow PO_2$ or to the decomposition of the excimer $(PO)_2^*$ [44]. Quantum yields for the reaction of PH_3 with O atoms were determined under varying experimental conditions and led to diverging results [38, 45]. This variation was judged to be a result of competing side reactions which do not lead to chemiluminescence [44]. The intensity of the continuous emission is only slightly influenced by the addition of diluents such as He, H_2, N_2, and NO. Added O_2 efficiently quenches the blue component of the emission and only slightly influences the green component [38, 44]. The blue component is also nearly extinguished if H atoms are present [44].

The reaction of PH_3 with O atoms was also investigated in an Ar matrix by cocondensing the reactants (or PD_3 or ^{18}O-enriched samples) at 12 to 14 K. When an excess of O atoms generated by a low power microwave discharge was used, IR bands of the products could be assigned to the phosphorus-containing species H_3PO, $H_3P\cdots O_3$, $(HO)_2PHO$, H_2POH, HPOH, H_2PO, HPO, HOPO, HPO_2 (probably), $HOPO_2$, $HP(O_2)O$, O_2POPO_2, PO_3, PO_2, PO, and $PO\cdots H_2$. Only the bands of the first three products were absent when the O atoms were generated by a strong MW discharge, where the accompanying vacuum UV radiation decomposed PH_3 almost completely. Additional IR bands were assigned tentatively to O_2PPH_2, $O_2PPH_2\cdots O_2$, and P_2O_3 [47]. In an earlier investigation, only the species HPO, PO, and possibly PO_2 were identified upon photolysis (121.6 nm) of cocondensed equimolar PH_3/N_2O mixtures in Ar or upon cocondensation of PH_3 (PD_3) in Ar with O_2 which had been passed through an MW discharge of moderate intensity [48].

The reaction of PH_3 with **O_2 molecules** yields H_3PO_3 and H_3PO_4. Earlier results on this reaction and fundamental investigations regarding the explosion limits of PH_3/O_2 mixtures were described in "Phosphor" C, 1965, pp. 35/9. The explosibility even of low concentrations of PH_3 in air makes such mixtures highly dangerous; see for example [49]. Most of the more recent publications on PH_3/O_2 mixtures describe refined investigations of the lower flammability (explosion) limit. Spontaneous ignition was found even for highly diluted PH_3 (4 vol% in N_2) when discharged into air through a nozzle [50]. Temperature-dependent experimental values of the lower explosion limit of PH_3 in air at 1 atm in the presence of 0.39 vol% water vapor were determined by spark discharge (values in parentheses were estimated) [42]:

t in °C	10	20	30	40	50	80	100
[PH_3] in vol%	2.10	2.04	1.98	1.92	1.85	(1.68)	(>1.5)

The presence of water vapor reduced the flammability in air, whereas NH_3 probably lowered the explosion limit [42]. Other explosion limits observed in air at ambient temperature and pressure were 1.6 vol% PH_3 [51, 52], 1.86 vol% in dry air, and 1.85 vol% in air saturated with humidity [53]. The last two values were constant at total pressures in the range of 150 to 760 Torr. The lower explosion limit of 2.18 vol% PH_3 in dry O_2 also remained constant in this pressure range at 25°C, whereas for 5 vol% O_2 in N_2 a lower explosion limit of 1.67 vol% PH_3 was measured [53].

The upper explosion limit of PH_3 in air at 1 atm and ambient temperature is believed to be close to 100 vol% of PH_3, as indicated by the explosions occurring on adding air to PH_3. These explosions yield small quantities of H_2 [51]. A formula for the upper explosion limit of PH_3 in the presence of varying concentrations of O_2 and N_2 at temperatures of 600 K or more was derived from a physicochemical model of the reaction [54].

The reaction of 0.1 to 0.6 vol% of PH_3 (in an inert gas) with O_2 in a concentration ratio of 1:5 began at 185 to 215°C; the complete conversion of PH_3 required temperatures of 320 to 360°C. The main volatile product, H_2, was formed in less than the stoichiometric amount and suggested that polyphosphanes were part of the solid product in addition to phosphorus oxides. The apparent activation energy of the reaction was determined to be 42 ± 13 kJ/mol [55, 56]. Flash photolysis of PH_3/O_2 mixtures was used in kinetic investigations of the reaction of PH_2 with O_2; see Section 1.2.1.5, p. 89, for details. The stability of PH_3 in air which contained various inorganic and organic compounds was studied at 20 to 50°C [57]. The catalytic oxidation of PH_3 by O_2 in the presence of MoO_3 is described in Section 1.3.1.5.7, p. 256.

A van der Waals complex of PH_3 with **ozone** forms upon cocondensation of equimolar amounts of PH_3/Ar and O_3/Ar mixtures or the isotopically substituted reactants at 12 to 18 K and additionally via diffusion at 30 K in the matrix. A minimal charge transfer to O_3 in $H_3P\cdots O_3$ is indicated by the moderate shifts of the product IR bands compared to those of the reactants [58]. Photolysis by red visible light gives a host of $H_xP_yO_z$ species as primary products [58, 59].

Reactions of diluted PH_3 with an excess of diluted O_3 in the gas phase are accompanied by a white glow with peach-colored fringes and yield a coating on the walls of the reactor. The coating consists of white and yellow phosphorus and an additional substance which is probably P_4O_{10} [60]. At normal pressure, the featureless continuum emission observed in reactions with O_3 in O_2 extends from 490 to above 810 nm and peaks at 670 nm at high concentrations of PH_3 (10 to 15%) in He, whereas for ~100 ppm of PH_3 in He, an emission range of 350 to 700 nm with a maximum at 530 nm was found [61]. At a PH_3 partial pressure of 8% in N_2 and a total pressure of 2 Torr, the chemiluminescence consisted of a continuum spanning the range 380 to 800 nm and peaking at ~610 nm; UV emissions were negligible [60]. An undefined weak system of bands at short wavelenghts from the maximum was observed only under special experimental conditions. In experiments with O_3 in N_2, the emission peaks at about 450 to 500 nm and the UV emission is greatly enhanced. Bands of the PO transitions $A\,^2\Sigma \rightarrow X\,^2\Pi$ and $B\,^2\Sigma \rightarrow X\,^2\Pi$ and of OH were identified in the UV range. The UV emissions of the reactions of PH_3 with O_3 in N_2 were stronger than those of reactions with O_3 in O_2; lower limits of the quantum yields were determined [45]. The chemiluminescence observable in reactions of PH_3 with O_3 in O_2 was used for the quantitative determination of phosphorus in phosphates [61].

Sulfur. Selenium. Tellurium. The reaction of PH_3 with $\mathbf{S_8}$ vapor can be induced by an argon discharge (which converts S_8 into smaller sulfur molecules). When the unstable, early products were trapped in an Ar matrix by condensation at 12 K, hydrogenated species of the general formula $H_xP_yS_z$ (x, y, z not specified) and the binary species P_2S, P_2S_4, PS_2, PS, PS_2^-, and *cyclo*-P_4S were identified by IR spectroscopy [62]. The removal of PH_3 impurities from H_2Se and H_2Te by using elementary Se and Te as sorbent was studied by a thermodynamic analysis. The standard Gibbs free energy change for the reaction $2\,PH_3+3\,\mathbf{Se}\rightleftharpoons 3\,H_2Se+0.5\,P_4$ was calculated from thermodynamic data to be $\Delta_rG^\circ_{298}=77.64$ kJ/mol which corresponds to an equilibrium constant of log $K_p=-13.62$. For the analogous reaction with **Te**, values of $\Delta_rG^\circ_{298}=$ 286.10 kJ/mol and log $K_p=-50.13$ were obtained [63].

Nitrogen. The reaction of PH_3 in Ar with atomic **N** is accompanied by a pale green luminescence. The bands observed in the range 6000 to 2300 cm^{-1} were assigned to PH, PH_2, PN, N_2^+, and NH. The final product is a solid [64]. The reaction of PH_3 in the carrier gas He with **N(^{4}S)** atoms was found to be very slow. A rate constant of $k\leqq 4.0\times10^{-14}$ $cm^3\cdot molecule^{-1}\cdot s^{-1}$ at ambient temperature was determined in a flow tube by molecular-beam sampling mass spectrometry. This rate constant is compatible with the slight endothermicity ($\Delta_rH_{298}=$ 4 kJ/mol) of the reaction $PH_3+N\rightarrow PH_2+NH$ [36].

Phosphorus. Silicon. The reaction of PH_3 with recoil ^{32}**P** atoms (from the irradiation of PH_3 with thermal neutrons) yields considerable quantities of $^{32}PH_3$ via H abstraction; details are given in Section 1.3.1.5.1.5, p. 215. The measured relative efficiency of the H abstraction in irradiated PH_3 and PF_3 was used to consider the likely reaction mechanism in [65]. A reaction of PH_3 (from adding Ca_3P_2 to the solution) with recoil ^{32}P was also considered in order to explain the formation of $30\pm10\%$ of $^{32}PH_3$ when KCl crystals [66] or KCl-$CaCl_2$ mixed crystals [67, 68] were irradiated with protons (635 to 660 MeV) and dissolved in water. The H abstraction between ^{32}P and PH_3 might have occurred at the surface of the dissolving crystals; however, a formation of $^{32}PH_3$ by secondary reactions was also considered to be conceivable [66 to 68].

The formation of SiH_4 from PH_3 and recoiling ^{31}**Si** atoms (generated by irradiation of PH_3 with fast neutrons) is described in Section 1.3.1.5.1.5, p. 215.

Boron. The reaction of PH_3 in Ar with amorphous **B** at 1000 to 1100°C yielded BP. This method is suitable for the preparation of BP powder because the reaction is relatively fast and proceeds almost quantitatively [69].

With Metals

Alkali Metals. The well-known reaction of PH_3 with solutions of alkali metals in liquid NH_3 leads to the formation of pure, crystalline dihydrogen phosphides, MPH_2, where M = Li, Na [70 to 72], K [70], Rb [70, 73], and Cs [70]. An excess of PH_3 is recommended to obtain phosphides which are free of alkali amides [70]. Sodium in diethyl ether/anthracene mixtures or potassium in THF/naphthalene mixtures also give high yields of the corresponding dihydrogen phosphides [74]. Additional details concerning the syntheses of solutions of PH_2^- and the isolation of solid phosphides under various experimental conditions are described in Section 1.2.4.1, p. 102. Earlier reactions of PH_3 with Na are described in "Natrium" Erg.-Bd. 4, 1967, pp. 1541/2.

Magnesium. Excited Mg atoms (3s 3p $^1P_1$) and PH_3 give MgH ($\nu=0$, 1); its nascent rotational quantum state distribution was determined [75].

Gallium. Mixtures of PH_3 and H_2 react with molten Ga forming a film of GaP on the surface of the melt. However, the film dissolves in the melt when the reaction is conducted at 1100°C, and crystals of GaP can be isolated after cooling the mixture [76].

Tantalum. The reaction of PH_3 with Ta commences above 200°C. A tantalum hydride probably forms above 300°C. Tantalum phosphides are obtained above 800°C and decompose above 900°C [77].

Chromium. Molybdenum. Tungsten. Cocondensation of PH_3 with chromium vapor at 60 K leads to the formation of a PH_3 complex of Cr [78]. Chromium powder with PH_3 forms phosphides of different composition at 700 to 950°C. The reaction at 850°C yields CrP, whereas at 950°C only the lowest phosphide, Cr_2P, is obtained; the latter dissociates at 1000°C to give Cr [79]. Gaseous PH_3 is inert in the presence of metallic Mo on an Al_2O_3 support at about 350°C [80]. The reaction of PH_3 with W at 850°C gives WP [81].

Iron. Ruthenium. Osmium. Iron (and steel) are rapidly corroded by PH_3 in the presence of air and humidity. The rate of corrosion increases with increasing temperature and relative humidity, and probably involves two or more stages. The initial formation of one or more acids of phosphorus is followed by an attack of these acids on iron [82].

Ruthenium powder and PH_3 in Ar yield RuP_2 at 1150°C. At 800°C the composition of the product is intermediate between RuP_2 and RuP. The course of the reaction can be described by a diffusion mechanism and obeys a parabolic law [83]. Osmium powder and PH_3 in Ar yield $OsP_{1.95}$ at 1150°C. The limiting composition OsP_2 cannot be reached because the compound decomposes under these reaction conditions. A diagram showing the consumption of Os as a function of the reaction time is given in the original paper [84].

Nickel. The corrosion of Ni by PH_3 in the presence of air and humidity proceeds as described for Fe (see above) [82]. The cocondensation at 10 K of nickel vapor with PH_3 in Ar yields $Ni(PH_3)_n$ with n=1 to 4, where n increases with an increasing concentration of PH_3 [85]. $Ni(PH_3)_4$ presumably forms when Ni vapor and pure PH_3 are cocondensed at 60 K [78]. An investigation of this reaction at 77 K showed H_2 to be the only volatile product being evolved both during the cocondensation and more rapidly on warming to room temperature. Cocondensation of nickel vapor with a mixture of equal volumes of PH_3 and PF_3 also frees some H_2. Warming of the frozen sample to ambient temperature yields $Ni(PF_3)_2(PH_3)_2$, $Ni(PF_3)_3PH_3$, and $Ni(PF_3)_4$ as volatile products [86].

Copper. Cocondensation of copper vapor and neat PH_3 or PD_3 at 60 K yields a phosphane complex of Cu. An investigation of the cocondensate in a Kr matrix at 12 K suggests the formation of $Cu(PH_3)_n$ with n=1 to 3 [78]. In a theoretical study of the electronic structure of octahedral copper(I) cluster complexes, the energy changes for reactions of PH_3 with Cu, Cu_6, and Cu^+ were calculated using the linear combination of Gaussian-type orbital local density functional (LCGTO-LDF) method. Values were computed and discussed for the formation of $[Cu(PH_3)]^{q+}$ with q=0 or 1 and of $[Cu_6(PH_3)_6]$ both from Cu and Cu_6 [87]. The reaction of Cu with PH_3 at 714 to 800°C gives H_2 and the molten Cu_3P–Cu eutectic. This process is claimed to be suitable for the recovery of phosphorus from sludges and of Cu from scrap [88]. The corrosion of Cu (and brass) by PH_3 is rapid in the presence of air and humidity [82, 89] (see Fe above) and is noticeable even at 10°C [90]. The corrosion rate increases with increasing temperature and increasing relative humidity [82, 90].

Lanthanides. The complete conversion of lanthanide metals by PH_3 in H_2 to the corresponding lanthanide phosphides was mentioned [91]. In more detailed studies reaction temperatures of 1000 to 1100°C for Nd to give NdP [92, 93] and of 900 to 950°C for Sm to give SmP [92, 94] were recommended. Solutions of Eu and Yb in liquid NH_3 react smoothly with PH_3, yielding H_2 and precipitates of $M(PH_2)_2 \cdot 7\,NH_3$ with M = Eu, Yb. The precipitates decompose upon warming with the formation of MP, a small amount of elemental phosphorus, and evolvement of PH_3, H_2, and NH_3 [95]; see also [96].

Actinides. Finely divided Th (obtained by decomposition of the hydride) reacts smoothly with PH_3 at 550°C to form Th_3P_4 as the major product containing some ThP, ThO_2, and thorium hydride as by-products [97]. Uranium powder (obtained analogously by decomposition of UH_3) yields a mixture of U_3P_4 and UP_2 in reactions with neat PH_3 at 385°C [98] and with PH_3 in Ar at 600°C [99]; see also [97, 100]. Reactions at 530°C yield U_3P_4 exclusively when undiluted PH_3 is used [98]. More detailed informations on the reactions of PH_3 with U are given in "Uranium" Suppl. Vol. C 14, 1981, pp. 2/3, 28. Finely divided Pu reacts with PH_3 in H_2 forming PuP [101].

References:

[1] Yee, D. S. C.; Stewart, W. B.; McDowell, C. A.; Brion, C. E. (J. Electron Spectrosc. Relat. Phenom. **7** [1975] 377/83).
[2] Latajka, Z.; Scheiner, S. (J. Mol. Struct. **198** [1989] 205/13).
[3] Chalasinski, G.; Cybulski, S. M.; Szczesniak, M. M.; Scheiner, S. (J. Chem. Phys. **91** [1989] 7809/17).
[4] Guenebaut, H.; Pascat, B. (J. Chim. Phys. Phys.-Chim. Biol. **61** [1964] 592/5).
[5] Guenebaut, H.; Pascat, B.; Berthou, J.-M. (J. Chim. Phys. Phys.-Chim. Biol. **62** [1965] 867/77).
[6] Guenebaut, H.; Pascat, B. (C. R. Hebd. Seances Acad. Sci. **259** [1964] 2412/5).
[7] Guenebaut, H.; Pascat, B. (C. R. Hebd. Seances Acad. Sci. **256** [1963] 677/80).
[8] Berkowitz, J.; Cho, H. (J. Chem. Phys. **90** [1989] 1/6).
[9] Guenebaut, H.; Pascat, B.; Brion, J. (C. R. Hebd. Seances Acad. Sci. **259** [1964] 3545/8).
[10] Hills, G. W.; McKellar, A. R. W. (J. Chem. Phys. **71** [1979] 1141/9).

[11] Aleksandrov, E. N.; Arutyunov, V. S.; Dubrovina, I. V.; Kozlov, S. N. (Fiz. Gorenya Vzryva **18** No. 4 [1982] 73/8; Combust. Explos. Shock Waves [Engl. Transl.] **18** [1982] 451/5; Oxid. Commun. **3** [1983] 327/39).
[12] Lee, J. H.; Michael, J. V.; Payne, W. A.; Whytock, W. A.; Stief, L. J. (J. Chem. Phys. **65** [1976] 3280/3; NBS Spec. Publ. [U.S.] No. 526 [1978] 345/7; C. A. **90** [1979] No. 95337).
[13] Stief, L. J.; Payne, W. A.; Nava, D. F. (J. Chem. Phys. **87** [1987] 2112/5).
[14] Leroy, G.; Sana, M. (J. Mol. Struct. **136** [1986] 283/301 [THEOCHEM **29**]).
[15] Janssen, R. A. J.; Visser, G. J.; Buck, H. M. (J. Am. Chem. Soc. **106** [1984] 3429/37).
[16] Trinquier, G.; Daudey, J.-P.; Caruana, G.; Madaule, Y. (J. Am. Chem. Soc. **106** [1984] 4794/9).
[17] Duewer, W. H.; Setser, D. W. (J. Chem. Phys. **58** [1973] 2310/20).
[18] Chowdhury, M. A.; Benard, D. J. (Proc. SPIE-Int. Soc. Opt. Eng. **875** [1988] 173/82; C. A. **109** [1988] No. 29694).
[19] Manocha, A. S.; Setser, D. W.; Wickramaaratchi, M. A. (Chem. Phys. **76** [1983] 129/46).
[20] Dyke, J. M.; Jonathan, N.; Morris, A. (Int. Rev. Phys. Chem. **2** [1982] 3/42, 22).

[21] Saito, S.; Endo, Y.; Hirota, E. (J. Chem. Phys. **82** [1985] 2947/50).
[22] Butcher, V.; Dyke, J. M.; Lewis, A. E.; Morris, A.; Ridha, A. (J. Chem. Soc. Faraday Trans. II **84** [1988] 299/310).
[23] Davies, P. B.; Russel, D. K.; Thrush, B. A. (Chem. Phys. Lett. **37** [1976] 43/6).
[24] Davies, P. B.; Russel, D. K.; Thrush, B. A.; Radford, H. E. (Chem. Phys. **44** [1979] 421/6).
[25] Pollock, T. L.; Jones, W. E. (Can. J. Chem. **51** [1973] 2041/6).
[26] Davis, S. J.; Hager, G.; Hadley, S. G. (IEEE J. Quantum Electron. **11** [1975] 693).
[27] Andrews, L.; Withnall, R. (Inorg. Chem. **28** [1989] 494/9).

[28] Wickramaaratchi, M. A.; Setser, D. W. (J. Phys. Chem. **87** [1983] 64/72).
[29] Iyer, R. S.; Rogers, P. J.; Rowland, F. S. (J. Phys. Chem. **87** [1983] 3799/801).
[30] Schlyer, D. J.; Wolf, A. P.; Gaspar, P. P. (J. Phys. Chem. **82** [1978] 2633/7).

[31] Azatyan, V. V.; Gagarin, S. G.; Zakhar'in, V. I.; Kalkanov, V. A.; Kolbanovskii, Yu. A. (Kinet. Katal. **26** [1985] 222/6; Kinet. Catal. [Engl. Transl.] **26** [1985] 191/4).
[32] Machara, N. P.; Ault, B. S. (J. Phys. Chem. **92** [1988] 73/7).
[33] Švehla, P. (Collect. Czech. Chem. Commun. **31** [1966] 4712/7).
[34] Horák, J.; Ettel, V. (Sb. Vys. Sk. Chem. Technol. Praze Org. Technol. **5** [1961] 93/7; C. A. **62** [1965] 12440).
[35] Schmidt, M.; Schröder, H. H. J. (Z. Anorg. Allg. Chem. **378** [1970] 192/209; Angew. Chem. **82** [1970] 808; Angew. Chem. Int. Ed. Engl. **9** [1970] 808).
[36] Hamilton, P. A.; Murrells, T. P. (J. Chem. Soc. Faraday Trans. II **81** [1985] 1531/41).
[37] Norrish, R. G. W.; Oldershaw, G. A. (Proc. R. Soc. [London] A **262** [1961] 10/8).
[38] Harris, D. G.; Chou, M. S.; Cool, T. A. (J. Chem. Phys. **82** [1985] 3502/15).
[39] Agrawalla, B. S.; Setser, D. W. (J. Chem. Phys. **86** [1987] 5421/32).
[40] Nava, D. F.; Stief, L. J. (J. Phys. Chem. **93** [1989] 4044/7).

[41] Endo, Y.; Saito, S.; Hirota, E. (J. Mol. Spectrosc. **97** [1983] 204/12).
[42] Green, A. R.; Sheldon, S.; Banks, H. J. (Dev. Agric. Eng. **5** [1984] 433/49; C. A. **102** [1985] No. 151587).
[43] Phillips, L. F. (Chem. Phys. Lett. **165** [1990] 545/50).
[44] Hamilton, P. A.; Murrells, T. P. (J. Phys. Chem. **90** [1986] 182/5).
[45] Fraser, M. E.; Stedman, D. H.; Dunn, T. M. (J. Chem. Soc. Faraday Trans. I **80** [1984] 285/95).
[46] Davies, P. B.; Thrush, B. A. (Proc. R. Soc. [London] A **302** [1968] 243/52).
[47] Withnall, R.; Andrews, L. (J. Phys. Chem. **92** [1988] 4610/9).
[48] Larzillière, M.; Jacox, M. E. (J. Mol. Spectrosc. **79** [1980] 132/50; NBS Spec. Publ. [U.S.] No. 561-1 [1979] 529/43; C. A. **92** [1980] No. 31431).
[49] Vogman, L. P. (Khim. Prom-st. [Moscow] **1984** 372/3 from C. A. **101** [1984] No. 116147).
[50] Urano, Y.; Horiguchi, S.; Tokuhashi, K.; Ohtani, H.; Iwasaka, M.; Kondo, S.; Hashiguchi, Y. (Kagaku Gijutsu Kenkyusho Hokoku **84** [1989] 585/93; C. A. **114** [1991] No. 87777).

[51] Ohtani, H.; Horiguchi, S.; Urano, Y.; Iwasaka, M.; Tokuhashi, K.; Kondo, S. (Combust. Flame **76** [1989] 307/10; C. A. **111** [1989] No. 25685).
[52] Ohtani, H.; Horiguchi, S.; Urano, Y.; Tokuhashi, K.; Iwasaka, M.; Kondo, S. (Anzen Kogaku **27** [1988] 96/8 from C. A. **109** [1988] No. 134304).
[53] Bond, E. J.; Miller, D. M. (J. Stored Prod. Res. **24** [1988] 225/8; C. A. **110** [1989] No. 2188231).
[54] Popov, V. P.; Rabinovich, O. S. (Fazovye Khim. Prevrashch. Vzaimodeistvii Tel Potokom Gaza **1975** 27/48; C. A. **85** [1976] No. 99805).
[55] Strater, K.; Mayer, A. (Semicond. Silicon 1st Int. Symp. Pap., New York City 1969, pp. 469/80, 472, 476/7; C. A. **71** [1969] No. 108630).
[56] Mayer, A.; Strater, K.; Puotinen, D. A. (AD-875321 [1970] 1/134, 26, 31, 32, 35; C. A. **78** [1973] No. 64078).
[57] Himi, K. (Kanagawa-ken Taiki Osen Chosa Kenkyu Hokoku **29** [1988] 58/63 from C. A. **111** [1989] No. 159408).
[58] Withnall, R.; Hawkins, M.; Andrews, L. (J. Phys. Chem. **90** [1986] 575/9).
[59] Withnall, R.; Andrews, L. (J. Phys. Chem. **91** [1987] 784/97).
[60] Fraser, M. E.; Stedman, D. H. (J. Chem. Soc. Faraday Trans. I **79** [1983] 527/42).

[61] Fujiwara, K.; Kanchi, T.; Tsumara, S.; Kumamaru, T. (Anal. Chem. **61** [1989] 2699/703).
[62] Mielke, Z.; Brabson, G. D.; Andrews, L. (J. Phys. Chem. **95** [1991] 75/9).
[63] Gladyshev, V. P.; Kovaleva, S. V.; Dubinina, L. K. (Izv. Vyssh. Uchebn. Zaved. Khim. Khim. Tekhnol. **24** [1981] 1462/4; C.A. **96** [1982] No. 111184).
[64] Guenebaut, H.; Pascat, B. (C. R. Hebd. Seances Acad. Sci. **256** [1963] 2850/3).
[65] Zeck, O. F.; Gennaro, G. P.; Tang, Y. N. (J. Am. Chem. Soc. **97** [1975] 4498/503; J. Chem. Soc. Chem. Commun. **1974** 52/3).
[66] Murin, A. N.; Banasevich, S. N.; Bogdanov, R. V. (Izv. Akad. Nauk SSSR Ser. Khim. **1961** 1433/7; Bull. Acad. Sci. USSR Div. Chem. Sci. [Engl. Transl.] **1961** 1334/7).
[67] Bogdanov, R. V.; Murin, A. N. (Zh. Obshch. Khim. **35** [1965] 916/23; J. Gen. Chem. USSR [Engl. Transl.] **35** [1965] 921/7).
[68] Bogdanov, R. V.; Murin, A. N. (Zh. Obshch. Khim. **37** [1967] 2425/31; J. Gen. Chem. USSR [Engl. Transl.] **37** [1967] 2308/12).
[69] Samsonov, G. V.; Titkov, Yu. B. (Zh. Prikl. Khim. [Leningrad] **36** [1963] 669/70; J. Appl. Chem. USSR [Engl. Transl.] **36** [1963] 636/7).
[70] Jacobs, H.; Hassiepen, K. M. (Z. Anorg. Allg. Chem. **531** [1985] 108/18).

[71] Wagner, R. I. (U.S. 3086053 [1957/63] 4 pp.; C.A. **59** [1963] 10124).
[72] Wagner, R. I. (U.S. 3086056 [1960/63] 4 pp.; C.A. **60** [1964] 559).
[73] Bergerhoff, G.; Schultze-Rhonhof, E. (Acta Crystallogr. **15** [1962] 420).
[74] Semenenko, K. N.; Taisumov, Kh. A.; Dubrovin, A. V.; Fofanov, A. A. (U.S.S.R. 497225 [1973/75] 2 pp.; C.A. **84** [1976] No. 124085).
[75] Breckenridge, W. H.; Umemoto, H. (J. Chem. Phys. **81** [1984] 3852/65).
[76] Yarmola, T. M. (Zh. Prikl. Khim. [Leningrad] **44** [1971] 2306/8; J. Appl. Chem. USSR [Engl. Transl.] **44** [1971] 2355/7).
[77] Chow, R.; Chai, Y. G. (J. Vac. Sci. Technol. [2] A **1** [1983] 49/54).
[78] Bowmaker, G. A. (Austral. J. Chem. **31** [1978] 2549/53).
[79] Vereikina, L. L.; Samsonov, G. V. (Ukr. Khim. Zh. **28** [1962] 441/3; C.A. **57** [1962] 16114).
[80] Paul, D. K.; Rao, L. F.; Yates, J. R., Jr. (J. Phys. Chem. **96** [1992] 3446/52).

[81] Samsonov, G. V.; Vereikina, L. L. (Fosfidy [Phosphides], Izd. Akad. Nauk SSSR; Kiev 1961, pp. 1/128, 52; C.A. **1961** 21958).
[82] Bond, E. J.; Dumas, T.; Hobbs, S. (J. Stored Prod. Res. **20** [1984] 57/63; C.A. **101** [1984] No. 95979).
[83] Chernogorenko, V. B.; Lynchak, K. A.; Kulik, L. Ya.; Shkaravskii, Yu. F.; Klochkov, L. A. (Zh. Neorg. Khim. **22** [1977] 1170/3; Russ. J. Inorg. Chem. [Engl. Transl.] **22** [1977] 639/41).
[84] Chernogorenko, V. B.; Solomatina, L. Ya. (Zh. Neorg. Khim. **28** [1983] 1940/3; Russ. J. Inorg. Chem. [Engl. Transl.] **28** [1983] 1100/2).
[85] Trabelsi, M.; Loutellier, A. J. (J. Mol. Struct. **43** [1978] 151/7).
[86] Timms, P. L. (J. Chem. Soc. A **1970** 2526/8).
[87] Bowmaker, G. A.; Pabst, M.; Rösch, N.; Schmidbaur, H. (Inorg. Chem. **32** [1993] 880/7).
[88] Buckholtz, H. E.; Miller, G. T. (U.S. 5019157 [1989/91] 4 pp.; C.A. **115** [1991] No. 54003).
[89] Pytlewski, L. L.; Iaconianni, F. J. (Proc. Symp. Explos. Pyrotech. **13** [1986] III 39/III 40 from C.A. **109** [1988] No. 152515).
[90] Moku, M.; Yasutomo, J.; Akiyama, H. (Shokubutsu Boekisho Chosa Kenkyu Hokoku No. 16 [1980] 99/100).

[91] Mirnov, K. E. (Izv. Akad. Nauk SSSR Neorg. Mater. **19** [1983] 714/7; Inorg. Mater. [Engl. Transl.] **19** [1983] 645/9).
[92] Samsonov, G. V.; Vereikina, L. L.; Endrzheevskaya, S. N.; Tikhonova, N. N. (Ukr. Khim. Zh. [Russ. Ed.] **32** [1966] 115/8; Sov. Prog. Chem. [Engl. Transl.] **32** No. 2 [1966] 89/91).
[93] Endrzheevskaya, S. N.; Samsonov, G. V. (Zh. Obshch. Khim. **35** [1965] 1983/4; J. Gen. Chem. USSR [Engl. Transl.] **35** [1965] 1973/4).
[94] Novruzly, I. M.; Tikhonova, N. N. (Uch. Zap. Azerb. Gos. Univ. Ser. Khim. Nauk **1965** No. 3, pp. 19/23; C.A. **65** [1966] 18134).
[95] Howell, J. K.; Pytlewski, L. L. (Inorg. Nucl. Chem. Lett. **6** [1970] 681/6).
[96] Pytlewski, L. L.; Howell, J. K. (Chem. Commun. **1967** 1280).
[97] Baskin, Y.; Dusek, J. T. (ANL-6868 [1963] 142/8; N.S.A. **18** [1964] No. 25891).
[98] Baskin, Y.; Shalek, P. D. (J. Inorg. Nucl. Chem. **26** [1964] 1679/84).
[99] Allbutt, M.; Junkison, A. R.; Carney, R. F. (Proc. Br. Ceram. Soc. No. 7 [1967] 111/26, 113/4; C.A. **67** [1967] No. 28348).
[100] Kruger, O. L.; Moser, J. B. (Chem. Eng. Prog. Symp. Ser. **63** No. 80 [1967] 1/10, 2/3).

[101] Moser, J. B.; Kruger, O. L. (ANL-6868 [1963] 149/56, 151; N.S.A. **18** [1964] No. 25891).
[102] Clyne, M. A. A.; Heaven, M. C. (Chem. Phys. **58** [1981] 145/50).

1.3.1.5.6 Reactions with Nonmetal Compounds

The results of earlier investigations are described in "Phosphor" C, 1965, pp. 40/3. Oxidation reactions of PH_3 with nonmetal compounds which can be used for the removal of small amounts of the compound, mainly from mixtures with other gases, are described in Section 1.3.1.5.4, pp. 222/33. Weakly bound complexes including a small number of PH_3 adducts were reviewed in [1].

Halogen Compounds. Weakly hydrogen-bound van der Waals complexes of PH_3 with **hydrogen halides** were generated from the diluted gases. The reactions were carried out either by pulsed-nozzle supersonic expansion with pulse-coupled microwave spectroscopic detection (MW) or by isolation in matrices at low temperature with identification by IR spectroscopy (IR). Experimentally observed complexes are $H_3P\cdots HF$ (MW [2], IR [8]), $H_3P\cdots HCl$ (MW [4], IR [3, 5]), and $H_3P\cdots HBr$ (MW [6], see also [4]).

An ab initio SCF MO calculation with a basis of moderate size predicts a decrease of the stabilization energy of $H_3P\cdots HX$ from X=F to I [7]. The formation of $H_3P\cdots HF$ and its geometry were investigated by ab initio calculations at the MP2 [8, 9] and the SCF level [10 to 12] using large basis sets (usually 6-31G**) and by SCF calculations using simpler basis sets [13 to 16]. A semiempirical PCILO calculation is described in [17]; the electrostatic interaction energy of the point multipoles PH_3 and HF was calculated in [18]. Ab initio MP2 MO calculations for $H_3P\cdots HCl$ with large basis sets [8, 9] and CEPA [19], CI [20, 21], and SCF MO calculations using smaller basis sets [13 to 16, 22] were described; the electrostatic interaction energy was calculated in [18]. Relatively simple basis sets were used in ab initio CI [21] and SCF MO calculations [22] for $H_3P\cdots HBr$.

The formation of $H_3P\cdots(HX)_2$ in cocondensed samples in matrices was deduced for X=F [3, 5] and Cl [3] from IR bands which were absent immediately after condensation, but which grew strongly in intensity on annealing the samples. The energetics and the geometry of $H_3P\cdots(HF)_2$ were computed in an ab initio MO calculation at the SCF/6-31G** level [12].

The dissolution of PH_3 in anhydrous HF leads to complete protonation [23]. The UV photolysis of PH_3/HI mixtures in Xe matrices at 4.2 to 100 K yields PH_4 via decomposition of HI into the atoms [103].

Matrix-isolated $H_3P\cdots$**ClF** was obtained by twin-jet deposition of the diluted reactants and identified by IR spectroscopy. Large shifts of the bands of the complex with respect to those of the isolated constituents indicate a quite strong bond between both reactants [24]. A complex of the molecules PH_3 and ClF without a hydrogen bond was also predicted to be favored by the product of the reference electrostatic potentials [14].

The oxidation of PH_3 by excess aqueous **ClO^-** in a homogenous solution at pH 12 to 13 formally follows the equation $PH_3+2ClO^-+OH^-\rightarrow H_2PO_2^-+2Cl^-+H_2O$. (Termination of the oxidation at the $H_2PO_2^-$ stage is consistent with the fact that further oxidation by ClO^- occurs only in acidic solution; see "Phosphor" C, 1965, p. 109). A kinetic investigation at 275 to 308 K gave the rate law $-d[ClO^-]/dt=k[PH_3][ClO^-]/[OH^-]$ with $k=3.47\times10^{10}\exp(-12.2\ kcal\cdot mol^{-1}/RT)$ and k in s^{-1}. The probable mechanism was discussed [25]. However, oxidation of gaseous PH_3 to H_3PO_4 by aqueous ClO^- at pH 9.4 to 13 was also described. The reaction was found to be first order in each reactant and for the rate constant at 301 K the equation $\log k=11.398-0.697\cdot pH$ with k in $L\cdot mol^{-1}\cdot s^{-1}$ was deduced. An experiment at pH 12.95 gave a numerical k value [26] which resembled the result obtained earlier under similar conditions but yielded a different product [25].

Oxygen Compounds. The interaction between PH_3 and **H_2O** was predicted to yield a van der Waals complex with an $HOH\cdots PH_3$ type of bonding based on an ab initio MO calculation at the MP4/6-31+G(2d, 2p) level. However, the bond energy was found to be too low to bind these two molecules together at ambient temperature. No equilibrium structure with PH_3 as the proton donor was found [27]. Calculations for the same complex at the SCF/4-31G level [15, 16] and at the HF/6-31G(d) level [28] were also published.

The reaction of PH_3 with H_2O yields H_3PO_3, H_3PO_4, and H_2 at elevated temperature and is already described in "Phosphor" C, 1965, p. 34. PH_3 in air and in N_2 forms only traces of H_3PO_4 when stored over water. The reaction becomes faster in the presence of moist oxide clay soils, but is still slow and incomplete [29]. The formation of the solid hydrate (clathrate) $PH_3\cdot5.9\,H_2O$ under pressure is also already covered in "Phosphor" C, 1965, p. 49. The entropy change for the reaction $PH_3(g)+6\,H_2O(l)\rightarrow PH_3\cdot6\,H_2O(s)$ was estimated from the entropies of solidification and condensation of the two reactants to be $\Delta_rS^\circ=-52.3\ cal\cdot mol^{-1}\cdot K^{-1}$. An entropy change of $\Delta_rS^\circ=-61.0\ cal\cdot mol^{-1}\cdot K^{-1}$ was calculated from the reaction enthalpy $\Delta_rH^\circ=-16.4$ kcal/mol and published thermodynamic data [30].

The bombardment of the mixture of PH_3 and H_2O vapor with 75 eV-electrons produced PO^+, PO_2^+, and PO_3^+ (and possibly PO_4^+) in decreasing ion intensity; the singly, doubly, and triply hydrogenated species of these ions were also observed [31]. The only cationic products observed in another experiment under similar conditions were POH^+ and POH_2^+ which originated from the reaction of the primary ion PH^+ with H_2O [32]; see Section 1.1.2, p. 42, for details. Only the products of proton transfer reactions were observed at an electron energy of 12 eV [32].

The reactions of PH_3 with water vapor and other compounds at 20 to 80°C under the influence of electric discharges were studied in model investigations of the primitive atmosphere of the earth. Mixtures of PH_3 and H_2O yielded insoluble polyphosphanes (probably P_nH_{n+2} containing elemental phosphorus), H_3PO_2, H_3PO_3, and H_3PO_4. The mixture $PH_3/H_2O/NH_3$ yielded as additional products diphosphoric, triphosphoric, and higher acids, and possibly polymeric derivatives of phosphinic acid. The system $PH_3/H_2O/NH_3/CH_4$ produced the compounds mentioned above and in addition aminoacids, ethanolamine, and unspecified aminoalkyl compounds of the types $H_2N(CR_2)_nPO_3H_2$ and $H_2N(CR_2)_nOPO_3H_2$ [33, 34]. In reducing planetary atmospheres like that of Jupiter, PH_3 is expected to be stable in the presence of H_2O only in the hot, lower layers. The actual observation of PH_3 in higher layers suggests that its transport is faster than its conversion to P_4O_6 and H_2 [35 to 37] and gives some indications on the structure and dynamics of the Jovian atmosphere [38].

Contrary to earlier findings that PH_3 is not oxidized by aqueous $\mathbf{H_2O_2}$ (cf. "Phosphor" C, 1965, p. 41), the formation of H_3PO_2 from these reactants was mentioned later (cf. "Phosphor" C, 1965, p. 94). The by-products consist of a moderate quantity of H_3PO_3 and very little H_3PO_4 [39, 40]. An acceptable rate of conversion and selectivity of the oxidation to H_3PO_2 was achieved by diluting PH_3 with N_2 and by using an excess of aqueous H_2O_2 at pH 3 to 6 and a temperature of about 20 to 45°C [40].

The reaction of PH_3 with the **OH** radical (from UV-irradiated HNO_3 in Ar) was found to be first order in both reactants. The temperature dependence of the rate constant of the reaction was found to be weak between 249 and 438 K and is given by the equation $k=(2.7\pm0.6)\times10^{-11}\exp(-155/T)$ (k in $cm^3 \cdot molecule^{-1} \cdot s^{-1}$). This rate constant implies that PH_3 has a lifetime of less than 1 day in the atmosphere. The primary reaction probably yields PH_2 and H_2O via H abstraction with $\Delta_rH=-35\pm2$ kcal/mol [41].

Sulfur Compounds. The formation of the van der Waals complex with $H_3P\cdots$**HSH** geometry was predicted to be energetically more favorable than $H_2PH\cdots SH_2$ in ab initio MO calculations at the SCF/4-31G level [15, 16]. The interaction of PH_3 with the **SH** radical is predicted at the HF/4-31G+P level to yield the radical H_3PSH. There seems to be only one stable configuration of H_3PSH, in which the HS group occupies an equatorial position in the pseudotrigonal bipyramidal radical. However, this species is higher in energy by 140.3 kJ/mol than the isolated constituents [42].

Bubbling PH_3 through liquid $\mathbf{SO_2}$ at −20°C yields crystalline H_3PO_3 and crystalline sulfur, probably by stepwise oxidation of PH_3 [43]. Nearly equimolar amounts of PH_3 and SO_2, sealed in a heavywall tube, reacted at room temperature in a 1:2 ratio, yielding a yellow to orange solid which was intractable to workup [44]. The calculation of the product of the electrostatic potentials of PH_3 and SO_2 indicated that the hypothetical complex of the molecules exhibits an $H_3P\cdots SO_2$ type of bonding [14]. The complex $H_3P\cdot\mathbf{SO_3}$ was obtained in an N_2 matrix by cocondensation of diluted PH_3 and SO_3. A fairly strong interaction of the constituents was indicated by the band shifts observed in the IR spectrum [45].

Solutions containing the PH_4^+ ion are formed from PH_3 and excess $\mathbf{H_2SO_4}$ [46] or $\mathbf{HSO_3F}$, when the acid is added dropwise to frozen PH_3 and the mixture is allowed to warm [47]. The experimental enthalpy of protonation of PH_3 in HSO_3F was determined at −52°C to be -14.0 ± 1.4 kcal/mol. At room temperature, HSO_3F vigorously decomposes PH_3 [48]. Bubbling PH_3 through a solution of $\mathbf{FSO_2NCO}$ in CH_2Cl_2 exothermally gives a high yield of $FSO_2N(H)C(O)PH_2$ [49].

The gamma irradiation of PH_3 (PD_3) in an $\mathbf{SF_6}$ matrix yields PH_3F (PD_3F); see Section 1.3.1.5.1.6, p. 216, for details. The decomposition of PH_3 in the presence of the photosensitizer SF_6 is described in Section 1.3.1.5.1.2, p. 204. The reaction of PH_3 with SiH_4 in the presence of SF_6 is described below (p. 251).

Nitrogen Compounds. Two equilibrium structures for a complex of PH_3 with $\mathbf{NH_3}$ were found in ab initio MO calculations of the intermolecular surface at the HF/6-31G(d) level; PH_3 acts either as the proton donor or the proton acceptor. However, both complexes are predicted to have low stabilization energies (MP4/6-31+G(2d, 2p)) and to be unstable at ambient temperature [50]. Calculations at the SCF/4-31G level [16] and semiempirical ones [17, 51] were also described.

PH_3 and NH_3 absorb UV irradiation at similar wavelengths and with similar absorption coefficients [52]; see [53] for details. However, even though about half of the light ($\lambda = 206$ nm) is absorbed by NH_3 in a 1:1 mixture of PH_3 and NH_3, the reaction products are the same as if only PH_3 absorbed light; consequently the initial quantum yields of P_2H_4 formation are the same for PH_3 and PH_3/NH_3. It seems that the reactions of PH_3 with the initial photoproducts of NH_3 according to $PH_3 + NH_2 \rightarrow PH_2 + NH_3$ and $PH_3 + H \rightarrow PH_2 + H_2$ are highly efficient. Prolonged UV photolysis of PH_3/NH_3 mixtures at ambient temperature yields P_2H_4, H_2, and red phosphorus; the steady-state concentration of P_2H_4 is greater and the yield of red phosphorus is smaller in the presence of NH_3 than in photolyses of pure PH_3. Addition of H_2 does not influence the course of the reaction [54]. The amount of P_2H_4 detected on extended irradiation at 175 K is much greater than at ambient temperature, because the smaller initial rate of P_2H_4 formation is outweighed by its slower decomposition due to condensation [55]. No N_2 is found which precludes an intermediate formation of N_2H_4 and suggests $PH_3 + NH_2 \rightarrow PH_2 + NH_3$ to be the dominant reaction of photochemically generated NH_2 [54]. The search for H_2PNH_2 and $(PN)_n$ after photolysis at 175 K failed [55]. The initial quantum yields Φ (with an absolute error of ~30%) were determined for the photolysis of PH_3/NH_3 mixtures at ambient temperature [54]:

	total pressure 11 Torr			total pressure 100 Torr			
PH_3 in vol%	10	50	90	1	5	10	50
$\Phi(P_2H_4)$	0.2	0.4[a)]	—[b)]	0.18	0.40	0.57	0.85
$\Phi(H_2)$	0.54	0.66	—	0.80	—	0.90	1.1

[a)] 0.23 and [b)] 0.37 at 175 K [55].

The initial quantum yield of P_2H_4 formation at ambient temperature was directly proportional to the partial pressure of PH_3 and independent of the partial pressure of NH_3 which suggests very efficient reactions of the initial NH_3 photolysis products with PH_3 as discussed above [54]. The initial quantum yield at 175 K reached only about half of the yield at ambient temperature [55]. The value of $\Phi(H_2)$ close to 1 at 100 Torr and ambient temperature is only slightly influenced by the concentration of PH_3 and demonstrates the marked enhancement of H_2 formation even in the presence of only 1% of PH_3 [54]. The photolysis of PH_3/NH_3 mixtures was investigated particularly in order to develop models of the atmospheric chemistry of Jupiter and Saturn [54, 55]. For discussions of possible reactions and products, see, e.g. [56 to 58].

The thermal decomposition of PH_3/NH_3 mixtures was investigated in a tantalum-based gas-injection cell at an input pressure of ~100 Torr and temperatures from 400 to 1100°C. The products were mainly H_2, N_2, P_2, and P_4 from the individual decomposition of each reactant. The minor product PN reached a maximum at 700°C; its yield increased with increasing excess of PH_3. Very low concentrations of P_2N_2 (observed at 400°C) and P_3N_5 (observed above 500°C) increased with increasing temperature [101].

Bombardment of gaseous PH_3/NH_3 mixtures with 70 eV electrons yielded the cationic products PHN_3^+, PNH_2^+, and a minor quantity of PNH^+. These ions originated most likely from the reaction of the primary ion PH^+ with NH_3 [32]; see Section 1.1.2, p. 42, for details. Only the products of proton transfer reactions were observed at an electron energy of 12 eV [32]. The reactions occurring in $PH_3/NH_3/H_2O$ mixtures when subjected to electric discharges in the gas phase, are described above on p. 245.

The reaction of PH_3 with $\mathbf{NH_2}$ was investigated by flash photolysis of highly diluted PH_3/NH_3 mixtures under conditions where PH_3 photolysis is minimal (175.0 or 206.0 nm interference filter, total pressure $\leqq 20$ Torr, PH_3 partial pressure $\leqq 0.12$ Torr, flash energy $\leqq 110$ J). For the reaction yielding PH_2 and NH_3 the enthalpy $\Delta_rH^\circ = -24.0$ kcal/mol [59] was deduced from $\Delta_fH_0^\circ$

values given in [60]. The kinetic investigation with highly diluted reactants showed a first-order dependence of the rate law for each reactant in the range 218 to 456 K; the rate constant k = $(1.52 \pm 0.16) \times 10^{-12} \exp(-928 \pm 56/T)$ with k in $cm^3 \cdot molecule^{-1} \cdot s^{-1}$ was determined. The experimental activation energy of $E_a = 1.84$ kcal/mol was found to be close to the value of $E_a = 2.05$ kcal/mol derived from a bond-energy-bond-order (BEBO) approach [59].

Gaseous PH_3 and **N_2O** do not react at ambient temperature [61]. PH_3 was burned with N_2O in a low-pressure discharge source to produce beams of PH_2^-, PH^-, and PO^- ions [102]. Reactions where N_2O serves as a source of oxygen are described in Section 1.3.1.5.5, pp. 236/7.

PH_3 reacts immediately with **NO** in the gas phase at ambient temperature [61], contrary to earlier results [62]. Upon warming, the condensed starting materials begin to react when the vaporizing PH_3 mixes with vaporized NO. The initial product is probably an adduct between the Lewis base NO and PH_3 which in this case acts as a Lewis acid. White clouds, formed at the beginning, settle on the walls of the reaction vessel as a nonvolatile yellow solid which upon contact with air is converted to a liquid mixture of H_3PO_3 and H_3PO_4 as determined by ^{31}P NMR spectra. N_2 and N_2O were identified as the gaseous products. The overall reaction ratio of PH_3 to NO is in the rage of 1:2.3 to 1:3. For each equivalent of PH_3 consumed, about 1 equivalent of N_2 and 0.8 to 0.9 equivalents of N_2O form [61]. A major quantity of N_2O was also detected when the reaction of PH_3 with NO was studied in a mass spectrometer; a peak attributable to H_3PO was not observed [62]. Mass spectra recorded during the photolysis of PH_3/NO mixtures at 147 nm indicated the formation of N_2O and a product which was thought to be H_2PNO. Compared to the photolysis of pure PH_3, the presence of NO reduces the quantum yield for the formation of P_2H_4 and increases the quantum yield for the depletion of PH_3 [63].

The oxygen transfer reaction $PH_3 + \mathbf{H_3NO} \rightarrow H_3PO + NH_3$ was predicted to have an activation energy of $E_a = 127$ kJ/mol and a reaction enthalpy of $\Delta_rH = -274$ kJ/mol from an ab initio MO calculation at the HF/6-31+G* level; the geometry of the transition state was calculated at various computational levels [64].

A fine particulate aerosol forms when PH_3 and **NF_3** are mixed in the gas phase at ambient temperature and is taken to indicate the formation of a weak complex. An explosion of the mixed reactants can be induced by irradiation with a CO_2 laser or by application of an electric discharge. Solid films result from mixtures rich in PH_3 (see Section 1.3.1.6, p. 295). When less than half the molar quantity of PH_3 reacts with NF_3, only the gaseous products N_2, HF, and PF_3 are obtained; as the excess of NF_3 increases, PF_3 diminishes in favor of PF_5. The explosion of the mixed starting materials is accompanied by an intense chemiluminescence, spanning the visible range, with a superimposed emission of PF(B $^3\Pi \rightarrow$ X $^3\Sigma$). The flame of diluted PH_3/NF_3 mixtures is greenish and less bright in mixtures richer in PH_3 and bright blue with a slight orange afterglow at a reactant ratio of about 1:4. In contrast to the continuum spanning the visible range due to the explosion, the intensity of the flame's chemiluminescence is reduced; principal, discrete emissions of PF(d $^1\Pi \rightarrow$ b $^1\Sigma$, d $^1\Pi \rightarrow$ a $^1\Delta$), NF(b $^1\Sigma^+ \rightarrow$ X $^3\Sigma^-$), and PN(A $^1\Pi \rightarrow$ X $^1\Sigma$) are observed [65].

Mixing of PH_3 and **N_2F_4** in the gas phase does not result in the formation of a complex. Explosions of N_2F_4-rich mixtures can be triggered as described for NF_3 (see above) and yield the same products [65].

The reaction of PH_3 with **NH_2Cl** proceeds with the formation of P–P bonds and NH_4Cl. A bright yellow solid forms from PH_3 and a benzene solution of NH_2Cl; the starting materials react in a ratio of 1:2.8. The product consists mainly (~85%) of NH_4Cl and contains polymeric phosphorus-hydrogen compounds and a small amount of HCl. In ether solution, the starting materials react in a ratio of 1:6 and yield a white solid; it consists mainly of NH_4Cl whereas the other components vary [66].

Phosphorus Compounds. The reactions of PH_3 with PH are described in Section 1.1.1.6, p. 32.

The direct reaction of PH_3 with $\mathbf{PI_3}$ in CS_2 solution at atmospheric pressure is prevented at ambient temperature by the rapid escape of PH_3 from the solution, whereas at −90 to −110°C the rate of the reaction of PH_3 in the liquid phase is too slow. However, equilibria containing the products PH_2I, PHI_2, and HI were established when the reaction of PI_3 in CS_2 solution with PH_3 (generated in situ from PH_4I) was carried out in closed vessels at ambient temperature. The reaction of PH_3 (from PH_4I) with $\mathbf{P_2I_4}$ under the same conditions led to equilibria containing the same products. The PH_2I/PHI_2 ratio in both reactions depended on the reactant ratio [67].

Arsenic and Antimony Compounds. The thermal decomposition of equimolar mixtures of PH_3 and $\mathbf{AsH_3}$ in H_2 at high temperatures yields complex mixtures of phosphorus- and arsenic-containing species. Their relative abundance in the gas phase was investigated by mass spectrometry [68, 69] and found to change with temperature as follows [70, 71]:

temperature in °C	abundance of the species relative to PH_3^+=100 at each temperature AsH_3^+	P_2^+	AsP^+	As_2^+	P_4^+	AsP_3^+	$As_2P_2^+$	As_3P^+	As_4^+
680	25	22	19	30	4	5	5	6	10
780	35	62	42	40	7	8	8	8	12
880	30	80	58	58	6	7	8	7	13

A mass spectrometric study of the decomposition of a mixture containing 77 mol% PH_3 and 23 mol% AsH_3 over a tantalum catalyst at 900 to 1100°C indicated the formation of P^+, As^+, $As_nP_{2-n}^+$ with n=0 to 2, $As_nP_{3-n}^+$ with n=0 to 3, and $As_nP_{4-n}^+$ with n=0 to 3. The relative abundance of the trimeric and tetrameric species decreased strongly with increasing temperature, whereas the abundance of the monomeric and dimeric species changed moderately and nonuniformly with changing temperature [72].

The explosive decomposition of the gaseous mixture of PH_3 with a large excess of $\mathbf{SbH_3}$ can be induced by heat; products are H_2, elemental antimony, and elemental phosphorus mainly in the P_4 modification [73].

Carbon Compounds. The results of earlier investigations are described in "Kohlenstoff" D 1, 1971, p. 25.

The cocondensation of Ar/PH_3 and Ar/**CO** mixtures with simultaneous vacuum UV irradiation (from an argon discharge) yields PCO and HPCO via the reaction of the intermediates P and PH with CO [74]. A computation of the product of the electrostatic potentials of PH_3 and $\mathbf{CO_2}$ predicted that the hypothetical complex $H_3P\cdots CO_2$ does not contain a hydrogen bond [14]. The UV photolysis of PH_3 and **COS** in inert gas matrices failed to yield SPH_3 [75, 76].

Diluted gas mixtures of PH_3 and **HCN** during pulsed-nozzle supersonic expansion form the hydrogen-bridged van der Waals complex $H_3P\cdots HCN$, which was identified by pulse-coupled microwave spectroscopy [77]. The formation of such a weakly bound complex was confirmed by an ab initio SCF MO calculation with a large GTO basis set [78] and a calculation of the electrostatic interaction energy for the point multipoles PH_3 and HCN [18]. Several attempts to induce a reaction of PH_3 with **(iso)cyanic acid**, liberated from KNCO by acids, were unsuccessful [79]. The reaction of PH_3 with **BrCN** yields monomeric $H_2NCP\cdot HBr$ (no experimental details given). The expected initial product from the addition of PH_3 to the C–N triple bond could not be detected by IR and ^{31}P NMR spectroscopic investigations [80].

Silicon Compounds. The results of earlier investigations are described for the parent silanes in "Silicon" Suppl. Vol. B 1, 1982, p. 152 (SiH_4) and p. 198 (Si_2H_6). The reactions of PH_3 with fluorosilanes are described in "Silicon" Suppl. Vol. B 7, 1992; while SiF_4 (p. 232) and Si_2F_6 (p. 331) do not react at or below ambient temperature, the cocondensation reaction of PH_3 and SiF_2 yields a considerable number of products (p. 81). The thermal decomposition of PH_3 in the presence of SiF_4 is described in Section 1.3.1.5.1.2, p. 204.

The formation of SiH_3PH_2 from PH_3 and **SiH_4** (see "Phosphor" C, 1965, p. 43) can be carried out at a lower temperature (300°C) when a trace of I_2 is added as a catalyst [81].

The reaction of PH_3 with SiH_4 can be induced by unfocused IR pulses at 944.19 cm^{-1} where SiH_4 absorbs practically all photons and decomposes to SiH_2 and H_2. The primary products SiH_3PH_2, Si_2H_6, and Si_3H_8 were identified by the time dependence of their partial pressures in mass spectra and result most likely from the insertion of **SiH_2** into P–H (see also p. 251) and Si–H bonds. Other products were small quantities of PSi_2H_7 and higher silanes and a brownish solid with the approximate composition $P_5Si_3H_3$. The rate constant for the formation of SiH_3PH_2 of $k = (8.8 \pm 1.5) \times 10^{-13}\ cm^3 \cdot molecule^{-1} \cdot s^{-1}$ at ambient temperature was determined relative to the rate of Si_2H_6 formation. A possible reaction scheme was discussed [82].

The UV photolysis (147 nm) of PH_3/SiH_4 mixtures was studied at a reactant ratio $\geqq 2$ where PH_3 (0.48 Torr) absorbs >90% of the total incident radiation. The volatile products consisted of SiH_3PH_2, P_2H_4, Si_2H_6, and H_2. Neither elemental phosphorus nor higher phosphanes, higher silanes, or higher silylphosphanes could be identified by mass spectrometry. The quantum yields of the products except for H_2 decreased with an increase in incident light intensity. Secondary reactions yielded a whitish solid containing phosphorus and P–Si hydrides. Similar results were obtained when appropriate deuterated substrates were used. The photolysis probably starts with formation of PH_2 radicals which react either by insertion into P–H or Si–H bonds or by combination with other radicals. Formed PH is mainly in the triplet state and does not react with insertion. Experimentally, the SiH_4 concentration decreases simultaneously and at rates comparable with those for the reaction of PH_3 in the early stages, until steady-state concentrations of the products are reached. A simplified reaction mechanism was discussed and a kinetic treatment of the chain reaction was given [83].

The mass spectrum of a PH_3/SiH_4 plasma, generated by an rf discharge in H_2, contained, besides the signals of the reactants, additional signals at m/z 65 and 95 which were probably due to $PSiH_6^+$ and $P_2SiH_5^+$ [84]. The mass spectrum of a PH_3/SiH_4 mixture after exposure to a dc multipole discharge exhibited signals of $PSiH_n^+$ in the range of m/z 58 to 67 along with signals of polysilanes and the starting materials [85].

The induction of the reaction of PH_3 with SiH_4 by unfocused laser IR pulses at 1025.3 cm^{-1} requires the presence of the photosensitizer **SiF_4** because the reactants do not absorb this wavelength. Major products in the presence of SiF_4 are H_2 and a solid brown deposit which differs from the deposit formed in absence of SiF_4 (see above) and exhibits P–H stretching and SiH_2 bending vibrations in the IR spectrum. Small quantities of SiH_3PH_2 and Si_2H_6 were found, whereas other silylphosphanes, elemental phosphorus, higher phosphanes, or higher silanes could not be detected among the volatiles, probably because all higher hydrides suffer photosensitized decomposition faster than SiH_4. The photosensitized reaction of the PH_3/SiH_4 mixture sets in at a pulse energy of ~1.5 J; the resulting product mixture depends strongly on the composition of the reactant mixture and the yield increases with the incident pulse energy. A minimum partial pressure of 2 Torr SiF_4 was required for the photolysis at a concentration ratio of $[PH_3]:[SiH_4]:[SiF_4] = 2:2:1$; the yield reaches a maximum at 7 Torr of SiF_4 and decreases slightly at higher partial pressures. At a constant pressure of SiF_4 the yield decreases with increasing pressures of PH_3 and SiH_4 and with increasing pressure of added He or N_2. The volatile products form simultaneously with the decomposition of PH_3 and SiH_4. While the yield of H_2 increases continuously during the photolysis, the yields of SiH_3PH_2 and Si_2H_6

pass through a maximum and almost reach zero at the end of the photolysis, their concentration ratio of about 3.2:1 remaining constant. Experiments with the addition of C_2H_2 showed that vibrationally excited SiF_4 transfers energy to PH_3 and SiH_4 which then decompose as the initial step of the reaction by the lowest-energy channels forming PH, SiH_2, and H_2 in their ground states. A kinetic treatment on the basis of a simplified reaction mechanism indicated that the energy transfer between SiF_4 and SiH_4 is more efficient than between SiF_4 and PH_3 [86].

The energy threshold of the decomposition of PH_3 by IR irradiation in the presence of the photosensitizer **SF_6** is lowered considerably by added SiH_4. The influence on the threshold is attributed to the enhancement of the PH_3 decomposition by the decomposition of SiH_4. The decomposition of the PH_3/SiH_4 mixture proceeds homogeneously in the gas phase via parallel loss of H_2 from both reactants and heterogeneously on the particles of the formed aerosol (the identification of products was not described) [87]. However, the photosensitizer SF_6 was mentioned as not inert in the presence of PH_3 without giving details [88].

An HF/6-31G* ab initio MO calculation predicted that PH_3 and ground-state singlet **SiH_2** form the complex $H_3P\cdots SiH_2$ with an interaction energy of $\Delta E = -17.5$ kcal/mol. Rearrangement of the complex by a proton shift yielding SiH_3PH_2 was found to be energetically favorable ($\Delta_rH^\circ = -53$ kcal/mol) [89]; for additional details see also [90].

No reaction was observed in the mixture of PH_3 and **SiH_3NCO** in the gas phase at room temperature during 1 h or at −78°C during 12 h [91].

The structures of the complexes in which PH_3 interacts with the hydrogen attached to the bridging oxygen in zeolite models of the type **$R_3SiO(H)AlR_3$** with R = H (HF/6-31G(d)) and OH (HF/6-31G) were calculated. For each zeolite model ZOH, the neutral adduct $ZO\text{–}H\cdots PH_3$ was found to be energetically more favorable than the ion pair $ZO^-HPH_3^+$ by about 57 kcal/mol [28].

Boron Compounds. The reactions of PH_3 with the boron compounds B_2H_6, BCl_3, and B_2S_3 at high temperatures lead to the formation of the boron phosphides $B_{13}P_2$ (from B_2H_6) and BP, see "Borverbindungen" 3, 1975, p. 93. The formation of BP from PH_3 and BCl_3 at 1000°C was also described [92].

Reactions of PH_3 with certain Lewis acid-type boron compounds at ambient temperature or below give adducts. References on the formation of individual adducts are listed in the following table:

boron compound	adduct	References ("Borverb." stands for "Borverbindungen")
B_2H_6	$H_3P\cdot BH_3$	"Bor" Erg.-Bd., 1954, pp. 102/3; "Borverb." 3, 1975, pp. 117/8; "Borverb." 18, 1978, p. 165; see also below
$B_3H_7\cdot THF$	$H_3P\cdot B_3H_7$	[93]
B_4H_{10}	$H_3P\cdot B_3H_7$, $H_3P\cdot BH_3$	[93]
B_5H_{11}	$H_3P\cdot B_5H_{11}$	[94]
BF_3	$H_3P\cdot BF_3$, $H_3P\cdot 2BF_3$	"Bor" Erg.-Bd., 1954, p. 179; "Borverb." 3, 1975, p. 129; see also below
B_8F_{12}	$H_3P\cdot B(BF_2)_3$	"Borverb." 19, 1978, pp. 338/9

boron compound	adduct	References ("Borverb." stands for "Borverbindungen")
BHF_2	no reaction	"Borverb." 9, 1976, p. 14
BCl_3	$H_3P \cdot BCl_3$	"Bor" 1926, p. 121; "Bor" Erg.-Bd., 1954, p. 201; "Borverb." 3, 1975, p. 130
B_2Cl_4	$B_2Cl_4 \cdot 2PH_3$	"Borverb." 3, 1975, p. 130; "Borverb." 19, 1978, p. 251
BBr_3	$H_3P \cdot BBr_3$	"Bor" 1926, p. 125; "Borverb." 3, 1975, p. 130
B_2H_5Br	$H_3P \cdot B_2H_5Br$	"Borverb." 9, 1976, p. 53
BI_3	$H_3P \cdot BI_3$	"Borverb." 3, 1975, p. 130

More recent ab initio calculations on the formation of the adduct $H_3P \cdot BH_3$ applied basis sets of double zeta [95], double zeta plus polarization [96], and 4-31G quality (including counterpoise corrections) [97]. A slightly exothermal decomposition of the adduct is predicted at the CEPA level and agrees with the observed instability of the adduct in the gas phase at ambient temperature [96]. A recent ab intio MO calculation for the 1:1 adduct $H_3P \cdot BF_3$ at the SCF and MP2 levels with basis sets of triple zeta quality gave a very large P–B distance. The adduct can thus be considered only a weak van der Waals complex which should not easily be observable in experiments. The calculated results therefore disagree [98] with the alleged experimental observation of this adduct in the gas phase [99].

The reactions of PH_3 with the adducts $BF_3 \cdot H_2O$ or $BF_3 \cdot CH_3OH$ give PH_4^+-containing solutions only in the presence of excess BF_3 [46]. Samples containing the partially deuterated $PH_{4-n}D_n^+$ were obtained by condensing PH_3 into a mixture of BF_3, CH_3OH, and CH_3OD [100].

References:

[1] Scheiner, S. (J. Mol. Struct. **200** [1989] 117/29 [THEOCHEM **59**]).
[2] Legon, A. C.; Willoughby, L. C. (Chem. Phys. **74** [1983] 127/36).
[3] Arlinghaus, R. T.; Andrews, L. (J. Chem. Phys. **81** [1984] 4341/51).
[4] Legon, A. C.; Willoughby, L. C. (J. Chem. Soc. Chem. Commun. **1982** 997/8).
[5] Andrews, L. (J. Phys. Chem. **88** [1984] 2940/9).
[6] Willoughby, L. C.; Legon, A. C. (J. Phys. Chem. **87** [1983] 2085/90).
[7] Müller, B.; Schüler, M.; Reinhold, J. (Chem. Phys. Lett. **172** [1990] 478/82).
[8] Szczęśniak, M. M.; Latajka, Z.; Scheiner, S. (J. Mol. Struct. **135** [1986] 179/88 [THEOCHEM **28**]).
[9] Latajka, Z.; Scheiner, S. (J. Chem. Phys. **81** [1984] 2713/6).
[10] Carroll, M. T.; Bader, R. F. W. (Mol. Phys. **65** [1988] 695/722).

[11] Carroll, M. T.; Chang, C.; Bader, R. F. W. (Mol. Phys. **63** [1988] 387/405).
[12] Kurnig, I. J.; Szczęśniak, M. M.; Scheiner, S. (J. Phys. Chem. **90** [1986] 4253/8).
[13] Hinchliffe, A. (J. Mol. Struct. **105** [1983] 335/41 [THEOCHEM **14**]).
[14] Kollman, P. (J. Am. Chem. Soc. **99** [1977] 4875/94).
[15] Kollman, P.; McKelvey, J.; Johansson, A.; Rothenberg, S. (J. Am. Chem. Soc. **97** [1975] 955/65).
[16] Topp, W. C.; Allen, L. C. (J. Am. Chem. Soc. **96** [1974] 5291/3).
[17] Weller, T. (Int. J. Quantum Chem. **12** [1977] 805/11).
[18] Buckingham, A. D.; Fowler, P. W. (Can. J. Chem. **63** [1985] 2018/25).

[19] Bacskay, G. B.; Kerdraon, D. I.; Hush, N. S. (Chem. Phys. **144** [1990] 53/69).
[20] De Almeida, W. B.; Hinchliffe, A. (Chem. Phys. **137** [1989] 143/56).

[21] Alabart, J. R.; Caballol, R. (Chem. Phys. Lett. **141** [1987] 334/8).
[22] Hinchliffe, A. (J. Mol. Struct. **121** [1985] 201/5 [THEOCHEM **22**]).
[23] Gut, R. (Inorg. Nucl. Chem. Lett. **12** [1976] 149/52).
[24] Machara, N. P.; Ault, B. S. (J. Phys. Chem. **92** [1988] 73/7).
[25] Lawless, J. J.; Searle, H. T. (J. Chem. Soc. **1962** 4200/5).
[26] Chandrasekaran, K.; Sharma, M. M. (Chem. Eng. Sci. **32** [1977] 275/80).
[27] Del Bene, J. E. (J. Phys. Chem. **92** [1988] 2874/80).
[28] Kassab, E.; Seiti, K.; Allavena, M. (J. Phys. Chem. **95** [1991] 9425/31).
[29] Hilton, H. W.; Robison, W. H. (J. Agric. Food Chem. **20** [1972] 1209/13; C.A. **78** [1973] No. 53809).
[30] Ionescu, L. G. (Rev. Roum. Chim. **23** [1978] 45/53; C.A. **88** [1978] No. 159456).

[31] Platzner, I. (Isr. J. Chem. **6** [1968] 34p).
[32] Holtz, D.; Beauchamp, J. L.; Eyler, J. R. (J. Am. Chem. Soc. **92** [1970] 7045/55).
[33] Rabinowitz, J. (Helv. Chim. Acta **53** [1970] 53/63).
[34] Rabinowitz, J.; Woeller, F.; Flores, J.; Krebsbach, R. (Nature **224** [1969] 796/8).
[35] Prinn, R. G. (Proc. Symp. Planet. Atmos., Ottawa 1977, pp. 103/4; C.A. **89** [1978] No. 62565).
[36] Lewis, J. S. (Icarus **10** [1969] 393/409; C.A. **71** [1969] No. 117944).
[37] Barshay, S. S.; Lewis, J. S. (Icarus **33** [1978] 593/611; C.A. **88** [1978] No. 156243).
[38] Drossart, P.; Encrenaz, T.; Tokunaga, A. T. (Icarus **60** [1984] 613/20; C.A. **102** [1985] No. 69791).
[39] Elsner, G.; Hack, H. (Ger. Offen. 3207716 [1982/83] 1/7; C.A. **99** [1983] No. 178343).
[40] Arzoumanidis, G. G.; Darragh, K. V. (U.S. 4265866 [1979/81] 1/12; C.A. **95** [1981] No. 27213).

[41] Fritz, B.; Lorenz, K.; Steinert, W.; Zellner, R. (EUR-7624 [1982] 192/202, 197/9; C.A. **96** [1982] No. 223993).
[42] Gonbeau, D.; Guimon, M.-F.; Ollivier, J.; Pfister-Guillouzo, G. (J. Am. Chem. Soc. **108** [1986] 4760/7).
[43] Fluck, E.; Binder, H. (Z. Anorg. Allg. Chem. **354** [1967] 139/48, 141).
[44] Chan, S.; Goldwhite, H. (Phosphorus Sulfur **4** [1978] 33/4).
[45] Sass, C. S.; Ault, B. S. (J. Phys. Chem. **91** [1987] 551/4).
[46] Sheldrick, G. M. (Trans. Faraday Soc. **63** [1967] 1077/84).
[47] Olah, G. A.; McFarland, C. W. (J. Org. Chem. **34** [1969] 1832/4).
[48] Arnett, E. M.; Wolff, J. F. (J. Am. Chem. Soc. **95** [1973] 978/80).
[49] Roesky, H. W.; Sidiropoulos, G. (Chem. Ber. **110** [1977] 3703/6).
[50] Del Bene, J. E. (J. Comput. Chem. **10** [1989] 603/15).

[51] Karelov, A. A.; Molostov, V. I. (Zh. Obshch. Khim. **55** [1985] 2233/7; J. Gen. Chem. USSR [Engl. Transl.] **55** [1985] 1982/5).
[52] di Stefano, G.; Lenzi, M.; Margani, A.; Mele, A.; Nguyen Xuan, C. (J. Photochem. **7** [1977] 335/44).
[53] Okabe, H. (Photochemistry of Small Molecules, Wiley, New York 1978, pp. 269/73).
[54] Ferris, J. P.; Bossard, A.; Khwaja, H. (J. Am. Chem. Soc. **106** [1984] 318/24).
[55] Ferris, J. P.; Khwaja, H. (Icarus **62** [1985] 415/23; C.A. **103** [1985] No. 218659).
[56] Strobel, D. F. (Astrophys. J. **214** [1977] L97/L99; C.A. **87** [1977] No. 56203).
[57] Strobel, D. F. (Int. Rev. Phys. Chem. **3** [1983] 145/76).

[58] Ferris, J. P.; Benson, R. (J. Am. Chem. Soc. **103** [1981] 1922/7).
[59] Bosco, S. R.; Brobst, W. D.; Nava, D. F.; Stief, L. J. (J. Geophys. Res. C **88** [1983] 8543/9; C.A. **99** [1983] No. 179403).
[60] Okabe, H. ([53], pp. 377/9).

[61] Odom, J. D.; Zozulin, A. J. (Phosphorus Sulfur Relat. Elem. **9** [1981] 299/305).
[62] Halmann, M.; Kugel, L. (J. Chem. Soc. **1962** 3272/3).
[63] Blazejowski, J.; Lampe, F. W. (J. Phys. Chem. **85** [1981] 1856/64).
[64] Bachrach, S. M. (J. Org. Chem. **55** [1990] 1016/9).
[65] McDonald, J. K.; Jones, R. W. (Proc. SPIE-Int. Soc. Opt. Eng. **669** [1986] 99/104; C.A. **106** [1987] No. 58762).
[66] Highsmith, R. E.; Sisler, H. H. (Inorg. Chem. **7** [1968] 1740/2).
[67] Schmidt, M.; Schröder, H. H. J. (Z. Anorg. Allg. Chem. **378** [1970] 192/209; Angew. Chem. **82** [1970] 808; Angew. Chem. Int. Ed. Engl. **9** [1970] 808).
[68] Ban, V. S. (J. Electrochem. Soc. **118** [1971] 1473/8).
[69] Tietjen, J. J.; Ban, V. S.; Enstrom, R. E.; Richman, D. (J. Vac. Sci. Technol. **8** [1971] S5/S11; C.A. **75** [1971] No. 134768).
[70] Ban, V. S.; Gossenberger, H. F.; Tietjen, J. J. (J. Appl. Phys. **43** [1972] 2471/2).

[71] Ban, V. S. (J. Cryst. Growth **17** [1972] 19/30).
[72] Panish, M. B.; Hamm, R. A. (J. Cryst. Growth **78** [1986] 445/52).
[73] Gunn, S. R.; Green, L. G. (J. Phys. Chem. **65** [1961] 779/83).
[74] Mielke, Z.; Andrews, L. (Chem. Phys. Lett. **181** [1991] 355/60).
[75] Mielke, Z.; Brabson, D.; Andrews, L. (J. Phys. Chem. **95** [1991] 75/9).
[76] Hawkins, M.; Almond, M. J.; Downs, A. J. (J. Phys. Chem. **89** [1985] 3326/34).
[77] Legon, A. C.; Willoughby, L. C. (Chem. Phys. **85** [1984] 443/50; Chem. Phys. Lett. **111** [1984] 566/70).
[78] Hinchliffe, A. (J. Mol. Struct. **136** [1986] 193/9 [THEOCHEM **29**]).
[79] Buckler, S. A. (J. Org. Chem. **24** [1959] 1460/2).
[80] Matveev, I. S. (Zh. Strukt. Khim. **15** [1974] 145/8; J. Struct. Chem. [Engl. Transl.] **15** [1974] 131/4).

[81] Sabherwal, I. H.; Burg, A. B. (Inorg. Nucl. Chem. Lett. **8** [1972] 27/30).
[82] Blazejowski, J.; Lampe, F. W. (J. Photochem. **20** [1982] 9/16).
[83] Blazejowski, J.; Lampe, F. W. (J. Photochem. **16** [1981] 105/20).
[84] Günzel, E. (Symp. Proc. 7th Int. Symp. Plasma Chem., Eindhoven 1985, Vol. 1, pp. 171/6; C.A. **104** [1986] No. 210005).
[85] Lloret, A.; De Rosny, G. (Int. J. Mass Spectrom. Ion Processes **62** [1984] 89/98; C.A. **102** [1985] No. 37958).
[86] Blazejowski, J.; Lampe, F. W. (J. Photochem. **24** [1984] 235/48).
[87] Adamova, Yu. A.; Kaganyuk, D. S.; Pimenov, V. P.; Skachkov, A. N.; Stolyarova, G. I.; Shmerling, G. V. (Khim. Fiz. **8** [1989] 1354/61; C.A. **112** [1990] No. 105876).
[88] Blazejowski, J.; Lampe, F. W. (J. Phys. Chem. **88** [1984] 1666/70).
[89] Raghavachari, K.; Chandrasekhar, J.; Gordon, M. S.; Dykema, K. (J. Am. Chem. Soc. **106** [1984] 5853/9).
[90] Dykema, K. J.; Truong, T. N.; Gordon, M. S. (J. Am. Chem. Soc. **107** [1985] 4535/41).

[91] Ebsworth, E. A. V.; Mays, M. J. (J. Chem. Soc. **1962** 4844/7).
[92] Williams, F. V.; Ruehrwein, R. A. (J. Am. Chem. Soc. **82** [1960] 1330/2).
[93] Bishop, V. L.; Kodama, G. (Inorg. Chem. **20** [1981] 2724/7).
[94] Parry, R. W.; Kodama, G. (AD-A 153443 [1984] 1/33; C.A. **104** [1986] No. 236197).

[95] Umeyama, H.; Kudo, T.; Nakagawa, S. (Chem. Pharm. Bull. **29** [1981] 287/92; C. A. **94** [1981] No. 163060).
[96] Ahlrichs, R.; Koch, W. (Chem. Phys. Lett. **53** [1978] 341/4).
[97] Cammi, R.; Tomasi, J. (Theor. Chim. Acta **69** [1986] 11/22).
[98] Ahlrichs, R.; Bär, M. R.; Häser, M.; Sattler, E. (Chem. Phys. Lett. **184** [1991] 353/8).
[99] Odom, J. D.; Kalasinsky, V. F.; Durig, J. R. (Inorg. Chem. **14** [1975] 2837/9).
[100] Wasylishen, R. E.; Burford, N. (Can. J. Chem. **65** [1987] 2707/12).

[101] Baillargeon, J. N.; Cheng, K. Y.; Jackson, S. L.; Stillman, G. E. (J. Appl. Phys. **69** [1991] 8025/30).
[102] Zittel, P. F.; Lineberger, W. C. (J. Chem. Phys. **65** [1976] 1236/43).
[103] Shimokoshi, K.; Nakamura, K.; Sato, S. (Mol. Phys. **53** [1984] 1239/49).

1.3.1.5.7 Reactions with Metal Compounds

The results of earlier investigations are described in "Phosphor" C, 1965, pp. 44/8. Reactions with metal compounds, in particular of Fe^{III} and Cu^{II}, which are used for the oxidative removal of PH_3 from mixtures of gases, are described in Section 1.3.1.5.4, pp. 224/5. Reactions leading to the formation of films or layers of phosphorus compounds on different substrates are described in Section 1.3.1.6, pp. 293/308.

Alkali Metal Compounds. The position of the deprotonation equilibrium of PH_3 and NaOH suspended in liquid NH_3 in a sealed tube at 31°C lies towards the side of the starting materials [1]. An excess of powdered KOH in dimethyl sulfoxide absorbs gaseous PH_3 at ambient temperature with formation of a phosphide solution [2]. The concentration of the PH_2^- ion in this solution decreases in the presence of increasing amounts of H_2O [3]. For the decomposition of PH_3 by aqueous KOH leading to the evolvement of H_2, see "Phosphor" C, 1965, p. 44. The reaction of excess, gaseous PH_3 with the amides of the alkali metals in liquid NH_3 at −35°C can be used for the preparation of MPH_2 with M = Li, Na, K, Rb, and Cs which could be isolated at room temperature as pure, crystalline products except for $LiPH_2$ [4]. Additional details concerning the syntheses of solutions of PH_2^- and the isolation of solid phosphides under various experimental conditions are described in Section 1.2.4.1, p. 102. Heating $LiBH_4$ in a PH_3 stream at 1200 to 1400°C yields Li_3P and BP [5]. The reaction of PH_3 with $MB(s\text{-}C_4H_9)_3H$ (M = Li, Na, K) at −80°C in THF is very slow and gives as yet undefined products [6].

Aluminium, Gallium, and Indium Compounds. PH_3 reacts with $LiAlH_4$ in diglyme or THF at ambient temperature yielding a solution of $LiAl(PH_2)_4$ according to $4PH_3 + LiAlH_4 \rightarrow LiAl(PH_2)_4 + 4H_2$ [7, 8]. In contrast, an insoluble product forms in ether, and almost three moles of H_2 evolve for every mole of PH_3 that reacts. No reaction was observed between liquid PH_3 and $LiAlH_4$ at the boiling point of PH_3. Gaseous PH_3 also does not react with AlH_3 solutions in ether or diglyme [7] and it does not react with dry Al_2O_3 at 350°C or less [33].

Excess PH_3 and $GaCl_3$ (at ambient temperature) and $GaBr_3$ (with gentle heating) form 1:1 adducts in reactions of the neat compounds [9, 10]. The same products are also obtained as precipitates from benzene solutions. The reaction of PH_3 with GaI_3 did not yield a pure product [10]. The reaction of PH_3 in H_2 with gaseous InCl evaporated at 600°C starts to yield crystalline InP at 350°C and proceeds efficiently at a reaction temperature of 400°C [11, 12]. The yield of InP from the reactions of PH_3 with $InCl_3$ or InBr is considerably lower [11].

Germanium Compounds. Circulating an equimolar mixture of PH_3 and GeH_4 through a silent electric discharge at −78°C and at total pressures between 0.25 and 0.5 atm yielded GeH_3PH_2, Ge_2PH_7, Ge_3PH_9, GeP_2H_6, Ge_2H_6, Ge_3H_8, Ge_4H_{10}, and P_2H_4 which were detected by mass spectrometry. The separation of the germanium–phosphorus compounds by distillation or gas-liquid chromatography failed [13]. The formation of $GeF_4 \cdot PH_3$ from the constituents was mentioned; the adduct is probably polymeric [14]. The other germanium halides do not react with PH_3; see "Phosphor" C, 1965, p. 45.

Transition Metal Compounds. The reaction of excess PH_3 with **TaF_5** suspended in anhydrous HF results in the gradual dissolution of the solid with formation of PH_4TaF_6 [15].

The oxidation of adsorbed PH_3 by **MoO_3** on an Al_2O_3 support starts at ~300°C and is accompanied by the consumption of isolated OH species on the solid surface. The complete conversion to the products H_2 and an adsorbed species containing the HPO moiety requires 350°C. The reaction can be used for the batchwise catalytic oxidation of PH_3; added O_2 regenerates MoO_3 and oxidizes the adsorbed HPO groups with formation of $(HO)_xPO$ which can be removed by heating in vacuum [33].

Dehydrated **Mn_3**$[Co(CN)_6]_2 \cdot x\,H_2O$ absorbs PH_3 with formation of $Mn_3[Co(CN)_6]_2 \cdot x\,PH_3$; the value of x was not given [17]. The reaction of excess PH_3, paraformaldehyde, and $(C_2H_5)_3N$ with the cage complex [**Co**(sen)]$^{3+}$ with sen = 4,4′,4″-ethylidynetris(3-azabutan-1-amine) in acetonitrile at ambient temperature yields a mixture of the phospha-capped cage complex [Co(Mephosphasar)]$^{3+}$ (Mephosphasar = 8-methyl-3,6,10,13,16,19,1-hexaazaphosphabicyclo-[6.6.6]icosane) and its phosphane oxide derivative [16].

In a theoretical study of the electronic structure of octahedral copper(I) cluster complexes, the energy changes for reactions of PH_3 with a series of copper and gold species was calculated using the linear combination of Gaussian-type orbital local density functional (LCGTO-LDF) method. Values were computed and discussed for the formation of [**Cu**$(PH_3)_{n+1}]^{q+}$ from PH_3 and $[Cu(PH_3)_n]^{q+}$, where n = 1 to 3 and q = 0 or 1, of $[H_6Cu_6(PH_3)_6]$ from PH_3 and CuH or $[H_6Cu_6]$, of $[CCu_4(PH_3)_4]$ from PH_3 and $[CCu_4]$, and of $[CM_6(PH_3)_6]^{2+}$ from PH_3 and $[CM_6]^{2+}$ where M = Cu or **Au** [34].

Bubbling PH_3 through a concentrated, aqueous solution of **Ag**NO_3 yields unstable, egg yellow $Ag_3P \cdot 3\,AgNO_3$; see "Phosphor" C, 1965, p. 48. The compound blackens upon addition of aqueous NH_3 [18]. Ag and/or Ag_3P precipitate when Ag^+ cations react with excess PH_3 in aqueous solution, but not in anhydrous HF. The precipitation is possibly prevented by the formation of $Ag(PH_3)_n^+$ with n = 1, 2 via PH_4^+ as an intermediate [19] (cf. Section 1.4.2.1, p. 318).

The reaction of PH_3 with **Hg**Cl_2 yields various precipitates depending on the experimental conditions. In aqueous solutions at pH < 1, in ether, dioxane, or THF, a yellow substance of the composition Hg_3PCl_3 is obtained which probably can be formulated as $(Hg_2P)(HgCl_3)$; see "Quecksilber" B 4, 1969, pp. 1366/7. In aqueous solution at pH > 5 or in pyridine, a black precipitate with Hg : P = 3 : 2 forms; it is frequently contaminated with elemental mercury. In aqueous solution at pH 2 to 3 as well as in methanol, ethanol, or isopropanol, brown mixtures of the two substances mentioned above are obtained [20]. The HCl liberated from the reaction of PH_3 with aqueous, slightly acidic $HgCl_2$ (1.5 to 5 wt%) can be used for the quantitative analysis of PH_3 [21 to 24]. The least amount of additional HCl from the decomposition of the thus formed mercury compound is released at an initial pH of 3 of the $HgCl_2$ solution [23]. The addition of ethanol in amounts up to half the volume of the aqueous solution enhances the solubility of PH_3 and leads to a more rapid attainment of steady-state conditions during titration [21]. Diglyme instead of ethanol was used in [25]. Further results of earlier investigations are given in "Phosphor" C, 1965, pp. 44/5 and in "Quecksilber" B 4, 1969, pp. 1357/8.

Lanthanide and Actinide Compounds. A hydrogen stream containing PH_3 converts M_2O_3 (M = Sc to Lu) into MP within 1 to 1.5 h at 1200 to 1300°C [26]. Individual preparations by this method were described for LaP [27, 28], NdP [28, 29], and SmP, pure PH_3 being used in the case of the last compound [28]. Reacting four equivalents of PH_3 in H_2 with $PrO_{1.83}$ at ~1300°C yields nonstoichiometric praseodymium oxides with a low phosphorus content. The formation of PrP sets in when a tenfold excess of PH_3 is used; the preparation of homogeneous PrP succeeds with a hundredfold excess of PH_3 [30]. The reaction of PH_3 in H_2 with Eu_2O_3 at 800 to 1400°C gives products containing Eu, O, and P. The maximum P:Eu ratio of 1.8 is reached at 1250 to 1300°C; X-ray and pycnometric studies suggest partial formation of EuP_2 [31].

For reactions of PH_3 with UO_2 and UF_6, see "Uranium" Suppl. Vol. C 14, 1981, p. 3. The reaction at 600°C of PH_3 with PuH_{2-x} (from the thermal decomposition of PuH_3 at 400°C) yields PuP after homogenizing the crude product at 1400°C [32].

References:

[1] Birchall, T.; Jolly, W. L. (Inorg. Chem. **5** [1966] 2177/80).
[2] Jolly, W. L. (Inorg. Synth. **11** [1968] 124/6).
[3] Langhans, K. P.; Stelzer, O.; Svara, J.; Weferling, N. (Z. Naturforsch. **45b** [1990] 203/11).
[4] Jacobs, H.; Hassiepen, K. M. (Z. Anorg. Allg. Chem. **531** [1985] 108/18).
[5] Kischio, W. (Z. Anorg. Allg. Chem. **349** [1967] 151/7).
[6] Snow, S. S.; Jiang, D.-X.; Parry, R. W. (Inorg. Chem. **24** [1985] 1460/3).
[7] Finholt, A. E.; Helling, C.; Imhof, V.; Nielsen, L.; Jacobson, E. (Inorg. Chem. **2** [1963] 504/7).
[8] Norman, A. D.; Wingeleth, D. C.; Heil, C. A. (Inorg. Synth. **15** [1974] 177/81).
[9] Taylor, M. J.; Riethmiller, S. (J. Raman Spectrosc. **15** [1984] 370/6).
[10] Balls, A.; Greenwood, N. N.; Straughan, B. P. (J. Chem. Soc. A **1968** 753/6).

[11] Born, P. J.; Robertson, P. S. (J. Mater. Sci. **11** [1976] 395/8).
[12] Born, P. J.; Robertson, P. S. (U.S. 3947549 [1976] 1/5; C.A. **85** [1976] No. 23259).
[13] Drake, J. E.; Jolly, W. L. (Chem. Ind. [London] **1962** 1470/1).
[14] Aggarwal, R. C.; Onyszchuk, M. (Proc. Chem. Soc. [London] **1962** 20).
[15] Gut, R. (Inorg. Nucl. Chem. Lett. **12** [1976] 149/52).
[16] Höhn, A.; Geue, R. J.; Sargeson, A. M.; Willis, A. C. (J. Chem. Soc. Chem. Commun. **1989** 1644/5).
[17] Beall, G. W.; Milligan, W. O.; Petrich, J. A.; Swanson, B. I. (Inorg. Chem. **17** [1978] 2978/81).
[18] Rangaswamy, J. R. (J. Assoc. Off. Anal. Chem. **67** [1984] 117/22).
[19] Gut, R.; Rueede, J. (J. Coord. Chem. **8** [1978] 47/53).
[20] Puff, H. (Angew. Chem. **74** [1962] 659; Angew. Chem. Int. Ed. Engl. **1** [1962] 411/2).

[21] Berck, B. (J. Agric. Food Chem. **16** [1968] 415/8; C.A. **69** [1968] No. 1856).
[22] Zugravescu, P. G.; Zugravescu, M. A. (Rev. Chim. [Bucharest] **17** [1966] 704/5; C.A. **66** [1967] No. 101326).
[23] Banks, H. J. (J. Stored Prod. Res. **23** [1987] 213/21; C.A. **108** [1988] No. 70521).
[24] Taylor, R. W. D. (Chem. Ind. [London] **1968** 1116).
[25] Römer, F. G.; Schimmelpenninck van der Oije, A. J. H.; Griepink, B. (Mikrochim. Acta I **1978** 185/91).
[26] Mirnov, K. E. (Izv. Akad. Nauk SSSR Neorg. Mater. **19** [1983] 714/7; Inorg. Mater. [Engl. Transl.] **19** [1983] 645/9).
[27] Samsonov, G. V.; Endrzheevskaya, S. N. (Zh. Obshch. Khim. **33** [1963] 2803/4; J. Gen. Chem. USSR [Engl. Transl.] **33** [1963] 2729/30).

[28] Samsonov, G. V.; Vereikina, L. L.; Endrzheevskaya, S. N.; Tikhonova, N. N. (Ukr. Khim. Zh. [Russ. Ed.] **32** [1966] 115/8; Sov. Prog. Chem. [Engl. Transl.] **32** [1966] 89/91).
[29] Endrzheevskaya, S. N.; Samsonov, G. V. (Zh. Obshch. Khim. **35** [1965] 1983/4; J. Gen. Chem. USSR [Engl. Transl.] **35** [1965] 1973/4).
[30] Mironov, K. E.; Vasil'eva, I. G.; Sinitsyna, E. D. (Izv. Akad. Nauk SSSR Neorg. Mater. **2** [1966] 1315/6; Inorg. Mater. [Engl. Transl.] **2** [1966] 1124/5).

[31] Mironov, K. E.; Brygalina, G. P.; Vasil'eva, I. G.; Popova, E. D. (Metalloterm. Protsessy Khim. Metall. Mater. Konf., Novosibirsk 1971, pp. 116/20; C. A. **78** [1973] No. 11059).
[32] Kruger, O. L.; Moser, J. B. (J. Inorg. Nucl. Chem. **28** [1966] 825/32).
[33] Paul, D. K.; Rao, L. F.; Yates, J. R., Jr. (J. Phys. Chem. **96** [1992] 3446/52).
[34] Bowmaker, G. A.; Pabst, M.; Rösch, N.; Schmidbaur, H. (Inorg. Chem. **32** [1993] 880/7).

1.3.1.5.8 Reactions with Main Group Organometallics, Organosilicon and Organophosphorus Compounds

Earlier published reactions of PH_3 with organometallic compounds are covered in "Phosphor" C, 1965, p. 48. A large series of reactions of PH_3 with certain organyl-substituted compounds of the main group elements which were claimed to be suitable for the deposition of a variety of films or layers mainly of phosphides on different substrates are covered in Section 1.3.1.6, pp. 293/308.

***n*-Butyllithium.** A selective monolithiation of phosphane to give crystalline $LiPH_2 \cdot 1$ monoglyme can be achieved by reacting PH_3 in monoglyme at −78°C with a solution of n-C_4H_9Li in hexane [1, 2]. Further lithiation of PH_3 with n-C_4H_9Li yielding Li_2PH and Li_3P, the latter in a hexane/toluene mixture, is described in detail, e.g. in [3].

Organogallium and Organoindium Compounds. Reactions of PH_3 with certain gallium and indium alkyls are very important in the MOCVD-processing of GaP and InP films or layers or more generally of III-V semiconductors (cf. 1.3.1.6, pp. 293/6). Although the deposition of such films is mostly carried out at relatively high temperatures, the formation of adducts between the Lewis base PH_3 and the organometallic Lewis acids as short-lived intermediates is discussed in many cases to play a major role in the mechanism.

PH_3 and $(CH_3)_3$**Ga** form the 1:1 adduct $(CH_3)_3Ga \cdot PH_3$, when equimolar quantities of the two compounds are condensed together at liquid nitrogen temperature and the mixture is allowed to warm up slowly to 0°C. At ambient temperature the adduct starts to decompose releasing a noncondensable gas. PD_3 reacts analogously [4]. In reactions of PH_3 with $(CH_3)_3Ga$ at 200 to 300°C, CH_4 and H_2 were found to be the only gaseous products. The solid deposit was suggested to have the composition $(CH_3)_{3-x}GaPH_{3-x}$, where x averaged 2.7 at 240°C and 2.9 at 270°C. The reaction orders with respect to both components, the reaction mechanism, and the activation energy of this heterogeneous reaction, which is catalyzed by the product surface, were determined. It was suggested that the corresponding MOCVD processes finally yielding GaP proceed in a similar manner [5, 6]. Reactions of excess PH_3 with $(CH_3)_3Ga$ within the pores of Na^+/H^+ exchanged zeolite Y between 200 and 400°C give microdisperse GaP and CH_4 [7]. Excess PH_3 and $(t$-$C_4H_9)_3Ga$ (neat or in benzene solution), when condensed together at −196°C and warmed up until the PH_3 melts and exerts a pressure of about 0.5 atm, react to yield the planar, six-membered Ga_3P_3 ring compound $[(t$-$C_4H_9)_2GaPH_2]_3$ [8]. The interaction of PH_3 with a benzene solution of $Ga[P(t$-$C_4H_9)_2]_3$ at ambient temperature overnight gave a precipitate which was claimed to be a precursor for the low-temperature (400°C) synthesis of GaP containing C and H [9]; see also the reaction of a PH_3/AsH_3 mixture on p. 259.

The formation of a 1:1 adduct between PH_3 and $(CH_3)_3$**In** at low temperatures is already described in "Phosphor" C, 1965, p. 48 and in "Organoindium" 1, 1991, pp. 38, 44. In the latter volume (pp. 14, 38, 44, 313) this adduct was suggested to be a short-lived intermediate in reactions between the two components at high temperatures, leading to the formation of InP films or layers on different substrates. Studies intended to obtain such films of InP from PH_3 and $(C_2H_5)_3In$ revealed that, upon mixing these two compounds at room temperature and atmospheric pressure, black nonvolatile deposits were formed on the walls of the reaction vessels. Since the black substances were found to be stable up to several hundred °C, it is assumed that they are polymers analogous to the decomposition products of the adduct with $(CH_3)_3In$. At low pressures the interaction between PH_3 and $(C_2H_5)_3In$ diminishes [10 to 13]. Treating a THF solution of $In[P(t\text{-}C_4H_9)_2]_3$ with PH_3 at ambient temperature for four hours yielded a precipitate which was claimed to be a precursor for the low-temperature (375°C) synthesis of (impure) InP. Analogously, a PH_3/AsH_3 mixture reacted with a THF solution of the above indium and the corresponding gallium compounds to give a precipitate which upon heating to 380°C formed a GaInAsP alloy contaminated with C and H [9].

Organosilicon Compounds. PH_3 can conveniently be silylated using trimethylsilyl triflate, $(CH_3)_3SiOSO_2CF_3$. Thus, introducing gaseous PH_3 into an equimolar mixture of $(CH_3)_3SiO$-SO_2CF_3 and $(C_2H_5)_3N$ in diethyl ether leads to the formation of about equal quantities of $[(CH_3)_3Si]_2PH$ and $[(CH_3)_3Si]_3P$. An excess of $(CH_3)_3SiOSO_2CF_3$ favors the formation of the trisilylated phosphane [14, 15]. The reaction of three equivalents of PH_3 with the adduct $[(CH_3)_3Si]_3Al \cdot O(C_2H_5)_2$ in hexane gave a polymeric product which could be pyrolized to yield AlP containing a small amount of SiC [16].

Organogermanium, Organotin, and Organolead Compounds. Passing PH_3 through benzene solutions of $(CH_3)_3$**Ge**Cl or $(C_6H_5)_3GeCl$ at room temperature while dropwise adding $(C_2H_5)_3N$ to these mixtures leads to the formation of $[(CH_3)_3Ge]_3P$ and $[(C_6H_5)_3Ge]_3P$, respectively [17 to 19]. The methylated compound is also formed when excess PH_3 is introduced into a solution of $(CH_3)_3GeN(CH_3)_2$ in ether, thereby cleaving the Ge–N bond and releasing $(CH_3)_2NH$ [20]. Reactions of PH_3 with diorganylgermanium dihalides, R_2GeX_2, give polymeric substances containing $R_2Ge{=}$ and $\equiv P$ structure units [19].

PH_3 reacts with $(CH_3)_3$**Sn**Cl or $(C_6H_5)_3SnCl$ in benzene at ambient temperature in the presence of $(C_2H_5)_3N$ yielding $[(CH_3)_3Sn]_3P$ and $[(C_6H_5)_3Sn]_3P$, respectively [17 to 19]. Diorganyltin dihalides, R_2SnX_2, and PH_3 generally react to give polymeric materials containing $R_2Sn{=}$ and $\equiv P$ groups. However, in the special case of the reaction of PH_3 with $(C_6H_5)_2SnCl_2$, small quantities of a compound could be isolated for which an adamantane structure, $[(C_6H_5)_2Sn]_6P_4$, was proposed, the main product again being a polymer [19]. CH_3SnCl_3 and $C_4H_9SnCl_3$ give only highly polymeric, insoluble reaction products with PH_3. On the other hand, when benzene solutions of $C_6H_5SnCl_3$ and $(C_2H_5)_3N$ were added simultaneously to benzene into which PH_3 is passed, a small amount of a compound could be obtained, to which the cubane structure, $(C_6H_5SnP)_4$, is attributed. Again, the major product consisted of a polymeric material [19, 21].

Introducing gaseous PH_3 into suspensions of $(CH_3)_3$**Pb**Cl or $(C_6H_5)_3PbCl$ in benzene in the presence of $(C_2H_5)_3N$ at room temperature yielded $[(CH_3)_3Pb]_3P$ and $[(C_6H_5)_3Pb]_3P$, respectively [17 to 19, 22]. It was briefly noted that diorganyllead dihalides, R_2PbX_2, and PH_3 react to give polymeric substances consisting of $R_2Pb{=}$ and $\equiv P$ structure units [19].

Organophosphorus Compounds. PH_3 and $(CH_3)_2PCl$, reacted together in a sealed tube, gave sublimable $(CH_3)_2PH_2Cl$ and yellow $(PH)_x$ [23]. The gas-phase reaction of a 2:1 mixture of PH_3 and $(CF_3)_2PI$ in a sealed ampule for 12 h at 20°C led to the formation of $(CF_3)_2PH$, white crystals of PH_4I, and a reddish brown deposit of phosphorus [24]. For reactions of PH_3 with phosphorus-containing alkenes, see Table 18, p. 274.

References:

[1] Schäfer, H.; Fritz, G.; Hölderich, W. (Z. Anorg. Allg. Chem. **428** [1977] 222/4).
[2] Baudler, M.; Glinka, K. (Inorg. Synth. **27** [1990] 227/35).
[3] Fritz, G.; Biastoch, R. (Z. Anorg. Allg. Chem. **535** [1986] 63/85).
[4] Odom, J. D.; Chatterjee, K. K.; Durig, J. R. (J. Mol. Struct. **72** [1981] 73/84).
[5] Schlyer, D. J.; Ring, M. A. (J. Electrochem. Soc. **124** [1977] 569/73).
[6] Newman, C. G.; Diel, B. N.; Paquin, D. P.; Ring, M. A. (J. Organomet. Chem. **137** [1977] 281/6).
[7] Mac Dougall, J. E.; Eckert, H.; Stucky, G. D.; Herron, N.; Wang, Y.; Moller, K.; Bein, T.; Cox, D. (J. Am. Chem. Soc. **111** [1989] 8006/7).
[8] Cowley, A. H.; Harris, P. R.; Jones, R. A.; Nunn, C. M. (Organometallics **10** [1991] 652/6).
[9] Baker, R. T. (U.S. 5084128 [1990/92] 6 pp.; C.A. **116** [1992] No. 164388).
[10] Larsen, C. A.; Stringfellow, G. B. (J. Cryst. Growth **75** [1986] 247/54).

[11] Ikeda, M.; Mori, Y.; Kawai, H. (Eur. Appl. 171242 [1984/86] 1/19; C.A. **105** [1986] No. 52049).
[12] Ogura, M.; Ban, Y.; Morisaki, M.; Hase, N. (J. Cryst. Growth **68** [1984] 32/8).
[13] Duchemin, J. P.; Hirtz, J. P.; Razeghi, M.; Bonnet, M.; Hersee, S. D. (J. Cryst. Growth **55** [1981] 64/73).
[14] Uhlig, W.; Tzach, A. (Z. Anorg. Allg. Chem. **576** [1989] 281/3).
[15] Uhlig, W.; Thust, U.; Tzach, A.; Gomille, R. (Ger. [East] 274626 [1988/89] 3 pp.; C.A. **113** [1990] No. 59546).
[16] Janik, J. F.; Duesler, E. N.; McNamara, W. F.; Westerhausen, M.; Paine, R. T. (Organometallics **8** [1989] 506/14).
[17] Schumann, H.; Schwabe, P.; Stelzer, O. (Chem. Ber. **102** [1969] 2900/13).
[18] Schumann, H.; Schwabe, P.; Schmidt, M. (Inorg. Nucl. Chem. Lett. **2** [1966] 309/10).
[19] Schumann, H. (Angew. Chem. **81** [1969] 970/83; Angew. Chem. Int. Ed. Engl. **8** [1969] 937/50).
[20] Schumann, I.; Blass, H. (Z. Naturforsch. **21b** [1966] 1105).

[21] Schumann, H.; Benda, H. (Angew. Chem. **80** [1968] 846; Angew. Chem. Int. Ed. Engl. **7** [1968] 813).
[22] Schumann, H.; Roth, A.; Stelzer, O.; Schmidt, M. (Inorg. Nucl. Chem. Lett. **2** [1966] 311/2).
[23] Seel, F.; Keim, H. (Chem. Ber. **112** [1979] 2278/81).
[24] Harris, G. S. (J. Chem. Soc. **1958** 512/9).

1.3.1.5.9 Reactions with Transition Metal Carbonyls and Carbonyl Derivatives

General Remarks and General References

The **ligand properties** of PH_3, especially in regards to transition metals in low oxidation states, have been the subject of extensive, mostly controversial discussions during the past decades. The relative donor-acceptor property and the steric influence (the latter generally expressed by Tolman's cone angle) of the PH_3 ligand are quite often compared with those of selected series of organyl-substituted phosphanes, PX_3 derivatives where X = F, Cl, Br, I, OR, SR, and NR_2 (R = alkyl or aryl), and other monodentate ligands with N, P, As, Sb, Bi, O, S, or Se donor atoms. Numerous attempts have been made to understand trends observed in chemical reactivity and in experimental or calculated physical, mainly spectroscopic, parameters. However, the numerous studies related to the separation of steric from electronic influences and of σ-donor from π-acceptor properties still give no conclusive picture of the bonding of the parent phosphane molecule towards transition metals. Therefore, the relative positions of PH_3 in the various series of ligands selected by different authors and assumed to represent the trend within a certain electronic effect and the role of the phosphorus d orbitals are still a matter of discussion, see e.g. [1 to 26, 66].

Most of the reactions of PH_3 with the carbonyls of the transition metals and with a variety of carbonyl derivatives can essentially be classified according to three main **types of reactions:** (i) substitution of CO ligands, of other neutral π-acceptor ligands like, e.g. PR_3, weak O-, S-, N-donor ligands, and alkenes, or of anionic ligands, (ii) cleavage of metal–metal bonds, and (iii) oxidative addition. Only in a few cases combinations of two or three of these reaction types or completely different reactions are observed. Most reviews dealing with the chemistry of phosphanes, with complexes of transition metals, or with transition metal carbonyls and derivatives cover only some aspects of the coordination chemistry of PH_3; see e.g. [1, 4, 14, 16, 22, 24, 27 to 29].

Reactions

Abbreviations used in the following subsections:

n-Bu	*n*-butyl = *n*-C_4H_9
CHT	cycloheptatriene = η^6-C_7H_8
COD	1,5-cyclooctadiene = η^4-C_8H_{12}
Cp	cyclopentadienyl = η^5-C_5H_5
Cp*	pentamethylcyclopentadienyl = η^5-$C_5(CH_3)_5$
DTO	2,2,7,7-tetramethyl-3,6-dithiaoctane = η^2-$(CH_3)_3CSCH_2CH_2SC(CH_3)_3$
DMPE	1,2-bis(dimethylphosphino)ethane = η^2-$(CH_3)_2PCH_2CH_2P(CH_3)_2$
DPPE	1,2-bis(diphenylphosphino)ethane = η^2-$(C_6H_5)_2PCH_2CH_2P(C_6H_5)_2$
Et	ethyl = C_2H_5
Me	methyl = CH_3
NBD	2,5-norbornadien = η^4-C_7H_8
Ph	phenyl = C_6H_5

Substitution of CO. The substitution of a CO by a PH_3 ligand in transition metal carbonyl compounds can be initiated thermally or by UV irradiation. In some cases, when such reactions are carried out in donor solvents, like THF or acetonitrile, complexes in which one CO ligand is replaced by a solvent molecule are assumed to be intermediates. These intermediates, however, generally can not be isolated or have not been isolated in practice; instead the resulting solution is reacted in situ with PH_3. Therefore, these "solvent stabilized 16-electron intermediates" were not included in the subsequent description of the various reactions. Attempts to replace more than one CO ligand in a certain complex directly with the same number of PH_3 molecules led mostly to the formation of a mixture of complexes containing different numbers of PH_3 ligands.

In a theoretical study of bimolecular nucleophilic substitutions, at six-, five-, and four-coordinate metal carbonyl radicals, Walsh diagrams, contour maps, and the atomic character and energies of the frontier MO's derived from SCF-discrete variational-Xα calculations were employed to deduce the most favorable mode of attack of PH_3 at the 17-electron complexes

$V(CO)_6$, **$Mn(CO)_5$**, and **$Co(CO)_4$**. The study, in which these homoleptic metal carbonyl radicals were constrained to octahedral, trigonal-bipyramidal, square-pyramidal, and tetrahedral geometries, showed that two-center, three-electron bonding may stabilize a hypervalent 19-electron transition state or intermediate formed during nucleophilic attack. Attack at a face, rather than at an edge of an octahedron, is predicted for $V(CO)_6$. Nucleophilic attack at the open face of the square-pyramidal $Mn(CO)_5$ is the only mode of attack (of six available) that can be stabilized by two-center, three-electron bonding. Preferred attack at an edge in the equatorial plane is predicted for the trigonal-bipyramidal $Mn(CO)_5$. Attack at a tetrahedral face is predicted for $Co(CO)_4$ [30]. Ab initio MO calculations with effective core potentials of the mechanisms of substitution reactions of 17-electron and 18-electron transition metal hexacarbonyl complexes predicted pseudo-C_{2v} transition states. In particular, the potential energy curves along the reaction coordinates for the reactions **$M(CO)_6$** + $PH_3 \rightarrow M(CO)_5PH_3 + CO$ where M = **Ta**, **W** were computed. Activation energies and the valence-electron charge distributions of the transition states were discussed [67]. The construction of ab initio potential energy surfaces of CO substitution reactions on ***trans*-$W(CO)_4(NO)Cl$** and **$Re(CO)_5Cl$** using RHF and CISD calculations revealed that the corresponding reactions with PH_3 proceed in both cases by dissociative or dissociative interchange (I_d) mechanisms [68].

In the reaction of PH_3 with Cp**V**$(CO)_4$ under UV irradiation in THF or benzene, a CO ligand is replaced yielding $CpV(CO)_3(PH_3)$ [31, 32].

The selective substitution of one CO ligand by reaction with PH_3 in $M(CO)_6$ where M = **Cr**, **Mo**, **W** can be achieved photochemically in ethanol [33] or in the case of Mo thermally in an autoclave in ether at 80°C with 10 atm PH_3 [34]. The photochemical substitution of CO by PH_3 in the complexes $M(CO)_6$ in hydrocarbon solvents like benzene or cyclohexane generally gives mixtures of compounds and, depending upon the reaction conditions, leads to the formation of $M(CO)_5(PH_3)$, $M(CO)_4(PH_3)_2$, $M(CO)_3(PH_3)_3$, and *cis*-$M(CO)_2(PH_3)_4$ [31, 35 to 38]. By reacting PH_3 with $Cr(CO)_5(PR_3)$ (R = *n*-Bu, Ph) in benzene under UV irradiation, both a CO and a PR_3 ligand are replaced to give a mixture of $Cr(CO)_4(PH_3)(PR_3)$, $Cr(CO)_3(PH_3)_3$, and $Cr(CO)_2(PH_3)_4$ [36, 37]. When the carbene complex $Cr(CO)_5[C(CH_3)OCH_3]$ is treated with PH_3 in diglyme at 65°C, even the carbene ligand is substituted along with CO to form *cis*-$Cr(CO)_4(PH_3)_2$ [36, 37].

One equatorial CO group is exchanged by PH_3 in the photochemical reaction with **Mn_2**-$(CO)_{10}$ in THF or in a benzene/hexane mixture [39]. Similarly, the substitution of only one CO ligand occurs, when PH_3 is reacted photochemically with $CpMn(CO)_3$ in benzene [31] or with $Cp^*Mn(CO)_3$ in THF [40]. Depending upon the mole ratio PH_3 : complex, the reaction with Mn-$(CO)_5Br$ in THF under reflux leads to *cis*-$Mn(CO)_4(PH_3)Br$ or $Mn(CO)_3(PH_3)_2Br$ [41 to 43]. The reaction with the corresponding iodo complex $Mn(CO)_5I$ was carried out in an autoclave at 50°C in hexane (10 atm PH_3) to give *cis*-$Mn(CO)_4(PH_3)I$ and *cis*-$Mn(CO)_3(PH_3)_2I$. The latter complex is also formed quantitatively from the former and PH_3 under similar conditions at 80°C [34].

Fe$(CO)_5$ and excess PH_3 react in an autoclave at 10 atm PH_3 to give $Fe(CO)_4(PH_3)$ (see also the cleavage of $Fe_2(CO)_9$ below) [34]. The treatment of *cis*-$Fe(CO)_4I_2$ with PH_3 in a 1:1 molar ratio in ether at 25°C in an autoclave (10 atm PH_3) yields *cis*-$Fe(CO)_3(PH_3)I_2$. The latter complex reacts with excess PH_3 at 80°C without solvent in an autoclave (10 atm PH_3) to give $Fe(CO)_2(PH_3)_2I_2$ in nearly quantitative yield [34, 44].

The reaction of **Co**$(CO)_3(NO)$ with PH_3 to give $Co(CO)_2(PH_3)(NO)$ can be carried out in sunlight at ambient temperature or in the dark at 60°C [45]. With $Co(CO)(PF_3)_3(H)$ phosphane reacts yielding $Co(PH_3)(PF_3)_3(H)$ [46]. In the course of the reaction of **Rh**$(CO)(PEt_3)_2(H)Cl_2$ with PH_3 in CH_2Cl_2 at 219 K (see below), NMR spectroscopic evidence indicated that PH_3 can replace the CO group in the intermediate cationic complex $[Rh(CO)(PH_3)(PEt_3)_2(H)Cl]^+$ forming the cation $[Rh(PH_3)_2(PEt_3)_2(H)Cl]^+$ [47].

In **Ni**$(CO)_4$, even with an excess of PH_3, only one CO ligand is replaced yielding Ni$(CO)_3(PH_3)$, when the reaction is carried out in sealed tubes at 38°C in the dark or at ambient temperature upon exposure to sunlight [48] or in an autoclave at room temperature with 10 atm PH_3 [34].

Substitution of Neutral P-, N-, O-, and S-Donor Ligands and Alkenes. The substitution of PBu_3 and PPh_3 ligands in the corresponding **Cr**$(CO)_5(PR_3)$ complexes by reaction with PH_3 was already mentioned in the previous subsection [36, 37]. When PH_3 is reacted with $Cr(CO)_3$-(η^6-$B_3N_3Me_6$) in cyclohexane at room temperature, the six-membered borazine ring ligand is replaced by three PH_3 ligands to give $Cr(CO)_3(PH_3)_3$ [49]. The substitution of three acetonitrile by the same number of phosphane molecules in $M(CO)_3(CH_3CN)_3$, M = Cr, **Mo**, **W**, to give *fac*-$M(CO)_3(PH_3)_3$ can be achieved by treating the complexes with PH_3 in the cases of Cr and Mo in THF at 0 to 25°C and in the case of W in the same solvent in an autoclave at 50°C [23, 41]. PH_3 and the complexes $M(CO)_4(DTO)$ react at 65°C in THF in an autoclave forming mixtures of $M(CO)_5(PH_3)$ and *cis*-$M(CO)_4(PH_3)_2$ [23, 41]. Versatile methods for introducing selectively two or three PH_3 ligands in Mo complexes are the reactions of PH_3 in pentane with $Mo(CO)_4(NBD)$ yielding *cis*-$Mo(CO)_4(PH_3)_2$ and with $Mo(CO)_3(CHT)$ yielding *cis*-$Mo(CO)_3(PH_3)_3$ [50].

In $M(CO)(PPh_3)_3(H)Cl$ where M = **Ru, Os** treatment with PH_3 leads to a selective replacement of one PPh_3 by a PH_3 ligand to give $M(CO)(PH_3)(PPh_3)_2(H)Cl$ [51, 64].

In **Co**$(PF_3)_4H$ a PF_3 ligand is substituted by PH_3 to give $Co(PH_3)(PF_3)_3H$ [46]. The reaction of PH_3 with **Rh**$(PPh_3)_3Cl$ was postulated to give "$Rh(PH_3)(PPh_3)_2Cl$" in spite of deviating IR spectroscopic observations. Similarly, **Ir**$(CO)(PPh_3)_2Cl$ and PH_3 were suggested to interact leading to the formation of "$Ir(CO)(PH_3)(PPh_3)Cl$" [42]. Both Ir^{III} complexes, $Ir(PPh_3)_3(H)Cl_2$ and $Ir(PPh_3)_3(H)_2Cl$, react with PH_3 yielding $Ir(PH_3)(PPh_3)_2(H)Cl_2$ and $Ir(PH_3)(PPh_3)_2(H)_2Cl$, respectively [51, 64]. The H_2O ligand was replaced in the cation *trans, mer*-$[Ir(H_2O)(PMe_2Ph)_3Cl_2]^+$, when PH_3 gas was bubbled through CH_2Cl_2 solutions of the BF_4^- or ClO_4^- salts of this complex, and the corresponding salts of *trans, mer*-$[Ir(PH_3)(PMe_2Ph)_3Cl_2]^+$ could be isolated [65].

At **Ni** centers the bulky PPh_3 and $P(OPh)_3$ ligands can also readily be substituted by PH_3 groups. Thus, reacting PH_3 with $Ni(PPh_3)_4$ in ether gives $Ni(PH_3)(PPh_3)_3$ from an equimolar mixture of the starting materials at 25°C and $Ni(PH_3)_2(PPh_3)_2$ when PH_3 is bubbled through the slurry of the tetrakis(triphenylphosphane) complex at 0 to 5°C. Bubbling PH_3 through a refluxing solution of $Ni[P(OPh)_3]_4$ in THF gives $Ni(PH_3)[P(OPh)_3]_3$ [41, 43]. The substitution of both diene ligands in $Ni(COD)_2$ by four PH_3 molecules to give $Ni(PH_3)_4$, the only homoleptic PH_3 complex known, so far, can be achieved by treating the nickel complex with a large excess of phosphane at −40°C in a sealed tube with or without ether as solvent [52] or by reacting the two components in THF at normal pressure at −55 to −25°C [43]. When $Ni(COD)_2$ is treated with both PH_3 and $P(OPh)_3$ in ether or THF, the complexes $Ni(PH_3)_2[P(OPh)_3]_2$ and/or $Ni(PH_3)_3[P(OPh_3)_3]$ are formed depending upon the exact reaction conditions [41, 43].

Substitution of Anionic Ligands. In several anionic complexes of **Cr**, **Mo**, and **W**, halide ions or the bidentate $B_3H_8^-$ ion can be substituted by PH_3 to give neutral mono- or disubstituted phosphane complexes. Thus, the salts $R_4N[M(CO)_5X]$ where R = Me, Et, M = Cr, Mo, W, and X = Cl, Br, I, can react with PH_3 eventually after adding a Lewis acid yielding the complexes $M(CO)_5(PH_3)$ with high yield under mild conditions [53]. In particular, $Et_4N[Cr(CO)_5Cl]$ was treated with PH_3 in methanol at 0°C [23, 41] or in CH_2Cl_2 with the addition of Et_3OBF_4 to form $Cr(CO)_5(PH_3)$ [41]. The reaction of $[Cr(CO)_5I]^-$ with PH_3 in THF at 45°C gives *cis*-$Cr(CO)_4(PH_3)_2$ [36]. Treating the salts $Me_4N[M(CO)_4B_3H_8]$ with PH_3 at 70°C in THF in a pressure bomb or in a shaker tube yields *cis*-$M(CO)_4(PH_3)_2$ [41, 42].

Replacing halide ligands in certain neutral **Rh**, **Ir**, and **Pt** complexes by PH_3 molecules may lead to the formation of cationic PH_3 complexes. Thus, *trans*-$Rh(CO)(PEt_3)_2X$, where X = Cl, Br,

I, NCS, reacts with PH_3 in the mole ratio 1:2 in dichloromethane at 183 K to give $[Rh(CO)(PH_3)_2(PEt_3)_2]^+$. In $Rh(CO)(PEt_3)_2(H)Cl_2$ both Cl^- ions can be substituted by PH_3 ligands depending upon the reaction conditions. With one equivalent PH_3 at 219 K in CH_2Cl_2, the monocation $[Rh(CO)(PH_3)(PEt_3)_2(H)Cl]^+$ is obtained. On warming the reaction mixture further with additional PH_3, first a CO group is replaced by phosphane (see above) and then the monocation $[Rh(PH_3)_2(PEt_3)_2(H)Cl]^+$ reacts with PH_3 yielding the dication $[Rh(PH_3)_3(PEt_3)_2(H)]^{2+}$ [47]. The product of the initial reaction of PH_3 with *trans*-$Ir(CO)(PEt_3)_2X$ (X = Cl, Br; molar ratio 2:1) in CD_2Cl_2 at 180 K is the cationic complex $[Ir(CO)(PH_3)_2(PEt_3)_2]^+$. Upon warming the reaction mixture in the presence of excess PH_3, oxidative addition reactions were observed as was the case when the same reaction was carried out in toluene (see below) [54, 55]. In the reaction of PH_3 with *trans*-$Pt(PEt_3)_2(H)X$, where X = Cl, I, in CD_2Cl_2, at −80°C, the formation of the cation $[Pt(PH_3)(PEt_3)_2(H)]^+$ was deduced from ^1H and ^{31}P NMR spectroscopic evidence. Above −30°C decomposition reactions were observed. Reactions with two equivalents PH_3 gave no definite five-coordinate cationic complex [56].

Cleavage of Metal–Metal Bonds. Only in a few cases PH_3 was found to cleave bonds between transition metals in its reactions with carbonyl derivatives. Treating $Cp_2\mathbf{V}_2(CO)_5$ with PH_3 in benzene at room temperature leads to the formation of some $CpV(CO)_3(PH_3)$ [31]. Reacting $\mathbf{Fe}_2(CO)_9$ with an excess of PH_3 in an autoclave (10 atm PH_3) or in $Fe(CO)_5$ as a solvent at room temperature under a PH_3 atmosphere gives $Fe(CO)_4(PH_3)$ [31, 34]. In the reactions of $Cp(CO)Fe[\mu\text{-}CO, \mu\text{-}P(SiMe_3)_2]ML_n$ ($ML_n = Fe(CO)_3$, **Cr**$(CO)_4$) with CH_3OH in the presence of PH_3, it is assumed that subsequent to the cleavage of the Si–P bonds by methanol PH_3 cleaves the Fe–Fe or Fe–Cr bonds, since they are labilized by the PH_2 bridges, finally yielding $Cp(CO)_2Fe(\mu\text{-}PH_2)Fe(CO)_3(PH_3)$ and $Cp(CO)_2Fe(\mu\text{-}PH_2)Cr(CO)_4(PH_3)$, respectively [57]. The addition of PH_3 to the metal–metal multiple bonds of coordinatively unsaturated di- and trinuclear complexes is described below.

Oxidative Addition Reactions. The so-called oxidative addition of PH_3 to transition metal centers was observed with a **Ta**, an **Os**, and some **Ir** complexes. The alkylidene hydride complex $Cp^*_2Ta(=CH_2)H$, which is in equilibrium with the unstable derivative $Cp^*_2TaCH_3$, reacts with PH_3 formally via an oxidative addition to the latter complex yielding $Cp^*_2Ta(CH_3)(PH_2)H$ [63]. PH_3 and $Os(CO)_2(PPh_3)_3$ react to give a mixture of $Os(CO)_2(PPh_3)_2(H)PH_2$, $Os(CO)_2(PPh_3)_2(H)_2$, and unidentified complexes [51, 64]. The initial reactions observed between PH_3 and *trans*-$Ir(CO)(PEt_3)_2X$ (X = Cl, Br) at low temperatures depend on the solvent. While in CD_2Cl_2 at 180 K the halide ligands are replaced by two equivalents of PH_3 to give cationic complexes in a first step (see above) followed by oxidative addition upon warming, in toluene (C_7D_8) even at 180 K one equivalent of PH_3 adds directly to the Ir center yielding $Ir(CO)(PEt_3)_2(H)(X)PH_2$ [55, 58]. Similarly, PH_3 simply adds to the chelate complexes $[Ir(DPPE)_2]X$ (X = Cl, BPh_4) at −78°C in CH_2Cl_2 to give $[Ir(PH_3)(DPPE)_2]X$ as shown by ^1H NMR spectra. Upon warming the solution to 25°C in CH_2Cl_2 or in CH_3CN in a closed system, oxidative addition of PH_3 gives *cis*-$[Ir(DPPE)_2(H)PH_2]X$. PH_3 and $[Ir(CO)(DMPE)_2]X$ react at 25°C in CH_3CN yielding *trans*-$[Ir(DMPE)_2(H)PH_2]X$. Both types of oxidative addition reactions were also carried out with PD_3 to give the corresponding complexes containing PD_2 and D ligands [43].

Miscellaneous Reactions. PH_3 and $\mathbf{V}(CO)_6$ react at room temperature in hexane yielding $[V(CO)_4PH_2]_2$, a complex with a formal V=V double bond [59]. When $[\mathbf{Re}(CO)_4I]_2$ is treated with PH_3 in $CHCl_3$ at room temperature for 10 d, both iodo bridges are cleaved forming $Re(CO)_4(PH_3)I$ [60]. The interaction of PH_3 with the central **Cr** atom in the heterogeneous, single-component, ethene polymerization catalyst $Cr(\eta^3\text{-}C_3H_5)_3$/silica gel before and after CO addition was investigated [62]. In the first step of the reaction of $Cp_2M_2(CO)_4$ (featuring an M≡M bond where M = **Mo**, **W**) with PH_3 in CD_2Cl_2 at 223 K, two phosphane molecules are added to the metal–metal triple bonds to give $Cp_2M_2(CO)_4(PH_3)_2$ (featuring an M–M bond). Above 233 K rearrangement reactions occur leading to the formation of μ-H, μ-PH_2 complexes. Similarly, PH_3 and H_2**Os**$_3(CO)_{10}$ react at room temperature yielding $H(\mu_2\text{-}H)Os_3(CO)_{10}(PH_3)$ [61]. PH_3 adds readily to the coordinatively unsaturated complexes $M(CO)(PPh_3)_2(R)Cl$ (where M = **Ru** and R = Ph or M = **Os** and R = *p*-tolyl) at room temperature yielding the six-coordinate complexes $M(CO)(PH_3)(PPh_3)_2(R)Cl$ [64]. Reactions of PH_3 with the complexes **Ru**$(CO)_2Cl_2$, **Rh**$_2(CO)_4Cl_2$, **Pd**$(PPh_3)_2Cl_2$, and **Pt**$(PPh_3)_2I_2$ were postulated to give clusters. However, the authors noted that a rigorous physical characterization of the clusters assumed to be formed (see the original paper) was impeded by their relative intractability and that the possibility of PH_2 bridging groups (from oxidative addition reactions) must be considered. The reaction of 10 atm of PH_3 with solid Cp_2**Ni** in a sealed tube was also investigated [42].

References:

[1] Levason, W. (in: Hartley, F. R.; The Chemistry of Organophosphorus Compounds, Vol. 1, John Wiley & Sons, Chichester 1990, pp. 567/641, 570/7).
[2] Gilhany, D. G. (in: Hartley, F. R.; The Chemistry of Organophosphorus Compounds, Vol. 1, John Wiley & Sons, Chichester 1990, pp. 9/49, 42/3).
[3] Bertholdt, U.; Wenschuh, E.; Reinhold, J. (Z. Anorg. Allg. Chem. **577** [1989] 59/73).
[4] McAuliffe, C. A. (in: Wilkinson, G.; Gillard, R. D.; McCleverty, J. A.; Comprehensive Coordination Chemistry, Vol. 2, Pergamon Press, Oxford 1987, pp. 989/1066).
[5] Blomberg, M. R. A.; Brandemark, U. B.; Siegbahn, P. E. M.; Mathisen, K. B.; Karlström, G. (J. Phys. Chem. **89** [1985] 2171/80).
[6] Tossel, J. A.; Moore, J. H.; Giordan, J. C. (Inorg. Chem. **24** [1985] 1100/3).
[7] Imyanitov, N. S. (Koord. Khim. **11** [1985] 1041/5, 1171/8; Sov. J. Coord. Chem. [Engl. Transl.] **11** [1985] 597/601, 663/70).
[8] Bartik, T.; Himmler, T.; Schulte, H.-G.; Seevogel, K. (J. Organomet. Chem. **272** [1984] 29/41).
[9] Marynick, D. S. (J. Am. Chem. Soc. **106** [1984] 4064/5).
[10] Xiao, S.-X.; Trogler, W. C.; Ellis, D. E.; Berkovitch-Yellin, Z. (J. Am. Chem. Soc. **105** [1983] 7033/7).

[11] Nakatsuji, H.; Onishi, Y.; Ushio, J.; Yonezawa, T. (Inorg. Chem. **22** [1983] 1623/30).
[12] DeSanto, J. T.; Mosbo, J. A.; Storhoff, B. N.; Bock, P. L.; Bloss, R. E. (Inorg. Chem. **19** [1980] 3086/92).
[13] Bodner, G. M.; May, M. P.; McKinney, L. E. (Inorg. Chem. **19** [1980] 1951/8).
[14] McAuliffe, C. A.; Levason, W. (Phosphine, Arsine, and Stibine Complexes of the Transition Elements, Elsevier, Amsterdam - Oxford - New York 1979, pp. 1/546, 66/72, 75, 78).
[15] Tolman, C. A. (Chem. Rev. **77** [1977] 313/48).
[16] Emsley, J.; Hall, D. (The Chemistry of Phosphorus, Harper & Row, London 1976, pp. 178/80, 191/207).
[17] Stelzer, O.; Unger, E. (Chem. Ber. **108** [1975] 1246/58).
[18] Bodner, G. M. (Inorg. Chem. **14** [1975] 2694/9).
[19] Lappert, M. F.; Pedley, J. B.; Wilkins, B. T.; Stelzer, O.; Unger, E. (J. Chem. Soc. Dalton Trans. **1975** 1207/14).
[20] Malatesta, L.; Cenini, S. (Zerovalent Compounds of Metals, Academic, London - New York - San Francisco 1974, pp. 5/67).

[21] Higginson, B. R.; Lloyd, D. R.; Connor, J. A.; Hillier, I. H. (J. Chem. Soc. Faraday Trans. II **70** [1974] 1418/25).
[22] Pidcock, A. (in: McAuliffe, C. A.; Transition Metal Complexes of Phosphorus, Arsenic, and Antimony Ligands, Macmillan, London 1973, pp. 1/31).

[23] Guggenberger, L. J.; Klabunde, U.; Schunn, R. A. (Inorg. Chem. **12** [1973] 1143/8).
[24] Robinson, S. D. (MTP Intern. Rev. Sci. Inorg. Chem. Ser. One **6** Pt. 2 [1972] 121/69, 123/4, 157/60).
[25] Tolman, C. A. (J. Am. Chem. Soc. **92** [1970] 2953/6).
[26] Tolman, C. A. (J. Am. Chem. Soc. **92** [1970] 2956/65).
[27] Stelzer, O. (Top. Phosphorus Chem. **9** [1977] 1/229, 3/5).
[28] Issleib, K. (Pure Appl. Chem. **44** [1975] 237/67).
[29] Fluck, E. (Fortschr. Chem. Forsch. **35** [1973] 1/64, 49/51).
[30] Therien, M. J.; Trogler, W. C. (J. Am. Chem. Soc. **110** [1988] 4942/53).

[31] Fischer, E. O.; Louis, E.; Bathelt, W.; Müller, J. (Chem. Ber. **102** [1969] 2547/56).
[32] Fischer, E. O.; Louis, E.; Schneider, R. J. J. (Angew. Chem. **80** [1968] 122/3; Angew. Chem. Int. Ed. Engl. **7** [1968] 136/7).
[33] Jeanne, C.; Pince, R.; Poilblanc, R. (Spectrochim. Acta A **31** [1975] 819/38).
[34] Bigorgne, M.; Loutellier, A.; Pankowski, M. (J. Organomet. Chem. **23** [1970] 201/7).
[35] Fischer, E. O.; Louis, E. (J. Organomet. Chem. **18** [1969] P26/P27).
[36] Fischer, E. O.; Louis, E.; Bathelt, W. (J. Organomet. Chem. **20** [1969] 147/52).
[37] Fischer, E. O.; Louis, E.; Bathelt, W.; Moser, E.; Müller, J. (J. Organomet. Chem. **14** [1968] P9/P12).
[38] Moser, E.; Fischer, E. O. (J. Organomet. Chem. **15** [1968] 157/63).
[39] Fischer, E. O.; Herrmann, W. A. (Chem. Ber. **105** [1972] 286/9).
[40] Herrmann, W. A.; Koumbouris, B.; Herdtweck, E.; Ziegler, M. L.; Weber, P. (Chem. Ber. **120** [1987] 931/6).

[41] Klanberg, F. K. (U.S. 3695853 [1968/72] 1/9; C.A. **79** [1973] No. 116689).
[42] Klanberg, F.; Muetterties, E. L. (J. Am. Chem. Soc. **90** [1968] 3296/7).
[43] Schunn, R. A. (Inorg. Chem. **12** [1973] 1573/9).
[44] Birck, J.-L.; Le Cars, Y.; Baffier, N.; Legendre, J.-J.; Huber, M. (C. R. Seances Acad. Sci. C **273** [1971] 880/3).
[45] Sabherwal, I. H.; Burg, A. B. (Chem. Commun. **1969** 853/4).
[46] Campbell, J. M.; Stone, F. G. A. (Angew. Chem. **81** [1969] 120; Angew. Chem. Int. Ed. Engl. **8** [1969] 140).
[47] Conkie, A.; Ebsworth, E. A. V.; Mayo, R. A.; Moretom, S. (J. Chem. Soc. Dalton Trans. **1992** 2951/4).
[48] Sabherwal, I. H.; Burg, A. B. (Inorg. Nucl. Chem. Lett. **5** [1969] 259/61).
[49] Fischer, E. O.; Louis, E.; Kreiter, C. G. (Angew. Chem. **81** [1969] 397/8; Angew. Chem. Int. Ed. Engl. **8** [1969] 377/8).
[50] Barlow, C. G.; Holywell, G. C. (J. Organomet. Chem. **16** [1969] 439/47).

[51] Bohle, D. S.; Clark, G. R.; Rickard, C. E. F.; Roper, W. R. (Chem. Austral. **54** [1987] 293/4).
[52] Trabelsi, M.; Loutellier, A.; Bigorgne, M. (J. Organomet. Chem. **40** [1972] C45/C46).
[53] Connor, J. A.; Jones, E. M.; McEwen, G. K. (J. Organomet. Chem. **43** [1972] 357/60).
[54] Ebsworth, E. A. V.; Mayo, R. A. (J. Chem. Soc. Dalton Trans. **1988** 477/84).
[55] Ebsworth, E. A. V.; Mayo, R. A. (Angew. Chem. **97** [1985] 65/6; Angew. Chem. Int. Ed. Engl. **24** [1985] 68).
[56] Ebsworth, E. A. V.; Edwards, J. M.; Reed, F. J. S.; Whitelock, J. D. (J. Chem. Soc. Dalton Trans. **1978** 1161/4).
[57] Schäfer, H.; Leske, W. (Z. Anorg. Allg. Chem. **552** [1987] 50/68).
[58] Ebsworth, E. A. V.; Gould, R. O.; Mayo, R. A.; Walkinshaw, M. (J. Chem. Soc. Dalton Trans. **1987** 2831/8).

[59] Hieber, W.; Winter, E. (Chem. Ber. **97** [1964] 1037/43).
[60] Moedritzer, K. (Synth. Inorg. Met.-Org. Chem. **2** [1972] 121/8).

[61] Ebsworth, E. A. V.; McIntosh, A. P.; Schröder, M. (J. Organomet. Chem. **312** [1986] C41/C43).
[62] Zakharov, V. A.; Bukatov, G. D.; Ermakov, Yu. I.; Demin, E. A. (Dokl. Akad. Nauk SSSR **207** [1972] 857/60; Dokl. Chem. [Engl. Transl.] **202/207** [1972] 913/6).
[63] Parkin, G.; Bunel, E.; Burger, B. J.; Trimmer, M. S.; Van Asselt, A.; Bercaw, J. E. (J. Mol. Catal. **41** [1987] 21/39; C.A. **109** [1988] No. 6661).
[64] Bohle, D. S.; Clark, G. R.; Rickard, C. E. F.; Roper, W. R.; Taylor, M. J. (J. Organomet. Chem. **348** [1988] 385/409).
[65] Deeming, A. J.; Doherty, S.; Marshall, J. E.; Powell, J. L.; Senior, A. M. (J. Chem. Soc. Dalton Trans. **1993** 1093/100).
[66] Pacchioni, G.; Bagus, P. S. (Inorg. Chem. **31** [1992] 4391/8).
[67] Lin, Z.; Hall, M. B. (Inorg. Chem. **31** [1992] 2791/7).
[68] Song, J.; Hall, M. B. (J. Am. Chem. Soc. **115** [1993] 327/36).

1.3.1.5.10 Reactions with Organic Compounds

Part of the results of earlier investigations are described in "Phosphor" C, 1965, p. 49.

The reactions of PH_3 with organic compounds have been described in detail in various handbooks [1 to 5] and reviews [6 to 10] during the last 30 years and to some extent in [11 to 17].

Weak interactions between PH_3 and $\mathbf{CH_4}$, $\mathbf{CH_2NH}$, $\mathbf{CHONH_2}$, $\mathbf{CHF_3}$, **HCP**, and the alkynes **HC≡C–R** (where R = H, t-C_4H_9, C_6H_5, C(O)CH_3, and CF_3) to give H-bonded complexes were investigated in ab initio MO [18, 19] and IR matrix isolation studies [20].

With Alkanes

Spark discharge of a mixture of PH_3 with $\mathbf{CH_4}$ yields CH_3PH_2, P_2H_4, and the hydrocarbons normally obtained in a CH_4 discharge. A yellowish solid is deposited on the wall of the reactor which has the analytical composition CH_2P_2 when an equimolar mixture is sparked. No $C_2H_5PH_2$ or $(CH_3)_2PH$ was detected. PH_3 is very unstable in a discharge; half of the initial PH_3 is destroyed after only one minute of sparking. Addition of hydrogen does not alter the qualitative composition, however, the partial pressure of CH_3PH_2 formed is roughly inversely proportional to the hydrogen pressure. A reaction mechanism is suggested in the original paper, which includes the insertion reaction $PH_3 + CH_2 \rightarrow CH_3PH_2$ [21]. A nonempirical, pseudopotential SCF calculation (DZ + d basis) for $PH_3 + CH_2(^1A_1) \rightarrow H_3P{=}CH_2$ giving $\Delta_rH = -36$ kcal/mol predicts a larger barrier for an approach of CH_2 being coplanar with the P atom lone pair, suggested by the product geometry. If the approach is perpendicular, the barrier disappears [22].

In addition to the reactions occurring in the pure components, new products with the empirical formulas CH_nP^+ (n = 2 to 4) and $CH_nP_2^+$ (n = 1, 3 to 5) were observed in a PH_3/CH_4 mixture (molar ratio 1:3.2) by ion-cyclotron resonance spectroscopy at 70 eV. Both CH_3^+ and fragment ions derived from PH_3 react to generate the CH_nP^+ species. At the highest pressure of about 10^{-3} Torr, for example, CH_5^+ and $C_2H_5^+$ react to form PH_4^+ [23]. For details of the reactions of PH_3 with the various cations, see Section 1.3.1.5.2, p. 219. Reactions observed upon neutron irradiation of PH_3 in the presence of a large amount of CH_4 are described in Section 1.3.1.5.1.5, p. 215. For an electric discharge in a $PH_3/H_2O/NH_3/CH_4$ mixture, see Section 1.3.1.5.6, pp. 245/6.

Sparking a mixture of PH_3 and $\mathbf{C_2H_6}$ (molar ratio 1:12) yields CH_3PH_2 and $C_2H_5PH_2$. The total quantity of the alkylphosphanes equals roughly that obtained under similar conditions from a PH_3/CH_4 mixture [21]. The gamma irradiation of PH_3 in a **neopentane** matrix is described in Section 1.3.1.5.1.6, p. 216.

With Alkenes and Alkynes

PH_3 adds, preferably under pressure, to compounds containing carbon–carbon multiple bonds, in particular alkenes to give primary, secondary, and/or tertiary phosphanes depending on the structure of the alkene and the reaction conditions. The reaction can be initiated by UV light, gamma irradiation, or peroxides or catalyzed by acids or bases. UV irradiation of PH_3 or the reaction of an initiator with PH_3 produce PH_2 which attacks the α-alkenes to give finally phosphanes with terminal phosphorus (anti-Markownikoff product) [1, 2], [4, pp. 61/8], [5, 6]. The reaction of linear α-alkenes usually gives a mixture of primary, secondary, and tertiary phosphanes. Using excess alkene favors the formation of the tertiary phosphane at high temperatures, high pressures, and short residence times, provided the alkene has no interfering steric or electronic properties. Reacting PH_3 with cyclic, branched, or β-alkenes leads to small amounts of tertiary phosphanes only, in some cases none is formed at all; instead, the corresponding primary and/or secondary phosphanes are obtained [26]. Basic catalysts are used for adding PH_3 to activated carbon–carbon double bonds to give phosphanes with terminal phosphorus like with the radical-initiated reactions [2], [4, pp. 62/3], [5, 6]. The acid-catalyzed reaction, which is a normal Markownikoff addition (carbenium-ion mechanism), yields only primary and some secondary phosphanes, but no tertiary phosphane [1, 2], [4, p. 62], [5, 6]. The use of zeolite catalysts leads to the formation of primary phosphanes with high selectivity [27, 28].

The activation energies for PH_3 addition to alkenes were calculated by a simple electrostatic model of point dipoles. Anti-Markownikoff addition was found always to require a higher activation energy than the normal Markownikoff addition [29].

Table 18, pp. 269/74, lists the reactions of linear, branched, and cyclic alkenes, haloalkenes, and oxygen-, sulfur-, nitrogen-, and phosphorus-containing compounds with C=C double bonds, the reaction conditions, and the product distribution of the phosphanes obtained (abbreviations are explained in the footnotes at the end of the table).

The influence of $\mathbf{C_2H_4}$ on the photolytic decomposition of PH_3 was studied [76 to 78]. Reactions of PH_3 with some **other alkenes** yielding mainly primary, not exactly identified phosphanes are given in [33].

Dicyclopentadiene reacts with PH_3 (mole ratio 3:1) in toluene at 90°C in the presence of AIBN to give a secondary phosphane (viscous yellow oil) of unknown constitution [79]. The addition of PH_3 successively to **bicyclo[2.2.1]hepta-2,5-diene** and 1-octadecene in the presence of AIBN to give a mixture of bicyclic tertiary phosphane isomers is described in [183]. **1,5,9-Cyclododecatriene** (90% trans,trans,trans isomer; 10% trans,trans,cis isomer) reacts with PH_3 on irradiation with ^{60}Co radiation at 100 to 140°C to give a mixture of cis and trans isomers of 1-cyclododeca-4,8-dienylphosphane, 13-phospha-tricyclo[6.4.1.$0^{4.13}$]tridecane, and 13-phospha-tricyclo[7.3.1.$0^{5.13}$]tridecane. The conversion of the cyclodecatriene was 53% [80].

Photochemical addition of PH_3 to a vinyl-substituted **furanose** of the type $RCH{=}CH_2$ produces a mixture of $RCH_2CH_2PH_2$ and $(RCH_2CH_2)_2PH$ which could not be separated [81]. The photochemical addition to **diketene** in toluene solution in the presence of AIBN gives methyl(2-oxo-oxetane-4-yl)phosphane and diastereo isomers of methylbis(2-oxo-oxetane-4-yl)phosphane which could not be isolated because of their limited stability [82].

Table 18
Reactions of PH_3 with Alkenes.

alkene	PH_3: alkene ratio	t in°C /time	catalyst or initiator	yield[1)] and/or product distribution of phosphorus-containing products RPH_2	R_2PH	R_3P	R	Ref.
with unsubstituted alkenes								
$CH_2{=}CH_2$	1:3.03	90 to 100	AIBN[2)]			90 to 95	C_2H_5	[30]
	1:0.93	250 h	UV	5	7	25		[31]
	1:1	200	Z 1	7(90)[13)]				[27, 32]
	1:0.28	60/16 h	CH_3SO_3H/BF_3[5)]	14	<10			[33]
$CH_2{=}CHCH_3$	1:3.13	90	AIBN[4)]			93	$n\text{-}C_3H_7$	[30]
	1:0.3	80/16 h	CH_3SO_3H	20[11)]	+		$i\text{-}C_3H_7$	[33, 34]
	1:1	200	Z 2	16(84)[13)]				[27, 32]
$CH_2{=}C{=}CH_2$	1:1	25/11 h	UV	3[9)]			$CH_2{=}CCH_3$	[35]
$CH_2{=}CHC_2H_5$	1:3.03	70 to 106	AIBN			<95	$n\text{-}C_4H_9$	[30, 36]
		90/9 h	AIBN			96		[37]
	1:3.03	122/16 h	DTBP	+	+	70		[38 to 41]
	1:3.03	20/6.5 h	UV			67		[38]
	1:1	20/2.5 h	UV	38	10	2		[38 to 41]
$CH_2{=}C(CH_3)_2$	1:0.91	80/4 h	AIBN	34	34	10	$i\text{-}C_4H_9$	[42]
	1:1	80	AIBN[2)]	18	52	21		[42]
	1:0.7	78 to 82/4 h	AIBN	30	35	7		[42]
	1:2.9	20/2.5 h	acetone/UV	+	+	+		[38 to 41]
	1:0.35	60/3 h	CH_3SO_3H	61[11)]	+		$t\text{-}C_4H_9$	[33, 34]
	2:1	200	Z 1	70(90)[13)]				[27, 32]
$CH_3CH{=}CHCH_3$	1:3.03	122/18 h	DTBP	+	+		$s\text{-}C_4H_9$	[39 to 41]

Table 18 (continued)

alkene	PH_3: alkene ratio	t in °C /time	catalyst or initiator	yield[1] and/or product distribution of phosphorus-containing products: RPH_2	R_2PH	R_3P	R	Ref.
$CH_2{=}C(CH_3)CH_2$-$C(CH_3)_3$[20]	1:3.03	60 to 100/5 h	AIBN	34.6[12]	22.4[12]	2[12]	$(CH_3)_3CCH_2CH$-$(CH_3)CH_2$	[28, 43]
$CH_2{=}C(CH_3)CH_2$-$CH_2C(CH_3){=}CH_2$	2:1	100	Z 1	19(79)[13]			$CH_2{=}C(CH_3)CH_2$-$CH_2C(CH_3)_2$[19]	[27, 32]
$CH_2{=}CHC_6H_{13}$	1:3.03	80 to 100/6 h	AIBN			83	C_8H_{17}	[42]
	1:0.91	~120 to 130/21 h	DTBP	23	28	22		[39 to 42]
	1:0.67	70 to 120/0.5 h	AIBN	36	32	14		[42]
	1:0.28	78 to 80/4 h	AIBN[5]	65	18	4		[42]
		120	AIBN[2]	80				[44, 45]
	1:0.25	90/5 h	AIBN[6]	70	20	10		[44, 45]
		85/24 h	DBP			+		[46]
		25/5 min	UV		+			[46]
	1:3.13	30	^{60}Co			68		[47]
$CH_2{=}C(C_2H_5)C_4H_9$	1:0.24	60/16 h	CH_3SO_3H	76[11]	+		$CH_3C(C_2H_5)C_4H_9$	[33, 34]
$CH_2{=}CHC_6H_5$	1:0.67	80/4 h	AIBN[5]	36	29	6	$C_6H_5CH_2CH_2$	[42]
	1:1	25/2 h	UV	++				[39 to 41]
	1:0.33	60/3 h	CH_3SO_3H	44				[33]
$CH_2{=}CHC_{10}H_{21}$	1:0.83	84 to 87/4 h	AIBN	23	20	26	$C_{12}H_{25}$	[42]
	1:0.36	60/16 h	CH_3SO_3H	14[11]	+			[34]
		90/16 h	CH_3SO_3H	98[12]	2[12]			[33, 34]

c-C_5H_8	2:1	100	Z 1		10(85)[13]			c-C_5H_9	[27, 32]
1-CH_3-c-C_5H_7-1	1:0.56	60/16 h	CH_3SO_3H		41[11](90)[12]	+	(10)[12]	1-CH_3-c-C_5H_8	[33, 34]
c-C_6H_{10}	1:2.2	80 to 84/6 h	AIBN		19	22		c-C_6H_{11}	[42]
	1:0.67	78 to 80/3.5 h	AIBN		49	29			[42]
		120	AIBN[4]		70				[44, 45]
	1:0.5	85/6 h	AIBN[3]		60	40			[44, 45]
	1:1	20/7 h	UV		34	+			[38 to 41]
	2:1	100	Z 1		12(81)[13]				[27, 32]
4-CH_3-c-C_6H_9-1	1:0.56	60/16 h	CH_3SO_3H		33(97)[12]	+	(3)[12]	1-CH_3-c-C_6H_{10}	[33, 34]
c-C_6H_8-2,5	2:1	100	Z 1		20(75)[13, 19]			c-C_6H_9-2	[27, 32]
c-C_8H_{12}-1,5		70 to 120	AIBN[2, 3, 4]			96; 57		$R_2 = C_8H_{14}$[22]	[44, 45, 49 to 53]
with halogen-containing alkenes									
$CH_2{=}CF_2$	1:1	230 to 240/5 d			40			CHF_2CH_2	[54]
	1:1	150/8 h			1				[55]
	1:1	200 h	UV		74	~28			[31]
$CH_2{=}CF_2$ } $CF_2{=}CF_2$ }	1:1	1 h	UV		{ 15 { 85			{ CHF_2CH_2 { CHF_2CF_2	[31]
$CHF{=}CF_2$	1:1	190/14 d		17	{ (78)[12] { (22)[12]			{ CHF_2CHF { CH_2FCF_2	[56]
	1:1	37/52 d	UV	75	{ (87)[12] { (13)[12]			{ CHF_2CHF { CH_2FCF_2	[56]

Table 18 (continued)

alkene	PH_3 : alkene ratio	t in °C /time	catalyst or initiator	yield[1] and/or product distribution of phosphorus-containing products RPH$_2$	R_2PH	R_3P	R	Ref.
$CF_2{=}CF_2$	1:3.03	150/8 h		33[12), 16)]	3[12)]		CHF_2CF_2	[57]
	1:1	150/8 h		53[16)]	7			[55]
	1:1	20/19 to 160 h	UV	84 to 86[16)]	2			[31]
$CF_2{=}CFCF_3$	1:1	150/8 h		36			$(CF_3)CHF_2CF$	[55]
	1:1	350 h	UV	97			{ CF_3CHFCF_2 and $(CF_3)CHF_2CF$	[31]
$CF_2{=}C(CH_3)_2$	1:1	150/8 h		6			$(CH_3)_2CHCF_2$	[55]
$CClF{=}CF_2$	1:3.03	150/8 h		45.4[12), 14)]	5.2[12)]		$CHClFCF_2$	[57]
	1:1	150/8 h		54[14)]	6			[55]
	1:1	12 to 55 h	UV	89 to 91				[31]
$CCl_2{=}CF_2$	1:1	150/8 h		30[15)]	3		$CHCl_2CF_2$	[55]
$CF_2{=}CClCClF_2$	1:0.76	160/96 h		+			$CF_2CHClCClF_2$	[58]
$CH_2{=}CHCH_2Cl$	1:2	20/6 h	UV	+	+	+	$CH_2ClCH_2CH_2$	[38 to 41]
with oxygen- and sulfur-containing alkenes								
$CH_2{=}CHOCH_3$	1:3.3	20/6 h	UV	+	+	+	$CH_3OCH_2CH_2$	[39, 40]
$(CH_2{=}CH)_2O$	~1:1	75/12 h	AIBN[3)]		7		$O(CH_2CH_2)_2$ (2R)	[59]
$CH_2{=}CHOC(O)CH_3$		25 to 80/4 h	AIBN			100	$CH_3C(O)OCH_2CH_2$	[60, 61]
$CH_2{=}CHOC_4H_9$	1:0.53	~80/4 h	AIBN[5)]	45	30	10	$C_4H_9OCH_2CH_2$	[42]

$CH_2{=}CHCH_2OH$ [21]	1:1	20/1 h	UV	26	5	3	$HO(CH_2)_3$	[38, 39, 41, 58, 62]
	1:>5.56	80/8 h	AIBN			93		[63]
$CH_2{=}CHC(O)CH_3$	1:1	25/1 h	UV	++			$CH_3C(O)CH_2CH_2$	[39, 40]
$CH_2{=}C(CH_3)OC(O)CH_3$		80/4 h	AIBN [6]			+	$CH_3C(O)OCH(CH_3)CH_2$	[60]
$CH_2{=}CHC(O)C_2H_5$	1:0.7	67 to 102/0.5 h	AIBN [5]	21	20	23 [17]	$C_2H_5C(O)OCH_2CH_2$	[42]
		54/1 h	HMBG [9]	25	6	42		[64]
$CH_2{=}CHC(O)OCH_3$	1:2	20/5 h	UV	++	+		$CH_3OC(O)CH_2CH_2$	[39, 40]
$CH_2{=}CHCH_2OC(O)CH_3$	1:1	20/5 h	UV	+	+		$CH_3C(O)O(CH_2)_3$	[39, 40]
$CH_2{=}CHSCH_3$	1:2	25/6 h	UV	+	+	+	$CH_3SCH_2CH_2$	[39 to 41]
with nitrogen- and phosphorus-containing alkenes								
$CH_2{=}CHCN$	1:3.3	105/4.5 h	AIBN [10]			91.7	$NCCH_2CH_2$	[65, 66]
	1:0.5	28 to 30	10 N aq KOH [9]	52	12			[67]
	1:1.52	30 to 35	KOH or Dowex-2 [9]	6	56	28		[67]
	1:1.67	8/2.5 h	10 N aq KOH [9]	48	12			[64]
	1:2	7/1.5 h	10 N aq KOH [9]	6	58	13		[64]
	1:2.63	15 to 20/1.5 h	10 N aq KOH [9]			80		[64, 67]
	1:2.5	30	$C_6H_4(OH)_2$/NaOH/($PtCl_4$)			80		[68]
	1:3.03	30	$NiCl_2$ in aq NH_3			76		[68]
	1:2.78	35	$[C_6H_5CH_2N(CH_3)_3]OH$			70		[68]
		20	$P(CH_2OH)_3$/Fe			83		[69]

Table 18 (continued)

alkene	PH_3: alkene ratio	t in °C /time	catalyst or initiator	yield[1] and/or product distribution of phosphorus-containing products RPH$_2$	R$_2$PH	R$_3$P	R	Ref.
$CH_2{=}CHCN$		20 to 50	$P[CHR'(OH)]_3/H_2PtCl_6$[18]			90 to 95		[70]
		20/>1 h	$Pt[P(NCCH_2CH_2)_3]_3$	+	+	+		[71]
$CH_2{=}CHCH_2NH_2$	1:2	20/5 h	UV	28	20	+	$H_2N(CH_2)_3$	[39 to 41]
$CH_2{=}CHCH_2N(C_2H_5)_2$		85/1 h	AIBN[8]	++	+	+	$(C_2H_5)_2N(CH_2)_3$	[63]
$CH_2{=}CHP(C_6H_5)_2$	1:2.38	100/36 h	LiC_6H_5[6]			24	$(C_6H_5)_2PCH_2CH_2$	[72, 73]
	1:0.53	65/3 h	$KOC(CH_3)_3$[7]			89		[73]
$CH_2{=}C[P(C_6H_5)_2]_2$			$KOC(CH_3)_3$			++	$[(C_6H_5)_2P]_2CHCH_2$	[74]
$CH_2{=}CHCH_2P(O)(OC_4H_9)_2$	1:1	25/1.5 h	UV	++			$(C_4H_9O)_2P(O)(CH_2)_3$	[39, 40]
$CH_2{=}CHP(S)(CH_2C(CH_3)_3)_2$	<1:0.56	~66/3 h	$KOC(CH_3)_3$			95	$((CH_3)_3CCH_2)_2P(S)(CH_2)_2$	[75]

AIBN = α,α′-azobisisobutyronitrile. – DTBP = di-*t*-butyl peroxide. – DBP = dibenzoyl peroxide. – HMBG = heptamethyl biguanide. – Z 1 = boron silicate zeolite impregnated with $Cr(NO_3)_3$. – Z 2 = boron silicate zeolite.

+ = corresponding phosphane is formed; ++ = corresponding phosphane is main product.

[1] The yields listed in % as in the original papers are based on the amount of PH_3 charged or consumed, or based on the amount of alkene charged or consumed. – [2] In toluene. – [3] In pentane. – [4] In hexane. – [5] In heptane. – [6] In benzene. – [7] In tetrahydrofuran. – [8] In 2-propanol. – [9] In acetonitrile. – [10] In molten $(NCCH_2CH_2)_3P$. – [11] % of RPH_2 and R_2PH formed. – [12] Product distribution. – [13] Conversion in %. Selectivity in parentheses in %. – [14] Little $H_2PCFClCF_2PH_2$ also forms. – [15] $H_2PCCl_2CF_2PH_2$ also forms. – [16] Some $H_2PCF_2CF_2PH_2$ also forms. – [17] 19% $(C_2H_5C(O)OCH_2CH_2)_2PCH_2CH(C(O)OC_2H_5)CH_2CH_2C(O)OC_2H_5$ also obtained. – [18] R′ = H or alkyl (C_1 to C_6). – [19] And other isomers. – [20] Diisobutylene (70% 2,4,4′-trimethylpentene-1 and 30% 2,4,4′-trimethylpentene-2). The latter isomer does not react. – [21] 3-Butene-2-ol and 3-butene-2-methyl-2-ol react similarly [62]. – [22] 9H-9-phosphabicyclononane-3.3.1 and its 4.2.1-isomer.

Tetracyanoethene is only reduced by PH_3 in an acetonitrile/ethanol mixture at 16 to 20°C to give *sym*-tetracyanoethane with 96% yield [83]. PH_3 reacts with **acrylamide** or **lauryl acrylate** in CH_3CN in the presence of pentamethylguanidine at room temperature to give mixed carbamoylethylphosphanes and 2-carblauroxyethylphosphanes, respectively [64].

Passing a mixture of PH_3 with a slight excess of $\mathbf{C_2H_2}$ at about 1 Torr through a silent electric discharge apparatus yields P_2H_4, $HC{\equiv}CPH_2$, and 1,3-butadiyne after low-temperature distillation [84]. The influence of C_2H_2 on the photolytic decomposition of PH_3 was studied [76, 85].

With Halogen-Containing Compounds

Reactions of PH_3 with halogen-containing alkenes are described in the subsection covering the reactions with alkenes; see Table 18, pp. 271/2.

PH_3 and $\mathbf{CF_4}$ react in a dc glow discharge to give PF, PF_2, and PF_3 [24]. The impact of an electron pulse on a mixture of PH_3 with $\mathbf{CHF_3}$ or $\mathbf{CHCl_3}$ induces reactions of CHF_2^+ and $CHCl_2^+$, respectively, with PH_3 [25].

The reaction of PH_3 with $\mathbf{CH_3Cl}$ (mole ratio 1:1) in a solid-bed reactor filled with active carbon (280°C, residence time (r.t.) 17 s) yields 25% CH_3PH_2 and small amounts of $(CH_3)_2PH$ and $(CH_3)_3P$ along with HCl and the corresponding phosphonium chlorides. The reaction at 200°C gives a lower yield, whereas the reaction at 325°C increases the yield of the primary phosphane to 29% (based on PH_3 conversion) and significantly also the yields of the secondary and tertiary phosphanes. A $PH_3:CH_3Cl$ mole ratio of 1:2 does not strongly influence the yield, but a 1:10 mole ratio causes a considerable increase in the yield of $(CH_3)_2PH$ and $(CH_3)_3P$ to 16 and 6%, respectively; however, the yield of CH_3PH_2 then drops to less than 1%. Metals like platinum, gold, or palladium were found not to catalyze this reaction [86]. Increasing the residence time to 210 s at 280°C (mole ratio 1:5) results in the formation of $(CH_3)_4PCl$ with 83% yield, based on PH_3 converted [87 to 89]. In this case, CH_3PH_2, $(CH_3)_2PH$, and $(CH_3)_3P$ are by-products [89]. Reacting PH_3 with CH_3Cl (mole ratio 1:1.21) mixed with 0.96 equivalents $(CH_3)_3N$ in a solid-bed reactor filled with active carbon (300°C, r.t. 31 s) gives 39.1% CH_3PH_2, 19.7% $(CH_3)_2PH$, 9% $(CH_3)_3P$, 18.6% $(CH_3)_4PCl$, and 10.6% phosphorus, the yield being based on PH_3 converted. By varying the quantities of the reactants and the reaction conditions, the yield of one or the other product can be maximized [90].

The analogous gas-phase reactions of PH_3 with $\mathbf{C_2H_5Br}$ (mole ratio 1:1, 250°C, r.t. 5 s) [86] and $\mathbf{C_3H_7Br}$ (mole ratio 1:0.7, in the presence of $(C_3H_7)_3N$, 320°C, r.t. 10 s) [90] yield the corresponding primary phosphanes with 2.8 [86] and 28.5% [90] yield based on PH_3 conversion.

In dimethyl sulfoxide or in two-phase systems employing toluene or an alkane as one of the phases, alkylation of PH_3 with **organyl halides** RX (R = CH_3 [91, 92], C_2H_5 [91], *n*-C_4H_9 [91, 92], *s*-C_4H_9 [91], $C_{16}H_{33}$, $CH_2{=}CHCH_2$ [91, 92], $C_6H_5CH_2$, C_5H_4N-2-CH_2CH_2 [91], $(C_6H_5)_2PCH_2$ [93]; X = Cl and/or Br) in the presence of concentrated aqueous KOH proceeds in the temperature range 0 to 80°C at normal or higher pressures to give predominantly the primary or secondary phosphanes with yields as high as 100% depending mainly on the molar ratio of the reactants. As phase-transfer catalysts in the two-phase system a tetraalkylammonium chloride or a tetraalkylphosphonium bromide was used [91 to 93]. Using 1,3-dichloropropane [92], 1,3-dibromopropane, or -butane yields the diprimary phosphanes, $H_2P(CH_2)_2CHRPH_2$ (R = H or CH_3), while 1,4-dibromopentane yields the heterocyclic 2-methylphospholane [91].

PH_3 in the presence of solid KOH reacts with $\mathbf{CH_3I}$ in dimethyl sulfoxide in the appropriate reactant ratios to give CH_3PH_2 or $(CH_3)_2PH$ [94].

Aroyl chlorides react with PH_3 in pyridine to give the corresponding aroyl phosphanes, $(Ar(O)C)_3P$, where Ar = C_6H_5, *m*-$CH_3C_6H_4$, *p*-$CH_3C_6H_4$, α-$C_{10}H_7$, β-$C_{10}H_7$ [177].

With Alcohols, Alkoxy Radicals, Alkoxides, and Ethers

The addition of PH_3 to **allyl alcohol** and various **vinyl ethers** is summarized in Table 18 on pp. 272/3.

At ambient temperature, PH_3 reacts with **methanol** in dimethylformamide in the presence of a base and CCl_4 according to $PH_3 + 3CH_3OH + 3CCl_4 + 3(C_2H_5)_3N \rightarrow P(OCH_3)_3 + 3CHCl_3 + 3(C_2H_5)_3NHCl$ [95].

The reaction with the alcohols **ROH** ($R = C_2H_5$ [96 to 98], C_3H_7 [96, 99], C_4H_9 [96 to 99], or *i*-C_4H_9 [98]) in alcoholic solutions of $CuCl_2$ [96, 97, 99], $Cu(CH_3COO)_2$ [96], or $FeCl_3$ [98] yields the corresponding alkyl phosphates $(RO)_3PO$ and dialkyl ethers along with the reduced metal salts and HCl; see also [100]. Mixed $CuCl_2/HgCl_2$ or $CuCl_2/PdCl_2$ catalysts accelerate the reaction rate. Mechanistic details are described in the original papers. The reduced catalysts can be reoxidized by O_2. Addition of H_2O, HCl, or pyridine decreases the rate of reaction and the yield of the alkyl phosphate in favor of inorganic phosphate [96, 97].

At high temperatures and in the presence of catalysts, PH_3 reacts with **butanol** (200°C, 5 Å molecular sieves) or **cyclohexanol** (300°C, hydrogen faujasites) to give the mono- and disubstituted phosphanes [101].

The hydrogen abstraction from PH_3 in the gas phase by *tert*-**butoxy radicals** formed by thermal (403 to 458 K, 20 to 350 Torr) [102] or photochemical (UV, ≥30 Torr) dissociation of di-*tert*-butyl peroxide [103] yields *i*-butanol and acetone (formed by $(CH_3)_3CO^\bullet \rightarrow CH_3C(O)CH_3 + \dot{C}H_3$) [102, 103]. The Arrhenius parameters for the reaction $PH_3 + (CH_3)_3CO^\bullet \rightarrow (CH_3)_3COH + \dot{P}H_2$ obtained by a competition method are log A = 9.0 (A in $L \cdot mol^{-1} \cdot s^{-1}$) and $E_a = 1.4$ kcal/mol (thermal reaction) [102] or $E_a = 5.0$ kcal/mol (photochemical reaction) [103]. The results are based on the assumption that hydrogen abstraction is the only route for the reaction of $(CH_3)_3CO^\bullet$ with PH_3 [103]. UV photolysis of a mixture of PH_3 and di-*tert*-butyl peroxide at −95°C in cyclopropane solution directly in the cavity of an ESR spectrometer gave the spectrum of two radical species, supposedly $H_3\dot{P}OC(CH_3)_3$ and $H_2\dot{P}(OC(CH_3)_3)_2$ [104, 105]. On raising the temperature to between 0 and 30°C, a third radical appeared which is thought to be a phosphinyl radical, potential candidates being $\dot{P}(OH)_2$, $(CH_3)_3CO\dot{P}OH$, and $\dot{P}(OC(CH_3)_3)_2$ [104].

Reaction of PH_3 with **$NaOCH_3$** in a CH_3OH/CCl_4 mixture yields $P(OCH_3)_3$ [106]. The analogous reaction of **$NaOC_6H_5$** in a propene carbonate/CCl_4 mixture at 50°C yields $P(OC_6H_5)_3$, $CH_2ClP(OC_6H_5)_2$, and $CHCl_2P(OC_6H_5)_2$ in equal amounts [107].

For the reactions in a PH_3/**ethene oxide** mixture under the influence of low-energy electrons [178], see 1.3.1.5.2, p. 219.

With Aldehydes, Ketones, and Carbonic Acids

The addition of PH_3 to ketones or carbonic acid esters containing carbon–carbon double bonds is described in Table 18 on pp. 272/3.

The addition reaction of PH_3 to **carbonyl compounds** has been described in detail in various reviews during the last three decades [7, 8] and to some extent in [13 to 15, 17].

In the absence of acids, PH_3 reacts with anhydrous **paraformaldehyde** (mole ratio 1:3) at 100°C and 35 to 40 atm to give tris(hydroxymethyl)phosphane, $(HOCH_2)_3P$ [108, 109]. For the reaction of paraformaldehyde pellets with gaseous PH_3 an activation energy of 9.7 kcal/mol between 90 and 110°C was found [110]. Reaction of paraformaldehyde with PH_3 in an aqueous medium in the presence of platinum salts (e.g., $PtCl_2$, $PtCl_4$, K_2PtCl_4) yields the same product. The reaction also proceeds in an acetonitrile suspension. The reaction ($PH_3 : CH_2O = 1:4$) in an acetonitrile/water mixture in the presence of $PtCl_4$ at ambient temperature yields $(HOCH_2)_4P$-OH, which further reacts with PH_3 to give $(HOCH_2)_3P$. In these reactions, paraformaldehyde can be replaced by trioxane [111].

Passing PH_3 at 150 to 300 Torr through an aqueous solution of **formaldehyde** at 40 to 80°C yields $(HOCH_2)_4POH$ [112], which is also obtained at ambient temperature in the presence of a small amount of $PdCl_2$ [113]. Reaction of PH_3 at 9.87 atm (145 psig) with a fresh aqueous solution of CH_2O at 40°C yields $(HOCH_2)_3P \cdot 0.82\,CH_2O$ [114]. $(HOCH_2)_3P$ is obtained in aqueous solution in the absence (100°C, 26.3 atm [115]; 65 to 70°C, 4.4 atm (50 psig) [116]) or presence of catalysts like Raney nickel [116], $FeCl_3$, $CoCl_3$ [111], $RhCl_3$ [111, 113], $Ni(CO)_4$ [116], $NiCl_2$ [117, 118], $PdCl_2$ [111, 113], H_2PtCl_6 [113, 116], $M[P(CH_2OH)_3]_4$ with M = Ni, Pd, Pt [119, 181] at 20 to 30°C, or other platinum compounds $(Pt(OH)_4$, $PtCl_2$, $PtCl_4$ [111], K_2PtCl_4 [111, 181, 182]; −3 to +5°C), $CdCl_2$, $HgCl_2$ (30°C) [111]. Hydroquinone (60°C) [116], or RNH_2 and R_2NH with $R = CH_3$ and C_2H_5, or $H_2NC_2H_4NH_2$ [118] have also been used as catalysts. The reaction of PH_3 with CH_2O in the presence of $CdCl_2$ at 25 to 75°C and pH 10.8 (NaOH) yields $(HOCH_2)_3PO$ [120]. The reaction also proceeds without catalysts in methanol (75 to 90°C, 31.6 atm (450 psig) to 33.3 atm (475 psig)) [121] or in xylene solution (75°C, 4.4 atm (50 psig)) [122].

Formaldehyde in concentrated aqueous hydrochloric acid solution reacts with PH_3 at 25°C [123, 124] or 80°C [125] to form $(HOCH_2)_4PCl$. This reaction explains the absorption in aqueous dioxane solution which is first-order with respect to PH_3, CH_2O, and HCl. The reaction rate is independent of the HCl concentration above 0.1 mol/L. The initial rates of reaction were determined at 25°C [126]:

$$\left.\begin{array}{l}[HCl] < 0.1\ \text{mol/L}\\ [CH_2O] = 0.42\ \text{to}\ 1.35\ \text{mol/L}\end{array}\right\}\ r_i = k_1[CH_2O][PH_3][HCl] \qquad k_1 = 12\ \text{to}\ 14\ L^2\cdot mol^{-2}\cdot h^{-1}$$

$$\left.\begin{array}{l}[HCl] > 0.1\ \text{mol/L}\\ [CH_2O] = 0.4\ \text{to}\ 1.2\ \text{mol/L}\end{array}\right\}\ r_i = k_2[CH_2O][PH_3] \qquad k_2 = 1.30\ \text{to}\ 1.45\ L\cdot mol^{-1}\cdot h^{-1}$$

An activation energy of 17.5 kcal/mol (10 to 40°C; [HCl] = 0.15 mol/L, $[CH_2O]_i = 1.01$ mol/L, $[PH_3]_i = 0.095$ mol/L) was determined [126]. The rate of absorption of PH_3 (with 40% H_2) in an aqueous solution of HCl (0.5 M) and CH_2O (0.5 M) was found to be in the order of 1×10^{-8} to 7×10^{-8} $mol \cdot cm^{-3} \cdot s^{-1}$ at 27°C depending on partial pressure of PH_3 and speed of agitation. Under these conditions, the reaction is approximately first-order with respect to PH_3 (0.80) and CH_2O (0.88), and approximately half-order with respect to HCl (0.42). An activation energy of 14.74 kcal/mol was determined between 4 and 40°C [127].

The reaction of PH_3 with CH_2O in aqueous solutions of H_2SO_4, H_3PO_4, CH_3COOH [128], or $(COOH)_2$ [128, 129] at 40 to 75°C gives the corresponding tetrakis(hydroxymethyl)phosphonium salts. Reaction of the phosphonium salt with additional PH_3 finally yields $(HOCH_2)_3P$ [128].

PH_3 reacts with a mixture of aqueous formaldehyde and diethylamine at ambient temperature to give tris(diethylaminomethyl)phosphane [130]. Reactions of PH_3 and CH_2O with urea in hydrochloric acid, with guanidinium sulfate or dicyandiamide, and with tri-*n*-butylamine yield viscous syrups with a phosphorus content of 4.6, 5.65, and 2.63%, respectively [131].

Electron bombardment of a mixture of PH_3 and **acetaldehyde** at low electron energies and low pressures causes a proton transfer reaction; see 1.3.1.5.2, p. 219 [23]. CH_3CHO reacts with PH_3 in isopropanol or H_2O in the presence of hydroquinone and $PtCl_4$ or $CdCl_2$ to form $(CH_3CHOH)_3P$. Reaction of **crotonaldehyde** (2-butenal) yields a product with the approximate composition $C_{32}H_{50}O_9P$ under similar conditions. Reaction of **butyraldehyde** (butanal) or **isovaleraldehyde** (3-methylbutanal) with PH_3 in the presence of $PtCl_4$ and hydroquinone in acetonitrile gives probably $(C_3H_7CHOH)_3P$ or $(i\text{-}C_4H_9CHOH)_3P$, respectively [133]. In an aqueous, THF-containing, acidic solution, **linear aliphatic aldehydes** react with PH_3 to give tetraalkylolphosphonium salts according to $PH_3 + 4\,RCHO \xrightarrow{HX} (RCHOH)_4PX$ where $R = CH_3$ ($X = Cl$) [134, 135], C_2H_5 ($X = Br$ in C_2H_5OH) [135], C_6H_{13} ($X = Cl$) [134], C_6H_{13} ($X = Cl, I, HSO_4$) [135], $C_{11}H_{23}$ ($X = Cl$) [134], and $C_{11}H_{23}$ ($X = Cl, NO_3, H_2PO_4$) [135].

The reaction of **α-branched aliphatic aldehydes** with PH_3 in hydrochloric acid does not yield hydroxyalkylphosphonium salts because of steric repulsion among the substituent groups. Instead, secondary phosphanes are obtained, in which phosphorus is part of heterocyclic 1,3-dioxa-5-phospha-cyclohexanes [134, 136 to 138, 179, 180].

PH_3 reacts with **fluoral**, CF_3CHO, at −70°C or with fluoral hydrate in concentrated hydrochloric acid to form tris(2,2,2-trifluoro-1-hydroxyethyl)phosphane. Formation of primary or secondary phosphanes has not been observed on passing excess PH_3 into the solution of fluoral hydrate [139]. However, the formation of $(CF_3CHOH)_2PH$ by reacting CF_3CHO with 4 atm PH_3 in a mixture of tetrahydrofuran and concentrated hydrochloric acid is claimed in a patent [140]. **2,2,3,3-Tetrafluoropropionaldehyde** or **2,2,3,3,4,4,5,5-octafluorovaleraldehyde** react with PH_3 in dioxane at ambient temperature to form the corresponding $(RCHOH)_3P$ [139].

It has been reported that in the reaction of PH_3 with **chloral**, CCl_3CHO, (molar ratio 1:2) in diethyl ether [141] or with a solution of chloral hydrate in a mixture of tetrahydrofuran or ethanol with concentrated hydrochloric acid the compound bis(2,2,2-trichloro-1-hydroxy-ethyl)phosphane, $(CCl_3CHOH)_2PH$, [134, 140] or its hydrate [141] form. This reaction has found its way into reference works [5], in which it is stated that the further addition of chloral to the secondary phosphane does not occur because of steric hinderance and the reduced nucleophilicity of the phosphorus atom. However, a patent [142] claims that the reaction of PH_3 with an aqueous hydrochloric acid solution of chloral hydrate at a reactant ratio of 1:3.5 causes a precipitate of tris(2,2,2-trichloro-1-hydroxyethyl)phosphane, $(CCl_3CHOH)_3P$, to form in accordance with [143]. However, with further reaction of PH_3 the composition of the product changes, gradually approaching the composition $(CCl_3CHOH)_2PH$ [143]. The reactions of PH_3 (4 atm) and aldehydes RCHO with $R = CHCl_2$ [134, 140], $CH_3CHClCCl_2$, or $C_2H_5CCl_2$ in a mixture of tetrahydrofuran and concentrated hydrochloric acid yielding the secondary phosphanes, $(RCHOH)_2PH$, are also described [140]. Reacting an aqueous solution of CH_2ClCHO with PH_3 in the presence of $PtCl_4$ yields an oily phosphorus compound, supposedly having the composition $(CH_2ClCHOH)_3P$. Phosphane and **2,2,3-trichlorobutanal** react in acetonitrile in the presence of $PtCl_4$ and hydroquinone to give an organophosphorus compound melting at 88°C [133].

Aromatic Aldehydes. PH_3 reacts with benzaldehyde in a 1:3 molar ratio in diethyl ether saturated with HCl [144, 145] or hydrochloric acid or dioxane solution (containing aqueous HCl) [146] to give benzyl-bis(α-hydroxybenzyl)phosphane oxide rather than the expected tris-(α-hydroxybenzyl)phosphane [147]. A by-product is supposed to be 1,3-dioxa-5-phospha-2,4,6-triphenylcyclohexane [147]. Other aldehydes, RCHO, react analogously in THF containing hydrochloric acid to give $(RCH_2)(RCHOH)_2PO$ with $R = 2\text{-}ClC_6H_4$ [146], $4\text{-}ClC_6H_4$, $4\text{-}CH_3C_6H_4$ [145, 146], $4\text{-}(n\text{-}C_3H_7)C_6H_4$, $4\text{-}(i\text{-}C_3H_7)C_6H_4$, $4\text{-}BrC_6H_4$, and $2,4\text{-}Cl_2C_6H_3$ [146]. $(C_6H_5CHOH)_3P$ was reported to form from C_6H_5CHO in isopropanol solution in the presence of $PtCl_4$ and hydroquinone [133]. In alcoholic solution containing hydrochloric acid, PH_3 reacts with C_6H_5CHO with participation of the alcohol as reactant to give $(C_6H_5CHOR)_3P$ where $R = CH_3$. In the case of $R = C_2H_5$ or $i\text{-}C_3H_7$, the secondary phosphanes $(C_6H_5CHOR)_2PH$ form [144, 147]. In *n*-propanol or *n*-butanol only sirupy products are obained decomposing on distillation [147].

The kinetics of the reaction of PH_3 with benzaldehyde was investigated in aqueous methanol (10 vol% H_2O) containing hydrochloric acid. The reaction $PH_3 + 3\,C_6H_5CHO + 3\,CH_3OH \rightarrow (C_6H_5(OCH_3)CH)_3P + 3\,H_2O$ is first-order with respect to both PH_3 and C_6H_5CHO

and is catalyzed by HCl. The reaction rate is independent of the hydrochloric acid concentration above 0.1 mol/L; below this concentration the reaction rate is proportional to the acid concentration. The initial rates of reaction were determined at 25°C [126]:

$$\left.\begin{array}{l}[HCl] < 0.1\ mol/L \\ [C_6H_5CHO] = 0.98\ to\ 1.75\ mol/L\end{array}\right\}\ r_i = k_1\ [C_6H_5CHO][PH_3][HCl] \quad k_1 = 0.4\ to\ 0.5\ L^2 \cdot mol^{-2} \cdot h^{-1}$$

$$\left.\begin{array}{l}[HCl] > 0.1\ mol/L \\ [C_6H_5CHO] = 0.8\ to\ 1.6\ mol/L\end{array}\right\}\ r_i = k_2\ [C_6H_5CHO][PH_3] \quad k_2 = 5.4\ to\ 6.8\ L \cdot mol^{-1} \cdot h^{-1}$$

An activation energy of 16.5 kcal/mol was determined in the temperature range 10 to 40°C ($[HCl] = 0.40$ mol/L, $[C_6H_5CHO]_i = 1.75$ mol/L, $[PH_3]_i = 0.092$ mol/L) [126].

Ketones. ICR spectroscopy of the ion-molecule reactions occurring in a $PH_3/CH_3C(O)CH_3$ mixture was used among other techniques to evaluate the proton affinity of PH_3 [23]; see Section 1.3.1.5.2, p. 217.

A semiempirical bond energy/bond order method was used to calculate the kinetic parameters for the photoreductions of carbonyl triplets by PH_3. Phosphorus–hydrogen bonds were found to be highly reactive for the hydrogen abstraction via $PH_3 + R^1R^2CO^\bullet \rightarrow \dot{P}H_2 + R^1R^2C^\bullet OH$ [148].

Ketones, RC(O)R′, react only in strongly acidic media to give a mixture of two types of products, the corresponding primary phosphane oxide, $RR'CHP(O)H_2$, and secondary 1-hydroxyalkyl- or 1-hydroxyaryl phosphane oxides, (RR′COH)RR′CHP(O)H. The primary phosphane oxides form in the first step of the reaction sequence by transfer of oxygen from carbon to phosphorus. The second step leading to the secondary phosphane oxides represents a normal carbonyl addition [149 to 151]. Ketones like acetophenone, 3-pentanone, 4-heptanone, cyclopentanone, and cyclohexanone, react in concentrated hydrochloric acid according to this reaction scheme to give the corresponding primary and secondary phosphane oxides [149, 151]. The reaction of acetone with PH_3 was found to be very slow in 8 N HCl, but fast and complete in 12 N HCl or 10 M H_2SO_4 solution [149 to 151]. There is no reaction in neutral solution [149, 152], even at 100°C under pressure [114]. However, due to the influence of the trifluoromethyl groups, 1,1,1-trifluoroacetone [153, 154] and hexafluoroacetone [154] react with PH_3 in the absence of a catalyst at higher temperatures (100 to 105°C for 6 h [153], 50 to 60°C for 2 h [154]) under pressure (≦30 atm) to give predominantly the primary phosphanes $CF_3(CH_3)(OH)CPH_2$ and $(CF_3)_2(OH)CPH_2$, respectively. According to a reinvestigation, $(CF_3)_2(OH)CPH_2$ forms as the sole reaction product at ambient temperature and pressure only if the reactant ratio is exactly 1:1. Further addition of hexafluoroacetone leads to the formation of $((CF_3)_2(OH)C)_2PH$ [155]. $CF_3C(O)CH_3$ reacts stereospecifically with PH_3 (mole ratio 2:1) at 100°C for 6 h to give $(CF_3(CH_3)(OH)C)_2PH$ in 78% yield [132]. Hexafluorocyclobutanone reacts with PH_3 at ambient temperature to give mono- and bis-(1-hydroxyhexafluorocyclobutyl)phosphane, the relative ratio of the two phosphanes being determined by the ratio of the reactants. Excess PH_3 favors the primary phosphane, whereas excess ketone favors the secondary phosphane [156].

Dicarbonyl Compounds. Glyoxal in neutral solution reacts with PH_3 in the presence of $PtCl_4$ and hydroquinone to give a compound whose elemental analysis corresponded approximately to the formula $(CH(OH)_2CHOH)_3P$ [133]. The reaction, carried out in a suitable mole ratio in a mixture of THF and concentrated hydrochloric acid, failed to give a spirocyclic phosphonium salt [134]. However, succinaldehyde and glutaraldehyde, reacted with PH_3 in a mixture of THF and concentrated hydrochloric acid at 20 to 30°C for 2 h, form the corresponding spirocyclic phosphonium salts [134, 138]. PH_3 reacts with 2,4-pentanedione in 4 to 6 N HCl to form 1,3,5,7-tetramethyl-2,4,8-trioxa-6-phospha-adamantane (20°C, 2 to 3 atm, 50 min) [157]. The reactions of PH_3 liberated in situ from Mg_3P_2 in a dioxane hydrochloric acid reaction mixture at 70°C with 1,5-diketones yield cyclic phosphane oxides with the general formula $HO\overline{CR_1CR_2CR_3CR_4CR_5P}(O)H$, where (a) $R_1=R_5=C_6H_5$, $R_2=R_3=R_4=H$ [158], (b) $R_1+R_2=-(CH_2)_4-$, $R_3=R_4=H$, $R_5=C_6H_5$ [158, 159]; (c) $R_1=C_6H_5$, $R_2=R_3=H$, $R_4+R_5=-(CH_2)_4-$ [158]; (d) $R_1+R_2=-(CH_2)_4-$, $R_3=R_5=C_6H_5$, $R_4=H$ [158, 159]; (e) $R_1=R_3=C_6H_5$, $R_2=H$, $R_4+R_5=-(CH_2)_4-$ [158]; (f) $R_1=R_2=-CH_2C(CH_3)_2OCH_2-$, $R_3=R_5=C_6H_5$, $R_4=H$ [158]; (g) $R_1+R_2=R_4+R_5=-(CH_2)_4-$, $R_3=C_6H_5$ [158], and (h) $R_1=R_3=C_6H_5$, $R_2=H$, $R_4+R_5=-CH_2OC(CH_3)_2CH_2-$ [158]. Analogously, PH_3 reacts with 2,2′-methylenebis(cyclohexanone) at 50 to 60°C in dioxane-containing hydrochloric acid to give 10-oxo-4a-hydroxy-10-phospha-perhydroanthracene [160].

The reaction of PH_3 with dialdehyde cellulose in different mole ratios in an organic solvent was claimed to give a PH_2-substituted, nonflammable cellulose derivative [161].

p-Benzoquinone barely reacts with PH_3 in the absence of catalysts. However, in the presence of iodide or bromide, it is reduced at a high rate to give p-$C_6H_4(OH)_2$ and H_3PO_4 [162, 163]. This reaction is related to the removal of PH_3 from C_2H_2/N_2 mixtures; see Section 1.3.1.5.4, p. 225.

The reaction of **pyruvic acid** (2-oxopropionic acid) with PH_3 in diethyl ether in the presence of dry HCl proceeds in two steps. First, the three P–H bonds each add to a keto group. Then, condensation leads to the formation of the tricyclic compound $((O)CO(CH_3)C)_3P$ in which the P atom functions as one of the bridgeheads [145].

PH_3 reacts with ***m*-chloroperoxybenzoic acid** to give $H_2P(O)OH$, $HP(O)(OH)_2$, and H_3PO_4. When the reaction was carried out in alcohol solution (e.g., methanol), $H_2P(O)OCH_3$, $HP(O)(OCH_3)_2$, $HP(O)(OCH_3)OH$, and $CH_3OP(O)(OCH_3)_2$ were also formed [164].

With Sulfur-Containing Compounds

The reaction of PH_3 with $CH_2{=}CHSCH_3$ is described in the section covering the reactions with alkenes; see Table 18, p. 273.

Aromatic sulfonyl chlorides react with PH_3 only in the presence of a base like pyridine or much better in pyridine as solvent. The products vary with reaction conditions. Benzene sulfonyl chloride is reduced to thiophenol and diphenylsulfane, and under certain conditions S,S,S-triphenyl trithiophosphate forms in addition to thiophenol. *p*-Toluene sulfonyl chloride reacts analogously. Reaction of p-$BrC_6H_4SO_2Cl$ with excess PH_3 yields p-BrC_6H_4SH. Reacting p-$NO_2C_6H_4SO_2Cl$ with excess PH_3 gives the corresponding thiophenol and disulfane. The oxidation products of PH_3 in the reactions with these sulfonyl chlorides are H_3PO_2, H_3PO_3, and H_3PO_4 [165].

With Nitrogen-Containing Compounds

Reactions of PH_3 with $CH_2{=}CHCN$, $CH_2{=}CHCH_2NH_2$, and $CH_2{=}CHCH_2N(C_2H_5)_2$ are described in the section covering the reactions with alkenes; see Table 18, pp. 273/4.

Reactions of PH_3 with $\mathbf{(CH_3)_3N}$ at an approximately 1:2 molar ratio at, e.g., 300°C in a solid-bed reactor filled with active carbon (residence time 64 s) yield CH_3PH_2 (67.5%), $(CH_3)_2PH$ (27.3%), $(CH_3)_3P$ (2.5%), and phosphorus (yields in parentheses refer to 90% PH_3 converted). Shorter residence times or a $PH_3:(CH_3)_3N$ mole ratio >1:2 favor formation of the primary phosphane. Longer residence times, higher temperatures, and a low $PH_3:(CH_3)_3N$ mole ratio favor the formation of the secondary and tertiary phosphane. In a fluid-bed reactor filled with finely dispersed SiO_2 covered with a layer of Cu or Ni at 350°C and 7 to 8 s residence time, mainly the primary phosphane is obtained in this reaction, but with reduced conversion and yield. Phosphorus is a significant by-product. Analogous reactions with **$(CH_3)_2NH$** or **CH_3NH_2** at 300°C and 10 s residence time in a solid-bed reactor yield mainly the primary phosphane. **$(C_2H_5)_3N$** and **$(C_3H_7)_3N$** react similarly with PH_3 at 320 to 350°C (17 to 25 s residence time) to give the primary phosphane, however, with low conversion [166]. PH_3 reacts very rapidly with **$C_6H_5CH_2NH_2$**, either neat or in pyridine solution, in the presence of Cu^I or Cu^{II} chlorides at ambient temperature to give the triaminophosphane, $(C_6H_5CH_2NH)_3P$ [167].

PH_3 and **$(CF_3)_2C{=}NH$** condensed together at −196°C react on warming to ambient temperature during 2 days to give $NH_2C(CF_3)_2PH_2$ [168].

In reactions with **aromatic nitro compounds** in alkaline solution, PH_3 functions as a reducing agent. In neutral solution nitrobenzene does not react even at 100 or 250°C nor in the presence of Cu^{II} or Fe^{III} ions. Nitrobenzene or ring-substituted nitrobenzenes, RNO_2 ($R=C_6H_5$, *o*-$CH_3C_6H_4$, *m*-$CH_3C_6H_4$, *p*-$CH_3C_6H_4$, *m*-ClC_6H_4, *p*-ClC_6H_4), are reduced in aqueous ethanol containing NaOH or KOH (RNO_2: NaOH ratio = 1:1 to 1:5) to the corresponding azoxy derivatives RN=N(O)R. PH_3 is oxidized in these reactions to a mixture of phosphinic acid, H_3PO_2, and phosphorous acid, H_3PO_3 [165]. Attempts to reduce 1-nitronaphthalene with PH_3 under similar conditions yielded a dark brown reduction product from which no crystalline compound could be isolated. Reduction of 2-nitronaphthalene, however, gave benzo[*f*]naphtho[2.1-*c*]cinnoline-N-oxide and a small amount of 2,2′-azoxynaphthalene. The 2-, 3- or 4-nitrobiphenyls or 4,4′-dinitrobiphenyl are readily reduced to the corresponding azoxy compounds. Reduction of 2,2′-dinitrobiphenyl yields benzo[*c*]cinnoline-3,4-dioxide [169]. Reaction of 6-nitroquinoline yields pyridazino[4.3-*f*:5.6-*f*′]diquinoline [170] along with pyridazino[4.3-f:5.6-f′]diquinoline 3-oxide [169, 170]. The reaction of **$(CF_3)_2NONO$** with PH_3 in a 1:1 molar ratio at room temperature in an evacuated glass ampule yielded $(CF_3)_2NON{=}PH$ with 60% yield [171].

PH_3 (2 to 4 atm) reacts with organic **isocyanates, XC_6H_4NCO** (X = H, *p*-Cl, *p*-NO_2 [172]; *m*-CH_3 [173]), in benzene in the presence of triethylamine or tributylamine to give tricarbamoylphosphanes, $(XC_6H_4NHC(O))_3P$. Ethyl, octyl (in the presence of tripropylamine at 100°C and 340 atm), or 1-naphthyl isocyanate react analogously [173]. The reaction of 2,4-tolylene diisocyanate with PH_3 is slow, and no definite product could be isolated. Phenyl isothiocyanate failed to react with PH_3 under reaction conditions analogous to those for the isocyanates [172].

The behavior of **heptafluoroazacyclopentane-2-one** towards PH_3 is quite different to that of hexafluorocyclobutanone (cf. p. 279). Ring expansion occurs instead of PH_3 addition to give a diazacyclodecanedione derivative, PF_3, and an unidentified orange solid [174].

Reaction of PH_3 with **dimethylchloramine** in diethyl ether yields red phosphorus and $(CH_3)_2NH_2Cl$ [175].

N-Bromosuccinimide reacts very slowly with PH_3 in alkaline solution with formation of hypophosphite, bromide, and succinimide [176].

References:

[1] Svara, J.; Weferling, N.; Hofmann, T. (Ullmann's Encycl. Ind. Chem. 5th Ed. A **19** [1991] 545/72, 546/8).
[2] Elsner, G. (Houben-Weyl Methoden Org. Chem. 4th Ed. E **1** [1982] 106/82, 113/6).
[3] Boeing, I. A.; Crutchfield, M. M.; Heitsch, C. W. (Kirk-Othmer Encycl. Chem. Technol. 3rd Ed. **17** [1982] 490/539, 527/31).
[4] Maier, L. (in: Kosolapoff, G. M.; Maier, L.; Organic Phosphorus Compounds, 2nd Ed., Vol. 1, Interscience, New York 1972, pp. 1/287).
[5] Sasse, K. (Houben-Weyl Methoden Org. Chem. 4th Ed. **12** Pt. 1 [1963] 17/73, 25/31).
[6] Wolfsberger, W. (Chem. Ztg. **112** [1988] 53/68).

[7] Wolfsberger, W. (Chem. Ztg. **109** [1985] 317/32).
[8] Fluck, E. (Fortschr. Chem. Forsch. **35** [1973] 1/64, 39/48).
[9] Walling, C.; Pearson, M. S. (Top. Phosphorus Chem. **3** [1966] 1/56, 4/10).
[10] Stacey, F. W.; Harris, J. F. (Org. React. [N.Y.] **13** [1963] 150/375, 218/20, 350/2).

[11] Kellner, K.; Tzschach, A. (Z. Chem. **24** [1984] 365/75).
[12] Gross, H.; Costisella, B. (Z. Chem. **17** [1977] 281/6).
[13] Harnisch, H. (Angew. Chem. **88** [1976] 517/24; Angew. Chem. Int. Ed. Engl. **15** [1976] 468).
[14] Staendeke, H. (Chem. Ztg. **96** [1972] 494/9).
[15] Bruker, A. B.; Grinshtein, E. I.; Raver, Kh. R.; Balashova, L. D.; Soborovskii, L. Z. (Khim. Primen. Fosfororg. Soedin. Tr. 3rd Vses. Konf., Moscow 1965 [1972], pp. 285/302; C.A. **77** [1972] No. 62048).
[16] Boyer, N. E.; Vajda, A. E. (SPE Trans. **4** [1964] 45/55, 46/7; C.A. **60** [1964] 9428).
[17] Horák, J. (Chem. Listy **55** [1961] 1278/91).
[18] Buckingham, A. D.; Fowler, P. W. (Can. J. Chem. **63** [1985] 2018/25).
[19] Kollmann, P.; McKelvey, J.; Johansson, A.; Rothenberg, S. (J. Am. Chem. Soc. **97** [1975] 955/65).
[20] Jeng, Mei-Lee H.; Ault, B. S. (J. Phys. Chem. **94** [1990] 1323/7).

[21] Bossard, A. R.; Kamga, R.; Raulin, F. (Icarus **67** [1986] 305/24; C.A. **105** [1986] No. 137394).
[22] Trinquier, G.; Malrieu, J. P. (J. Am. Chem. Soc. **101** [1979] 7169/72).
[23] Holtz, D.; Beauchamp, J. L.; Eyler, J. R. (J. Am. Chem. Soc. **92** [1970] 7045/55, 7051).
[24] Saito, S.; Endo, Y.; Hirota, E. (J. Chem. Phys. **82** [1985] 2947/50).
[25] Lias, S. G.; Ausloos, P. (Int. J. Mass Spectrom. Ion Phys. **22** [1976] 135/45).
[26] Weferling, N. (Phosphorus Sulfur Relat. Elem. **30** [1987] 641/4).
[27] Hoelderich, W.; Hesse, M.; Sattler, E. (Proc. 9th Int. Congr. Catal., Calgary, Alberta, 1988, Vol. 1, pp. 316/23; C.A. **112** [1990] No. 77339).
[28] Rickelton, W. A.; Boyle, R. J. (Eur. Appl. 299169 [1988/89] 1/14; C.A. **110** [1989] No. 235058).
[29] Haugen, G. R.; Benson, S. W. (Int. J. Chem. Kinet. **2** [1970] 235/55).
[30] Elsner, G.; Vollmer, H.; Reutel, E. (Ger. Offen. 2936210 [1979/81] 1/8; C.A. **95** [1981] No. 43345).

[31] Burch, G. M.; Goldwhite, H.; Haszeldine, R. N. (J. Chem. Soc. **1963** 1083/91).
[32] Hoelderich, W.; Hesse, M.; Schwarzmann, M. (Eur. Appl. 300331 [1988/89] 1/11; C.A. **110** [1989] No. 173472).
[33] Brown, H. C. (U.S. 2584112 [1950/52] 1/7; C.A. **1952** 9580).
[34] Hoff, M. C.; Hill, P. (J. Org. Chem. **24** [1959] 356/9).
[35] Goldwhite, H. (J. Chem. Soc. **1965** 3901/2).
[36] Nippon Chemical Industrial Co., Ltd. (Jpn. Kokai Tokkyo Koho 83-222097 [1982/83] 1/6; C.A. **100** [1984] No. 175071).
[37] Weferling, N.; Elsner, G.; Stephan, H. W.; Frorath, F. K. (Ger. Offen. 3629189 [1986/88] 1/5 from C.A. **109** [1988] No. 57041).
[38] Stiles, A. R.; Rust, F. F.; Vaughan, W. E. (J. Am. Chem. Soc. **74** [1952] 3282/4).
[39] N. V. De Bataafsche Petroleum Maatschappij (Br. 673451 [1952] 1/6; C.A. **1953** 5426).
[40] Stiles, A. R.; Rust, F. F.; Vaughan, W. E. (Ger. 899040 [1949/50] 1/5).

[41] Stiles, A. R.; Rust, F. F.; Vaughan, W. E. (U.S. 2803597 [1949/57] 1/5; C.A. **1958** 2049).
[42] Rauhut, M. M.; Currier, H. A.; Semsel, A. M.; Wystrach, V. P. (J. Org. Chem. **26** [1961] 5138/45).
[43] Robertson, A. J. (U.S. 4374780 [1981/83] 1/3; C.A. **98** [1983] No. 198447).
[44] Elsner, G.; Heymer, G.; Stephan, H.-W. (Br. 1561874 [1977/80] 1/5; C.A. **94** [1981] No. 84294).
[45] Elsner, G.; Heymer, G.; Stephan, H.-W. (Ger. Offen. 2703802 [1977/78] 1/15; C.A. **89** [1978] No. 180154).
[46] Hamilton, L. A.; Williams, R. H. (U.S. 2957931 [1949/60] 1/29, 27; C.A. **1961** 10317).
[47] Ross, A. (Can. 1057296 [1973/79] 1/12; C.A. **91** [1979] No. 140999).
[48] Nippon Chemical Industrial Co., Ltd. (Jpn. Kokai Tokkyo Koho 80-122792 [1979/80] 1/6; C.A. **95** [1981] No. 7451).
[49] Hoye, P. A. T. (Eur. Appl. 281311 [1988] 1/12; C.A. **110** [1989] No. 193116).
[50] Harris, T. V.; Pretzer, W. R. (Inorg. Chem. **24** [1985] 4437/9).

[51] Nippon Chemical Industrial Co., Ltd. (Jpn. Kokai Tokkyo Koho 80-122790 [1979/80] 1/4; C.A. **95** [1981] No. 43338).
[52] Van Winkle, J. L.; Morris, R. C.; Mason, R. F. (Ger. Offen. 1909620 [1968/69] 1/13; C.A. **72** [1970] No. 3033).
[53] Mason, R. F.; Van Winkle, J. L. (U.S. 3400163 [1965/68] 1/5).
[54] Green, M.; Haszeldine, R. N.; Iles, B. R.; Rowsell, D. G. (J. Chem. Soc. **1965** 6879/82).
[55] Parshall, G. W.; England, D. C.; Lindsey, R. V., Jr. (J. Am. Chem. Soc. **81** [1959] 4801/2).
[56] Fields, R.; Goldwhite, H.; Haszeldine, R. N.; Kirman, J. (J. Chem. Soc. C **1966** 2075/80).
[57] England, D. C.; Parshall, G. W. (U.S. 2879302 [1957/59] 1/3; C.A. **1959** 15978).
[58] Bissey, J. E.; Goldwhite, H.; Rowsell, D. G. (J. Org. Chem. **32** [1967] 1542/6).
[59] Tavs, P. (Ger. Offen. 1900706 [1969/70] 1/8; C.A. **73** [1970] No. 77287).
[60] Hechenbleikner, I.; Eulow, W. P. (Ger. Offen. 2601520 [1975/76] 1/17; C.A. **85** [1976] No. 192883).

[61] Hechenbleikner, I. (Ger. Offen. 2605307 [1975/76] 1/17; C.A. **85** [1976] No. 192890).
[62] Drucker, A.; Grayson, M. (U.S. 3489811 [1967/70] 1/4).
[63] Hinman, L. M.; Miller, L. S. (Eur. Appl. 339217 [1989] 1/17; C.A. **112** [1990] No. 158982).
[64] Hechenbleikner, I.; Rauhut, M. M. (U.S. 2822376 [1957/58] 1/3; C.A. **1958** 10147).
[65] Robertson, A. J.; Oppelt, J. C. (Can. 1151212 [1981/83] 1/6; C.A. **100** [1984] No. 6852).
[66] Oppelt, J. C.; Robertson, A. J. (Br. Appl. 2092590 [1981/82] 1/3; C.A. **98** [1983] No. 16846).
[67] Rauhut, M. M.; Hechenbleikner, I.; Currier, H. A.; Schaefer, F. C.; Wystrach, V. P. (J. Am. Chem. Soc. **81** [1959] 1103/7).
[68] Reuter, M.; Wolf, E. (Ger. 1078574 [1960] 1/2; C.A. **1961** 16427).
[69] Kuznetsov, E. V.; Valetdinov, R. K.; Sharifullin, A. Sh. (U.S.S.R. 941382 [1978/82] 1/2; C.A. **97** [1982] No. 216466).
[70] Kuznetsov, E. V.; Voskresenskii, V. A.; Valetdinov, R. K.; Sharifullin, A. Sh.; Pavlov, V. Ya.; Gol'tser, S. I.; Karpov, V. S.; Murdasov-Murda, B. D.; Zuikova, A. N. (U.S.S.R. 941381 [1978/82] 1/3; C.A. **97** [1982] No. 216465).

[71] Pringle, P. G.; Smith, M. B. (J. Chem. Soc. Chem. Commun. **23** [1990] 1701/2).
[72] King, R. B.; Kapoor, P. N. (J. Am. Chem. Soc. **93** [1971] 4158/66).
[73] King, R. B.; Kapoor, P. N. (U.S. 3657298 [1970/72] 1/13; C.A. **77** [1972] No. 75338).
[74] Bookham, J. L.; McFarlane, W.; Colquhoun, I. J. (J. Chem. Soc. Chem. Commun. **1986** 1041/2).

[75] King, R. B.; Cloyd, J. C., Jr.; Reimann, R. H. (J. Org. Chem. **41** [1976] 972/7).
[76] Ferris, J. P.; Khwaja, H. (Icarus **62** [1985] 415/23; C.A. **103** [1985] No. 218659).
[77] Blazejowski, J.; Lampe, F. W. (J. Phys. Chem. **85** [1981] 1856/64).
[78] Ruiz, H. G. V.; Rowland, F. S. (Geophys. Res. Lett. **5** [1978] 407/10; C.A. **89** [1978] No. 92778).
[79] Elsner, G.; Vollmer, H.; Heymer, G. (Ger. Offen. 2939588 [1979/81] 1/6; C.A. **95** [1981] No. 150883).
[80] Mason, R. F. (U.S. 3435076 [1966/69] 1/4; C.A. **71** [1969] No. 22187).

[81] Whistler, R. L.; Wang, C.-C.; Inokawa, S. (J. Org. Chem. **33** [1968] 2495/7).
[82] Dingwall, J. G.; Tuck, B. (J. Chem. Soc. Perkin Trans. I **1986** 2081/90).
[83] Nasakin, O. E.; Petrov, G. N.; Alekseev, V. V.; Promonenkov, V. K.; Sukhobokov, A. V. (Zh. Prikl. Khim. [Leningrad] **55** [1982] 1399/402; J. Appl. Chem. USSR [Engl. Transl.] **55** [1982] 1399/402).
[84] Albrand, J. P.; Anderson, S. P.; Goldwhite, H.; Huff, L. (Inorg. Chem. **14** [1975] 570/3).
[85] Noy, N.; Podolak, M.; Bar-Nun, A. (J. Geophys. Res. C **86** [1981] 11985/8; C.A. **96** [1982] No. 43568).
[86] Hestermann, K.; Lippsmeier, B.; Heymer, G. (Ger. Offen. 2407461 [1974/75] 1/13; C.A. **84** [1976] No. 17558).
[87] Staendeke, H.; Hestermann, K.; Lippsmeier, B. (Ger. Offen. 2522021 [1975/76] 1/10; C.A. **86** [1977] No. 72879).
[88] Staendeke, H.; Hestermann, K.; Lippsmeier, B. (Ger. Offen. 2511933 [1975/76] 1/9; C.A. **86** [1977] No. 90017).
[89] Hestermann, K.; Staendeke, H.; Lippsmeier, B. (Ger. Offen. 2457442 [1974/76] 1/11; C.A. **85** [1976] No. 124146).
[90] Hestermann, K.; Jödden, K. (Ger. Offen 2727390 [1977/79] 1/17; C.A. **90** [1979] No. 138022).

[91] Langhans, K. P.; Stelzer, O. (Z. Naturforsch. **45b** [1990] 203/11).
[92] Stelzer, O.; Langhans, K. P.; Svara, J.; Weferling, N. (Eur. Appl. 307702 [1988/89] 1/6; C.A. **111** [1989] No. 97500).
[93] Langhans, K. P.; Stelzer, O.; Weferling, N. (Chem. Ber. **123** [1990] 995/9).
[94] Jolly, W. L. (Inorg. Synth. **11** [1968] 124/6, 126/8).
[95] Kant, M.; Riesel, L.; Kochmann, W.; Oertel, M. (Ger. [East] 231074 [1984/85] 1/4; C.A. **105** [1986] No. 209205).
[96] Dorfman, Ya. A.; Levina, L. V.; Aibasov, E. Zh.; Tungatarov, S. A.; Polimbetova, G. S. (Zh. Obshch. Khim. **61** [1991] 1987/2002; J. Gen. Chem. USSR [Engl. Transl.] **61** [1991] 1840/53).
[97] Dorfman, Ya. A.; Levina, L. V.; Petrova, T. V.; Aleshkova, M. M.; Abdreimova, R. R.; Polimbetova, G. S.; Emel'yanova, V. S. (Zh. Obshch. Khim. **60** [1990] 1275/87; J. Gen. Chem. USSR [Engl. Transl.] **60** [1990] 1137/440).
[98] Dorfman, Ya. A.; Levina, L. V.; Petrova, T. V.; Emel'yanova, V. S.; Polimbetova, G. S. (Zh. Obshch. Khim. **60** [1990] 840/7; J. Gen. Chem. USSR [Engl. Transl.] **60** [1990] 736/42).
[99] Dorfman, Ya. A.; Levina, L. V.; Polimbetova, G. S.; Emel'yanova, V. S.; Kel'man, I. V.; Karinskaya, A. S. (Koord. Khim. **17** [1991] 280/7; Sov. J. Coord. Chem. [Engl. Transl.] **17** [1991] 145/51).
[100] Dorfman, Ya. A.; Levina, L. V.; Petrova, T. V.; Emel'yanova, V. S.; Polimbetova, G. S. (Zh. Obshch. Khim. **59** [1989] 1454/5; J. Gen. Chem. USSR [Engl. Transl.] **59** [1989] 1292/3).

[101] Hamilton, L. A. (U.S. 3352925 [1964/67] 1/3; C.A. **68** [1968] No. 49762).
[102] Lee, Y. E.; Choo, K. Y. (Int. J. Chem. Kinet. **18** [1986] 267/79).
[103] Park, C. R.; Choo, K. Y. (Bull. Korean Chem. Soc. **6** [1985] 206/9; C.A. **104** [1986] No. 129219).
[104] Ingold, K. U. (J. Chem. Soc. Perkin Trans. II **1973** 420/4).
[105] Krusic, P. J.; Mahler, W.; Kochi, J. K. (J. Am. Chem. Soc. **94** [1972] 6033/41).
[106] Lehmann, H.-A.; Schadow, H.; Richter, H.; Kurze, R.; Oertel, M. (Ger. Offen. 2643282 [1976/77] 1/4; C.A. **87** [1977] No. 184678).
[107] Lehmann, H.-A.; Schadow, H.; Pfuetzner, L.; Kurze, R.; Richter, H.; Ober, D. (Ger. [East] 206897 [1981/84] 1/7; C.A. **101** [1984] No. 152075).
[108] Grinshtein, E. I.; Bruker, A. B.; Soborovskii, L. Z. (Dokl. Akad. Nauk SSSR **139** [1961] 1359/62; Proc. Acad. Sci. USSR Chem. Sect. [Engl. Transl.] **136/141** [1961] 841/3).
[109] Grinshtein, E. I.; Bruker, A. B.; Soborovskii, L. Z. (U.S.S.R. 138617 [1960/61] 1/2; C.A. **56** [1962] 6002).
[110] Bruker, A. B.; Baranaev, M. K.; Grinshtein, E. I.; Novoselova, R. I.; Prokhorova, V. V.; Soborovskii, L. Z. (Zh. Obshch. Khim. **33** [1963] 1919/23; J. Gen. Chem. USSR [Engl. Transl.] **33** [1963] 1866/9).

[111] Reuter, M.; Orthner, L. (Ger. 1035135 [1957/59] 1/3; C.A. **1960** 14125).
[112] Raver, Kh. R.; Bruker, A. B.; Soborovskii, L. Z. (Zh. Obshch. Khim. **32** [1962] 588/90; J. Gen. Chem. USSR [Engl. Transl.] **32** [1962] 578/80).
[113] Valetdinov, R. K.; Kuznetsov, E. V.; Kozin, N. P.; Sharifullin, A. Sh.; Kiselev, I. N. (Tr. Kazan. Khim. Tekhnol. Inst. **40** No. 2 [1969] 107/12; C.A. **75** [1971] No. 89084).
[114] Carlson, R. H. (U.S. 3666817 [1970/72] 1/3; C.A. **77** [1972] No. 101878).
[115] American Cyanamid Co. (Fr. 2156972 [1971/73] 1/8; C.A. **80** [1974] No. 4880).
[116] Lin, K. C. (U.S. 3660495 [1969/72] 1/3; C.A. **77** [1972] No. 48635).
[117] Valetdinov, R. K.; Dorfman, A. Ya.; Kutuev, A. A.; Shaikhutdinova, I. T. (Khim. Tekhnol. Elementoorg. Soedin. Polim. **1984** 22/5; C.A. **102** [1985] No. 132147).
[118] Dorfman, Ya. A.; Levina, L. V.; Grekov, L. I.; Korolev, A. V. (Kinet. Katal. **30** [1989] 662/7; Kinet. Catal. [Engl. Transl.] **30** [1989] 578/83).
[119] Harrison, K. N.; Hoye, P. A. T.; Orpen, A. G.; Pringle, P. G.; Smith, M. B. (J. Chem. Soc. Chem. Commun. **1989** 1096/7).
[120] Lippsmeier, B.; Hestermann, K. (Ger. Offen. 2511932 [1975/76] 1/14; C.A. **86** [1977] No. 90016).

[121] Stockel, R. F.; Herbes, W. F. (Ger. Offen. 2158823 [1970/72] 1/6; C.A. **78** [1973] No. 43702).
[122] Stockel, R. F.; Herbes, W. F. (U.S. 3704325 [1970/72] 1/2; C.A. **78** [1973] No. 43703).
[123] Kuznetsov, E. V.; Valetdinov, R. K.; Roitburd, Ts. Ya.; Zakharova, L. B. (Tr. Kazakhsk. Khim. Tekhnol. Inst. **1960** No. 29, pp. 20/1; C.A. **58** [1963] 547).
[124] Reeves, W. A.; Flynn, F. F.; Guthrie, J. D. (J. Am. Chem. Soc. **77** [1955] 3923/4).
[125] Hoffman, A. (J. Am. Chem. Soc. **43** [1921] 1684/7).
[126] Horák, J.; Ettel, V. (Collect. Czech. Chem. Commun. **26** [1961] 2401/9; C.A. **56** [1962] 5995).
[127] Chaudhari, R.; Doraiswamy, L. K. (Chem. Eng. Sci. **29** [1974] 129/33).
[128] Katz, D. S. (Ger. Offen. 2648249 [1975/77] 1/16; C.A. **87** [1977] No. 135931).
[129] Leavitt, J. J. (U.S. 3835194 [1973/74] 1/3; C.A. **81** [1974] No. 171298).
[130] Coates, H.; Hoye, P. A. T. (Br. 854182 [1960] 1/4; C.A. **56** [1962] 1482).

[131] Coates, H.; Lawless, J. J. (U.S. 3050522 [1958/62] 1/2; C.A. **58** [1963] 6975).
[132] Francke, R.; Röschenthaler, G. V. (Chem. Ztg. **113** [1989] 320/1).

[133] Reuter, M.; Orthner, L. (Ger. 1075610 [1958/60] 1/2; C.A. **1961** 13316).
[134] Buckler, S. A.; Wystrach, V. P. (J. Am. Chem. Soc. **83** [1961] 168/73).
[135] Buckler, S. A.; (U.S. 3013085 [1959/61] 1/3; C.A. **57** [1962] 11240).
[136] Rauhut, M. M. (U.S. 3261857 [1965/66] 1/6; C.A. **65** [1966] 15424).
[137] Rauhut, M. M. (U.S. 3238248 [1962/66] 1/6; C.A. **64** [1966] 17639).
[138] Buckler, S. A.; Wystrach, V. P. (J. Am. Chem. Soc. **80** [1958] 6454/5).
[139] Shermolovich, Yu. G.; Danchenko. E. A.; Solov'ev, A. V.; Markovskii, L. N. (Zh. Obshch. Khim. **55** [1985] 2218/26; J. Gen. Chem. USSR [Engl. Transl.] **55** [1985] 1969/76).
[140] Buckler, S. A.; Doll, L. (U.S. 2999882 [1959]; C.A. **56** [1962] 2473).

[141] Ettel, V.; Horák, J. (Collect. Czech. Chem. Commun. **26** [1961] 2087/9; C.A. **56** [1962] 5994).
[142] Gordon, I.; Baranauckas, C. F. (U.S. 3054718 [1959/62] 1/3; C.A. **57** [1962] 17139).
[143] Kozlov, E. S.; Solov'ev, A. V.; Markovskii, L. N. (Zh. Obshch. Khim. **48** [1978] 2437/42; J. Gen. Chem. USSR [Engl. Transl.] **48** [1978] 2212/6).
[144] Ettel, V.; Horák, J. (Collect. Czech. Chem. Commun. **26** [1961] 1949/57; C.A. **56** [1962] 5994).
[145] Buckler, S. A. (J. Am. Chem. Soc. **82** [1960] 4215/20).
[146] Buckler, S. A.; Day, N. E. (U.S. 2927945 [1959/60] 1/2; C.A. **1960** 15316).
[147] Ettel, V.; Horák, J. (Collect. Czech. Chem. Commun. **25** [1960] 2191/5; C.A. **1961** 5402).
[148] Previtali, C. M.; Scaiano, J. C. (J. Chem. Soc. Perkin Trans. II **1975** 934/8).
[149] Buckler, S. A.; Epstein, M. (Tetrahedron **18** [1962] 1211/9).
[150] Buckler, S. A.; Epstein, M. (J. Am. Chem. Soc. **82** [1960] 2076/7).

[151] Buckler, S. A.; Epstein, M. (U.S. 3005029 [1959/61] 1/5; C.A. **56** [1962] 6003).
[152] Bruker, A. B.; Grinshtein, E. I.; Soborovskii, L. Z. (Zh. Obshch. Khim. **36** [1966] 1133/8; J. Gen. Chem. USSR [Engl. Transl.] **36** [1966] 1146/50).
[153] Grinshtein, E. I.; Bruker, A. B.; Soborovskii, L. Z. (Zh. Obshch. Khim. **36** [1966] 1138/41; J. Gen. Chem. USSR [Engl. Transl.] **36** [1966] 1151/4).
[154] Grinshtein, E. I.; Bruker, A. B.; Soborovskii, L. Z. (U.S.S.R. 170498 [1964/65] 1/2; C.A. **63** [1965] 13319).
[155] Röschenthaler, G. V. (Z. Naturforsch. **33b** [1978] 311/5).
[156] Parshall, G . W. (Inorg. Chem. **4** [1965] 52/4).
[157] Epstein, M.; Buckler, S. A. (J. Am. Chem. Soc. **83** [1961] 3279/92).
[158] Vysotskii, V. I.; Pavlycheva, E. V.; Tilichenko, M. N. (Zh. Obshch. Khim. **45** [1975] 1466/8; J. Gen. Chem. USSR [Engl. Transl.] **45** [1975] 1434/5).
[159] Vysotskii, V. I.; Kalinov, S. M.; Tilichenko, M. N. (Zh. Obshch. Khim. **50** [1980] 1707/12; J. Gen. Chem. USSR [Engl. Transl.] **50** [1980] 1383/7).
[160] Vysotskii, V. I.; Pavlycheva, E. V.; Tilichenko, M. N. (U.S.S.R. 445670 [1973/74] 1/2; C.A. **82** [1975] No. 73169).

[161] Khardin, A. P.; Tuzhikov, O. I.; Lemasov, A. I.; Martynova, E. N. (U.S.S.R. 717066 [1977/80] 1/4; C.A. **93** [1980] No. 9830).
[162] Sokol'skii, D. V.; Dorfman, Ya. A.; Evtikov, N. I. (Zh. Org. Khim. **9** [1973] 735/7; J. Org. Chem. USSR [Engl. Transl.] **9** [1973] 757/8).
[163] Sokol'skii, D. V.; Dorfman, Ya. A.; Evtikov, N. I. (Zh. Org. Khim. **8** [1972] 2566/9; J. Org. Chem. USSR [Engl. Transl.] **8** [1972] 2615/8).
[164] Lam, W. W.; Toia, R. F.; Casida, J. E. (J. Agric. Food Chem. **39** [1991] 2274/8; C.A. **115** [1991] No. 272985).
[165] Buckler, S. A.; Doll, L.; Lind, F. K.; Epstein, M. (J. Org. Chem. **27** [1962] 794/8).

[166] Hestermann, K.; Vollmer, H.; Heymer, G.; Schlosser, E.-G. (Ger. Offen. 2636558 [1976/78] 1/20; C.A. **88** [1978] No. 191053).
[167] Dorfman, Ya. A.; Levina, L. V.; Pakhorukova, O. M. (Kinet. Katal. **32** [1991] 617/22; Kinet. Catal. [Engl. Transl.] **32** [1991] 550/5).
[168] Kischkel, H.; Röschenthaler, G. V. (Z. Naturforsch. **39b** [1984] 356/8).
[169] Bellaart, A. C. (Tetrahedron **21** [1965] 3285/8).
[170] Bellaart, A. C. (Recl. Trav. Chim. Pays-Bas **83** [1964] 718/22).

[171] Ang, H. G.; Lee, F. K. (Polyhedron **8** [1989] 379/80).
[172] Buckler, S. A. (J. Org. Chem. **24** [1959] 1460/2).
[173] Buckler, S. A. (U.S. 2969390 [1959/61] 1/3; C.A. **1961** 14381).
[174] Takashima, M.; Shreeve, J. M. (Inorg. Chem. **18** [1979] 3281/3).
[175] Highsmith, R. E.; Sisler, H. H. (Inorg. Chem. **7** [1968] 1740/2).
[176] Jaura, K. L.; Maini, B. K.; Kaushik, R. L. (Res. Bull. Panjab Univ. [2] **18** [1967] 165/70; C.A. **69** [1968] No. 26738).
[177] Tyka, R.; Plazek, E. (Bull. Acad. Pol. Sci. Ser. Sci. Chim. **9** [1961] 577/84; C.A. **60** [1964] 1482).
[178] Staley, R. H.; Corderman, R. R.; Foster, M. S.; Beauchamp, J. L. (J. Am. Chem. Soc. **96** [1974] 1260/1).
[179] Peters, G. A. (U.S. 3159667 [1959/64] 1/3; C.A. **62** [1965] 6515).
[180] American Cyanamid Co. (Fr. 1333818 [1962/63] 1/7; C.A. **60** [1964] 30136).

[181] Hoye, P. A. T.; Pringle, P. G.; Smith, M. B.; Worboys, K. (J. Chem. Soc. Dalton Trans. **1993** 269/74).
[182] Ellis, J. W.; Harrison, K. N.; Hoye, P. A. T.; Orpen, A. G.; Pringle, P. G.; Smith, M. B. (Inorg. Chem. **31** [1992] 3026/33).
[183] Li, Dagang; Wang, Zhongheng; Liu, Shufa; et al. (Faming Zhuanli Shenqing Gongkai Shuomingshu 1043640 [1988/90] 17 pp. from C.A. **114** [1991] No. 166754).

1.3.1.5.11 Sorption

This section summarizes the investigations dealing with purely physical adsorption of PH_3 on adsorbents like activated carbon or silica gel, dissociative adsorption on various adsorbents which can already occur at low temperatures, and irreversible chemisorption involving the decomposition of PH_3 on certain surfaces at higher temperatures. The products formed by PH_3 reacting with the surface atoms were generally not studied. This section also includes studies dealing with the coadsorption of PH_3 with H_2, D_2, O_2, H_2O, D_2O, or CO on various surfaces.

On Aluminium and Gallium Compounds. It was mentioned that PH_3 is irreversibly adsorbed on **alumina gels**. However, no details were given [1]. An adsorption isotherm for PH_3 adsorption ($p < 35$ Torr) on **Al_2O_3** at 20°C is displayed in [2, 3]. The percentage of PH_3 (500 pg in N_2) sorbed on Al_2O_3 between 50 and 120°C is given in [4].

The adsorption of PH_3 on Cr-doped ***n*-GaAs**(100)-(4×1) was studied at 140 and 300 K using temperature-programmed desorption (TPD), high-resolution electron energy loss spectroscopy (HREELS), and isotope exchange reactions. No saturation coverage was observed indicative of multilayer formation. PH_2 species were produced on the surface during adsorption either at 120 or 300 K. It was suggested that PH_3 desorbs either molecularly or by recombination of PH_2 and hydrogen atoms [5].

Ellipsometric studies at $p=2\times10^{-4}$ Torr showed that PH_3 was adsorbed on the (110) plane of **GaP** single crystals. The fractional coverage (number of PH_3 molecules per surface atom) was 0.2 at room temperature [6]. The adsorption process on GaP powder (specific surface area 1 m^2/g) in the pressure range 5×10^{-3} to 5×10^{-2} Torr involved a fast uptake followed by a slow continuous logarithmic adsorption to a fractional coverage exceeding about 0.2 [7].

On Carbon. A gas-chromatographic study revealed that the adsorption of PH_3 on activated carbon (AGN-2), like that of other volatile hydrides, is physical in character and obeys Langmuir-type adsorption. The limiting monolayer adsorption was found to be 0.87 mmol/g. A mean enthalpy of adsorption of 5.1 kcal/mol was determined in the temperature range 30 to 60°C. The ratio of the enthalpy of adsorption to the enthalpy of evaporation was calculated to be 1.5. In the series NH_3–PH_3–AsH_3, a linear relation between the enthalpy of adsorption and the boiling points of the hydrides was found; see figure 2 in the original paper [8]. Adsorption of PH_3 (80 to 120 Torr) by BAU-type activated carbon in the temperature range 0 to −80°C can be described by the Freundlich equation. The enthalpy of adsorption varies from 6.38 to 6.58 kcal/mol (0.8 to 2.3 mmol PH_3/g carbon), and the ratio to the enthalpy of evaporation was calculated to be 1.91 [9]. Other adsorption measurements at 10, 20, and 50°C in the pressure range 8.5 to 511 Torr yielded adsorption enthalpies of 6.5 to 7.2 kcal/mol for surface coverages of 3.0 to 0.05 mmol PH_3/g carbon [10]; see also [1]. The adsorption on SKT-type activated carbon at 20°C can be described by the Freundlich equation as well. Adsorption isotherms of PH_3 up to 35 Torr at 20°C on other activated carbons (BAU-, AR-3-, and AG-5-type) are also displayed in [2, 3]. For a quantum-chemical study of the interaction of PH_3 with a molecular fragment of a carbon surface, see [11]. The adsorption of PH_3 on activated carbon was investigated with respect to its removal by oxidation on the adsorbent at higher temperatures; see Section 1.3.1.5.4, p. 223 [12].

On Silicon. B-doped single-crystal *p*-Si(111)–(7×7) surfaces were used as substrates in more recent investigations dealing with the adsorption of PH_3 on silicon single crystals [13 to 16]; see also [17 to 19]. Partly dissociative adsorption of PH_3 was observed as low as 80 K [13], at higher temperatures, e.g., 120 K [15, 16], and at 300 K [13]. SiH, PH_2, and a minority of PH_3 species were identified on the surface by HREELS [13]. The surface phosphorus concentration was determined by Auger electron spectroscopy (AES) [13, 15, 16]. Mass-spectrometric studies of the sorption at 120 K suggested that PH_3 adsorption proceeds with a constant sticking probability (sticking coefficient $s_0=1$) up to a surface coverage of $(1.5\pm0.2)\times10^{14}$ molecules·cm^{-2}, corresponding to about 75% of the saturation coverage. This behavior indicates a mobile precursor adsorption mechanism. Additional PH_3 was adsorbed with reduced sticking probability until saturation was reached at $(1.9\pm0.3)\times10^{14}$ molecules·cm^{-2} [13, 15, 16]. At 300 K the initial sticking coefficient was determined to be $s_0=0.26$ [13]. Adsorption of atomic H or D prior to PH_3 exposure blocks the adsorption of PH_3, indicating that the surface dangling bonds are the active sites [15, 16]. Previous investigations of the adsorption on Si(111) planes by AES and low-energy electron spectroscopy (LEES) are interpreted by the formation of a P:Si(111)–(7×7) structure [20, 21]. The surface coverage was found to be 2.6×10^{14} molecules·cm^{-2}, corresponding approximately to one PH_3 molecule per three silicon surface atoms [21]; see also [22]. The PH_3 adsorption was studied as a function of both PH_3 pressure and surface temperature. The p/T diagram of the interaction between PH_3 and an Si(111) surface exhibits several Si–P surface structures in the temperature range up to 1070 K [20, 21]. The structure and chemical reactivity of the P-doped surfaces obtained after adsorption of PH_3 on *n*-Si(111)–(7×7) planes was also investigated [14].

Thermal activation of the adsorbate resulted in PH_3 desorption up to 550 K with a maximum at about 180 K. The amount of PH_3 released corresponded to a small fraction of a monolayer. Further heating caused desorption of H_2 between 650 and 850 K with a maximum at about

740 K and of P_2 between 900 and 1100 K with a maximum at 1010 K [13, 15, 16]. If an adsorbed PH_3 layer is heated below the phosphorus desorption temperature, additional PH_3 can be adsorbed onto the P:Si(111) layer after removing the surface hydrogen [15].

The saturation phosphorus coverage on *n*-Si(100)–(2×1) surfaces at room temperature was obtained within 100 s by exposing the surface to 2×10^{-7} Torr PH_3 [23, 24] as shown by low-energy electron diffraction (LEED) [25] or AES [23, 24, 26]. A simple Langmuir model fits the adsorption [24]. The PH_3 molecules were assumed to be chemisorbed nondissociatively at room temperature with a high sticking coefficient of about one. P(2p) X-ray photoemission spectra suggested the PH_3 saturation coverage to be about 2.5×10^{14} molecules·cm^{-2}. The TPD spectrum exhibited a peak at about 275°C for the desorption of PH_3. Annealing of the PH_3 overlayer at 500°C or adsorption at 200 to 400°C caused dissociation of the PH_3 molecules, leading to the formation of Si–P and Si–H bonds. A surface coverage of 1.9×10^{14} molecules·cm^{-2} found at 400°C corresponds to about 1/4 of the number of silicon surface atoms. Hydrogen was desorbed from the silicon surface above 400°C offering more silicon sites for phosphorus chemisorption. The maximum phosphorus coverage achieved was close to the density of silicon surface atoms (6.8×10^{14} atoms·cm^{-2}) and was attained at about 550°C. Beyond 550°C phosphorus was thermally desorbed as P_2 [26]. Other results of previous work are given in [22 to 24], [27, p. 279].

Adsorption studies of PH_3 on silicon powder at room temperature predict a surface coverage of about one PH_3 molecule per six silicon surface atoms. On heating hydrogen evolves, and at 500°C nearly all hydrogen is desorbed [28, pp. 91/2].

On Silicon Compounds. Experimental results of the adsorption of PH_3 on three different **silica gel** samples between −80 and 20°C can be described by the Langmuir equation. The values of the isosteric enthalpies of adsorption indicate the physical nature of the adsorption in these cases [29]. The adsorption on silica gel of the KSM-6-, KSS-3-, and KSK-2-type, on the other hand, are described by the Freundlich equation [2, 29] and the adsorption on silica gel of the ASM-type can be expressed by the Dubinin-Radushkevich equation [1]. The isotherms at 10 [1] and 20°C [2, 3] are displayed in the original papers. The adsorptive capacity of the KSM silica gel impregnated by acetonitrile was found to be less than that of a silica gel activated in a specific manner [30].

The adsorption on **silicon carbide** is described in "Silicon" Suppl. Vol. B 3, Pt. 2, 1986, p. 397.

The adsorption of PH_3 by **zeolites** (NaX [1 to 3], CaX, NaA [2, 3], or CaA [1 to 3]) at 20 [1 to 3], 50 or 70°C [1] can be described either by the Freundlich equation [2, 3] or by the Dubinin-Radushkevich equation [1]. The amount of PH_3 adsorbed on the zeolite NaX is minute as shown by the limiting value of adsorption. The adsorption obviously takes place only on the external surface of the zeolite crystal. Thus, molecular sieves of this type can be used for drying PH_3; see Section 1.3.1.3.9, p. 128 [1]. Differing values for the heat of adsorption of PH_3 on zeolites were given (10.4 kcal/mol [1], 10.5 kcal/mol [31], 6.4 kcal/mol [2, 3]). A quantum-chemical treatment of intermolecular interactions in zeolite cavities, whereby the H^+ donor sites are simulated by $H_3SiOHAlH_3$, is given in [32].

The percentage of PH_3 (500 pg in N_2) sorbed on **glass beds** or various **soils** (e.g. montmorillonite, etc.) between 45 and 120°C is given in [4].

On Germanium. Adsorption of PH_3 on Ge(111) surfaces and heating to 300°C led to the loss of H_2 and formation of a Ge(111)-P-1 structure with one or three germanium atoms per phosphorus atom [20, p. 396]. A more recent investigation of the exposure of a c(2×8) Ge(111) surface to 10^{-6} Torr PH_3 at 200°C for several minutes revealed a change of the c(2×8) reconstruction to a (1×1) pattern. After annealing to 330 to 350°C, a (1×1)PH_x/Ge(111) geometry with x probably two was deduced from angle-resolved photoemission extended fine structure (P 1s level) [33]. The Auger electron yield for phosphorus on Ge(100) planes in the form of a monolayer of PH_3 at room temperature indicated that more than 4×10^{14} molecules $\cdot$ cm^{-2} were adsorbed. It was assumed that the atoms lie atop or are within the first layer of germanium atoms [27, pp. 274/6].

On clean germanium powder, 1.4×10^{14} molecules $\cdot$ cm^{-2} of PH_3 were found to be adsorbed at room temperature after 10 min [28, pp. 55/7]. This corresponds to about one PH_3 molecule per six germanium surface atoms [28, 34]. On heating, desorption of hydrogen took place [28]. On an oxygen-covered germanium surface (one oxygen atom per germanium surface atom), 0.8×10^{14} molecules $\cdot$ cm^{-2} of PH_3 adsorbed after 10 min at room temperature, representing about 10% of a monolayer [28, pp. 59/60].

On Zirconium. AES indicated that PH_3 can be adsorbed on clean Zr(0001) surfaces at room temperature, though only diffuse (1×1) LEED patterns were found for exposures to 10^{-5} to 10^{-6} Torr $\cdot$ s. Desorption, probably of PH_3, started around 100°C and further desorption, probably of H_2 and/or P–H-containing species, started at 360°C [35].

On Molybdenum. The adsorption of PH_3 on Mo(100) surfaces, which were chemically modified by 1.5, 1.2 [36] or 1.0 monolayers of oxygen [36, 37], 0.8 monolayer of sulfur [37], or 1.0 monolayer of carbon [37], was investigated. In the case of the oxygen- and carbon-modified surfaces, some of the adsorbed molecules decomposed, while others desorbed intact as shown by the thermal decomposition spectra [37]. The maximum of the desorption from the oxygen-modified surface was above 300 K (1.5 monolayer of oxygen) or below 300 K (1.2 or 1.0 monolayer of oxygen) [36]. From the sulfur-modified surface, PH_3 desorbed above 100°C and from the carbon-modified surface at about 350°C [37]. The activation energy of desorption decreases along with the desorption temperature. In the case of the various oxygen-modified surfaces, the activation energy decreases with the decrease of oxygen coverage [36]. The trends in desorption energy are consistent with the expectation that the more electronegative oxygen creates more acidic adsorption sites where the soft Lewis base PH_3 can be adsorbed. The desorption energy in a series of Lewis bases including PH_3 can be correlated with the proton affinity [36, 37].

On Tungsten. The adsorption of PD_3 on and its interaction with (100)-oriented tungsten foils was studied by TPD. On clean surfaces at 140 K, both molecular and dissociative adsorption were found. At low PD_3 exposures, complete dissociation of PD_3 occurred already below 170 K. D_2 formed by recombination of D atoms desorbed above 400 K. At higher exposures, PD_3 and D_2 desorbed with phosphorus left on the surface. Under these conditions, D_2 mainly desorbed between 200 and 400 K (H_2 in the temperature interval from PH_3) and PD_3 desorbed as a single broad peak at 230 K independent of PD_3 exposure (first-order desorption). From the onset of D_2 desorption it was concluded that PD_3 decomposition starts at 170 K. At saturation coverage of PD_3, the D_2 desorption can be resolved into two peaks at 290 and 350 K, indicating a two-step mechanism for the decomposition of PD_3 into D_2 (gas) and PD (ads) which decomposes further to P (ads) and D_2 (gas). Energies of decomposition of 17.2 kcal/mol for PD_3 and 20.8 kcal/mol for PD were determined. No isotope effect could be detected. Preadsorbed hydrogen is not displaced by PD_3 at 140 K and no isotope exchange between PD_3 and surface H was observed. Preadsorbed phosphorus causes the PH_3 decomposition to decrease and the concentration of reversibly adsorbed PD_3 to increase. At a surface coverage of 0.5 monolayer, no decomposition of PD_3 could be observed [38].

On Iron. PH_3 adsorption on clean polycrystalline iron at 100 K was suggested to be both molecular and dissociative. Near saturation coverage, molecularly adsorbed PH_3 desorbs around 170 K as shown by thermal desorption spectra. An activation energy for PH_3 desorption of 9.6 kcal/mol was calculated assuming a simple first-order desorption kinetics. The hydrogen desorption due to decomposition of PH_3 depends on the PH_3 exposure. At low PH_3 exposure, maximum hydrogen desorption occurs around 415 K and with increasing exposure shifts to lower temperatures by as much as 100 K between 0.1×10^{-6} Torr·s and 0.4×10^{-6} Torr·s of PH_3 exposure. A second-order H–H recombination-desorption kinetics was suggested. Preadsorbed D_2 decreases the fraction of PH_3 decomposed by about 10%. There is no detectable exchange between PH_3 and D_2. This suggests that PH_3 decomposition is irreversible [39]. Sorption is presumably the reason for a small loss of PH_3 (2.44 $\mu g \cdot cm^{-2} \cdot d^{-1}$), when the gas (28 mg/L) is kept in stainless steel tubes [40].

On Rhodium. Adsorption of PH_3 on Rh(100) planes occurs with a high sticking coefficient. Saturation was achieved at 100 K at 3×10^{-6} Torr·s and at 300 K at 5×10^{-6} Torr·s. Warming the crystal above 100 K desorbed molecular PH_3 and H_2 and left adsorbed phosphorus on the surface [41, 42]. Molecular PH_3 and atomic phosphorus can be readily distinguished by XPS on the basis of their P(2p) binding energies [41]. The thermal desorption of PH_3 was found to be a function of PH_3 exposure ((0.05 to 3)$\times10^{-6}$ Torr·s, 325 to 450 K) suggesting that the binding energy is coverage-dependent. Heating above 450 K completely desorbed PH_3 and H_2 and left phosphorus atoms adsorbed on the surface. Previous partial passivation of the surface raises the temperature of the PH_3 decomposition [42]. The reversible adsorption of PH_3 molecules is only slightly affected by preadsorbed H_2, D_2, or O_2. Preadsorbed O_2, however, reduces the amount of H_2 desorption resulting from PH_3 decomposition and increases the H_2 desorption temperature. A significant fraction of the PH_3 desorption, reduced by the preadsorbed O_2, resulted from the formation of water [43]. Preexposure of Rh(100) surfaces to H_2O [43] or D_2O [41] decreases the extent of PH_3 adsorption and PH_3 decomposition. Experiments with coadsorbed D_2 [42, 43] or D_2O [41] indicated negligible isotope exchange.

Transmission infrared spectroscopic investigations of the interaction of PH_3 with Al_2O_3-supported Rh surfaces indicated that PH_3 initially adsorbs in a molecular state at 90 K and decomposes to PH_2 and PH at 100 K, probably via an intermediate hydrogen-bonded chemisorbed state of molecular PH_3 [44]. PH_3 can displace CO on these surfaces, and the formation of $Rh(CO)PH_3$ was verified spectroscopically [45]. Displacement of CO or NO by PH_3 was also found on Rh(100) surfaces [43].

On Nickel. Multilayer adsorption of 300 K gas phase PH_3 molecules on 25 K Ni(100) planes was accompanied by some PH_3 decomposition as shown by UV photoelectron spectroscopy [41]. No hydrogen evolution was observed mass spectrometrically when PH_3 was adsorbed on Ni(100) planes at 100 K. However, on warming, H_2 and trace amounts of PH_3, PH_2, and PH were observed [46]. Room-temperature adsorption of PH_3 (2×10^{-8} Torr) on Ni(100) planes produced disordered surface structures, which on heating above 550°C gave a three-dimensionally ordered structure of a phosphide [47].

On Platinum. The adsorption of PH_3 on Pt(111) surfaces was investigated by TPD, AES, and HREELS. PH_3 adsorbed molecularly at 100 K with a saturation coverage of 0.5 monolayer. Molecular desorption was not observed, but PH_3 started to decompose upon warming above 160 K to give adsorbed phosphorus and hydrogen atoms. The temperature of hydrogen desorption ranged from about 200 (0.5 monolayer) to about 330 K (0.06 monolayer). No loss in phosphorus was observed initially during PH_3 decomposition; however, in the temperature range 200 to 400 K some desorption of P_4 was observed. Changes in the AES observed in the temperature range 400 to 900 K are best interpreted by diffusion of phosphorus into and out of the bulk material. Preadsorbed D_2 had little effect on the adsorption of PH_3, but adsorbed PH_3 blocked the adsorption of D_2 and also the adsorption of CO substantially. Slow exchange of D for H into PH_3 to give HD occurred at 150 K, but not at 100 K [48].

On Silver. According to TPD studies, PH_3 adsorbs molecularly through the P atom on Ag(111) planes at 105 K and desorbs without any detectable decomposition. The coverage dependence of the desorption energy is interpreted by repulsive lateral interactions between the adsorbed PH_3 molecules. Annealing to 250 K yielded a clean surface again. PH_3 also adsorbs molecularly on Cl-covered Ag(111) surfaces and desorbs reversibly; but partial decomposition takes place on K-covered Ag(111) planes [49].

On Cadmium Sulfide. PH_3 was found to adsorb on CdS powder (mean diameter 0.5 μm, specific surface area 3 m^2/g). The fractional coverage at the beginning at 20°C and 0.6 Torr was found to be 0.09; after evacuation it subsequently dropped to 0.01 [50].

On Organic Polymers. Sorption on PVC, nylon, and Teflon is assumed to be the reason for a small loss of PH_3 (including diffusion) when the gas is kept in tubes of these materials [40]. The percentage of PH_3 (500 pg in N_2) sorbed on various sorts of flours and on starch, etc., are given in [4].

References:

[1] Efremov, A. A.; Morozov, V. I.; Fedorov, V. A. (Vysokochist. Veshchestva **1991** No. 3, pp. 115/21; High Purity Subst. **5** [1991] 468/74).
[2] Tkach, O. D.; Davydov, L. G. (Khim. Khim. Tekhnol. Drev. No. 3 [1975] 23/8; C.A. **86** [1977] No. 57500).
[3] Tkach, O. D. (Izv. Vyssh. Uchebn. Zaved. Khim. Khim. Tekhnol. **17** [1974] 1653/4; C.A. **82** [1975] No. 103609).
[4] Berck, B.; Gunther, F. A. (J. Agric. Food Chem. **18** [1970] 148/53; C.A. **72** [1970] No. 65449).
[5] Singh, N. K.; Murrell, A. J.; Harrison, D.; Foord, J. S. (J. Phys. Condens. Matter **3** [1991] S167/S172).
[6] Morgan, A. E. (Surf. Sci. **43** [1974] 150/72).
[7] van Velzen, W. J. M.; Morgan, A. E. (Surf. Sci. **39** [1973] 255/9).
[8] Karabanov, N. T.; Zorin, A. D. (Zh. Fiz. Khim. **50** [1976] 180/2; Russ. J. Phys. Chem. [Engl. Transl.] **50** [1976] 97/9).
[9] Zorin, A. D.; Chesnokova, S. G. (Tr. Khim. Khim. Tekhnol. **1968** No. 2, pp. 71/5; C.A. **71** [1969] No. 64504).
[10] Morozov, V. I.; Efremov, A. A.; Zel'venskii, Ya. D. (Tr. Inst. Mosk. Khim. Tekhnol. Inst. im. D.I. Mendeleeva No. 81 [1974] 52/3; C.A. **84** [1976] No. 35676).

[11] Lavrinenko-Ometsinskaya, E. D. (Teor. Eksp. Khim. **27** [1991] 488/90; Theor. Exp. Chem. [Engl. Transl.] **27** [1991] 426/8).
[12] Colabella, J. M.; Stall, R. A.; Sorenson, C. T. (J. Cryst. Growth **92** [1988] 189/95).
[13] Chen, P. J.; Colaianni, M. L.; Wallace, R. M.; Yates, J. T., Jr. (Surf. Sci. **244** [1991] 177/84).
[14] Bozso, F.; Avouris, P. (Phys. Rev. B Condens. Matter **43** [1991] 1847/50).
[15] Taylor, P. A.; Wallace, R. M.; Choyke, W. J.; Yates, J. T., Jr. (Surf. Sci. **238** [1990] 1/12).
[16] Wallace, R. M.; Taylor, P. A.; Choyke, W. J.; Yates, J. T., Jr. (J. Appl. Phys. **68** [1990] 3669/78).
[17] Chen, P. J.; Colaianni, M. L.; Wallace, R. M.; Yates, J. T. (AD-225313 [1990] 23 pp.; C.A. **115** [1991] No. 240460).
[18] Taylor, P. A.; Wallace, R. M.; Choyke, W. J.; Yates, J. T. (AD-A222417 [1990] 42 pp.; C.A. **115** [1991] No. 190433).
[19] Wallace, R. M.; Taylor, P. A.; Choyke, W. J.; Yates, J. T. (AD-A221508 [1990] 43 pp.; C.A. **114** [1991] No. 129942).
[20] van Bommel, A. J.; Meyer, F. (Surf. Sci. **8** [1967] 381/98).

[21] van Bommel, A. J.; Crombeen, J. E. (Surf. Sci. **36** [1973] 773/7).
[22] Bootsma, G. A.; Meyer, F. (Surf. Sci. **14** [1969] 52/76, 64).

[23] Meyerson, B. S.; Yu, M. L. (J. Electrochem. Soc. **131** [1984] 2366/8).
[24] Meyerson, B. S.; Yu, M. L. (Proc. Electrochem. Soc. **84** Pt. 6 [1984] 287/94).
[25] Yu, M. L.; Meyerson, B. S. (J. Vac. Sci. Technol. [2] A **2** [1984] 446/9; C.A. **101** [1984] No. 12640).
[26] Yu, M. L.; Vitkavage, D. J.; Meyerson, B. S. (J. Appl. Phys. **59** [1986] 4032/7).
[27] Meyer, F.; Vrakking, J. J. (Surf. Sci. **33** [1972] 271/94).
[28] Boonstra, A. H. (Philips Res. Rep. Suppl. **1968** No. 3, pp. 1/106; C.A. **69** [1968] No. 81708).
[29] Zorin, A. D.; Dudorov, V. Ya.; Rogozhnikova, T. S.; Ryabenko, E. A. (Zh. Fiz. Khim. **44** [1970] 717/9; Russ. J. Phys. Chem. [Engl. Transl.] **44** [1970] 398/400).
[30] Bochkarev, E. P.; Khripunov, V. M. (Poluch. Anal. Veshchestv Osoboi Chist. Mater. Vses. Konf., Gorki, USSR, 1963 [1966], pp. 287/9; C.A. **67** [1967] No. 26019).

[31] Potolokov, N. A.; Kolganov, V. P.; Efremov, A. A.; Grinberg, E. E. (Tr. IREA No. 46 [1984] 90/2; C.A. **102** [1985] No. 226424).
[32] Kassab, E.; Seiti, R.; Allavena, M. (Bul. Shkencave Nat. **44** [1990] 76/82 from C.A. **114** [1991] No. 214636).
[33] Terminello, L. J.; Leung, K. T.; Hussain, Z.; Hayashi, T.; Zhang, X. S.; Shirley, D. A. (Phys. Rev. B Condens. Matter **41** [1990] 12787/98).
[34] Boonstra, A. H.; van Ruler, J. (Surf. Sci. **4** [1966] 141/9, 145).
[35] Lou, J. R.; Wong, P. C.; Mitchell, K. A. R. (Can. J. Chem. **66** [1988] 3157/61).
[36] Deffeyes, J. E.; Smith, A. H.; Stair, P. C. (Surf. Sci. **163** [1985] 79/98).
[37] Deffeyes, J. E.; Horlacher, S. A.; Stair, P. C. (Appl. Surf. Sci. **26** [1986] 517/33).
[38] Zhou, X. L.; Yoon, C.; White, J. M. (Appl. Surf. Sci. **44** [1990] 103/14).
[39] Hegde, R. I.; White, J. M. (J. Phys. Chem. **90** [1986] 2159/63).
[40] Waterford, C. J.; Winks, R. G. (J. Stored Prod. Res. **22** [1986] 25/7; C.A. **104** [1986] No. 181595).

[41] Greenlief, C. M.; Hegde, R. I.; White, J. M. (J. Phys. Chem. **89** [1985] 5681/5).
[42] Hegde, R. I.; Tobin, J.; White, J. M. (J. Vac. Sci. Technol. [2] A **3** [1985] 339/45; C.A. **102** [1985] No. 191805).
[43] Hegde, R. I.; White, J. M. (Surf. Sci. **157** [1985] 17/28).
[44] Lu, G.; Crowell, J. E. (J. Phys. Chem. **94** [1990] 5644/6).
[45] Lu, G.; Darwell, J. E.; Crowell, J. E. (J. Phys. Chem. **94** [1990] 8326/8).
[46] Kiskinova, M.; Goodman, D. W. (Surf. Sci. **108** [1981] 64/76).
[47] Matsudaira, T.; Onchi, M. (J. Phys. C **12** [1979] 3381/7).
[48] Mitchell, G. E.; Henderson, M. A.; White, J. M. (Surf. Sci. **191** [1987] 425/48).
[49] Zhou, X. L.; White, J. M. (Surf. Sci. **221** [1989] 534/52).
[50] Bootsma, G. A. (Surf. Sci. **9** [1968] 396/406).

1.3.1.6 Applications and Uses

Films and Layers

General Remarks. Methods. Phosphane can be used for the manufacture of a variety of films or layers usually containing phosphorus. In particular, the use of PH_3 together with an organometallic main group 3 compound for the high-temperature deposition of III-V semiconducting materials on different substrates is of great industrial interest for producing many kinds of electronic devices. Quite often the phosphorus-containing films deposited on the various substrates consist of metastable phases. In the paragraphs below the films or layers that have been obtained by using PH_3 as one of the starting materials are mainly arranged by their overall chemical composition without differentiating between (i)PH_3 being originally a component of the mixture of gases to be decomposed directly or (ii)PH_3 being precracked before reacting its gaseous decomposition products with other compounds. Thus, **the formulas without any subscripts** of the films and layers given below, especially for the ternary and quarternary phases, **indicate only qualitatively the components**, out of which the different materials are composed and are not intended to indicate either the exact or the idealized stoichiometric composition.

A variety of growth technologies for obtaining such layered materials have been used. The references cited below for the most important methods are simply provided to facilitate the access to these topics without claiming to be even close to completeness. The methods used include, e.g., chemical vapor deposition (CVD) [1]; vapor phase epitaxy (VPE) [2, 3], in particular, metalorganic chemical vapor deposition (MOCVD), also called organometallic chemical vapor deposition (OMCVD); metalorganic chemical vapor epitaxy (MOCVE) or organometallic vapor phase epitaxy (OMVPE) [2 to 8]; hydride vapor phase epitaxy (HVPE) [9]; as well as molecular beam epitaxy (MBE) [2, 10, 11], and chemical beam epitaxy (CBE) [2]. The other components which are reacted with PH_3 or its decomposition products are in most cases organometallic compounds, or rarely their adduct with tertiary phosphanes or amines [2, 8]. The gaseous elements, their hydrides, or their halides have also been used. The deposition on the desired substrate at high temperatures is usually achieved by using carrier gases like H_2 and N_2. The following paragraphs concentrate only on the chemical formation of the various films or layered substances from PH_3 and other components without referring either to the method used or to the physical properties of the material obtained or the resulting electronic device.

Al. The vapor-phase deposition of a metallic aluminium film containing small amounts of carbide impurities by reacting PH_3 with $(CH_3)_3Al$ was claimed in a patent [12].

Mo. The reduction of molybdenum fluoride with PH_3 to give films of metallic molybdenum was patented [13].

W. Similarly, films of metallic tungsten can be vapor-phase grown by reacting PH_3 with WF_6 [13 to 15].

BP. Films or layers of boron phosphide were obtained by reacting PH_3 with B_2H_6 [16, 17], BCl_3 [18], $(CH_3)_3B$ or $(C_2H_5)_3B$ [19 to 22].

AlP. The implantation of phosphorus by treating Al-based wires with PH_3 was patented [23]. A $PH_3/H_2/HCl$ mixture reacts with Al at high temperatures to form AlCl as an intermediate which is subsequently converted to an AlP layer [24]. The growth of semiconducting AlP layers by decomposing the adducts $AlH_3 \cdot PH_3$ or $AlCl_3 \cdot PH_3$ was mentioned in two early patents [25, 26]. For attempts to obtain AlP from PH_3 and $(CH_3)_3Al$, see [27].

GaP. The onset of the reaction between PH_3 and liquid metallic Ga at 950°C is limited by the formation of a GaP film on the surface of the liquid [28]; see also [29]. Also a $PH_3/H_2/HCl$ mixture reacts with gallium vapor to finally give a GaP layer [30, 31], see also [32]. In a similar reaction PCl_3 was used instead of HCl [33]. The reaction of PH_3 with a gallium oxide layer was also claimed in a patent to give a GaP film [34]. The most versatile and most often used methods for the manufacture of GaP films are based on the reactions of PH_3 with the organometallic gallium compounds $(CH_3)_3Ga$ [35 to 46, 48 to 52] (see also [53 to 55]), $(C_2H_5)_3Ga$ [40, 50] (see also [56]), and $Ga(P(t\text{-}C_4H_9)_2)_3$ [57]. Treating GaP with a PH_3/He mixture can be used to maintain a stoichiometric 1:1 composition [58].

InP. Reactions of $PH_3/H_2/HCl$ mixtures with indium vapor have been widely used to obtain indium phosphide layers [31, 59 to 73], see also [74, 75]. The interaction of PH_3 with an In film

[76], In vapor [77 to 79], In^{III} in a zeolite [80], or an indium oxide surface was also reported to yield InP films [34]. The most thoroughly investigated processes for the deposition of InP films or layers on different substrates comprise reactions of PH_3 with the organometallic indium compounds $(CH_3)_3In$ [81 to 116] (see also [53, 117 to 122]), its adducts $(CH_3)_3In \cdot PH_3$ [123], $(CH_3)_3In \cdot P(C_2H_5)_3$ [124] (see also [125]), and $(CH_3)_3In \cdot (C_3H_7)_2NH$ [82], $(C_2H_5)_3In$ [99, 101, 126 to 144], and $In(P(t\text{-}C_4H_9)_2)_3$ [57]. It was claimed that PH_3 reacts with diethylcyclopentadienylindium to give InP layers with a low carbon content [145]. The treatment of InP films with PH_3, e.g., to maintain a certain stoichiometry on the surface, was studied in great detail [146 to 162]. For the deposition of two-layered heterostructures including InP layers as one component, see [163 to 167].

SiP. An amorphous Si-P solid solution was obtained by decomposing a PH_3/Si_2H_6 mixture [168].

ScP. PH_3 reacts with intermediately formed scandium chlorides (from Sc and HCl) yielding a scandium phosphide layer [169].

WP. Heating peroxopolytungstic acid on InP in a PH_3/H_2 ambient gives a β-WP_2 layer [170].

ZnP. Radiofrequency sputtering of Zn in a PH_3 atmosphere leads to the deposition of ZnP [171] or Zn_3P_2 [172] films. Films of the latter composition can also be grown by reactions of PH_3 with Zn vapor [173].

PN. The thermal [174 to 177] or photolytic reaction of PH_3 with NH_3 [178 to 180] and the thermal reaction of PH_3 with NF_3 [181] yield phosphorus nitride films.

PO. The oxidation of PH_3 with molecular oxygen at high temperatures can be used to deposit P_4O_{10} glass films [182, 183].

AlPAs. The decomposition of a gaseous mixture containing PH_3, AsH_3, and $(CH_3)_3Al$ gives a layer consisting of this ternary phase [184].

GaPAs. Analogously, reactions of PH_3/AsH_3 mixtures with $(CH_3)_3Ga$ [39, 40, 50, 51, 184 to 190] or $(C_2H_5)_3Ga$ [50] lead to the deposition of GaPAs films. Phosphidation of GaAs with PH_3 also yields GaPAs [191 to 198], see also [199]. The growth of layers composed of the same three components can be achieved by reacting PH_3/AsH_3 [29] or $PH_3/AsH_3/H_2/HCl$ mixtures with elemental gallium [200 to 205]. A two-layered superlattice featuring this phase as one component is described in [206].

InPAs. Films of ternary phases containing these three elements can be obtained by reacting $PH_3/AsH_3/H_2/HCl$ mixtures with elemental indium [201, 207, 208] or PH_3 and AsH_3 with $(C_2H_5)_3In$ [144, 209, 210].

GaPSb. The decomposition of PH_3 in the presence of Ga and Sb [211] or their trimethyl derivatives yields GaPSb layers [212].

InPSb. Films of this qualitative composition are deposited when PH_3 is pyrolized together with the trimethyl derivatives of In and Sb [212].

AlGaP. The growth of such a film from PH_3, R_3Al, and R_3Ga ($R = CH_3$, C_2H_5) was claimed in a patent [50]. A double heterostructure containing this phase as one component was obtained using the trimethyl derivatives of the two metals [213].

AlInP. Reacting PH_3 with the trimethyl or triethyl derivatives of aluminium and indium leads to the formation of ternary phosphide films of the two metals [50, 214 to 218], see also [219, 220].

GaInP. Of the ternary phosphides of main group 3, films and layers comprising the metals gallium and indium have been investigated most thoroughly. These films are quite often deposited by pyrolyzing PH_3 along with both trialkylgallium and trialkylindium, the alkyl groups generally being CH_3 or C_2H_5 [129, 214 to 218, 221 to 231], see also [220, 232]. The formation of GaInP phases by reacting the two metals with a $PH_3/H_2/HCl$ mixture is described in [31, 233 to 235]. Double heterostructures containing GaInP layers as a component have also been obtained [213, 236].

InPO. PH_3, $(C_2H_5)_3In$, and O_2 were the starting materials for depositing such films [237, 238].

SiPO. Phosphosilicate glass films can be vapor-phase deposited when mixtures of PH_3 with certain silicon compounds are pyrolyzed in the presence of oxygen or selected oxygen compounds. As silicon compounds SiH_4 [239 to 257], Si_2H_6, SiH_2Cl_2 [258], $CH_3Si(OCH_3)_3$ [259, 260], and $Si(OC_2H_5)_4$ [261] were used. In addition to molecular oxygen [239, 242, 244, 245, 247 to 249, 251 to 253, 255 to 257, 259, 261], the compounds H_2O [246, 253], N_2O [258], NO [243], or NO_2 [254] were used as an oxygen source.

AlGaPAs. It was claimed in a patent that layers containing these four elements can be obtained by decomposing the two methyl- or ethyl-substituted main group 3 organometallics together with a PH_3/AsH_3 mixture [50].

AlInPAs. Analogously, it was patented that when the corresponding organoindium compounds replaced the gallium derivatives in the four-component system, AlInPAs films can be grown [50].

GaInPAs. The most thoroughly investigated quarternary phase featuring these four elements was first obtained by reacting $PH_3/AsH_3/H_2/HCl$ mixtures with the vapors of both metals [262 to 265], see also [266]. More commonly applied procedures now use the decomposition of PH_3/AsH_3 mixtures in the presence of both trimethyl- [50, 99, 100, 187, 267 to 272] or triethylmetal compounds [50, 99, 100, 141, 143, 216, 269, 270, 273 to 281], see also [282 to 284]. The use of both $Ga(PR_2)_3$ and $In(PR_2)_3$ ($R = t\text{-}C_4H_9$) [50] or of the adduct $(CH_3)_3In \cdot P(CH_3)_3$ for the deposition of such layers was also mentioned [187]. For the manufacture of a double heterostructure containing a GaInPAs layer as component, see [167].

InPAsSb. PH_3, AsH_3, $(C_2H_5)_3Sb$, and $(C_2H_5)_3In$ were the starting materials for depositing films containing the three different main group 5 elements [285 to 287].

AlGaInP. Layers of this quarternary phosphide containing three different main group 3 metals can be grown by reacting PH_3 in the presence of $(CH_3)_3Al$ with Ga and In vapor [288]. Decomposing PH_3 in the presence of a mixture of all three trimethyl- [217, 226, 227, 289 to 294] or triethylmetal compounds [295] also yields films of this qualitative composition, and it is similarly obtained using a mixture of $(CH_3)_3Al$ with $(C_2H_5)_3Ga$ and $(C_2H_5)_3In$ [215, 216, 230]; see also [232]. For the manufacture of a double heterostructure containing an AlGaInP layer as component, see [236].

BSiPO. Borophosphosilicate glass films can be vapor-phase grown by adding the volatile boron compounds B_2H_6 [258, 296], BCl_3 [297 to 299], or $B(OCH_3)_3$ [296] to the gaseous mixtures suitable for the deposition of phosphosilicate glass films (see above SiPO).

FePCO. Films containing about 67 at% Fe, 21 at% P, 10 at% C, and 2 at% O were deposited when a mixture of PH_3 and $Fe(CO)_5$ in an H_2 stream was decomposed by a radiofrequency discharge [300].

NiPCO. Analogously, the plasma decomposition by a radiofrequency discharge of $PH_3/Ni(CO)_4$ mixtures in Ar as the carrier gas yielded black or silvery films containing about 68 at% Ni, 13 at% P, 13 at% C, and 4 at% O. Hydrogen as the carrier gas caused an increase of the P content up to 18 to 25 at% [300, 301].

Doping

The doping of all kinds of predominantly semiconducting films or layers with specially purified PH_3 is a highly important process used in the microelectronics industry for gradually changing the electronic properties of certain microelectronic devices. The doping can be achieved either by adding very small amounts of PH_3 to the mixture of volatile compounds used for the chemical vapor deposition of some layer or by subsequently treating the surface of the grown film with PH_3 at high temperatures (cf. 1.3.1.5.1.2, pp. 201/6). The detailed coverage of the several hundred papers dealing with this subject was considered to be beyond the scope of this handbook volume.

Organophosphorus Compounds

PH_3 is used as a starting material in the production of a variety of organophosphorus compounds. The most important reaction types consist of the addition of a P–H bond of PH_3 to the C=C bond of alkenes or to the C=O bond of aldehydes and ketones. These reactions and other reactions with organic compounds are described in Section 1.3.1.5.10, pp. 267/87, including a large number of related patents.

Fumigation

PH_3 is widely applied for pest control where grain, other dry foodstuffs, or animal feeds are to be stored. The fumigation of, e.g., grain silos is mainly based on a slow and controlled hydrolysis of metal phosphide formulations evolving PH_3 (cf. 1.3.1.3.10, p. 131). Thus, several hundred papers describe the use of PH_3 as a fumigant, its manner of action, its toxicity towards all stages of the life cycle (eggs, pupae, larvae, and adults) of insect pests, and its metabolism. A critical coverage of all these papers was considered to be beyond the scope of this handbook volume; the reviews [302 to 305] are intended to allow a ready access to this topic.

Miscellaneous

The preparation of pure elemental phosphorus for semiconductor materials by thermal decomposition of PH_3 in H_2 at about 900°C on quartz surfaces [306 to 308] or in a column filled with FeNiP alloy powder was described [309] (see also 1.3.1.5.1.2, pp. 201/6). PH_3 was proposed as a versatile component of gaseous fill mixtures suitable for the manufacture of high-temperature tungsten-halogen lamps [310 to 312]. PH_3 was suggested in a patent to be a promoter for the preparation of alkylhalosilanes from silicon and alkyl halides [313]. The treatment with PH_3 of a solid catalyst whose surface was covered with nickel sulfide was claimed to improve its hydrodenitrogenation and hydrodesulfurization catalytic activity towards hydrocarbons [314]. The continuous addition of small amounts of PH_3 to the vapor phase of the cracking zone can suppress certain side reactions in the thermal cracking of hydrocarbons [315, 316]. A PH_3/pyrogallol mixture was claimed to inhibit the oxidative thermal degradation of cis-1,4-polyisoprene rubber vulcanizates [47].

References:

[1] Holstein, W. L. (Prog. Cryst. Growth Charact. **24** [1992] 111/211).
[2] Razeghi, M. (The MOCVD Challenge, Vol. 1, Hilger, Bristol 1989, 328 pp., pp. 3/28).
[3] Dupuis, R. D. (Science **226** [1984] 623/9).
[4] Stringfellow, G. B. (Organometallic Vapor-Phase Epitaxy: Theory and Practice, Academic, Boston 1989, 345 pp.).
[5] Smith, F. T. J. (Prog. Solid State Chem. **19** [1989] 111/64).

[6] Stringfellow, G. B. (Semicond. Semimet. **22** [1985] 209/59; C.A. **103** [1985] No. 62742).
[7] Razeghi, M. (Semicond. Semimet. **22** [1985] 299/378; C.A. **103** [1985] No. 62744).
[8] Dapkus, P. D. (Annu. Rev. Mater. Sci. **12** [1982] 243/69; C.A. **97** [1982] No. 118612).
[9] Beuchet, G. (Semicond. Semimet. **22** [1985] 261/98; C.A. **103** [1985] No. 62743).
[10] Herman, M. A.; Sitter, H. (Molecular Beam Epitaxy, Springer, Berlin 1989).

[11] Panish, M. B. (Science **208** [1980] 916/22).
[12] Kiyota, H. (Jpn. Kokai Tokkyo Koho 01-252776 [89-252776] [1988/89] 1/4 from C.A. **113** [1990] No. 136941).
[13] Mitani, K.; Kobayashi, N.; Suzuki, M.; Masuda, H.; Kusano, C.; Takahashi, S. (Jpn. Kokai Tokkyo Koho 02-109327 [90-109327] [1988/90] 1/3 from C.A. **113** [1990] No. 124786).
[14] Sekine, M. (Jpn. Kokai Tokkyo Koho 03-191520 [91-191520] [1989/91] 1/4 from C.A. **115** [1991] No. 267496).
[15] Van der Putte, P. (Philips J. Res. **42** [1987] 608/26; C.A. **109** [1988] No. 12659).
[16] Lund, J. C.; Olschner, F.; Ahmed, F.; Shah, K. S. (Mater. Res. Soc. Symp. Proc. **162** [1990] 601/4 from C.A. **114** [1991] No. 131421).
[17] Kumashiro, Y.; Hirabayashi, M.; Koshiro, T.; Okada, Y. (J. Less-Common Met. **143** [1988] 159/65).
[18] Williams, F. V.; Ruehrwein, R. A. (J. Am. Chem. Soc. **82** [1960] 1330/2).
[19] Manasevit, H. M.; Hewitt, W. B.; Nelson, A. J.; Mason, A. R. (J. Electrochem. Soc. **136** [1989] 3070/6).
[20] Manasevit, H. M. (NASA-CR-181622 [1988] 1/35 from C.A. **110** [1989] No. 48746).

[21] Hatano, G.; Izumitani, T.; Ooba, Y. (Jpn. Kokai Tokkyo Koho 03-34551 [91-34551] [1989/91] 1/6 from C.A. **114** [1991] No. 219629).
[22] Semiconductor Energy Research Institute Co., Ltd. (Jpn. Kokai Tokkyo Koho 57-196710 [82-196710] [1981/82] 1/4 from C.A. **98** [1983] No. 163232).
[23] Ozawa, K. (Jpn. Kokai Tokkyo Koho 03-003243 [91-003243] [1989/91] 1/4 from C.A. **114** [1991] No. 155255).
[24] Richman, D. (J. Electrochem. Soc. **115** [1968] 945/7).
[25] Fischer, H.; Wiberg, E. (Ger. 1042539 [1954/58] 1/2; C.A. **1960** 20519).
[26] Siemens & Halske AG (Br. 812818 [1959] from C.A. **1959** 18647).
[27] Manasevit, H. M. (J. Electrochem. Soc. **118** [1971] 647/50).
[28] Yarmola, T. M. (Zh. Prikl. Khim. [Leningrad] **44** [1971] 2306/8; J. Appl. Chem. USSR [Engl. Transl.] **44** [1971] 2355/7).
[29] Morris, F. J.; Fukui, H. (J. Vac. Sci. Technol. **11** [1974] 506/10).
[30] Seifert, W.; Jacobs, K.; Pickenheim, R.; Biehne, G. (Cryst. Res. Technol. **20** [1985] 625/33; C.A. **103** [1985] No. 46474).

[31] Ban, V. S. (J. Cryst. Growth **17** [1972] 19/30).
[32] Shalumov, B. Z.; Shaulov, Yu. Kh.; Ryabenko, E. A.; Mosin, A. M.; Pesotskii, G. S. (Zh. Fiz. Khim. **48** [1974] 1064; Russ. J. Phys. Chem. [Engl. Transl.] **48** [1974] 622).
[33] Dement'ev, Yu. S.; Buzyhin, A. N.; Bletskan, N. I.; Sokolov, E. B.; Fedorov, V. A. (Kristallografiya **25** [1980] 1267/72; Sov. Phys. Crystallogr. [Engl. Transl.] **25** [1980] 722/5).
[34] Nagao, H.; Takemura, K.; Misonou, M.; Kawahara, H. (Ger. Offen. 3817733 [1987/88] 1/7 from C.A. **110** [1989] No. 145496).
[35] Leys, M. R.; Pistol, M. E.; Titze, H.; Samuelson, L. (J. Electron. Mater. **18** [1989] 25/31 from C.A. **110** [1989] No. 163757).
[36] MacDougall, J. E.; Eckert, H.; Stucky, G. D.; Herron, N.; Wang, Y.; Moller, K.; Bein, T.; Cox, D. (J. Am. Chem. Soc. **111** [1989] 8006/7).

[37] Kobayashi, N.; Makimoto, T.; Horikoshi, Y. (Br. Appl. 2192198 [1987/88] 1/60 from C.A. **108** [1988] No. 230063).
[38] Morrow, B. A.; McFarlane, R. A. (J. Phys. Chem. **90** [1986] 3192/7).
[39] Olson, J. M.; Al-Jassim, M. M.; Kibbler, A.; Jones, K. M. (J. Cryst. Growth **77** [1986] 515/23).
[40] Fraas, L. M.; McLeod, P. S.; Partain, L. D.; Weiss, R. E.; Cape, J. A. (J. Cryst. Growth **77** [1986] 386/91).

[41] Biefeld, R. M. (J. Cryst. Growth **56** [1982] 382/8).
[42] Beneking, H.; Roehle, H. (J. Cryst. Growth **55** [1981] 79/86).
[43] Pogge, H. B.; Kemlage, B. M.; Broadie, R. W. (J. Cryst. Growth **37** [1977] 13/22).
[44] Schlyer, D. J.; Ring, M. A. (J. Electrochem. Soc. **124** [1977] 569/73).
[45] Pogge, H. B.; Kemlage, B. M.; Broadie, R. W. (Thin Solid Films **36** [1976] 147/50; C.A. **85** [1976] No. 151933).
[46] André, J. P.; Hallais, J.; Schiller, C. (J. Cryst. Growth **31** [1975] 147/57).
[47] Sokolov, A. N.; Kozlov, N. S.; Shcherbina, E. I. (Dokl. Akad. Nauk BSSR **25** [1981] 534/5; C.A. **95** [1981] No. 26368).
[48] Wang, C. C.; Ladany, I.; McFarlane, S. H., III; Dougherty, F. C. (J. Cryst. Growth **24/25** [1974] 239/43).
[49] Wang, C. C.; McFarlane, S. H., III (J. Cryst. Growth **13/14** [1972] 262/7).
[50] Manasevit, H. M. (Br. 1319311 [1970/73] 1/7; C.A. **79** [1973] No. 84821).

[51] Manasevit, H. M.; Simpson, W. I. (J. Electrochem. Soc. **116** [1969] 1725/32).
[52] McFarlane, S. H., III; Wang, C. C. (J. Appl. Phys. **43** [1972] 1724/31).
[53] Kodama, K. (Jpn. Kokai Tokkyo Koho 03-204922 [91-204922] [1990/91] 1/6 from C.A. **116** [1992] No. 74076).
[54] Naida, G. A.; Ivanyutin, L. A.; Sokolov, E. B.; Kul'chitskii, A. A.; Arendarenko, N. A. (Sint. Rost Soversh. Krist. Plenok Poluprovodn. Mater. Simp., Novosibirsk 1978 [1981], pp. 105/7 from C.A. **95** [1981] No. 106521).
[55] Sokolov, E. B. (Tezisy Dokl. 5th Vses. Soveshch. Rostu Krist., Tiflis 1977, Vol. 2, pp. 27/8 from C.A. **93** [1980] No. 85370).
[56] Ivanyutin, L. A.; Naida, G. A.; Sokolov, E. B. (Tezisy Dokl. 5th Vses. Soveshch. Rostu Krist., Tiflis 1977, Vol. 2, p. 51 from C.A. **93** [1980] No. 85379).
[57] Baker, R. T. (U.S. 5084128 [1990/92] 1/6; C.A. **116** [1992] No. 164388).
[58] Ishizuka, F.; Yoshizawa, H.; Itoh, T. (J. Appl. Phys. **63** [1988] 2091/3).
[59] Banvillet, H.; Gil, E.; Vasson, A. M.; Cadoret, R.; Tabata, A.; Benyattou, T.; Guillot, G. (Proc. SPIE-Int. Soc. Opt. Eng. **1361** [1991] 972/9 from C.A. **115** [1991] No. 101890).
[60] Kol'chenko, T. I.; Moroz, S. E. (Vysokochist. Veshchestva **1989** No. 2, pp. 64/7; High Purity Subst. [Engl. Transl.] **3** [1989] 227/30).

[61] Karlicek, R. F., Jr.; Mitcham, D.; Ginocchio, J. C.; Hammarlund, B. (J. Electrochem. Soc. **134** [1987] 470/4).
[62] Nishibe, T.; Takena, M. (Jpn. Kokai Tokkyo Koho 63-207119 [88-207119] [1987/88] 1/3 from C.A. **110** [1989] No. 86066).
[63] Parfitt, H. T.; Robertson, D. S.; Wilson, A. R. (J. Mater. Sci. **19** [1984] 2211/8; C.A. **101** [1984] No. 82272).
[64] Jürgensen, H.; Korec, J.; Heyen, M.; Balk, P. (J. Cryst. Growth **66** [1984] 73/82).
[65] Yanase, T.; Kato, Y. (Jpn. Kokai Tokkyo Koho 60-191100 [85-191100] [1984/85] 1/4 from C.A. **104** [1986] No. 100035).
[66] Kato, Y. (Jpn. Kokai Tokkyo Koho 60-223129 [85-223129] [1984/85] 1/3 from C.A. **104** [1986] No. 160541).

[67] Nippon Electric Co., Ltd. (Jpn. Kokai Tokkyo Koho 57-145313 [82-145313] [1981/82] 1/4 from C.A. **98** [1983] No. 10201).
[68] Erstfeld, T. E.; Quinlan, K. P. (J. Electron. Mater. **11** [1982] 647/62; C.A. **97** [1982] No. 64853).
[69] Quinlan, K. P. (U.S. Appl. 437655 [1982/83] 1/22; C.A. **99** [1983] No. 14448).
[70] Mizutani, T.; Yoshida, M.; Usui, A.; Watanabe, H.; Yuasa, T.; Hayashi, I. (Jpn. J. Appl. Phys. **19** [1980] L113/L116; C.A. **92** [1980] No. 172534).

[71] Born, P. J.; Robertson, D. S. (J. Mater. Sci. **11** [1976] 395/8).
[72] Born, P. J.; Robertson, D. S. (U.S. 3947549 [1973/76] 1/5; C.A.**85** [1976] No. 23259).
[73] Harrous, M.; Laporte, J. L.; Cadoret, M.; Pariset, C.; Cadoret, R. (J. Cryst. Growth **83** [1987] 279/85).
[74] Braun, I.; Klíma, P.; Stejskal, J.; Cerný, C.; Vonka, P.; Holub, R. (Collect. Czech. Chem. Commun. **51** [1986] 1213/21).
[75] Jones, K. A. (J. Cryst. Growth **60** [1982] 313/20).
[76] Fujimoto, K. (Jpn. Kokai Tokkyo Koho 03-201428 [91-201428] [1989/91] 1/3 from C.A. **116** [1992] No. 13768).
[77] Chin, T. P.; Liang, B. W.; Hou, H. Q.; Tu, C. W. (Mater. Res. Soc. Symp. Proc. **216** [1991] 517/22 from C.A. **115** [1991] No. 194418).
[78] Chow, R.; Chai, Y. G. (Appl. Phys. Lett. **42** [1983] 383/5).
[79] Panish, M. B. (J. Electrochem. Soc. **127** [1980] 2729/33).
[80] Uchida, H.; Ogata, T.; Yoneyama, H. (Chem. Phys. Lett. **173** [1990] 103/6).

[81] Stringfellow, G. B. (AFOSR-TR-90-0950 [1990] 1/13; C.A. **116** [1992] No. 21168).
[82] Hövel, R.; Brianese, N.; Brauers, A.; Balk, P.; Zimmer, M.; Hostalek, M.; Pohl, L. (J. Cryst. Growth **107** [1991] 355/9).
[83] Rudra, A.; Carlin, J. F.; Proctor, M.; Ilegems, M. (J. Cryst. Growth **111** [1991] 589/93).
[84] Iga, R.; Sugiura, H.; Yamada, T. (Jpn. Kokai Tokkyo Koho 03-68129 [91-68129] [1989/91] 1/4 from C.A. **115** [1991] No. 83585).
[85] Butler, B. R.; Briggs, A. T. R.; Kitching, S. A.; Chew, A. (J. Cryst. Growth **102** [1990] 393/7).
[86] Sakuma, Y. (Jpn. Kokai Tokkyo Koho 02-12814 [90-12814] [1988/90] 1/6 from C.A. **112** [1990] No. 243589).
[87] Hallock, R. B.; Manzik, S. J.; Mitchell, T.; Hui, B. C. (U.S. 4847399 [1987/89] 1/10 from C.A. **112** [1990] No. 56270).
[88] Lueckerath, R.; Richter, W.; Jensen, K. F. (NATO ASI Ser. B **198** [1989] 157/67; C.A. **113** [1990] No. 29957).
[89] Rose, B.; Kazmierski, C.; Robein, D.; Gao, Y. (J. Cryst. Growth **94** [1989] 762/6).
[90] Vernon, S. M. (ARO-25536.1-EL-SBI [1988] from C.A. **111** [1989] No. 106738).

[91] Olson, J. M.; Kurtz, S. R.; Kibbler, A. E. (J. Cryst. Growth **89** [1988] 131/6).
[92] Buchan, N. I.; Larsen, C. A.; Stringfellow, G. B. (J. Cryst. Growth **92** [1988] 605/15).
[93] Harrous, M.; Chaput, L.; Bendraoui, A.; Cadoret, M.; Pariset, C.; Cadoret, R. (J. Cryst. Growth **92** [1988] 423/31).
[94] Buchan, N. I.; Larsen, C. A.; Stringfellow, G. B. (Appl. Phys. Lett. **51** [1987] 1024/6).
[95] Naitoh, M.; Umeno, M. (Jpn. J. Appl. Phys. **26** Pt. 2 [1987] L1538/L1539; C.A. **107** [1987] No. 226963).
[96] Dentai, A. G.; Joyner, C. H., Jr.; Weidman, T. W.; Zilko, J. L. (PCT Int. Appl. 8804830 [1987/88] 1/18 from C.A. **110** [1989] No. 86051).
[97] Nelson, A. W.; Westbrook, L. D. (U.S. 4734387 [1985/88] 1/17 from C.A. **110** [1989] No. 16525).

[98] Sakuma, Y.; Kodama, K.; Ozeki, M. (Jpn. J. Appl. Phys. **27** Pt. 2 [1988] L2189/L2191; C.A. **110** [1989] No. 85820).
[99] Maurel, P.; Defour, M.; Grattepain, C.; Omnes, F.; Acher, O.; Timms, G.; Razeghi, M.; Portal, J. C. (Chemtronics **4** [1989] 40/3; C.A. **111** [1989] No. 68795).
[100] Campbell, J. C.; Tsang, W. T.; Qua, G. J.; Johnson, B. C. (IEEE J. Quantum Electron. **24** [1988] 496/500).

[101] Andrews, D. A.; Davey, S. T.; Tuppen, C. G.; Wakefield, B.; Davies, G. J. (Appl. Phys. Lett. **52** [1988] 816/8).
[102] Eguchi, K.; Ohba, Y.; Kushibe, M.; Funamizu, M.; Nakanisi, T. (J. Cryst. Growth **93** [1988] 88/92).
[103] Clawson, A. R.; Hanson, C. M.; Vu, T. T. (J. Cryst. Growth **77** [1986] 334/9).
[104] Olson, J. M.; Kibbler, A. (J. Cryst. Growth **77** [1986] 182/7).
[105] Hsu, C. C.; Yuan, J. S.; Cohen, R. M.; Stringfellow, G. B. (J. Cryst. Growth **74** [1986] 535/42).
[106] Chen, C. H.; Kitamura, M.; Cohen, R. M.; Stringfellow, G. B. (Appl. Phys. Lett. **49** [1986] 963/5).
[107] Mircea, A.; Mellet, R.; Rose, B.; Robein, D.; Thibierge, H.; Leroux, G.; Daste, P.; Godefroy, S.; Ossart, P.; Pougnet, A.-M. (J. Electron. Mater. **15** [1986] 205/13; C.A. **105** [1986] No. 123966).
[108] Yuan, J. S.; Gal, M.; Taylor, P. C.; Stringfellow, G. B. (Appl. Phys. Lett. **47** [1985] 405/7).
[109] Zhu, L. D.; Chan, K. T.; Ballantyne, J. M. (J. Cryst. Growth **73** [1985] 83/95).
[110] Zhu, L. D.; Chan, K. T.; Ballantyne, J. M. (Appl. Phys. Lett. **47** [1985] 47/8).

[111] Zilko, J. L.; van Haren, D. L.; Lu, P. Y.; Schumaker, N. E.; Leung, S. Y. (J. Electron. Mater. **14** [1985] 563/72; C.A. **103** [1985] No. 151093).
[112] Moss, R. H. (J. Cryst. Growth **68** [1984] 78/87).
[113] Stringfellow, G. B. (J. Cryst. Growth **68** [1984] 111/22).
[114] Sacilotti, M.; Mircea, A.; Azoulay, R. (J. Cryst. Growth **63** [1983] 111/5).
[115] Bass, S. J.; Pickering, C.; Young, M. L. (J. Cryst. Growth **64** [1983] 68/75).
[116] Hsu, C. C.; Cohen, R. M.; Stringfellow, G. B. (J. Cryst. Growth **63** [1983] 8/12).
[117] Tsang, W. T.; Choa, F. S.; Ha, N. T. (J. Electron. Mater. **20** [1991] 541/4 from C.A. **115** [1991] No. 83124).
[118] Davies, G. J.; Scott, E. G.; Lyons, M. H.; Rejman-Greene, M. A. Z.; Andrews, D. A. (NATO ASI Ser. B **206** [1989] 45/63 from C.A. **114** [1991] No. 91535).
[119] Stringfellow, G. B.; Buchan, N. I.; Larsen, C. A. (Mater. Res. Soc. Symp. Proc. **94** [1987] 245/53; C.A. **107** [1987] No. 243779).
[120] Larsen, C. A.; Buchan, N. I.; Stringfellow, G. B. (J. Cryst. Growth **85** [1987] 148/53).

[121] Nelson, A. W.; Cole, S.; Harlow, M. J.; Wong, S. Y. K. (Eur. Appl. 242084 [1986/87] 1/6 from C.A. **113** [1990] No. 242988).
[122] Larsen, C. A.; Stringfellow, G. B. (J. Cryst. Growth **75** [1986] 247/54).
[123] Karlicek, R.; Long, J. A.; Donnelly, V. M. (J. Cryst. Growth **68** [1984] 123/7).
[124] Chatterjee, A. K.; Faktor, M. M.; Moss, R. H.; White, E. A. D. (J. Phys. Colloq. [Paris] **43** [1982] C5-491/C5-503).
[125] Moss, R. H.; Evans, J. S. (J. Cryst. Growth **55** [1981] 129/34).
[126] Ando, H.; Okamoto, N.; Sandhu, A.; Fujii, T. (Jpn. J. Appl. Phys. **30** Pt. 2 [1991] 1696/8; C.A. **115** [1991] No. 245111).
[127] Yamamoto, N.; Uwai, K.; Takahei, K. (J. Appl. Phys. **65** [1989] 3072/5).
[128] Uwai, K.; Yamada, S.; Takahei, K. (J. Appl. Phys. **61** [1987] 1059/62).

[129] Razeghi, M. (NATO ASI Ser. B **163** [1987] 151/69; C.A. **109** [1988] No. 161434).
[130] DiForte-Poisson, M. A.; Brylinski, C.; Duchemin, J. P. (Appl. Phys. Lett. **46** [1985] 476/8).

[131] Zhu, L. D.; Chan, K. T.; Wagner, D. K.; Ballantyne, J. M. (J. Appl. Phys. **57** [1985] 5486/92).
[132] Nippon Telegraph and Telephone Public Corp. (Jpn. Kokai Tokkyo Koho 58-156592 [83-156592] [1982/83] 1/4 from C.A. **100** [1984] No. 28351).
[133] Nippon Telegraph and Telephone Public Corp. (Jpn. Kokai Tokkyo Koho 58-208200 [83-208200] [1982/83] 1/4 from C.A. **100** [1984] No. 149519).
[134] Nippon Telegraph and Telephone Public Corp. (Jpn. Kokai Tokkyo Koho 58-209118 [83-209118] [1982/83] 1/4 from C.A. **100** [1984] No. 113416).
[135] Uwai, K.; Susa, N.; Mikami, O.; Fukui, T. (Jpn. J. Appl. Phys. **23** Pt. 2 [1984] 121/3; C.A. **100** [1984] No. 165598).
[136] Bass, S. J.; Young, M. L. (J. Cryst. Growth **68** [1984] 311/8).
[137] Sugou, S.; Kameyama, A.; Katsuda, H.; Miyamoto, Y.; Furuya, K.; Suematsu, Y. (Electron. Lett. **19** [1983] 1036/7; C.A. **100** [1984] No. 15010).
[138] Kasemset, D.; Hess, K. L.; Mohammed, K.; Merz, J. L. (J. Electron. Mater. **13** [1984] 655/71; C.A. **101** [1984] No. 141822).
[139] Uwai, K.; Mikami, O.; Susa, N. (Ext. Abstr. Conf. Solid State Devices Mater. **16** [1984] 667/70; C.A. **101** [1984] No. 238673).
[140] Matsushita Electric Industrial Co., Ltd. (Jpn. Kokai Tokkyo Koho 58-125698 [83-125698] [1982/83] 1/4 from C.A. **99** [1983] No. 204027).

[141] Razeghi, M.; Poisson, M. A.; Larivain, J. P.; Duchemin, J. P. (J. Electron. Mater. **12** [1983] 371/95; C.A. **98** [1983] No. 153398).
[142] Matsushita Electric Industrial Co., Ltd. (Jpn. Kokai Tokkyo Koho 58-132921 [83-132921] [1982/83] 1/4 from C.A. **99** [1983] No. 204033).
[143] Duchemin, J. P. (J. Vac. Sci. Technol. **18** [1981] 753/5; C.A. **95** [1981] No. 53520).
[144] Manasevit, H. M.; Simpson, W. I. (J. Electrochem. Soc. **120** [1973] 135/7).
[145] Sugawara, S.; Sato, K.; Sukegawa, T. (Jpn. Kokai Tokkyo Koho 03-88324 [91-88324] [1989/91] 1/6 from C.A. **115** [1991] No. 171513).
[146] Sugino, T.; Itoh, H.; Boonyasirikool, A.; Shirafuji, J. (Electron. Mater. **21** [1992] 99/104 from C.A. **116** [1992] No. 136839).
[147] Sugino, T.; Yamamoto, H.; Yamada, T.; Ninomiya, H.; Sakamoto, Y.; Matsuda, K.; Shirafuji, J. (J. Electron. Mater. **20** [1991] 1001/6 from C.A. **116** [1992] No. 73465).
[148] Sugino, T.; Yamamoto, H.; Sakamoto, Y.; Ninomiya, H.; Shirafuji, J. (Jpn. J. Appl. Phys. **30** Pt. 2 [1991] L1439/L1442; C.A. **115** [1991] No. 171844).
[149] Ito, H.; Sugino, T.; Shirafuji, J. (Technol. Rep. Osaka Univ. **41** [1991] 59/65 from C.A. **115** [1991] No. 171801).
[150] Mihailovic, M.; Banvillet, H.; Gruzza, B.; Gil, E.; Cadoret, M. (Cryst. Prop. Prep. **36/38** Pt. 2 [1991] 480/2 from C.A. **116** [1992] No. 13952).

[151] Sugino, T.; Yamamoto, H.; Shirafuji, J. (Jpn. J. Appl. Phys. **30** Pt. 2 [1991] L948/L951; C.A. **115** [1991] No. 39886).
[152] Glew, R. W.; Adams, A. R.; Crookes, C. G.; Greene, P. D.; Holmes, S. N.; Kitching, S. A.; Klipstein, P. C.; Lancefield, D.; Stradling, R. A.; Woolley, R. A. (Semicond. Sci. Technol. **6** [1991] 1088/92; C.A. **115** [1991] No. 29207).
[153] Sugino, T.; Ito, H.; Shirafuji, J. (Jpn. J. Appl. Phys. **29** Pt. 2 [1990] L1771/L1774; C.A. **114** [1991] No. 92677).

[154] Murata, M. (Jpn. Kokai Tokkyo Koho 03-236219 [91-236219] [1990/91] 1/4 from C.A. **116** [1992] No. 97359).
[155] Sugino, T.; Boonyasirikool, A.; Shirafuji, J. (Defect Control Semicond, Proc. Int. Conf. Sci. Technol. Defect Control Semicond., Yokohama 1989 [1990], Vol. 1, pp. 873/7 from C.A. **114** [1991] No. 219017).
[156] Sugino, T.; Boonyasirikool, A.; Hashimoto, H.; Shirafuji, J. (Proc. SPIE-Int. Soc. Opt. Eng. **1144** [1989] 224/32 from C.A. **112** [1990] No. 208801).
[157] Fujieda, S.; Akimoto, K.; Hirosawa, I.; Mizuki, J; Matsumoto, Y.; Matsui, J. (Jpn. J. Appl. Phys. **28** Pt. 2 [1989] L16/L18; C.A. **110** [1989] No. 183965).
[158] Bendraoui, A.; El Younoussi, Y.; Cadoret, M. (Rev. Phys. Appl. **24** [1989] 351/6 from C.A. **111** [1989] No. 31471).
[159] Cole, S.; Evans, J. S.; Harlow, M. J.; Nelson, A. W.; Wong, S. (J. Cryst. Growth **93** [1988] 607/12).
[160] Fletcher, J. C. Q. (S. Afr. J. Sci. **84** [1988] 699/701 from C.A. **110** [1988] No. 105321).

[161] Ishikawa, H.; Kamata, M.; Kobayashi, T. (Jpn. Kokai Tokkyo Koho 01-57711 [89-57711] [1987/89] 1/3 from C.A. **111** [1989] No. 88887).
[162] Masut, R. A.; Sacilotti, M. A.; Roth, A. P.; Williams, D. F. (Can. J. Phys. **65** [1987] 1047/52; C.A. **108** [1988] No. 140932).
[163] Jeong, Y. H.; Jeong, D. H.; Hong, W. P.; Caneau, C.; Bhat, R.; Hayes, J. R. (Jpn. J. Appl. Phys. **31** Pt. 2 [1992] L66/L67; C.A. **116** [1992] No. 118314).
[164] Tran, C. A.; Masut, R. A.; Cova, P.; Brebner, J. L. (Appl. Phys. Lett. **60** [1992] 589/91).
[165] Kurishima, K.; Makimoto, T.; Kobayashi, T.; Ishibashi, T. (Jpn. J. Appl. Phys. **30** Pt. 2 [1991] L258/L261; C.A. **114** [1991] No. 197310).
[166] Wang, T. Y.; Reihlen, E. H.; Jen, H. R.; Stringfellow, G. B. (J. Appl. Phys. **66** [1989] 5376/83).
[167] Panish, M. B.; Temkin, H. (Appl. Phys. Lett. **44** [1984] 785/7).
[168] Ogino, T. (Jpn. J. Appl. Phys. **30** Pt. 1 [1991] 1585/90; C.A. **115** [1991] No. 171846).
[169] Yim, W. M.; Stofko, E. J.; Smith, R. T. (J. Appl. Phys. **43** [1972] 254/6).
[170] Karlicek, R. F., Jr.; Williams, K.; Baiocchi, F. A.; Thomas, P. M.; Nakahara, S. (Appl. Phys. Lett. **59** [1991] 2832/4).

[171] Suda, T.; Miyakawa, T.; Kurita, S. (J. Cryst. Growth **86** [1988] 423/9).
[172] Weber, A.; von Känel, H. (Springer Proc. Phys. **54** [1991] 463/7; C.A. **116** [1992] No. 185738).
[173] Chu, T. L.; Chu, S. S. (J. Appl. Phys. **54** [1983] 2063/8).
[174] Hirota, Y.; Kobayashi, T.; Furukawa, Y. (Jpn. J. Appl. Phys. **22** [1983] Suppl. 22-1, pp. 385/8).
[175] Hirota, Y.; Kobayashi, T. (J. Appl. Phys. **53** [1982] 5037/43).
[176] Kobayashi, T.; Hirota, Y. (Electron. Lett. **18** [1982] 180/1).
[177] Nippon Telegraph and Telephone Public Corp. (Jpn. Kokai Tokkyo Koho 58-56425 [83-56425] [1981/83] 1/13 from C.A. **99** [1983] No. 62711).
[178] Hirota, Y.; Hisaki, T.; Mikami, O. (Proc. Electrochem. Soc. **88**-15 [1988] 263/70; C.A. **109** [1988] No. 201142).
[179] Hirota, Y.; Hisaki, T.; Mikami, O. (Inst. Phys. Conf. Ser. No. 79 [1986] 313/8; C.A. **105** [1986] No. 89365).
[180] Hirota, Y.; Mikami, O. (Electron. Lett. **21** [1985] 77/8).

[181] Furukawa, Y. (Jpn. J. Appl. Phys. I **23** [1984] 376/7).
[182] Popov, V. P.; Rabinovich, O. S. (Fazovye Khim. Prevrashch. Vzaimodeistvii Tel Potokom Gaza **1975** 27/48; C.A. **85** [1976] No. 99805).

[183] Henning, W.; Exner, D.; Herrmann, K.; Yücelen, Y. (Z. Angew. Phys. **29** [1970] 114/7).
[184] Kobayashi, N.; Fukui, T. (J. Cryst. Growth **67** [1984] 513/20).
[185] Leys, M. R.; Titze, H.; Samuelson, L.; Petruzzello, J. (J. Cryst. Growth **93** [1988] 504/11).
[186] Fritz, I. J.; Biefeld, R. M.; Osburn, G. C. (Solid State Commun. **45** [1983] 323/5; C.A. **98** [1983] No. 117836).
[187] Ludowise, M. J.; Dietze, W. T.; Lewis, C. R. (Inst. Phys. Conf. Ser. No. 65 [1983] 93/100; C.A. **100** [1984] No. 111533).
[188] Saitoh, T.; Minagawa, S. (J. Electrochem. Soc. **120** [1973] 656/9).
[189] Manasevit, H. M. (J. Cryst. Growth **13/14** [1972] 306/14).
[190] Inoue, M.; Asahi, K. (Jpn. J. Appl. Phys. **11** [1972] 919/20; C.A. **77** [1972] No. 53911).

[191] Viktorovitch, P.; Gendry, M.; Krawczyk, S. K.; Krafft, F.; Abraham, P.; Bekkaoui, A.; Menteil, Y. (Appl. Phys. Lett. **58** [1991] 2387/9).
[192] Park, M. P.; Nitta, M.; Itoh, T. (Solid State Commun. **78** [1991] 569/72; C.A. **115** [1991] No. 124372).
[193] Sugino, T.; Yamada, T.; Shirafuji, J. (Defect Control Semicond. Proc. Int. Conf. Sci. Technol. Defect Control Semicond., Yokohama 1989 [1990], Vol. 1, pp. 849/53 from C.A. **115** [1991] No. 103606).
[194] Yamada, T.; Sugino, T.; Shirafuji, J. (Technol. Rep. Osaka Univ. **40** [1990] 57/62 from C.A. **113** [1990] No. 89147).
[195] Masut, R. A.; Sacilotti, M. A.; Roth, A. P.; Williams, D. F. (NATO ASI Ser. B **198** [1989] 75/84 from C.A. **113** [1990] No. 104111).
[196] Tachi, K.; Sugioka, K.; Toyoda, K.; Ohtsuka, M. (Reza Kagaku Kenkyu No. 11 [1989] 155/7 from C.A. **113** [1990] No. 15645).
[197] Bugge, F.; Diegner, B.; Jacobs, K.; Kloth, B.; Lehmann, L.; Strehmel, G. (Ger. [East] 258679 [1984/88] 1/6 from C.A. **110** [1989] No. 86096).
[198] Burmeister, R. A., Jr.; Pighini, G. P.; Greene, P. E. (Trans. Metall. Soc. AIME **245** [1969] 587/92; C.A. **70** [1969] No. 91431).
[199] Smeets, E. T. J. M. (J. Cryst. Growth **82** [1987] 385/95).
[200] Beccard, R.; Pelzer, B.; Heime, K.; Schreiner, R.; Deschler, M. (Inst. Phys. Conf. Ser. No. 112 [1990] 161/6 from C.A. **115** [1991] No. 38833).

[201] Koukitu, A.; Saegusa, A.; Seki, H. (J. Cryst. Growth **99** [1990] 556/9).
[202] Hitachi Ltd. (Br. 1308790 [1969/73] 1/5; C.A. **78** [1973] No. 129743).
[203] Ban, V. S.; Gossenberger, H. F.; Tietjen, J. J. (J. Appl. Phys. **43** [1972] 2471/2).
[204] Ban, V. S. (J. Electrochem. Soc. **118** [1971] 1473/8).
[205] Tietjen, J. T.; Amick, J. A. (J. Electrochem. Soc. **113** [1966] 724/8).
[206] Bedair, S. M.; Katsuyama, T.; Chiang, P. K.; El-Masry, N. A.; Tischler, M.; Timmons, M. (J. Cryst. Growth **68** [1984] 477/82).
[207] Buckmelter, J. R.; Kennedy, J. K. (J. Electrochem. Soc. **120** [1973] 133/4).
[208] Tietjen, J. J.; Maruska, H. P.; Clough, R. B. (J. Electrochem. Soc. **116** [1969] 492/4).
[209] Fukui, T.; Kobayashi, N. (J. Cryst. Growth **71** [1985] 9/11).
[210] Matsushita Electric Industrial Co., Ltd. (Jpn. Kokai Tokkyo Koho 58-140391 [83-140391] [1982/83] 1/3 from C.A. **99** [1983] No. 222887).

[211] Loualiche, S.; Le Corre, A.; Salaun, S.; Caulet, J.; Lambert, B.; Gauneau, M.; Lecrosnier, D.; Deveaud, B. (Appl. Phys. Lett. **59** [1991] 423/4).
[212] Jou, M. J.; Stringfellow, G. B. (J. Cryst. Growth **98** [1989] 679/89).

[213] Wang, T. Y.; Kimball, A. W.; Chen, G. S.; Birkedal, D.; Stringfellow, G. B. (J. Cryst. Growth **109** [1991] 285/91).
[214] Ozasa, K.; Yuri, M.; Tanaka, S.; Matsunami, H. (J. Cryst. Growth **95** [1989] 171/5).
[215] Ban, Y.; Ogura, M.; Morisaki, M.; Hase, N. (Ext. Abstr. Conf. Solid State Devices Mater. **16** [1984] 679/82; C.A. **101** [1984] No. 238320).
[216] Ogura, M.; Ban, Y.; Morisaki, M.; Hase, N. (J. Cryst. Growth **68** [1984] 32/8).
[217] Hino, I.; Suzuki, T. (J. Cryst. Growth **68** [1984] 483/9).
[218] Suzuki, T.; Hino, I.; Gomyo, A.; Nishida, K. (Jpn. J. Appl. Phys. **21** [1982] L731/L733; C.A. **98** [1983] No. 81216).
[219] Takamori, A.; Yokozuka, T.; Uchama, K.; Nakajima, M. (Jpn. Kokai Tokkyo Koho 03-50882 [91-50882] [1989/91] 1/4 from C.A. **115** [1991] No. 102497).
[220] Takamori, A.; Yokotsuka, T.; Uchiyama, K.; Nakajima, M. (Inst. Phys. Conf. Ser. No. 106 [1990] 229/34 from C.A. **114** [1990] No. 133518).

[221] Omnes, F.; Razeghi, M. (Rev. Tech. Thomson-CSF **23** [1991] 571/83 from C.A. **116** [1992] No. 141124).
[222] Feng, S. L.; Bourgin, J. C.; Omnes, F.; Razeghi, M. (Appl. Phys. Lett. **59** [1991] 941/3).
[223] Ota, T. (Jpn. Kokai Tokkyo Koho 03-293721 [91-293721] [1990/91] 1/9 from C.A. **116** [1992] No. 205158).
[224] Gomyo, A.; Kobayashi, K.; Kawata, S.; Hino, I.; Suzuki, T.; Yuasa, T. (J. Cryst. Growth **77** [1986] 367/73).
[225] Schaus, C. F.; Schaff, W. J.; Shealy, J. R. (J. Cryst. Growth **77** [1986] 360/6).
[226] Ishikawa, M.; Ohba, Y.; Sugawara, H.; Yamamoto, M.; Nakanisi, T. (Appl. Phys. Lett. **48** [1986] 207/8).
[227] Ishikawa, M.; Ohba, Y.; Sugawara, H.; Yamamoto, M.; Nakanisi, T. (Electron. Lett. **21** [1985] 1084/5; C.A. **103** [1985] No. 224079).
[228] Hsu, C. C.; Cohen, R. M.; Stringfellow, G. B. (J. Cryst. Growth **62** [1983] 648/50).
[229] Yoshino, J.; Iwamoto, T.; Kukimoto, H. (J. Cryst. Growth **55** [1981] 74/8).
[230] Hino, I.; Gomyo, A.; Kobayashi, K.; Suzuki, T.; Nishida, K. (Appl. Phys. Lett. **43** [1983] 987/9).

[231] Yoshino, J.; Iwamoto, T.; Kukimoto, H. (Jpn. J. Appl. Phys. **20** [1981] L290/L292; C.A. **94** [1981] No. 200104).
[232] Kobayashi, K. (Jpn. Kokai Tokkyo Koho 03-89583 [91-89583] [1989/91] 1/3 from C.A. **115** [1991] No. 170561).
[233] Ban, V. S.; Ettenberg, M. S. (J. Phys. Chem. Solids **34** [1973] 1119/29; C.A. **79** [1973] No. 24055).
[234] Sigai, A. G.; Nuese, C. J.; Enstrom, R. E.; Zamerowski, T. (J. Electrochem. Soc. **120** [1973] 947/55).
[235] Nuese, C. J.; Richman, D.; Clough, R. B. (Metall. Trans. **2** [1971] 789/94; C.A. **74** [1971] No. 148009).
[236] Ohba, Y.; Yamamoto, M.; Ishikawa, M.; Iwamoto, M.; Nakanisi, T. (Inst. Phys. Conf. Ser. No. 79 [1986] 679/84; C.A. **105** [1986] No. 69740).
[237] Tokuda, H.; Ishimura, H.; Sasaki, K.; Sasaki, F.; Yoshida, T. (Inst. Phys. Conf. Ser. No. 106 [1990] 689/93 from C.A. **114** [1991] No. 238427).
[238] Chang, H. L.; Meiners, L. G.; Sa, C. J. (Appl. Phys. Lett. **48** [1986] 375/7).
[239] Serghi, D.; Pavelescu, C. (Thin Solid Films **186** [1990] L25/L28; C.A. **113** [1990] No. 33076).
[240] Akimoto, T. (Jpn. Kokai Tokkyo Koho 02-170430 [90-170430] [1988/90] 1/4 from C.A. **113** [1990] No. 222875).

[241] Takamatsu, A.; Shibata, M.; Sakai, H.; Yoshimi, T. (J. Electrochem. Soc. **131** [1984] 1865/70).
[242] Hitachi Ltd., Hitachi Microcomputer Engineering Ltd. (Jpn. Kokai Tokkyo Koho 60-34021 [85-34021] [1983/85] 1/4 from C.A. **103** [1985] No. 46862).
[243] Misumi, T.; Ogawa, K.; Kaube, J.; Seitoh, K.; Osato, Y.; Shirai, S. (Ger. Offen. 3309627 [1982/83] 1/108 from C.A. **100** [1984] No. 94513).
[244] Matsushita Electronics Corp. (Jpn. Kokai Tokkyo Koho 59-100542 [84-100542] [1982/84] 1/4 from C.A. **101** [1984] No. 202743).
[245] Matsushita Electric Industrial Co., Ltd. (Jpn. Kokai Tokkyo Koho 59-68921 [84-68921] [1982/84] 1/4 from C.A. **101** [1984] No. 121535).
[246] Wichmann, B.; Doering, E.; Herbst, J. (Ger. Offen. 3133516 [1981/83] 1/8; C.A. **98** [1983] No. 208598).
[247] Lehrer, W. I. (Eur. Appl. 60783 [1981/82] 1/10 from C.A. **98** [1983] No. 131150).
[248] Toshiba Corp. (Jpn. Kokai Tokkyo Koho 58-93274 [83-93274] [1981/83] 1/3 from C.A. **99** [1983] No. 185908).
[249] Nippon Electric Co., Ltd. (Jpn. Kokai Tokkyo Koho 58-63138 [83-63138] [1981/83] 1/3 from C.A. **99** [1983] No. 62736).
[250] Fujitsu Ltd. (Jpn. Kokai Tokkyo Koho 58-56325 [83-56325] [1981/83] 1/2 from C.A. **99** [1983] No. 46402).

[251] Fujitsu Ltd. (Jpn. Kokai Tokkyo Koho 58-03633 [83-03633] [1981/83] 1/3 from C.A. **98** [1983] No. 226440).
[252] Mitsubishi Electric Corp. (Jpn. Kokai Tokkyo Koho 57-202741 [82-202741] [1981/82] 1/3 from C.A. **98** [1983] No. 82366).
[253] Fujitsu Ltd. (Jpn. Kokai Tokkyo Koho 57-180135 [82-180135] [1981/82] 1/4 from C.A. **98** [1983] No. 64193).
[254] Kokusai Electric Co., Ltd. (Jpn. Kokai Tokkyo Koho 57-128038 [82-128038] [1981/82] 1/3 from C.A. **98** [1983] No. 26319).
[255] Kern, W.; Schnable, G. L.; Fischer, A. W. (RCA Rev. **37** [1976] 3/54; C.A. **85** [1976] No. 50802).
[256] Shibata, M.; Yoshimi, T.; Sugawara, K. (J. Electrochem. Soc. **122** [1975] 157/8).
[257] Shibata, M.; Sugawara, K. (J. Electrochem. Soc. **122** [1975] 155/6).
[258] Inoue, S. (Jpn. Kokai Tokkyo Koho 01-303726 [89-303726] [1988/89] 1/4 from C.A. **112** [1990] No. 244341).
[259] Shin-Etsu Chemical Industry Co., Ltd. (Jpn. Kokai Tokkyo Koho 60-90838 [85-90838] [1983/85] 1/7 from C.A. **103** [1985] No. 109041).
[260] Shin-Etsu Chemical Industry Co., Ltd. (Jpn. Kokai Tokkyo Koho 60-90837 [85-90837] [1983/85] 1/9 from C.A. **103** [1985] No. 109042).

[261] Levin, R. M.; Adams, A. C. (J. Electrochem. Soc. **129** [1982] 1588/92).
[262] NEC Corp. (Jpn. Kokai Tokkyo Koho 60-30121 [85-30121] [1983/85] 1/3 from C.A. **103** [1985] No. 79965).
[263] Hyder, S. B.; Saxena, R. R.; Hooper, C. C. (Appl. Phys. Lett. **34** [1979] 584/6).
[264] Enda, H. (Jpn. J. Appl. Phys. **18** [1979] 2167/8; C.A. **92** [1980] No. 13779).
[265] Sugiyama, K.; Kojima, H.; Enda, H.; Shibata, M. (Jpn. J. Appl. Phys. **16** [1977] 2197/203; C.A. **88** [1978] No. 57045).
[266] NEC Corp. (Jpn. Kokai Tokkyo Koho 59-84417 [84-84417] [1982/84] 1/4 from C.A. **101** [1984] No. 162236).
[267] Wiedemann, P.; Klenk, M.; Koerber, W.; Koerner, U.; Weinmann, R.; Zielinski, E.; Speier, P. (J. Cryst. Growth **107** [1991] 561/6).
[268] Knight, D. G.; Miner, C. J.; Watt, B. (J. Cryst. Growth **107** [1991] 221/5).

[269] Ludowise, M. J.; Biswas, D.; Bhattacharya, P. K. (Appl. Phys. Lett. **56** [1990] 958/60).
[270] Campbell, J. C.; Tsang, W. T.; Qua, G. J.; Bowers, J. E. (Appl. Phys. Lett. **51** [1987] 1454/6).

[271] Miller, B. I.; Koren, U.; Capik, R. J.; Su, Y. K. (Appl. Phys. Lett. **51** [1987] 2260/2).
[272] Duchemin, J. P.; Hirtz, J. P.; Razeghi, M.; Bonnet, M.; Hersee, S. D. (J. Cryst. Growth **55** [1981] 64/73).
[273] Tsang, W. T.; Schubert, E. F.; Chiu, T. H.; Cunningham, J. E.; Burkhart, E. G.; Ditzenberger, J. A.; Agyekum, E. (Appl. Phys. Lett. **51** [1987] 761/3).
[274] Sugou, S.; Kameyama, A.; Miyamoto, Y.; Furuya, K.; Suematsu, Y. (Jpn. J. Appl. Phys. **23** [1984] 1182/9; C.A. **100** [1984] No. 15010).
[275] Iwamoto, T.; Mori, K.; Mizuta, M.; Kukimoto, H. (Jpn. J. Appl. Phys. **22** [1983] L191/L193; C.A. **98** [1983] No. 170488).
[276] Sugou, S.; Kameyama, A.; Katsuda, H.; Miyamoto, Y.; Furuya, K.; Suematsu, Y. (Electron. Lett. **19** [1983] 1036/7).
[277] Nippon Telegraph and Telephone Public Corp. (Jpn. Kokai Tokkyo Koho 59-03099 [84-03099] [1982/84] 1/8 from C.A. **100** [1984] No. 201406).
[278] Nippon Telegraph and Telephone Public Corp. (Jpn. Kokai Tokkyo Koho 58-209117 [83-209117] [1982/83] 1/4 from C.A. **100** [1984] No. 113417).
[279] Hirtz, J. P.; Razeghi, M.; Larivain, J. P.; Hersee, S.; Duchemin, J. P. (Electron. Lett. **17** [1981] 113/4; C.A. **94** [1981] No. 93273).
[280] Hirtz, J. P.; Beuchet, G. (Rev. Tech. Thomson-CSF **13** [1981] 263/92 from C.A. **95** [1981] No. 196048).

[281] Hirtz, J. P.; Duchemin, J. P.; Hirtz, P.; De Cremoux, B.; Pearsall, T.; Bonnet, M. (Electron. Lett. **16** [1980] 275/7; C.A. **93** [1980] No. 34603).
[282] Uchida, T.; Kuramata, A. (Jpn. Kokai Tokkyo Koho 03-237712 [91-237712] [1990/91] 1/6 from C.A. **116** [1992] No. 118946).
[283] Schmitz, D.; Strauch, G.; Juergensen, H.; Heyen, M.; Harde, P. (Proc. SPIE-Int. Soc. Opt. Eng. **1144** [1989] 55/60 from C.A. **112** [1990] No. 208793).
[284] Hersee, S. D.; Duchemin, J. P. (Annu. Rev. Mater. Sci. **12** [1982] 65/80; C.A. **97** [1982] No. 118609).
[285] Fukui, T.; Horikoshi, Y. (Jpn. J. Appl. Phys. **20** [1981] 587/91; C.A. **94** [1981] No. 131125).
[286] Fukui, T.; Horikoshi, Y. (Jpn. J. Appl. Phys. **19** [1980] L551/L554; C.A. **94** [1981] No. 10071).
[287] Fukui, T.; Horikoshi, Y. (Jpn. J. Appl. Phys. **19** [1980] L395/L397; C.A. **93** [1980] No. 105761).
[288] Nippon Telegraph and Telephone Publ. Corp. (Jpn. Kokai Tokkyo Koho 59-57991 [84-57991] [1982/84] 1/6 from C.A. **101** [1984] No. 101831).
[289] Hamada, H.; Shono, M.; Honda, S.; Hiroyama, R.; Matsukawa, K.; Yodoshi, K.; Yamaguchi, T. (Electron. Lett. **27** [1991] 1713/5 from C.A. **115** [1991] No. 218098).
[290] Murata, H.; Terasaki, R.; Ihana, T.; Shichizawa, A. (Jpn. Kokai Tokkyo Koho 02-102200 [90-102200] [1988/90] 1/5 from C.A. **113** [1990] No. 88784).

[291] Bour, D. P.; Shealy, J. R. (Appl. Phys. Lett. **51** [1987] 1658/60).
[292] Ohba, Y.; Ishikawa, M.; Sugawara, H.; Yamamoto, M. ; Nakanisi, T. (J. Cryst. Growth **77** [1986] 374/9).
[293] Ikeda, M.; Nakano, K.; Mori, Y.; Kaneko, K.; Watanabe, N. (J. Cryst. Growth **77** [1986] 380/5).

[294] Ohba, Y.; Yamamoto, M.; Ishikawa, M.; Iwamoto, M.; Nakanisi, T. (Inst. Phys. Conf. Ser. No. 79 [1986] 679/84 from C.A. **105** [1986] No. 69740).
[295] Ikeda, M.; Mori, Y.; Kawai, H. (Eur. Appl. 171242 [1984/86] 1/19; C.A. **105** [1986] No. 52049).
[296] Law, K.; Wong, J.; Leung, C.; Olsen, J.; Wang, D. (Solid State Technol. **32** [1989] 60/2 from C.A. **111** [1989] No. 107011).
[297] Raley, N. F.; Losee, D. L. (J. Electrochem. Soc. **135** [1988] 2640/3).
[298] O'Hanlon, J. F.; Fraser, D. B. (J. Vac. Sci. Technol. [2] A **6** [1988] 1226/54, 1249/50; C.A. **109** [1988] No. 75834).
[299] Foster, T. C.; Goldman, J. C.; Hoeye, G. W. (Ger. Offen. 3515135 [1985] 1/26 from C.A. **104** [1986] No. 80552).
[300] Bourcier, R. J.; Nelson, G. C.; Hays, A. K.; Romig, A. D., Jr. (J. Vac. Sci. Technol. [2] A **4** [1986] 2943/8; C.A. **106** [1987] No. 54258).

[301] Hays, A. K. (Mater. Res. Soc. Symp. Proc. **38** [1985] 337/42; C.A. **103** [1985] No. 79707).
[302] Bond, E. J. (FAO Plant Protec. Bull. **54** [1984]).
[303] Halliday, D.; Harris, A. H.; Taylor, R. W. D. (Chem. Ind. [London] **1983** 468/71).
[304] Monro, H. A. U. (FAO Agric. Stud. No. 79 [1969] 379 pp., pp. 8, 40, 145/55, 216/7, 222, 228).
[305] Cabrol, A. M.; De Saint Blanquat, G.; Derache, R. (Microbiol. Aliments Nutr. **4** [1986] 241/6 from C.A. **107** [1986] No. 95476).
[306] Iso, A. (Jpn. Kokai Tokkyo Koho 60-186408 [85-186408] [1984/85] 1/5 from C.A. **104** [1986] No. 52958).
[307] Iso, A. (Jpn. Kokai Tokkyo Koho 60-215510 [85-215510] [1984/85] 1/5 from C.A. **105** [1986] No. 26564).
[308] Frolov, I. A. (Zh. Prikl. Khim. [Leningrad] **51** [1978] 721/3; J. Appl. Chem. USSR [Engl. Transl.] **51** [1978] 709/11).
[309] Matsubara, H.; Tabei, S.; Ichimura, S.; Iso, A. (Jpn. Kokai Tokkyo Koho 01-313309 [89-313309] [1988/89] 1/5 from C.A. **112** [1990] No. 237709).
[310] Yu, T. H. C.; Olwert, R. J.; Bergman, R. S. (Eur. Appl. 420547 [1990/91] 1/6 from C.A. **115** [1991] No. 40206).

[311] Yu, T. H. C.; Holcomb, R. H. (Eur. Appl. 370318 [1989/90] 1/7 from C.A. **113** [1990] No. 162234).
[312] Weld, T. G.; Beschle, M. D. (Can. 1271513 [1986/90] 1/16 from C.A. **114** [1991] No. 91638).
[313] Degen, B.; Feldner, K.; Kaiser, H. J.; Schulze, M. (Ger. Offen. 3910665 [1989/90] 1/4 from C.A. **114** [1991] No. 62348).
[314] Akzo N. V. (Neth. Appl. 8900688 [1989/90] 1/17 from C.A. **114** [1991] No. 46349).
[315] Koszman, I. (Br. 1307542 [1970/73] 1/5; C.A. **78** [1973] No. 126575).
[316] Koszman, I. (Ger. Offen. 2026319 [1970/72] 1/14; C.A. **76** [1972] No. 74537).

1.3.2 PH_3^+ and PH_3^{2+}

1.3.2.1 PH_3^+

CAS Registry Numbers: PH_3^+ *[29724-05-8]* Phosphoniumyl; deleted *[97419-07-3]* Phosphine, radical ion(1+); PD_3^+ *[64782-73-6]* Phosphoniumyl-d_3

Formation

PH_3^+ ions were obtained by electron impact (see p. 212) and photoabsorption (see p. 193) of gaseous phosphane. A survey [1] of experimental appearance potentials for the production of charged species from phosphane shows that the ion in the electronic ground state $\tilde{X}\ ^2A_1$ appears in the range 9.97 to 10.4 eV [1], for instance in the mass spectrum of phosphane, PH_3^+ at 10.1±0.1 eV and PD_3^+ at 10.15±0.1 [2] or 10.1±0.2 eV [3]. The appearance potentials of the electronically excited species PH_3^+ ($\tilde{A}\ ^2E$) and PD_3^+ ($\tilde{A}\ ^2E$) were determined to be 12.5±0.2 and 12.75±0.2 eV, respectively [2]. See also Section 1.3.1.4.1.3 on the ionization potentials of PH_3 (p. 149). – For formation of PH_3^+ by reaction of PH_3 with electronically excited He atoms, see p. 233.

The PH_3^+ ion was ESR-spectroscopically detected in the γ radiolysis (4 Mrad/h, 2 to 4 h) of sulfuric acid solutions of phosphane [4].

Physical Properties

In the electronic ground state $\tilde{X}\ ^2A_1$ with the electron configuration ... $(4a_1)^2\,(2e)^4\,(5a_1)$, the cation forms a very flat trigonal pyramid (point group C_{3v}), as indicated experimentally by an analysis of the ESR spectrum [4] and a broadening of the vibrational fine structure in the first photoelectron band of PH_3 (see p. 197) [5] and theoretically by ab initio calculations on PH_3^+ [6 to 12]; for further calculations, see the bibliography cited on p. 176. The electronically excited state $\tilde{A}\ ^2E$ of the ion with the electron configuration ... $(4a_1)^2\,(2e)^3\,(5a_1)^2$ was shown to be the convergence limit of a Rydberg series (above the first ionization potential of PH_3) in the photoabsorption spectrum of PH_3 between 1130 and 1250 Å [13].

Ab initio calculations (multireference CI) gave the adiabatic ionization potential E_i=18.0 eV [14].

Ab initio calculations on PH_3^+ ($\tilde{X}\ ^2A_1$) gave the following values for the internuclear distance r_e, bond angle α_e (HPH) or angle β_e (angle between the P–H bond and the plane perpendicular to the C_3 axis passing through the P atom) and inversion barrier B (PTCI: second-order perturbation CI, CAS: complete active space):

parameter	method				
	CI [6]	CI [12]	CI [8]	CAS SCF [15]	SCF [10]
r_e in Å	1.39	1.392	1.402	1.416	1.381
α_e (β_e)	(15.5°)	(16.1°)	113.3°(15.0°)	113.0	112.6°
B in kJ/mol	15	25	13	—	—

From the vibrational structure in the photoelectron spectrum of the PH_3 frequencies of the symmetric out-of-plane (inversion) vibration of the ion, ν_2=450 [16], 500±20 [17, 18], and 530±80 cm^{-1} [5], were obtained. The so-called "frequency halving" (compared to $\nu_2\approx$ 900 cm^{-1} for PH_3) can be explained by a double minimum potential with a low inversion barrier which allows the "left" and "right" vibrational energy levels to interact and to split into equally spaced doublets; see e.g. [18]. Ab initio calculated harmonic vibrational frequencies were reported [7].

The analysis of the ESR spectrum of PH_3^+ in an H_2SO_4 matrix at 77 K gave the g values $g_\perp = 2.014$ and $g_\parallel = 1.993$ and the ^{31}P hyperfine coupling constants (in mT) $A_{iso} = 51.7$, $A_\perp = 4.23$ and $A_\parallel = 7.06$ [4].

The bond dissociation energies $D(PH_2^+-H) = 337.6$ [19] and 308.8 kJ/mol [20] were obtained from electron impact studies on PH_3. The proton detachment energy, $D(PH_2-H^+) =$ 709 kJ/mol, was reported [21].

The enthalpy of formation, $\Delta_fH^\circ_{298.15} = 967.3$ kJ/mol, was derived from $E_i(PH_3) = 9.97$ eV and $\Delta_fH^\circ_{298.15}(PH_3) = 5.4$ kJ/mol [19]. Similarly, $\Delta_fH = 992$ kJ/mol was obtained from $E_i(PH_3) =$ 10.12 eV [3]. $\Delta_fH^\circ_{298.15} = 972.8$ kJ/mol and $\Delta_fH^\circ_0 = 974.5$ kJ/mol were given in [22]. Semiempirical (MNDO) calculations yielded $\Delta_fH = 1030$ kJ/mol [23, 24].

Chemical Reactions

Selected-ion flow tube (SIFT) [21, 25] and ion cyclotron resonance (ICR) [25, 26] investigations showed that no measurable reaction occurs at 80 K with H_2 and CO_2 [25] and at room temperature with H_2 [21, 26], O_2, H_2O, CO_2, CO [19], and CH_4 [19, 21]. SIFT experiments yielded rate constants k_{298} (in $cm^3 \cdot molecule^{-1} \cdot s^{-1}$) for the protonation of the following compounds by reaction with PH_3^+ (for $CH_3C{\equiv}CH$ an additional minor channel to PCH_5^+ was found) [21]:

compound	CH_3NH_2	NH_3	H_2S	HCN	CH_3OH	$CH_3C{\equiv}CH$
10^9 k	1.9	2.3	1.0	2.6	1.9	1.6

ICR investigations of the $PH_3^+ + NH_3$ reaction gave $k = 1.9 \times 10^{-9}$ $cm^3 \cdot molecule^{-1} \cdot s^{-1}$. The branching ratio was found to be 0.98 for the proton transfer channel and 0.02 for the channel forming $PNH_5^+ + H$ [26]. The reaction of PH_3^+ with COS, C_2H_2, and C_2H_4 resulted in the formation of H_3PS^+, $PC_2H_3^+$, and $C_2H_4 \cdot PH_3^+$ as shown by SIFT experiments [21].

Room-temperature rate constants (in 10^{-10} $cm^3 \cdot molecule^{-1} \cdot s^{-1}$) measured for the $PH_3^+ + PH_3$ reaction are k = 11 [21], 10.8 [27], 9.8 [28], and 7.2 [25]. The proton transfer channel dominates, while minor channels lead to $P_2H_4^+ + H_2$ and $P_2H_5^+ + H$ [26, 27].

References:

[1] Zarate, E. B.; Cooper, G.; Brion, C. E. (Chem. Phys. **148** [1990] 277/88).
[2] Märk, T. D.; Egger, F. (J. Chem. Phys. **67** [1977] 2629/35).
[3] Wada, Y.; Kiser, R. W. (Inorg. Chem. **3** [1964] 174/7).
[4] Begum, A.; Lyons, A. R.; Symons, M. C. R. (J. Chem. Soc. A **1971** 2290/3).
[5] Maier, J. P.; Turner, D. W. (J. Chem. Soc. Faraday Trans. II **68** [1972] 711/9).
[6] Maripuu, R.; Reineck, I.; Ågren, H.; Nian-Zu, Wu; Rong, Ji Ming; Veenhuizen, H.; Al-Shamma, S. H.; Karlsson, L.; Siegbahn, K. (Mol. Phys. **48** [1983] 1255/67).
[7] Berkowitz, J.; Curtiss, L. A.; Gibson, S. T.; Green, J. P.; Hillhouse, G. L.; Pople, J. A. (J. Chem. Phys. **84** [1986] 375/84).
[8] Marynick, D. S. (J. Chem. Phys. **74** [1981] 5186/9).
[9] Yates, B. F.; Bouma, W. J.; Radom, L. (J. Am. Chem. Soc. **108** [1986] 6445/54).
[10] Carmichael, I. (Inorg. Chim. Acta **117** [1986] 75/9).

[11] Aarons, L. J.; Guest, M. F.; Hall, M. B.; Hillier, I. H. (J. Chem. Soc. Faraday Trans. II **69** [1973] 643/7).
[12] Müller, J.; Ågren, H.; Canuto, S. (J. Chem. Phys. **76** [1982] 5060/8).
[13] Xia, T. J.; Wu, C. Y. R.; Judge, D. L. (Phys. Scr. **41** [1990] 870/3).

[14] Pope, S. A.; Hillier, I. H.; Guest, M. F.; Kendric, J. (Chem. Phys. Lett. **95** [1983] 247/9).
[15] Pope, S. A.; Hillier, I. H.; Guest, M. F. (Faraday Symp. Chem. Soc. No. 19 [1984] 109/23).
[16] Xia, T. J.; Chien, T. S.; Wu, C. Y. R.; Judge, D. L. (J. Quant. Spectrosc. Radiat. Transfer **45** [1991] 77/91).
[17] Branton, G. R.; Frost, D. C.; McDowell, C. A.; Stenhouse, I. A. (Chem. Phys. Lett. **5** [1970] 1/2).
[18] Bock, H. (Pure Appl. Chem. **44** [1975] 343/72, 350).
[19] McAllister, T.; Lossing, F. P. (J. Phys. Chem. **73** [1969] 2996/8).
[20] Fehlner, T. B.; Callen, R. B. (Adv. Chem. Ser. **72** [1968] 181/90).

[21] Smith, D.; McIntosh, B. J.; Adams, N. G. (J. Chem. Phys. **90** [1989] 6213/9).
[22] Wagman, D. D.; Evans, W. H.; Parker, V. B.; Schumm, R. H.; Halow, I.; Bailey, S. M.; Churney, K. L.; Nutall, R. L. (J. Phys. Chem. Ref. Data **11** Suppl. No. 2 [1982] 2-1/2-392, 2-73).
[23] Glidewell, C. (Inorg. Chim. Acta **97** [1985] 173/8).
[24] Glidewell, C. (J. Chem. Soc. Perkin Trans II **1985** 551/5).
[25] Adams, N. G.; McIntosh, B. J.; Smith, D. (Astron. Astrophys. **232** [1990] 443/6).
[26] Thorne, L. R.; Anicich, V. G.; Huntress, W. T. (Chem. Phys. Lett. **98** [1983] 162/6).
[27] Holtz, D.; Beauchamp, J. L.; Eyler, J. R. (J. Am. Chem. Soc. **92** [1970] 7045/55).
[28] Halmann, M.; Platzner, I. (J. Phys. Chem. **71** [1967] 4522/6).

1.3.2.2 PH_3^{2+}

CAS Registry Numbers: PH_3^{2+} *[64782-74-7]* Phosphorus(2+), trihydro-; PD_3^{2+} *[64782-75-8]* Phosphorus(2+), tri(hydro-*d*)-

The ions PH_3^{2+} and PD_3^{2+} were detected in the mass spectrum of phosphane at 30.0±1.0 (PH_3^{2+}) and 29.8±0.5 eV (PD_3^{2+}) [1]. A surprisingly low value of 15.6 eV [2] was critically discussed [1]. Traces of the ion were observed in the time-of-flight mass spectrum of phosphane at photon energies of 132.5 eV (P 2p resonant excitation) and 135.5 eV (P 2p Rydberg excitation) [3].

The electronic ground state is expected to be ...$(1a')^2\,(1e')^4$, $\tilde{X}\ {}^1A_1'$ with a bond angle α of 120° (D_{3h} symmetry) [4]; see also [5]. A complete active space (CAS) SCF calculation gave the bond distance r(P–H)=1.451 Å [4]. Semiempirical (MINDO) calculations of the enthalpy of formation were reported [6].

Multireference CI calculations yielded for the exothermic process of deprotonation via $PH_3^{2+} \rightarrow PH_2^{+} + H^+$ an activation barrier of 272 kJ/mol and a deprotonation enthalpy of −85.8 kJ/mol [7]. Ab initio (SCF MO) calculations on the tetra-coordinated phosphorus species $PH_nF_{4-n}^{+}$ gave ΔH = −2091 kJ/mol for the exothermic process $PH_3^{2+} + F^- \rightarrow PH_3F^+$ [5].

References:

[1] Märk, T. D.; Egger, F. (J. Chem. Phys. **67** [1977] 2629/35).
[2] Fischler, J.; Halmann, M. (J. Chem. Soc. **1964** 31/6).
[3] Zarate, E. B.; Cooper, G.; Brion, C. E. (Chem. Phys. **148** [1990] 289/97).
[4] Pope, S. A.; Hillier, I. H.; Guest, M. F. (Faraday Symp. Chem. Soc. No. 19 [1984] 109/23).
[5] Deiters, J. A.; Holmes, R. R. (J. Am. Chem. Soc. **112** [1990] 7197/202).
[6] Bews, J. R.; Glidewell, C. (J. Mol. Struct. **94** [1983] 305/18 [THEOCHEM **11**]).
[7] Pope, S. A.; Hillier, I. H.; Guest, M. F.; Kendric, J. (Chem. Phys. Lett. **95** [1983] 247/9).

1.3.3 PH_3^- and PH_3^{2-}

CAS Registry Number PH_3^{2-} *[76211-33-1]* Phosphate(2–), trihydro-

Ab initio calculations (unrestricted Møller-Plesset perturbation theory of 2nd order) revealed $\mathbf{PH_3^-}$ not to be stable with respect to electron detachment. The anion lies at least $\Delta E = 0.5$ eV above PH_3 [1]. $\Delta E = 1.5$ eV was derived from electron transmission spectra of PH_3 [2]. The formation of a negative ion state by Feschbach resonance (temporary binding of an electron to PH_3 in a Rydberg excited state) was discussed [3].

Semiempirical (MINDO/3) calculations on $\mathbf{PH_3^{2-}}$ showed that the planar T-shaped structure (C_{2v}) is more stable than the D_{3h} structure. The unique P–H distance in the C_{2v} species, r(P–H) = 1.439 Å, is considerably shorter than the other P–H distance, r(P–H′) = 1.833 Å. The H–P–H′ angle was found to be 88.27°. The enthalpy of formation $\Delta H = 852.28$ kJ/mol was calculated [4].

References:

[1] Nguyen, M. T. (J. Mol. Struct. **180** [1988] 23/9 [THEOCHEM **49**]).
[2] Tosselli, J. A.; Moore, J. H.; Giordan, J. C. (Inorg. Chem. **24** [1985] 1100/3).
[3] Ben Arfa, M.; Tronc, M. (Chem. Phys. **155** [1991] 143/8).
[4] Glidewell, C. (J. Mol. Struct. **67** [1980] 121/32).

1.4 PH_4 and Ions

1.4.1 PH_4 λ^5-Phosphanyl, Phosphoranyl

CAS Registry Numbers: PH_4 *[25530-87-4]* Phosphoranyl; PH_3D *[55130-06-8]*; PD_4 *[55130-07-9]*

Formation

The ESR-spectroscopically detected radical was formed by the reaction $PH_3 + H \rightarrow PH_4$ during γ radiolysis of PH_3 (1 to 5%) in matrices of xenon [1, 2] or neopentane [2, 3] at 77 K. PH_4 radicals were similarly generated by UV photolysis ($\lambda > 250$ nm) of a mixture of PH_3, HI, and Xe (1:0.1:100) at 4.2 to 100 K (HI as source of H atoms) [2]. PD_4 radicals were formed by γ radiolysis of PD_3 in xenon or neopentane-d_{12} matrices [1, 3]. The much lower stability of PH_4 in neopentane (half-life ~1.7 s at 122 K) compared to that of PD_4 in neopentane-d_{12} (~100 s) was interpreted by H-atom transfer between PH_4 and trapped alkyl radicals. The mixed species PH_3D and PHD_3 were generated by γ radiolysis of PH_3 in a neopentane-d_{12} matrix or of PD_3 in a neopentane matrix [2].

Molecular Properties. ESR Spectrum

The quantum-chemically calculated PH_4 structure of minimum energy (see below) is a distorted trigonal bipyramid (TBP) with two axial (ax) and two equatorial (eq) hydrogen atoms, the unpaired electron placed at the position of the missing equatorial ligand (TBP-eq, symmetry C_{2v}), see structure A in **Fig. 12** from [4]. From quantum-chemical calculations [5 to 8] (for further calculations, see the bibliography cited on p. 176) revealing significant odd-electron density to reside on the axial position it was shown that the metastable structure B in Fig. 12 is equally correct, with dashed lines denoting three-electron three-center bonding [1, 4]. The

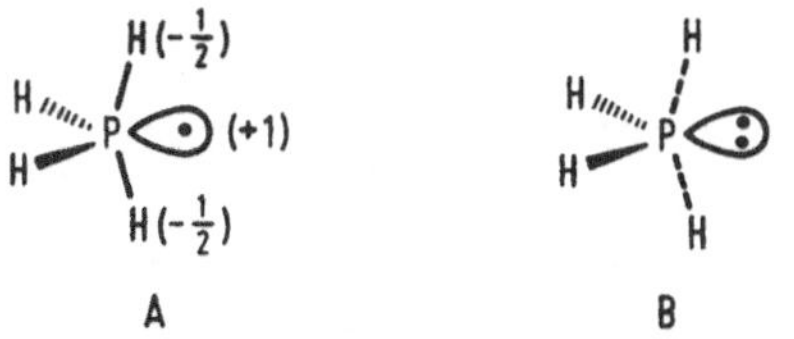

Fig. 12. Structure of PH_4.

following table summarizes optimized P–H bond lengths (in Å) and angles from ab initio calculations (UMP2: unrestricted second-order Møller-Plesset perturbation, total molecular energy E_t):

method	basis set	$-E_t$ in au	r_{eq}	r_{ax}	$\alpha(H_{eq}PH_{eq})$	$\alpha(H_{ax}PH_{ax})$	Ref.
UMP2	6-311 G**	343.2392	1.407	1.521	100.2°	171.0°	[9]
SCF MO	4-31 G*	342.6138	1.423	1.566	97°	172°	[10]
SCF MO	4-31 G*	342.5801	1.39	1.53	99.15°	171.3°	[7]

Further ab initio [5, 6, 11, 12] and semiempirical [8, 12, 13] calculations gave somewhat differing geometrical parameters.

Ab initio calculations showed that the TBP-eq form is energetically favored by about 80 kJ/mol over the TBP-ax structure (C_{3v} symmetry, missing ligand in axial position) [5, 6, 11]. An even higher instability relative to the TBP-eq form resulted for square pyramid (C_s or C_{4v}) [11], square-planar (D_{4h}) [11], and tetrahedral (T_d) [11, 14] structures of PH_4.

An ab initio study of the pathway of ligand scrambling of PH_4 indicated a turnstile mechanism involving a TBP-ax transition state with a lower barrier than a Berry pseudorotation process with a C_{4v} transition state [11]; see [7].

Ab initio calculations [11, 15] of the potential energy surface for $PH_3 + H \rightarrow PH_4$ (TBP-eq) showed the reaction to be a wholly downhill process (energy barrier of zero) with $PH_3 + H$ lying by some 46 [15] or 54 kJ/mol [11] lower in energy than PH_4. These findings are in conflict with the experimental detection of the radical [11]. A dissociation of PH_4 without a barrier was also predicted by ab initio calculations assuming the TBP-ax structure for PH_4 [5]. Similar results concerning the stability of PH_4 were obtained from an extrapolated potential energy curve (using the Z+1 core analogy model) [16]. A semiempirical calculation (MINDO) showed by contrast the radical to be more stable (by ~58 kJ/mol) than $PH_3 + H$ [17].

The electron affinity of PH_4 predicted by ab initio calculations (UMP2 procedure) to be A = 0.60 ± 0.05 eV [18].

An analysis of the **ESR** spectra of PH_4 (and deuterated species) in xenon and neopentane matrices confirmed the theoretically obtained TBP-eq structure of PH_4 [1 to 3, 23]. The best resolved ESR spectrum of PH_4 was observed in a neopentane matrix [2]. The spectrum at 100 K consists of eight 0.6 mT triplets (1:2:1). The large interaction of ~50 mT was due to a ^{31}P nucleus, and the smaller (~0.6 and 20 mT) interactions were due to pairs of equivalent protons. The ESR spectrum of PD_4 in neopentane-d_{12} consisted of a pair of 3 mT quintets (1:2:3:2:1) arising from hyperfine interactions with a ^{31}P nucleus and two equivalent deuterons [3]. Anisotropic features in the spectrum of PH_4 in a xenon matrix at 4.2 K were interpreted in terms of fast torsional oscillations about the symmetry axis of PH_4 [2]. The following g factors and hyperfine coupling constants A (in mT) were obtained (for values of PH_3D and PD_3H, see [3]).

species	matrix	T in K	g factor	A(P)	$A(H_{eq})$	$A(H_{ax})$	Ref.
PH_4	$C(CH_3)_4$	4.2, 77	2.0023(10)	51.81(10)	0.65(10)	20.05(10)	[2]
		100	2.0030(2)	51.93(3)	0.60(2)	19.87(3)	[3]
	Xe	4.2	2.0015(10)	49.76(10)	0.65(10)	20.86(10)	[2]
		77	2.0010(10)	48.99(10)	*)	20.70(10)	[2]
			—	48.2	*)	20.0	[1, 23]
PD_4	$C(CH_3)_4$	100	2.0030(2)	51.60	<0.1	0.302	[3]

*) Not resolved.

Hyperfine coupling constants were obtained by ab initio [9, 12, 19] and semiempirical [13, 20] calculations.

The assignment of an ESR spectrum, recorded during vacuum UV photolysis of PH_3 in a krypton matrix at 4.2 K, to PH_4 [21] was questioned [3, 19, 22]; a reanalysis of the spectrum showed that the spectrum presumably arises from $P_2H_6^+$ rather than PH_4 [19].

References:

[1] Sogabe, K.; Hasegawa, A.; Komatsu, T.; Miura, M. (Chem. Lett. **1975** 663/6).
[2] Shimokoshi, K.; Nakamura, K.; Sato, S. (Mol. Phys. **53** [1984] 1239/49).
[3] Colussi, A. J.; Morton, J. R.; Preston, K. F. (J. Chem. Phys. **62** [1975] 2004/6).
[4] Demolliens, A.; Eisenstein, O.; Hiberty, P. C.; Lefour, J. M.; Ohanessian, G.; Shaik, S. S.; Volatron, F. (J. Am. Chem. Soc. **111** [1989] 5623/31).
[5] Janssen, R. A. J.; Visser, G. J.; Buck, H. M. (J. Am. Chem. Soc. **106** [1984] 3429/37).
[6] Janssen, R. A. J.; Buck, H. M. (J. Mol. Struct. **110** [1984] 139/53 [THEOCHEM **19**]).
[7] Gonbeau, D.; Guimon, M.-F.; Ollivier, J.; Pfister-Guillouzo, G. (J. Am. Chem. Soc. **108** [1986] 4760/7).
[8] Colussi, A. J.; Morton, J. R.; Preston, K. F. (J. Phys. Chem. **79** [1975] 651/4).
[9] Cramer, C. J. (J. Am. Chem. Soc. **113** [1991] 2439/47).
[10] Kutzelnigg, W.; Wasilewski, J. (J. Chem. Soc. **104** [1982] 953/60).

[11] Howell, J. M.; Olsen, J. F. (J. Am. Chem. Soc. **98** [1976] 7119/27).
[12] Hudson, A.; Treweek, R. F. (Chem. Phys. Lett. **39** [1976] 248/9).
[13] Zuev, M. B.; Morozowa, I. D.; Charkin, O. P.; Klimenko, N. M.; Il'yasov, A. V. (Dokl. Akad. Nauk SSSR **254** [1980] 396/400; Dokl. Phys. Chem. [Engl. Transl.] **250/255** [1980] 756/9).
[14] Pope, S. A.; Hillier, I. A.; Guest, M. F. (Faraday Symp. Chem. Soc. No. 19 [1984] 109/23).
[15] Kutzelnigg, W.; Wallmeier, H.; Wasilewski, J. (Theor. Chim. Acta **51** [1979] 261/73).
[16] Schwarz, W. H. E. (Chem. Phys. **11** [1975] 217/28).
[17] Baird, N. C.; Hadley, G. C. (Chem. Phys. Lett. **128** [1986] 31/7).
[18] Ngyuen, M. T. (J. Mol. Struct. **180** [1988] 23/9 [THEOCHEM **49**]).
[19] Claxton, T. A.; Fullam, B. W.; Platt, E.; Symons, M. C. R. (J. Chem. Soc. Dalton Trans. **1975** 1395/7).
[20] Colussi, A. J.; Morton, J. R.; Preston, K. F. (J. Phys. Chem. **79** [1975] 1855/8).

[21] McDowell, C. A.; Mitchell, K. A. R.; Raghunathan, P. (J. Chem. Phys. **57** [1972] 1699/703).
[22] Griller, D.; Roberts, B. P. (J. Chem. Soc. Perkin Trans. II **1973** 1339/42).
[23] Sogabe, K. (J. Sci. Hiroshima Univ. A **39** [1975] 11/25).

1.4.2 PH_4^+ and PH_4^{2+}

1.4.2.1 PH_4^+

CAS Registry Numbers: PH_4^+ *[16749-13-6]* Phosphonium; PH_3D^+ *[28518-13-0]*; PD_4^+ *[18590-13-1]*; PT_4^+ *[31717-55-2]*

Results of earlier investigations are given in "Phosphor" C, 1965, pp. 6/7.

Formation

PH_4^+ ions were generated at ambient [1 to 4] and higher temperatures (up to 160°C) [5] in a low-pressure ion source by electron impact on PH_3. The main formation channel is the exothermic (secondary) reaction $PH_3^+ + PH_3 \rightarrow PH_4^+ + PH_2$; see for example [1 to 5]. Additional protonation reactions of PH_3 in the gas phase forming PH_4^+ are given on p. 217.

PH_4^+ ions were identified by NMR after slowly warming a mixture of P_2H_6, HF, and CS_2 in an NMR tube from 77 to 205 K [6].

Conductivity measurements indicated complete protonation of PH_3 in liquid anhydrous HF to give PH_4^+ [7].

Phosphonium ions were detected at room temperature by ^1H and ^{31}P NMR spectroscopy in solutions of PH_3 in the strong acids BF_3–H_2O and BF_3–CH_3OH (both containing excess BF_3). The ^1H NMR signals of PH_4^+ have also been observed in 96% sulfuric acid solutions of PH_3 at lower temperatures (–29 to –46°C) [8]. Similarly, PH_3 is protonated in HSO_3F at ambient temperature to give PH_4^+ [9]. For further protonation reactions of PH_3, see p. 246.

A sample of the isotopic species $PH_{4-n}D_n^+$ ($n \leq 4$) (~0.7 M) was prepared by condensing PH_3 (2.7 mmol) into a mixture of BF_3 (2.5 mL), CH_3OH (0.5 mL), and CH_3OD (1.5 mL) [10].

Physical Properties

The free PH_4^+ ion shows tetrahedral (T_d) symmetry (see "Phosphor" C, 1965, p. 6). The electronic ground state configuration is $(1a_1)^2 (2t_2)^6 (3a_1)^2 (3t_2)^6$, $\tilde{X}\ ^1A_1$ [11, 12]. Numerous ab initio calculations on the electronic structure and the properties of the ion have been performed: SCF calculations [11 to 29, 66] and SCF procedures including electron correlation [30 to 36] (see also the bibliography cited on p. 176).

Ab initio CI calculations yielded the bond length r(P–H) = 1.392 [31] or 1.397 Å [33]. A lower value, r = 1.384 Å, was obtained by an MP2 procedure [33]. Further values from SCF calculations range from 1.38 to 1.45 Å [12, 13, 15 to 21]. Neutron diffraction studies of the salts PH_4Br [37] and PH_4I [38] yielded for the PH_4^+ ions, distorted only slightly from tetrahedral symmetry, $r = 1.42 \pm 0.02$ Å (corrected for thermal motion) [37, 38]. From the ^1H NMR spectra of the salts (at 90 K), a value of $r = 1.42 \pm 0.02$ Å was derived [39]. A thermochemical radius of 1.57 Å was calculated from the lattice energy of the phosphonium halides [40].

Multireference CI calculations gave the adiabatic ionization potential $E_i = 20.7$ eV [30, 33] in good agreement with $E_i = 20.5$ eV from a one-center SCF MO investigation [20]. The deuteron quadrupole coupling constant in PD_4^+ [22], multipole moments, the magnetic susceptibility, and the polarizability of PH_4^+ [12, 21] were calculated by one-center SCF MO procedures.

The barrier height B = 643.5 kJ/mol at 0 K for the inversion through the planar D_{4h} transition state was calculated by an SCF MO procedure [41]; see also [11].

NMR Spectrum. The spectrum of PH_4^+ was recorded in highly acid solutions of PH_3 which suppress the rapid proton exchange with PH_3. The ^{31}P NMR spectrum consists of the anticipated 1:4:6:4:1 quintet according to four equivalent protons bonded to phosphorus

[8, 9]. Measurements in BF_3/H_2O [8], BF_3/CH_3OH [8, 10], fluorosulfuric acid [9], and HF [6] at room temperature gave the chemical shifts $\delta(^{31}P) = -217 \pm 1$ ppm (high-field from the external standard P_2O_4) [8] and −105.3 [10], −101.0 [9], −99.9 ppm [6] (from external 85% H_3PO_4). The proton shift $\delta(^1H)$, measured in concentrated sulfuric acid below −30°C [8] or in fluorosulfuric acid [9] and referenced to external $(CH_3)_4Si$ is between 6.0 and 6.4 ppm [8, 9]. The dependance of $\delta(^1H)$ on the composition and acidity of the solvent and on the temperature was investigated [7]. The P–H spin-spin coupling constant J [6, 8, 9, 10] is in the range 547.98 [10] to 548.5 [6], varying by less than 0.5% with change of conditions [8].

In the series PH_4^+, PH_3D^+, $PH_2D_2^+$, and PD_4^+, the isotope effect on the ^{31}P NMR shift is essentially zero. The primary isotope effect on J(P–H) is negative [10]. The two-bond coupling H–P–D in PH_3D^+ could not be resolved. Only an upper limit of $J < |0.2|$ Hz was given [42].

Ab initio calculations of phosphorus NMR chemical shifts in the gauge including atomic orbital (GIAO) were reported for 17 phosphorus-containing molecules, among them PH_4^+.

The anisotropy of the nuclear spin-spin coupling tensor relative to P, H and H, H interacting nuclei have been investigated by ab initio calculations (coupled perturbed Hartree-Fock). Discrepancies between calculated and experimental values of J(P–H) were interpreted in terms of neglected electron correlation contribution [14].

MO calculations predicted the magnetic screening of ^{31}P to decrease with increasing distortion of the ideal PH_4^+ tetrahedron [23].

The four **fundamental vibrations** $\nu_1(A_1)$, $\nu_2(E)$, $\nu_3(F_2)$, and $\nu_4(F_2)$ of the free ion (T_d symmetry) were determined from the vibrational spectra of the crystalline phosphonium halides PH_4X (X = Cl [43 to 46], Br [45 to 47], I [45, 46, 48]) and PD_4X [43, 47, 48] which show a splitting of the fundamentals due to the lowering of the symmetry of the ion in the crystal. Fundamentals of PH_4^+ (in cm^{-1}) are given in the following table:

species	X	ν_1	ν_2	ν_3	ν_4	Ref.
PH_4^+	Cl	2345	1132	2398, 2428	964, 994	[43]
	Br	2325	1120, 1100	2412, 2385	975, 950	[47]
	I	2295	1086, 1026	2366, 2272	974, 919	[48, 49]
PD_4^+	I	1654	772, 725	1732	677	[48, 49]

The values $\nu_1 = 2295$, $\nu_2 = 1086$, $\nu_3 = 2366$, and $\nu_4 = 994$ cm^{-1} for PH_4^+ [50] were used to calculate thermodynamic functions of PH_4^+ [50, 51].

Bands at 2530, 2486 (ν_1), 1134, and 990 cm^{-1} in the Raman spectrum of $PH_3F_2/PH_4^+HF_2^-$, dissolved in hydrogen fluoride, were assigned to PH_4^+ [6].

Harmonic frequencies of PH_4^+ were obtained by scaled SCF MO calculations [19] and from the ab initio calculated geometry combines with a force field (transferred from SiH_4) [31].

The frequencies of PD_4^+ and PT_4^+ were calculated with the Redlich-Teller rule [67].

Thermodynamic functions of PH_4^+ [51] in the ideal gas state, calculated in the rigid rotor-harmonic oscillator approximation, are presented in Table 19. They are based on the fundamental frequencies derived from the vibrational spectra of phosphonium halides (see above) and the calculated bond length [10] r = 1.382 Å [50].

Table 19
Thermodynamic Functions of PH_4^+ [51].

T in K	C_p° in $J \cdot mol^{-1} \cdot K^{-1}$	C_p°/C_0°	S° in $J \cdot mol^{-1} \cdot K^{-1}$	$H^\circ - H_0^\circ$ in kJ/mol	$-(G^\circ - H_0^\circ)/T$ in $J \cdot mol^{-1} \cdot K^{-1}$
100	33.26	1.333	165.1	3.33	131.9
200	34.67	1.315	188.4	6.69	154.9
273.15	38.75	1.273	199.7	9.36	165.5
298.15	40.57	1.258	203.2	10.35	168.5
300	40.71	1.257	203.5	10.43	168.7
400	48.54	1.207	216.2	14.89	179.0
500	55.98	1.174	227.9	20.12	187.6
600	62.65	1.153	238.7	26.06	195.3
700	68.54	1.138	248.8	32.62	202.2
800	73.67	1.127	258.3	39.74	208.6
900	78.08	1.119	267.2	47.33	214.6
1000	81.85	1.113	275.7	55.34	220.3

The enthalpy of formation of PH_4^+ in the gaseous state, $\Delta_f H = 746$ kJ/mol, was derived from a proton affinity $A_p = 789$ kJ/mol for PH_3 [52]. The Gibbs enthalpy of formation of PH_4^+ in aqueous solution (ideal solution at unit molality) is $\Delta_f G_{298.15}^\circ = 92.1$ kJ/mol [53].

Chemical Reactions

The thermochemistry of the process $PH_4^+ \rightarrow PH_3 + H^+$ is treated in the section on the proton affinity of PH_3.

Ab initio calculations (procedure in parentheses) yielded the reaction enthalpies $\Delta H^\circ = -698$ (CI) [34] and -835 (SCF) kJ/mol [25] for the processes $PH_4^+ + H^- \rightarrow PH_5$ [33] and $PH_4^+ + F^- \rightarrow PH_4F$, respectively [26].

Cross section measurements of the proton transfer reaction of PH_4^+ with Ca atoms have been carried out for ion kinetic energies E between 1 to 6 eV. The cross section drop sharply with increasing E, indicating exothermic behavior. The bimolecular rate constant is $k = 13.9 \times 10^{-10}$ at E=1 eV and 7.3×10^{-10} $cm^3 \cdot molecule^{-1} \cdot s^{-1}$ at E=2 eV [54].

The experimental hydration enthalpy of PH_4^+ has not been precisely determined. For the reaction $PH_4^+ + H_2O \rightarrow PH_4^+ \cdot H_2O$, $\Delta H = -(54)$ kJ/mol was cited [51]. Ab initio calculations on binary complexes formed from PH_4^+ and H_2O [18, 29, 56, 57], HF, HCl, and H_2S [18] were performed.

Selected-ion flow tube (SIFT) measurements showed that the PH_4^+ ion reacts extremely slowly with H_2 and CO at 80 [58] and 300 K [59]. The ion reacts with NH_3 and with CH_3NH_2 by fast proton transfer [59]. An ion cyclotron resonance (ICR) study at room temperature showed the reaction $PH_4^+ + NH_3 \rightarrow PH_3 + NH_4^+$ to be the only channel with a rate constant of $k = 2.1 \times 10^{-9}$ $cm^3 \cdot molecule^{-1} \cdot s^{-1}$. The reaction enthalpy was determined to be $\Delta H = -70$ [30] and -62 kJ/mol (at 320 K) [60, 61]. Ab initio studies on the hydrogen-bonded complex $PH_4^+ \cdot NH_3$ were performed [29, 62].

Investigations of ion-molecule reactions of PH_3 in a drift tube ion source revealed that PH_4^+ ions react above a source temperature of 25°C with PH_3 via a solvation equilibrium reaction followed by dissociation. The first step is $PH_4^+ + PH_3 \rightleftharpoons P_2H_7^+$, followed by the decomposition $P_2H_7^+ \rightarrow P_2H_5^+ + H_2$. The rate constant of the first step, $k_1 = 3.8 \times 10^{-12}$ $cm^3 \cdot molecule \cdot s^{-1}$, was determined. In the following analogous steps (e.g. $P_2H_5^+ + PH_3$) $P_3H_6^+$ and $P_4H_7^+$ were formed. At source temperatures below 25°C, ions of the general formula $PH_4^+(PH_3)_n$ with $n \leq 6$ were observed. Thermochemical data for the successive solvation are: for the first step, $PH_4^+ + PH_3$, the enthalpy of solvation is $\Delta H^\circ_{298} = -48$ kJ/mol and the entropy of solvation $\Delta S^\circ_{298} = -105$ $J \cdot mol^{-1} \cdot K^{-1}$ [5]. Ab initio calculations gave $\Delta H = -50$ (SCF) [18], −39 [29], −32 [36], and −31 kJ/mol [16] (MP4 procedures).

A potentiometric titration of PH_4^+ with AgF in a basic anhydrous HF solution showed an initial uptake of Ag^+ with an end point at [Ag]/[P] = 0.5 (formation of $Ag(PH_3)_2^+$) followed by a further uptake of Ag^+ forming $AgPH_3^+$. The complex-formation constants for the stepwise coordination were derived [63].

Proton transfer reactions of PH_4^+ with c-C_6H_{10}, $(CH_3)_2CO$ [58], CH_3COOH, C_2H_5OH, CH_3CN [60], and $Fe(C_6H_5)_2$ [65] were investigated by ion cyclotron resonance.

References:

[1] Holtz, D.; Beauchamp, J. L.; Eyler, J. R. (J. Am. Chem. Soc. **92** [1970] 7045/55).
[2] Eyler, J. R. (Inorg. Chem. **9** [1970] 981/2).
[3] Thorne, L. R.; Anicich, V. G.; Huntress, W. T. (Chem. Phys. Lett. **98** [1983] 162/6).
[4] Halman, M.; Platzner, I. (J. Phys. Chem. **71** [1967] 4522/6).
[5] Long, J. W.; Franklin, J. L. (J. Am. Chem. Soc. **96** [1974] 2320/7).
[6] Minkwitz, R.; Liedtke, A. (Inorg. Chem. **28** [1989] 4238/42).
[7] Gut, R. (Inorg. Nucl. Chem. Lett. **12** [1976] 149/52).
[8] Sheldrick, G. M. (Trans. Faraday Soc. **63** [1967] 1077/82).
[9] Olah, G. H.; McFarland, C. W. (J. Org. Chem. **34** [1969] 1832/4).
[10] Wasylishen, R. E.; Burford, N. (Can. J. Chem. **65** [1987] 2707/12).

[11] Krogh-Jespersen, M.-B.; Chandrasekhar, J.; Würthwein, E.-U.; Collins, J. B.; Schleyer, P. v. R. (J. Am. Chem. Soc. **102** [1980] 2263/8).
[12] Albasiny, E. L.; Cooper, J. R. A. (Proc. Phys. Soc. [London] **88** [1966] 315/23).
[13] Hendewerk, M. L.; Frey, R.; Dixon, D. A. (J. Phys. Chem. **87** [1983] 2026/32).
[14] Lazzeretti, P.; Rossi, E.; Taddei, F.; Zanasi, R. (J. Chem. Phys. **77** [1982] 408/14).
[15] Magnusson, E. (J. Comput. Chem. **5** [1984] 612/28).
[16] Del Bene, J. E.; Frisch, M. J.; Pople, J. A. (J. Phys. Chem. **89** [1985] 3669/74).
[17] Ikuta, S.; Kebarle, P. (Can. J. Chem. **61** [1983] 97/102).
[18] Desmeules, P. J.; Allen, L. C. (J. Chem. Phys. **72** [1980] 4731/48).
[19] Latajka, Z.; Schreiner, S. (J. Chem. Phys. **81** [1984] 2713/6).
[20] Moccia, R. (J. Chem. Phys. **40** [1964] 2176/85).

[21] Banyard, K. E.; Hake, R. B. (J. Chem. Phys. **43** [1965] 2684/9, **44** [1966] 3150).
[22] Pyykkö, P. (Proc. Phys. Soc. [London] **92** [1967] 841/2).
[23] Tarasevich, A. S. (Ukr. Khim. Zh. **54** [1988] 866/9; Sov. Progr. Chem. [Engl. Transl.] **54** No. 2 [1988] 86/8).
[24] Kutzelnigg, W.; Wasilewski, J. (J. Am. Chem. Soc. **104** [1982] 953/60).
[25] Deiters, J. A.; Holmes, R. R. (J. Am. Chem. Soc. **112** [1990] 7197/202).
[26] Pettitt, B. A.; Danchura, W. (J. Phys. B **20** [1987] 1899/907).
[27] Korkin, A. A.; Tsvetkov, E. N. (Zh. Neorg. Khim. **34** [1989] 290/4; Russ. J. Inorg. Chem. [Engl. Transl.] **34** [1989] 161/4).

[28] Choi, S. C.; Boyd, R. J.; Knop, O. (Can. J. Chem. **66** [1988] 2465/75).
[29] Evleth, E. M.; Hamou-Tahra, Z. D.; Kassab, E. (J. Phys. Chem. **95** [1991] 1213/20).
[30] Pope, S. A.; Hillier, I. H.; Guest, M. F. (Faraday Symp. Chem. Soc. No. 19 [1984] 103/23).

[31] Marynick, D. S.; Scanlon, K.; Eades, R. A.; Dixon, D. A. (J. Phys. Chem. **85** [1981] 3364/6).
[32] De Frees, D. J.; McLean, A. D. (J. Chem. Phys. **82** [1985] 333/41).
[33] Pope, S. A.; Hillier, I. H.; Guest, M. F.; Kendric, J. (Chem. Phys. Lett. **95** [1983] 247/9).
[34] Trinquier, G.; Daudey, J.-P.; Caruana, G.; Madaule, Y. (J. Am. Chem. Soc. **106** [1984] 4794/9).
[35] Del Bene, J. E.; Shavitt, I. (J. Phys. Chem. **94** [1990] 5514/8).
[36] Ikuta, S. (J. Mol. Struct. **152** [1987] 89/100 [THEOCHEM **37**]).
[37] Schroeder, L. W.; Rush, J. J. (J. Chem. Phys. **54** [1971] 1968/73).
[38] Sequeira, A.; Hamilton, W. C. (J. Chem. Phys. **47** [1967] 1818/22).
[39] Pratt, L.; Richards, R. E. (Trans. Faraday Soc. **50** [1954] 670/4).
[40] Jenkins, H. D. B.; Thakur, K. P. (J. Chem. Educ. **56** [1979] 576/7).

[41] Zahradník, R.; Havlas, Z.; Hess, B. A., Jr.; Hobza, Y. (Collect. Czech. Chem. Commun. **55** [1990] 869/89).
[42] Manatt, S. L.; Cohen, E. A.; Cowley, A. H. (J. Am. Chem. Soc. **91** [1969] 5919/20).
[43] Durig, J. R.; Antion, D. J.; Pate, C. B. (J. Chem. Phys. **52** [1970] 5542/8).
[44] Heinemann, A. (Ber. Bunsen-Ges. Phys. Chem. **68** [1964] 280/6).
[45] Rush, J. J.; Melvegger, A. J.; Lippincott, E. R. (J. Chem. Phys. **51** [1969] 2947/55).
[46] Rush, J. J.; Melvegger, A. J. (Chem. Phys. Lett. **2** [1968] 621/4).
[47] Durig, J. R.; Antion, D. J.; Pate, C. B. (J. Chem. Phys. **51** [1969] 4449/56).
[48] Durig, J. R.; Antion, D. J.; Baglin, F. G. (J. Chem. Phys. **49** [1968] 666/74).
[49] Nakamoto, K. (Infrared and Raman Spectra of Inorganic and Coordination Compounds, 4th Ed., Wiley, New York - Chichester - Brisbane - Toronto - Singapore 1986, pp. 1/484, 131).
[50] Loewenschuss, A.; Marcus, Y. (Chem. Rev. **84** [1984] 89/115, 99).

[51] Loewenschuss, A.; Marcus, Y. (J. Phys. Chem. Ref. Data **16** [1987] 61/89, 77).
[52] Lias, S. G.; Barteness, J. E.; Liebman, J. F.; Holmes, J. L.; Levin, R. D.; Mallard, W. G. (J. Phys. Chem. Ref. Data **17** Suppl. No. 1 [1988] 1/872, 623).
[53] Wagman, D. D.; Evans, W. H.; Parker, V. B.; Schumm, R. H.; Halow, I.; Bailey, S. M.; Churney, K. L.; Nuttall, R. L. (J. Phys. Chem. Ref. Data **11** Suppl. No. 2 [1982] 2-1/2-392, 2-73).
[54] Eslava, L. A.; Porter, R. F. (Chem. Phys. Lett. **52** [1977] 368/70).
[55] Keesee, R. G.; Castleman, A. W., Jr. (J. Phys. Chem. Ref. Data **15** [1986] 1011/71, 1016).
[56] Sreerama, N.; Vishveshwara, S. (J. Mol. Struct. **133** [1985] 139/46 [THEOCHEM **26**]).
[57] Del Bene, J. E. (J. Phys. Chem. **92** [1988] 2874/80).
[58] Adams, N. G.; McIntosh, B. J.; Smith, D. (Astron. Astrophys. **232** [1990] 443/6).
[59] Smith, D.; McIntosh, B. J.; Adams, N. G. (J. Chem. Phys. **90** [1989] 6213/9).
[60] Wolf, J. F.; Staley, R. H.; Koppel, I.; Taagepera, M.; McIver, R. T., Jr.; Beauchamp, J. L.; Taft, R. W. (J. Am. Chem. Soc. **99** [1979] 5417/29).

[61] Taft, R. W. (Footnote 3f in [64]).
[62] Del Bene, J. E. (J. Comput. Chem. **10** [1989] 603/15).
[63] Gut, R.; Ruede, J. (J. Coord. Chem. **8** [1978] 47/53).
[64] Lias, S. G.; Shold, D. M.; Ausloos, P. (J. Am. Chem. Soc. **102** [1980] 2540/8).

[65] Foster, M. S.; Beauchamp, J. L. (J. Am. Chem. Soc. **97** [1975] 4814/7).
[66] Müller, B.; Schüler, M.; Reinhold, J. (Chem. Phys. Lett. **172** [1990] 478/82).
[67] Fulea, A. O. (An. Univ. Bucuresti Chim. **18** [1969] 161/5 from C. A. **74** [1971] No. 93101).
[68] Chesnut, D. B.; Rusiloski, B. E. (Chem. Phys. **157** [1991] 105/10).

1.4.2.2 PH_4^{2+}

CAS Registry Number *[85420-12-8]* Phosphorus(2+), tetrahydro-

Complete active space (CAS) SCF calculations on this species with seven valence electrons showed it to be distorted from tetrahedral (T_d) symmetry. The energy minimum for C_{3v} symmetry in the electronic ground state $^2A_1$ predicts one very long P–H bond. The geometric parameters were calculated to be r(P–H)=1.438 and 1.877 Å and α(H–P–H)=95.5° [1]. Multireference CI calculations yielded a relatively high barrier of 83 kJ/mol towards proton loss. The energy of the strongly exothermic deprotonation was found to be −276 kJ/mol [2].

References:

[1] Pope, S. A.; Hillier, I. H.; Guest, M. F. (Faraday Symp. Chem. No. 19 [1984] 109/23).
[2] Pope, S. A.; Hillier, I. H.; Guest, M. F.; Kendric, J. (Chem. Phys. Lett. **95** [1983] 247/9).

1.4.3 PH_4^-

CAS Registry Number *[20774-06-5]* Phosphate(1−), tetrahydro-

SCF MO [1, 2] and MINDO/3 [3] calculations for PH_4^- (based on the valence shell electron pair repulsion theory) revealed the most stable structure to be a trigonal bipyramid in which one of the equatorial sites is occupied by a lone electron pair instead of a P–H bond. Geometry optimization of this C_{2v} structure gave the bond lengths r(P–H_{ax})=1.671 [1], 1.716 Å [2] and r(P–H_{eq})=1.402 [1], 1.409 Å [2]. The bond angles are α(H_{ax}–P–H_{ax})=166.4° [1], 165.6° [2] and α(H_{eq}–P–H_{eq})=103.3° [1], 102.8° [2]. For similar values, see [3]. Vibrational frequencies [1], ionization energies (from electron propagator calculations) [1], and an enthalpy of formation of Δ_fH=192 kJ/mol [3] were reported.

A less stable (by 1.453 eV [1]) structure of PH_4^- with T_d symmetry [1, 2] results from the model of double Rydberg anions which consist of a closed-shell cation core and two Rydberg-like diffuse electrons [1].

The entropy S°=213 $J \cdot mol^{-1} \cdot K^{-1}$ for gaseous PH_4^- was estimated with the Yatsimirskii rule [4].

References:

[1] Ortiz, J. V. (J. Phys. Chem. **94** [1990] 4762/3).
[2] Trinquier, G.; Daudey, J.-P.; Caruana, G.; Madaule, Y. (J. Am. Chem. Soc. **106** [1984] 4794/9).
[3] Glidewell, C. (J. Mol. Struct. **67** [1980] 121/32).
[4] Krestov, G. A. (Zh. Fiz. Khim. **42** [1968] 866/73; Russ. J. Phys. Chem. [Engl. Transl.] **42** [1968] 452/5).

1.5 PH_5 λ^5-Phosphane, Phosphorane; PH_5^{2-}

CAS Registry Numbers: PH_5 *[131232-65-0], [131232-51-4], [13769-19-2]* Phosphorane; PH_5^{2-} *[76009-54-6]* Phosphate(2−), pentahydro-

PH_5

The species PH_5, the prototype of phosphoranes and moreover of systems with pentavalent phosphorus, has not been detected experimentally. The formation via $PH_3 + H_2 \rightarrow PH_5$ is not feasible due to the high barrier of reaction (about 335 kJ/mol). The preparation via $PH_4^+ + H^-$ is expected to lead to $PH_3 + H_2$ rather than to PH_5 [1, 2] (see below).

Numerous ab initio calculations with at least split-valency quality basis sets have been performed. The results are extremely sensitive to the choice of the basis set, especially the inclusion of d orbitals, and to the procedure applied, such as nth-order Møller-Plesset theory (MPn, up to $n = 4$) [3 to 7], coupled electron pair approximation (CEPA) [1, 2, 8, 9], configuration interaction (CI) [8, 10, 11], or SCF MO calculations [12 to 19]. For semiempirical calculations, see [20, 21] (EHMO), [22] (PRDDO, partial retention of diatomic differential overlap), and [23] (INDO, intermediate neglect of differential overlap).

The calculations show the equilibrium structure of PH_5 in the electronic ground state to be a trigonal bipyramid of D_{3h} symmetry. The electronic structure of the molecule is discussed in terms of a set of five valence orbitals according to $\ldots(1a_1')^2\ (1e')^2\ (2e')^2\ (1a_2'')^2\ (2a_1')^2$ [5]. Bonding in PH_5 is decribed by three equatorial (eq) bonds and two weaker axial (ax), four-electron three-center bonds [5, 9, 14, 24, 25]. Several sets of optimized bond lengths, r_{eq} and r_{ax}, were calculated [1, 3, 5, 7 to 11, 14, 17, 18]. MP2 calculations yielded the bond lengths $r_{eq} = 1.414$ and $r_{ax} = 1.463$ Å [7], which are comparable to the MP2 values 1.424 and 1.474 Å (somewhat smaller basis) [3], to $r_{eq} = 1.41$ and $r_{ax} = 1.47$ Å given in CEPA [1, 8], CI [1, 8] and SCF MO [17, 18] studies as well as to $r_{eq} = 1.43$ Å and $r_{ax} = 1.48$ Å obtained in an MRD-CI (multireference double excitation) investigation [10].

Harmonic force fields and harmonic frequencies of PH_5 were obtained by SCF MO [8, 13, 17] and CEPA calculations [8]; for the transfer of force fields of other penta-coordinated molecules to PH_5, see [26, 27]. A predicted IR gas-phase spectrum with rotational fine structure (at 300 K) is displayed in [17]. Anharmonicity constants and vibration-rotation interaction constants were derived from an anharmonic force field. From a scaling procedure, making use of harmonic and anharmonic force fields of PH_5 and PH_3 (as reference molecule) the following wave numbers (in cm^{-1}) of the PH_5 fundamentals were predicted [18] (ν = stretching, δ = deformation, s = symmetric, as = antisymmetric, op = out-of-plane):

$\nu_1(A_1')$	$\nu_2(A_1')$	$\nu_3(A_2'')$	$\nu_4(A_2'')$	$\nu_5(E')$	$\nu_6(E')$	$\nu_7(E')$	$\nu_8(E'')$	Ref.
2232	1731	1844	1122	2239	1229	534	1381	[18]
ν_s eq	ν_s ax	ν_{as} ax	δ_{as} op	ν eq	δ ax	δ eq	δ op	[17]

The isomerization of PH_5 was decribed as Berry pseudorotation (BPR) or as turnstile rotation (TR); see **Fig. 13**, p. 322, from [16].

Both processes are shown to be topologically equivalent and while pseudorotation goes through a valley of the potential energy surface, turnstile proceeds on the slope of the same valley with no stationary properties of the turnstile transition state [1, 8]; for an analysis of the structures and for the arrangement of phosphoranes in terms of the BPR and TR processes, see [28]. Ab initio calculations on the BPR transition state (square-pyramidal C_{4v} structure) gave barrier heights ranging from about 4 to 8 kJ/mol [1, 2, 4 to 6, 8, 10, 11, 16] which are lower than the TR activation energies [10, 16, 21]. Since the zero-point energy of the vibration initiating the pseudorotation ($\nu_7(E')$) is of the same order [2, 8] (see also [5]), PH_5 is expected to be a markedly nonrigid molecule with all hydrogen atoms being dynamically equivalent [2, 24].

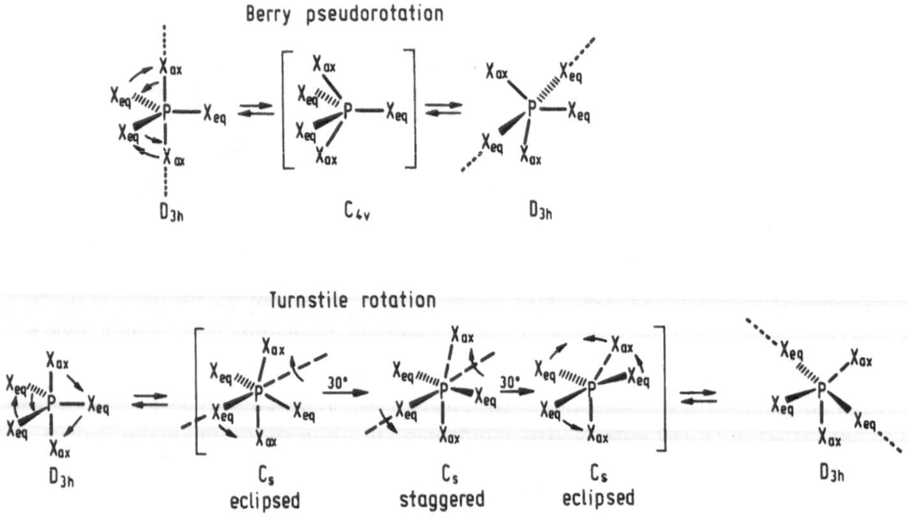

Fig. 13. Berry pseudorotation and turnstile rotation as mechanism of the intramolecular ligand exchange in phosphoranes PX_5 (X = H, halogen, aryl) [16].

The reaction $PH_5 \rightarrow PH_3 + H_2$ was investigated by ab initio calculations in terms of PH_5 transition states with C_{2v} and C_s symmetry. The results are very sensitive to the level of computational sophistication [1]. On an intermediate level of calculation (SCF MO + polarization functions), two saddle points on the potential energy surface (PES) were found, one corresponding to a concerted (equatorial-equatorial) H_2 abstraction reaction, the other to a zwitterionic reaction via $PH_4^+ + H^-$. The barrier for the concerted reaction is slightly smaller than that for the zwitterionic reaction, but the PES between the two saddle points is extremely flat. In higher level calculations (including electron correlation) only the lower barrier was found [1, 2, 8]. Calculated barrier heights of 151 [1, 2, 8] and 130 kJ/mol [3] (above PH_5) are such that PH_5 should be metastable in spite of the exoergicity of the reaction $PH_5 \rightarrow PH_3 + H_2$ (−159 to −243 kJ/mol [1 to 4, 7, 8, 12]). Catalytic amounts of acids (even another PH_5 molecule may serve as a Lewis acid) are expected to lower the barrier for decomposition of PH_5 considerably [1, 2].

PH_5^{2-}

MINDO/3 calculations on PH_5^{2-} yielded very similar enthalpies of formation, $\Delta_fH = 879.34$ kJ/mol for D_{3h} symmetry and 868 kJ/mol for C_{4v} symmetry. No activation barrier was found for the pseudorotation pathway from the most stable C_{4v} structure via D_{3h} to C_{4v} [29].

References:

[1] Kutzelnigg, W.; Wasilewski, J. (J. Am. Chem. Soc. **104** [1982] 953/60).
[2] Kutzelnigg, W.; Wasilewski, J.; Wallmeier, H. (Energy Storage Redistrib. Mol. **1983** 203/17).
[3] Reed, A. E.; Schleyer, P. v. R. (Chem. Phys. Lett. **133** [1987] 553/61).

[4] Wang, P.; Zhang, Y.; Glaser, R.; Reed, A. E.; Schleyer, P. v. R.; Streitwieser, A. (J. Am. Chem. Soc. **113** [1991] 55/64).
[5] Wasada, H.; Hirao, K. (J. Am. Chem. Soc. **114** [1992] 16/27).
[6] Wang, P.; Agrafiotis, D. K.; Streitwieser, A.; Schleyer, P. v. R. (J. Chem. Soc. Chem. Commun. **1990** 201/3).
[7] Ewig, C. S.; Van Wazer, J. R. (J. Am. Chem. Soc. **111** [1989] 1552/8).
[8] Kutzelnigg, W.; Wallmeier, H.; Wasilewski, J. (Theor. Chim. Acta **51** [1979] 261/73).
[9] Keil, F.; Kutzelnigg, W. (J. Am. Chem. Soc. **97** [1975] 3623/32).
[10] Shih, S.-K.; Peyerimhoff, S. D.; Buenker, R. J. (J. Chem. Soc. Faraday Trans. II **75** [1979] 379/89).

[11] Trinquier, G.; Daudry, J. P.; Caruana, G.; Madaule, Y. (J. Am. Chem. Soc. **106** [1984] 4794/9).
[12] Rauk, A.; Leland, C. A.; Kislow, K. (J. Am. Chem. Soc. **94** [1972] 3035/40).
[13] Walker, W. (J. Mol. Spectrosc. **43** [1972] 411/5).
[14] McDowell, R. S.; Streitwieser, A., Jr. (J. Am. Chem. Soc. **107** [1985] 5849/55).
[15] Howell, J. (J. Am. Chem. Soc. **99** [1977] 7447/52).
[16] Altmann, J. A.; Yates, K.; Csizmadia, I. H. (J. Am. Chem. Soc. **98** [1976] 1450/4).
[17] Breidung, J.; Thiel, W.; Komornicki, A. (J. Phys. Chem. **92** [1988] 5603/11).
[18] Breidung, J.; Schneider, W.; Thiel, W.; Schaefer, H. F., III (J. Mol. Spectrosc. **140** [1990] 226/36).
[19] Magnusson, E. (J. Am. Chem. Soc. **112** [1990] 7940/51).
[20] Issleib, K.; Gründler, W. (Theor. Chim. Acta **8** [1967] 70/2).

[21] Hoffmann, R.; Howell, J. M.; Muetterties, E. M. (J. Am. Chem. Soc. **94** [1972] 3047/58).
[22] Smolyar, A. E.; Zubyn, A. S.; Khaikina, E. A.; Charkin, O. P. (Zh. Neorg. Khim. **25** [1980] 313/7; Russ. J. Inorg. Chem. [Engl. Transl.] **25** [1980] 167/70).
[23] Colussi, A. J.; Morton, J. R.; Preston, K. F. (J. Phys. Chem. **79** [1975] 651/4).
[24] Kutzelnigg, W. (Angew. Chem. **96** [1984] 262/86; Angew. Chem. Int. Ed. Engl. **23** [1984] 272/95).
[25] Strich, A.; Veillard, A. (J. Am. Chem. Soc. **95** [1973] 5574/81).
[26] Holmes, R. R.; Deiters, R. M.; Golen, J. A. (Inorg. Chem. **8** [1969] 2612/20).
[27] Holmes, R. R.; Deiters, R. M. (J. Chem. Phys. **51** [1969] 4043/54).
[28] Lemmen, P.; Baumgartner, R.; Ugi, I.; Ramirez, F. (Chem. Scr. **28** [1988] 451/64).
[29] Glidewell, C. (J. Mol. Struct. **67** [1980] 121/32).

1.6 PH_6^- and PH_6^{3-}

CAS Registry Numbers: PH_6^- *[79839-88-6]* Phosphate(1–), hexahydro-; PH_6^{3-} *[76009-52-4]* Phosphate(3–), hexahydro-

Quantum-chemical calculations showed the ion $\mathbf{PH_6^-}$ to have O_h symmetry. The electronic ground-state configuration is $\ldots(1a_g)^2(1t_{2u})^6(1e_g)^4$. The first vertical ionization potential, E_i = 2.4 eV, was calculated by a Green's function method [1]. Ab inito SCF MO studies gave a P–H bond length of 1.494 Å [2] or 1.48 Å [3]. CI [2] and SCF MO [3] calculations revealed PH_6^- to be not stable with respect to $PH_4^- + H_2$ ($\Delta E = -45$ kJ/mol [2]), but stable with respect to $PH_5 + H^-$ ($\Delta E = 133$ [2], 142 kJ/mol [3]). The energy for the reaction $PH_6^- + PH_4^+ \rightarrow 2\,PH_5$ is −554 kJ/mol (CI calculation) [2]. MINDO calculations were performed for the minimum energy path of the attack of an H^- ion on PH_5 forming PH_6^- [4].

MINDO/3 calculations (constraining all P–H bonds to be equal) gave for **PH_6^{3-}** (O_h symmetry) a bond length of 1.83 Å. An enthalpy of formation $\Delta_fH = 1954$ kJ/mol was calculated [5].

References:

[1] Boldyrev, A. I.; Niessen, W., von (Chem. Phys. **155** [1991] 71/8).
[2] Trinquier, G.; Daudry, J. P.; Caruana, G.; Madaule, Y. (J. Am. Chem. Soc. **106** [1984] 4794/9).
[3] Kutzelnigg, W.; Wasilewski, J. (J. Am. Chem. Soc. **104** [1982] 953/60).
[4] Minaev, R. M.; Minkin, V. I. (Zh. Strukt. Khim. **20** [1979] 842/53; J. Struct. Chem. [USSR] [Engl. Transl.] **20** [1979] 715/25).
[5] Glidewell, C. (J. Mol. Struct. **67** [1980] 121/32).

Physical Constants and Conversion Factors

Avogadro constant N_A (or	$L) = 6.02214 \times 10^{23}\ mol^{-1}$	Planck constant	$h = 6.62608 \times 10^{-34}\ J \cdot s$
Faraday constant	$F = 9.64853 \times 10^{4}\ C/mol$	elementary charge	$e = 1.60218 \times 10^{-19}\ C$
molar gas constant	$R = 8.31451\ J \cdot mol^{-1} \cdot K^{-1}$	electron mass	$m_e = 9.10939 \times 10^{-31}\ kg$
molar volume (ideal gas) (273.15 K, 101 325 Pa)	$V_m = 2.24141 \times 10^{1}\ L/mol$	proton mass	$m_p = 1.67262 \times 10^{-27}\ kg$

1 kg = 2.205 pounds

1 m = 3.937×10^{1} inches = 3.281 feet

1 m³ = 2.642×10^{2} gallons (U.S.)

1 m³ = 2.200×10^{2} gallons (Imperial)

Force	N	dyn	kp
1 N	1	10^{5}	1.019716×10^{-1}
1 dyn	10^{-5}	1	1.019716×10^{-6}
1 kp	9.80665	9.80665×10^{5}	1

Pressure	Pa	bar	kp/m²	at	atm	Torr	lb/in²
1 Pa = 1 N/m²	1	10^{-5}	1.019716×10^{-1}	1.019716×10^{-5}	9.86923×10^{-6}	7.50062×10^{-3}	1.450378×10^{-4}
1 bar = 10^{6} dyn/cm²	10^{5}	1	1.019716×10^{4}	1.019716	9.86923×10^{-1}	7.50062×10^{2}	1.450378×10^{1}
1 kp/m² = 1 mm H_2O	9.80665	9.80665×10^{-5}	1	10^{-4}	9.67841×10^{-5}	7.35559×10^{-2}	1.422335×10^{-3}
1 at (technical)	9.80665×10^{4}	9.80665×10^{-1}	10^{4}	1	9.67841×10^{-1}	7.35559×10^{2}	1.422335×10^{1}
1 atm = 760 Torr	1.01325×10^{5}	1.01325	1.033227×10^{4}	1.033227	1	7.60×10^{2}	1.469595×10^{1}
1 Torr = 1 mm Hg	1.333224×10^{2}	1.333224×10^{-3}	1.359510×10^{1}	1.359510×10^{-3}	1.315789×10^{-3}	1	1.933678×10^{-2}
1 lb/in² = 1 psi	6.89476×10^{3}	6.89476×10^{-2}	7.03069×10^{2}	7.03069×10^{-2}	6.80460×10^{-2}	5.17149×10^{1}	1

Work, Energy, Heat	J	kW·h	kcal	Btu	eV
1 J = 1 W·s = 1 N·m = 10^7 erg	1	2.778×10^{-7}	2.39006×10^{-4}	9.4781×10^{-4}	6.242×10^{18}
1 kW·h	3.6×10^{6}	1	8.604×10^{2}	3.41214×10^{3}	2.247×10^{25}
1 kcal	4.1840×10^{3}	1.1622×10^{-3}	1	3.96566	2.6117×10^{22}
1 Btu (British thermal unit)	1.05506×10^{3}	2.93071×10^{-4}	2.5164×10^{-1}	1	6.5858×10^{21}
1 eV	1.602×10^{-19}	4.450×10^{-26}	3.8289×10^{-23}	1.51840×10^{-22}	1

$1\ cm^{-1} \triangleq 1.239842\times10^{-4}$ eV $1\ Hz \triangleq 4.135669\times10^{-15}$ eV

2 rydberg = 1 hartree = 27.2114 eV $1\ eV \triangleq 96.485$ kJ/mol

Power	kW	hp	kp·m·s^{-1}	kcal/s
1 kW = 10^3 J/s	1	1.35962	1.01972×10^{2}	2.39006×10^{-1}
1 hp (horsepower, metric)	7.3550×10^{-1}	1	7.5×10^{1}	1.7579×10^{-1}
1 kp·m·s^{-1}	9.80665×10^{-3}	1.333×10^{-2}	1	2.34384×10^{-3}
1 kcal/s	4.1840	5.6886	4.26650×10^{2}	1

References:

Mills, I. (Ed.), International Union of Pure and Applied Chemistry, Quantities, Units and Symbols in Physical Chemistry, Blackwell Scientific Publications, Oxford 1988.

The International System of Units (SI), National Bureau of Standards Spec. Publ. 330 [1972].

Landolt-Börnstein, 6th Ed., Vol. II, Pt. 1, 1971, pp. 1/14.

ISO Standards Handbook 2, Units of Measurement, 2nd Ed., Geneva 1982.

Cohen, E. R., Taylor, B. N., Codata Bulletin No. 63, Pergamon, Oxford 1986.